Applied Evolutionary Psychology

Applied Evolutionary Psychology

Edited by

S. Craig Roberts
Department of Psychology,
University of Stirling,
UK

OXFORD
UNIVERSITY PRESS

Great Clarendon Street, Oxford ox2 6dp

Oxford University Press is a department of the University of Oxford.
It furthers the University's objective of excellence in research, scholarship,
and education by publishing worldwide in

Oxford New York

Athens Auckland Bangkok Bogotá Buenos Aires Cape-Town
Chennai Dar-es-Salaam Delhi Florence Hong-Kong Istanbul Karachi
Kolkata Kuala-Lumpur Madrid Melbourne Mexico-City Mumbai Nairobi
Paris São-Paulo Shanghai Singapore Taipei Tokyo Toronto Warsaw
with associated companies in Berlin Ibadan

Oxford is a registered trade mark of Oxford University Press
in the UK and in certain other countries

Published in the United States
by Oxford University Press Inc., New York

© Oxford University Press, 2012

The moral rights of the authors have been asserted
Database right Oxford University Press (maker)

First edition published 2012

All rights reserved. No part of this publication may be reproduced,
stored in a retrieval system, or transmitted, in any form or by any means,
without the prior permission in writing of Oxford University Press,
or as expressly permitted by law, or under terms agreed with the appropriate
reprographics rights organization. Enquiries concerning reproduction
outside the scope of the above should be sent to the Rights Department,
Oxford University Press, at the address above

You must not circulate this book in any other binding or cover
and you must impose the same condition on any acquirer

British Library Cataloguing in Publication Data

Data available

Library of Congress Cataloging in Publication Data
Applied evolutionary psychology / edited by S. Craig Roberts. — 1st ed.
    p. ; cm.
 Includes bibliographical references and index.
 ISBN 978-0-19-958607-3 (alk. paper)
 I. Roberts, S. Craig.
 [DNLM: 1. Adaptation, Psychological. 2. Biological Evolution. BF 698.95]
 LC classification not assigned
 616.89—dc23
                                                                2011035841

Typeset in Minion by Cenveo
Printed and bound by
CPI Group (UK) Ltd, Croydon, CR0 4YY

ISBN 978–0–19–958607–3

10 9 8 7 6 5 4 3 2 1

Whilst every effort has been made to ensure that the contents of this book are as complete,
accurate and up-to-date as possible at the date of writing, Oxford University Press is not
able to give any guarantee or assurance that such is the case. Readers are urged to take
appropriately qualified medical advice in all cases. The information in this book is intended to
be useful to the general reader, but should not be used as a means of self-diagnosis or for the
prescription of medication.

Acknowledgements

I would like to express my deep thanks to all the authors. It has been a pleasure to interact with them and read their fascinating work more closely.

I'm also grateful to my former students at the University of Liverpool, who over the years stimulated me to think about how we might put all our ideas into practice. I also thank my former colleague John Lycett, who over several years challenged me to think more critically about them. Together, we hosted a desk at many a university Open Day—now, finally, we might have something to offer prospective students who are interested in the practical value of their evolutionary psychology courses.

I sincerely thank Robert Burriss, Vicky Mileva, Alice Murray, and, especially, Juan David Leongómez for their valuable help in the final editing of the volume.

The staff at OUP have been extremely supportive throughout the process. Grateful thanks go to Martin Baum for his enthusiasm about the idea, and to Charlotte Green for her efficient and friendly help throughout.

I didn't quite believe those who tried to warn me about how time-consuming it is to produce an edited book, but they were right. So, for putting up with my distracted state, I'm indescribably grateful to Isla, Struan, and Steph.

What Next? is an oil painting on canvas by Lilia Mazurkevich. More of her work can be seen at www.liliamazurkevich.com. Lilia's paintings are inspired by the unpredictability and fragility of human nature, and tend to explore the narrow line between humour and sadness. Most of the paintings in her *Ambition* series depict a twisted and tangled mass of human bodies, in beautiful and varied detail, each trying to climb over the rest to find some space. The images are often claustrophobic and disturbing. While the man in this painting is also confined, he is hopeful as he peers through the peephole to a happier place (or so it seems to me, at least). I like to think this characterization of the human condition—constrained and unfulfilled, yet with room for optimism—befits a book such as this.

Contents

List of abbreviations *ix*
List of contributors *xi*

1 Applying evolutionary psychology *1*
S. Craig Roberts

Section 1 **Business**

2 The evolutionary psychology of economics *7*
Paul H. Rubin and C. Monica Capra

3 The evolution of business and management *16*
Nigel Nicholson

4 The social animal within organizations *36*
Abraham P. Buunk and Pieternel Dijkstra

Section 2 **Family**

5 The evolved child: adapted to family life *55*
David F. Bjorklund and Patrick Douglas Sellers II

6 Application of evolutionary psychology to academic learning *78*
David C. Geary

7 Serial monogamy and clandestine adultery: evolution and consequences of the dual human reproductive strategy *93*
Helen E. Fisher

Section 3 **Society**

8 The evolutionary psychology of mass politics *115*
Michael Bang Petersen

9 Gender equity issues in evolutionary perspective *131*
Bobbi S. Low

10 The evolution of charitable behaviour and the power of reputation *149*
Pat Barclay

11 Altruism as showing off: a signalling perspective on promoting green behaviour and acts of kindness *173*
Wendy Iredale and Mark van Vugt

12 Evolutionary perspectives on intergroup prejudice: implications for promoting tolerance *186*
Justin H. Park

13 The evolutionary psychology of criminal behaviour *201*
   Aurelio José Figueredo, Paul Robert Gladden, and Zachary Hohman

14 War, martyrdom, and terror: evolutionary underpinnings of the moral imperative to extreme group violence *222*
   Scott Atran

15 Evolutionary theory and behavioural biology research: implications for law *239*
   David J. Herring

Section 4 Health

16 Motivational mismatch: evolved motives as the source of—and solution to—global public health problems *259*
   Valerie Curtis and Robert Aunger

17 Mental health and well-being: clinical applications of Darwinian psychiatry *276*
   Alfonso Troisi

18 Evolutionary perspectives on sport and competition *290*
   Diana Wiedemann, Robert A. Barton, and Russell A. Hill

Section 5 Marketing and communication

19 Why we buy: evolution, marketing, and consumer behaviour *311*
   Vladas Griskevicius, Joshua M. Ackerman, and Joseph P. Redden

20 Evolutionary psychology and perfume design *330*
   S. Craig Roberts and Jan Havlicek

21 Television programming and the audience *349*
   Charlotte De Backer

22 News as reality-inducing, survival-relevant, and gender-specific stimuli *361*
   Maria Elizabeth Grabe

Section 6 Technology

23 Media naturalness theory: human evolution and behaviour towards electronic communication technologies *381*
   Ned Kock

24 Evolutionary psychology, demography, and driver safety research: a theoretical synthesis *399*
   David L. Wiesenthal and Deanna M. Singhal

25 Evolutionary robotics *414*
   Dylan Evans and Walter de Back

Index *427*

List of abbreviations

BP	before present
CEO	Chief Executive Officer
DALY	disability-adjusted life year
EPC	extra-pair copulation
GDI	Gender-related Development Index
GEM	Gender Empowerment Measure
HDI	Human Development Index
HWWS	handwashing with soap
LH	life history
MHC	major histocompatibility complex
MIT	Massachusetts Institute of Technology
my	million years
OCB	Organizational Citizen Behaviours
PTSD	post-traumatic stress disorder
SES	socioeconomic status
TB	tuberculosis
UK	United Kingdom
UN	United Nations
UNDP	United Nations Development Programme
US	United States
WEF	World Economic Forum
WHO	World Health Organization
WTA	willingness to accept
WTP	willingness to pay

List of contributors

Joshua M. Ackerman
Sloan School of Management
Massachusetts Institute of Technology
Cambridge, MA, USA

Scott Atran
Research Center for Group Dynamics
University of Michigan
Ann Arbor, MI, USA

Robert Aunger
London School of Hygiene & Tropical Medicine
London, UK

Pat Barclay
Department of Psychology
University of Guelph
Guelph, ON, Canada

Robert A. Barton
Department of Anthropology
Durham University
Durham, UK

David F. Bjorklund
Department of Psychology
Florida Atlantic University
Boca Raton, FL, USA

Abraham P. Buunk
Department of Social and Organizational Psychology
University of Groningen
AD Groningen, The Netherlands

C. Monica Capra
Department of Economics
Emory University
Atlanta, GA, USA

Valerie Curtis
London School of Hygiene & Tropical Medicine
London, UK

Walter de Back
Center for High Performance Computing
Dresden University of Technology
Dresden, Germany

Charlotte De Backer
Department of Communication Sciences
University of Antwerp
Antwerpen, Belgium

Pieternel Dijkstra
Department of Social and Organizational Psychology
University of Groningen
AD Groningen, The Netherlands

Dylan Evans
School of Medicine
University College Cork
Cork, Ireland

Aurelio J. Figueredo
Department of Psychology
University of Arizona
Tucson, AZ, USA

Helen E. Fisher
Department of Anthropology
Rutgers University
New York City, NY, USA

David C. Geary
Department of Psychological Sciences
University of Missouri at Columbia
Columbia, MO, USA

Paul Robert Gladden
Department of Psychology
University of Arizona
Tucson, AZ, USA

Maria Elizabeth Grabe
Department of Telecommunications
University of Indiana
Bloomington, IN, USA

LIST OF CONTRIBUTORS

Vladas Griskevicius
Carlson School of Management
University of Minnesota
Minneapolis, MN, USA

Jan Havlicek
Department of Anthropology
Charles University
Prague, Czech Republic

David J. Herring
School of Law
University of Pittsburgh
Pittsburgh, PA, USA

Russell A. Hill
Department of Anthropology
Durham University
Durham, UK

Zachary Hohman
Department of Psychology
University of Arizona
Tucson, AZ, USA

Wendy Iredale
Department of Psychology
University of Kent
Kent, UK

Ned Kock
Division of International Business and Technology Studies
Texas A&M University
Laredo, TX, USA

Bobbi S. Low
School of Natural Resources and Environment
University of Michigan
Ann Arbor, MI, USA

Nigel Nicholson
London Business School
London, UK

Justin H. Park
Department of Experimental Psychology
University of Bristol
Bristol, UK

Michael Bang Petersen
Department of Political Science
Aarhus University
Aarhus C, Denmark

Joseph P. Redden
Carlson School of Management
University of Minnesota
Minneapolis, MN, USA

S. Craig Roberts
Department of Psychology
University of Stirling
Stirling, Scotland, UK

Paul H. Rubin
Department of Economics
Emory University, USA
Atlanta, GA, USA

Patrick Douglas Sellers II
Department of Psychology
Florida Atlantic University
Boca Raton, FL, USA

Deanna M. Singhal
Department of Psychology
Grant MacEwan University
Edmonton, AB, Canada

Alfonso Troisi
Department of Neurosciences
School of Medicine
Rome, Italy

Mark van Vugt
Department of Social and Organizational Psychology
University of Amsterdam
Amsterdam, The Netherlands

Diana Wiedemann
Department of Anthropology
Durham University
Durham, UK

David L. Wiesenthal
Department of Psychology
York University
Toronto, ON, Canada

Chapter 1

Applying evolutionary psychology

S. Craig Roberts

A new foundation

Evolutionary psychology aims to understand and describe human behaviour in the light of past and continuing selection and adaptation. For a relatively young discipline, considerable progress has been made over the past few decades. Onto the conceptual framework provided by evolutionary theory, evolutionary psychologists hang observations of human behaviour and, on the whole, there they hang very well. Whether we want to understand the nature of sex differences in psychological attributes, of dynamics within and between groups, or of variability and commonalities in how individuals choose mates, evolutionary theory imparts a rigorous, informative, and cohesive structure.

It is not surprising that many notable scholars in the field originally come from biological rather than psychological disciplines, since the approaches used in the study of animal behaviour and behavioural ecology are readily transferred to investigations of our own species. What is perhaps more remarkable is the extent to which many with a psychology background have come to recognize the explanatory power of evolution in their own research, and embrace it, even though they often lack formal exposure to evolutionary reasoning because many psychology departments have, at least until recently, failed to offer it. It is true that there remain numerous critics and sceptics, or worse, those that simply do not see how evolution is relevant to psychology. Increasingly, however, and as misconceptions about what evolutionary psychologists do and think are rebutted and clarified (for recent examples, see Dunbar 2008; Confer et al. 2010), ideas about selection and adaptation are taking root in all of the various psychological subdisciplines (Fitzgerald and Whitaker 2010). As Dunbar (2008) has argued, this is because evolutionary theory is a 'single seamless framework' capable of spanning disciplinary divides, and it is the only such framework we have. To many, then, Darwin's (1859, p. 449) prediction that 'Psychology will be built on a new foundation' appears to have been realized; at the very least, the foundation stone is laid and the builders have been booked.

Evolutionary psychologists argue that a comprehensive understanding of any aspect of human behaviour cannot be achieved without due consideration of the selective forces that have shaped that behaviour in our evolutionary past, and which may continue to do so in the present. One way to characterize their approach is to say that they are particularly interested in the *ultimate* explanations for behaviour, according to the scheme set out by the renowned ethologist Niko Tinbergen (1963). Mechanistic, or *proximate*, explanations for the problem of why we need to eat food (e.g. 'I eat because low blood sugar levels activate neurons in the lateral hypothalamus, producing the perception of hunger') provide part of the answer, but are incomplete without information about the evolutionarily functional significance of eating (e.g. 'I eat because food provides the energy I need to survive and reproduce'). To take another example, recently expounded by Nettle (2011), we can attempt to understand why women tend to vary in timing of the onset of their reproductive career depending upon the harshness of the environment in which they live. A functional (or ultimate) reason appears to be that delaying reproduction allows them to divert nutritional resources towards somatic development, which may result in children who survive and grow better; however, this comes at the risk of dying before they reproduce—a classic trade-off between the benefits and costs of fast or slow life cycles. In contrast, the question of how individual women negotiate this trade-off, that is how they come to make decisions that characterize the population from which they belong, is answered differently—the proximate explanation. Proximate explanations include the presence or absence of events that trigger the adoption of a particular life history trajectory, such as early separation from the mother, or not being breastfed.

The distinction between the proximate and ultimate explanations of behaviour is an important one, because psychologists may often be interested in the former, while evolutionary psychologists will stress the latter. However, it should be clear that the approaches are inextricably linked; each provides insights and defines constraints on possible outcomes, and it is to the detriment of our science if we focus on one to the exclusion of the other.

Evolutionary psychology applied

Psychologists have a long tradition of applying basic research findings to develop practical ways of dealing with problematic issues. This is why many were drawn to psychology in the first place, even if they go on to do basic psychological research—it might be said that this is psychology's *raison d'être*. In contrast, evolutionary psychology is at its heart a basic science, and few proponents are initially drawn to the field with ambitions to apply their research findings.

Again, the distinction between proximate and ultimate causation of behaviour is relevant here, as proximate explanations will usually lead more directly to identification of specific interventions. For example, if we can understand the specific triggers that cause an individual to behave in a particular way, perhaps to its own cost, then we may be able to find a way to change behaviour—as health psychologists might aim to do. Or if we understand the mechanisms underlying expression of a psychopathy, clinical psychologists may be able to ameliorate its effects. It might be argued that ultimate approaches to understanding behaviour are unlikely to contribute practical solutions to such problems and that evolutionary psychologists thus have nothing to offer in this regard. Perhaps this argument is justified to an extent, but as we have seen, knowledge about any behaviour is incomplete in the absence of due regard to both proximate and ultimate reasoning. Furthermore, having a fuller understanding of the evolutionary history or likely adaptive value of particular behaviours might help to identify which of a range of possible interventions is likely to be the most successful in achieving particular outcomes.

To illustrate this, consider the case of evolutionary medicine, where a similar debate concerning the utility of evolutionary approaches has been ongoing over the past 20 years. Following the paper in 1991 by Williams and Nesse, which heralded 'the Dawn of Darwinian Medicine' and which kick-started the field, increasing numbers of researchers have become interested in this area and it has had some considerable success (see reviews in Stearns and Koella 2007; Trevathan et al. 2007; Nesse and Stearns 2008). One example is the case of nausea and vomiting in pregnancy. Until recently, this was considered a condition for which treatment was desirable—as epitomized by its epithets 'pregnancy sickness' or 'morning sickness'—which are now generally considered to be misnomers. In fact, symptoms of nausea and vomiting in pregnancy are now known to be indicative of a healthy embryo and are associated with a range of positive pregnancy outcomes (Weigel and Weigel 1989). According to one functional explanation, the symptoms arise as a result of a classic evolutionary struggle—parent–offspring conflict, or the non-overlapping interests of mother and embryo (Haig 1993). Rather than being an illness to be treated, the evolutionary perspective has changed the perception of this condition experienced by up to 90% of pregnant women in modern societies (Pepper and Roberts 2006). Despite this, it remains the case that at most one or two lectures are devoted to evolutionary medicine during training of medical practitioners and that most researchers have an evolutionary background, rather than a clinical one. It appears that a body of specialists working within an evolutionary framework, time, and some convincing empirical demonstrations, are required to persuade practitioners of the merits of evolutionary thinking and to get them to incorporate this into their own work.

There are obvious parallels between the aims of evolutionary medicine and an applied evolutionary psychology. Researchers in the two fields use the same approaches and speak the same language, but often face antipathy, scepticism, or indifference from practitioners. However, evolutionary psychology perhaps lags slightly behind in the extent to which there is a core body of researchers who actively attempt to apply their research. Of course there are some notable exceptions, many of whom have contributed to this book, but concerted effort to apply principle to practice remains patchy and limited in extent. Perhaps this is appropriate to a young discipline, but one would expect that as it matures, its scope should inevitably begin to broaden towards application, and tackling contemporary issues in human society.

The idea for the book arose from discussions with graduate students about precisely this: what we felt to be an incipient focus on applied evolutionary psychology, founded upon the large body of theoretical work that has accumulated over recent years. Credit should be placed where it is due, however, and in this regard it is well worth noting that this book is not the first to highlight the potential for applications of evolutionary psychology—interested readers are encouraged to read some fascinating accounts of several applications in books by Beckstrom (1993), Crawford and Krebs (1997), and by Crawford and Salmon (2004). Furthermore, Saad (2007) and Miller (2009) have focused specifically on evolutionary insights into consumerism.

Evolutionary psychologists are often accused of (and, very occasionally, possibly guilty of) arguing that our underlying human nature dictates how things 'should' be. To do this is to commit what is known as the *naturalistic fallacy*. The danger of doing this, or at least being perceived to do it, is perhaps never higher than when we attempt to employ our understanding of human behaviour and its evolutionary origins to tackle specific issues, as this book explicitly sets out to do. In particular circumstances, it may turn out that what is natural is, in fact, good—the example of nausea and vomiting in pregnancy appears to be a case in point from evolutionary medicine—but this will by no means always be the case. Thus, in the main, evolutionary psychologists generally aim to map and describe the evolutionary forces at work while remaining detached from judgements of value and morality.

Here, in view of the theme of the book, the contributors spend more effort than is typically the case in speculating on the practical possibilities that these insights lend, but nonetheless they remain studiously aware of the pitfalls of attaching moral value to their findings. In each of the chapters, they: 1) describe theoretical and empirical research on evolutionary aspects of human behaviour in the relevant area, 2) elucidate how evolutionary thinking lends particular insight into applied issues that are missed by standard sociological or social psychological approaches, and 3) explore, highlight, and analyse ways in which these insights are already being or could potentially be used in practical and beneficial ways within applied settings.

Many of the authors are evolutionary psychologists who are interested in understanding how basic findings can be used for practical benefit. However, a little less than half are best described as specialists in other disciplines who have come to recognize the potential of evolutionary theory as it pertains to their existing research interests. In addition to providing an interesting blend of approaches, it is enormously encouraging that these contributors have, in their own areas, already initiated the process of introducing evolutionary perspectives to their research and to their colleagues. Persuading not just other psychologists, but also practitioners, is perhaps the greatest testament of all to the explanatory power of the evolutionary framework.

References

Beckstrom, J.H. (1993). *Darwinism applied: evolutionary paths to social goals.* Praeger, Westport, CT.

Confer, J.C., Easton, J.A., Fleischman, D.S., et al. (2010). Evolutionary psychology: controversies, questions, prospects, and limitations. *American Psychologist,* **65**, 110–26.

Crawford, C. and Krebs, D.L. (1997). *Handbook of evolutionary psychology: ideas, issues, and applications.* Lawrence Erlbaum Associates, Mahwah, NJ.

Crawford, C. and Salmon, C. (2004). *Evolutionary psychology, public policy and personal decisions.* Lawrence Erlbaum Associates, Mahwah, NJ.

Dunbar, R. (2008). Taking evolutionary psychology seriously. *The Psychologist,* **21**, 304–6.

Fitzgerald, C.J. and Whitaker, M.B. (2010). Examining the acceptance of and resistance to evolutionary psychology. *Evolutionary Psychology,* **8**, 284–96.

Haig, D. (1993). Genetic conflicts in human pregnancy. *Quarterly Review of Biology,* **68**, 495–532.

Miller, G. (2009). *Spent: Sex, Evolution and Consumer Behavior.* Viking Books, New York.

Nesse, R.M. and Stearns, S.C. (2008). The great opportunity: evolutionary applications to medicine and public health. *Evolutionary Applications,* **1**, 28–48.

Nettle, D. (2011). Flexibility in reproductive timing in human females: integrating ultimate and proximate explanations. *Philosophical Transactions of the Royal Society B,* **366**, 357–65.

Pepper, G.V. and Roberts, S.C. (2006). Rates of nausea and vomiting in pregnancy and dietary characteristics across populations. *Proceedings of the Royal Society B,* **273**, 2675–9.

Saad, G. (2007). *The evolutionary bases of consumption.* Lawrence Earlbaum Associates/Psychology Press, Mahwah, NJ.

Stearns, S.C. and Koella, J.C. (2007). *Evolution in health and disease, 2nd Edition.* Oxford University Press, Oxford.

Tinbergen, N. (1963). On aims and methods in ethology. *Zeitschrift für Tierpsychologie,* **20**, 410–33.

Trevathan, W.R., Smith, E.O., and McKenna, J.J. (2007). *Evolutionary medicine and health: new perspectives.* Oxford University Press, New York.

Weigel, R.M. and Weigel, M.M. (1989). Nausea and vomiting of early pregnancy and pregnancy outcome. A metaanalytical review. *British Journal of Obstetrics and Gynecology,* **96**, 1312–18.

Williams, G.C. and Nesse, R.M. (1991). The dawn of Darwinian medicine. *Quarterly Review of Biology,* **66**, 1–22.

Section 1

Business

Chapter 2

The evolutionary psychology of economics

Paul H. Rubin and C. Monica Capra

Introduction

The standard economic model of consumer behaviour has a consumer maximizing a 'utility' function subject to constraints. As far as economists are concerned, the utility function is arbitrary. That is, economics itself says nothing about the shape of the utility function or the elements that create utility. This has led some to say that economics is vacuous, because virtually any behaviour can be consistent with some utility function. This is not true. Economic theory does derive testable implications from the classical model of utility maximization, but they are fairly weak, dealing mostly with consistency of choices. Nonetheless, modern behavioural economics, descending from the pioneering work of Tversky and Kahneman (Tversky and Kahneman 1974, 1991, 1992; Kahneman and Tversky 1979), has demonstrated that, in experimental situations where theory has the best chance to work, subjects often violate even these weak restrictions. That is, human behaviour seems inconsistent with even the weak predictions of economic theory (for extended discussions, see Thaler 1980, 1981, 1985, 1992).

The consistency violations observed in laboratory experiments have led many to believe that a new paradigm of choice is required. Such a dramatic change may not be needed. Indeed, in this chapter, we argue that applying evolutionary psychology to economics can explain observed 'anomalies' in decision-making (Capra and Rubin (2011) provide a detailed analysis of this issue), and put some substance in the utility function itself. For instance, we could reasonably see 'utility' as just another name for 'evolved fitness'. To take the classic example, it is not arbitrary that we like to eat foods containing sugar and fat. In the Malthusian world of our predecessors, those who liked sugar and fat survived to become our ancestors. In contrast, those who liked only lettuce and carrots probably did not. This sort of analysis has been applied to issues relevant to economic choice such as time preference (Rogers 1994) and risk preference (Rubin and Paul 1979).[1]

In this chapter, we provide an analysis of the relationships between economics and the evolutionary and biological origin of economic choice. By no means do we cover all possible topics. Instead, we cover some selected topics that are of special interest to us. These are: 1) rationality and biases, especially as related to the endowment effect, 2) the extent to which individuals are selfish or pro-social, 3) individuality versus heterogeneity, and 4) perceptions of economics.

---

[1] For a general discussion, see Robson (2001). Robson has been a leader in applying evolution to economics. See his webpage for many links to relevant research: www.sfu.ca/~robson

Rationality and biases

Economics has always assumed that individuals are fully rational. However, experimental research has indicated that in many circumstances rationality is violated. From a biological perspective, these violations are not surprising. Traditional economics assumes that the brain is a maximizing general-purpose computer. But, in fact, the brain is made of several 'modules', each specialized for some particular purpose (Cosmides and Tooby 1994; Gigerenzer and Goldstein 1996; Gigerenzer, 2002; Gigerenzer and Selten 2001.) This is because evolving a general-purpose computer would itself have been too costly, and such a brain would not have created any evolutionary advantage. Rather, our brains have evolved to solve particular problems that our ancestors met in their day-to-day lives. 'Failures' of rationality, which behavioural economists call biases, occur when we use the wrong module to solve a problem, or when the structure of a problem is different from problems the brain evolved to solve.

One interesting example is the status quo bias. A perfectly rational maximizing individual would treat all wealth allocations symmetrically. This implies that whether or not an individual owns something would have no effect on its value. In other words, if an individual would be willing to pay, say, $10 for an item, then she should be willing to sell the item for $10 if she possesses it. That is, willingness to pay (WTP) to obtain an item and willingness to accept (WTA) to sell the item should be the same. However, a result in experimental economics is that these are not the same (e.g. Knetch 1989; Thaler 1992).[2] It seems, then, that people place a higher value on things that they already own, resulting in WTA always being greater than WTP. This bias is related to (and may be the same as) the utility difference between losses and gains. Basically, experimental results (and our own intuitions) indicate that losses are more highly valued—about twice as high—than gains, and giving up what we already possess may be seen as a loss. Similarly, in choice under risk, there is an asymmetry in how people behave with respect to losses and gains. Whereas people usually prefer a sure gain to a likely gain with same expected value, they prefer a likely loss to a sure loss of the same expected value.

All of these behaviours have in common that they privilege the status quo. That is, the current endowment is a baseline from which all changes are measured, and we are reluctant to move away from this baseline—particularly in a negative direction. This is inconsistent with a world of rational maximizers, for whom the status quo is just another possible state. However, all of these observations are consistent with a Malthusian world—a world in which living entities are at the edge of subsistence (McDermott et al. 2008).[3] Modern Western humans do not live at subsistence, but our ancestors did. It is believed that the essence of our utility functions was formed in this long period before modern humans populated our planet. At subsistence, any reduction in wealth or income will lead to death. Therefore, there is no evolutionary incentive to develop a utility function that will maximize over all states of the world if the individual will be dead in those where wealth is reduced. Rather, evolution would put a very high weight on maintaining current wealth. Thus, a bias towards the status quo and an aversion towards losses can be understood as evolved adaptations.

The endowment effect may have significant implications for markets (and other issues, for example, see Chapter 15, section 'The value of possession and the endowment effect'). For example, a rational individual would sell any stock that had lost value, because there are tax gains from

---

[2] For criticisms of the experimental design of these and other studies, see Plott et al. (2005) and Plott and Zeiler (2009).

[3] Interestingly, it was reading Malthus on life at subsistence that led Darwin to the fundamental insight of competition as driving natural selection.

such a sale. On the other hand, such an individual would resist selling stocks that had appreciated, again for tax reasons. But common behaviour is to resist taking losses and rather to hold a depreciated stock until it returns to the initial buying price. This is apparently in order to avoid the 'loss' associated with selling at a price below that which was paid even though, in fact, the loss occurs as soon as the stock falls in value. This phenomenon is called the 'disposition effect' and has been observed in several experiments. Clearly, the disposition effect is not rational, but is common. Similar behaviours delay housing markets in adapting to a new equilibrium as homeowners resist selling depreciated houses because of an unwillingness to realize the losses that have already occurred.

It is of course possible for rational individuals to learn how to maximize long-term wealth and ignore the endowment effect and related effects, but this must be learned. It is not our natural behaviour. Moreover, it is not only humans who exhibit this sort of biases. For example, the endowment effect has also been observed in chimpanzees (Brosnan et al. 2007) and capuchin monkeys (Lakshminaryanan 2008), and other species.

Selfish and pro-social?

Economic theory basically assumes that individuals are selfish. Economists allow for some altruism, but a desire for altruism is viewed as another good attached to an essentially individualistic utility function (see, for example, Andreoni 1995). But people are not as selfish as theory would suggest. Experimental evidence indicates that many people are more cooperative than theory would predict (e.g. Guth et al. 1982; Rabin 1993; Fehr and Schmidt 1999; see also Camerer 2003 for a review). People do not always cheat in the prisoner's dilemma game; in the ultimatum game people generally give away something; and many individuals contribute to public goods regardless of the private incentives not to do so. Finally, people also reciprocate, even in situations where there is little chance of repeated interaction.

Evolutionary theory can explain pro-social behaviour (see also Frank 1988). One possibility (although controversial among biologists) is group selection. That is, those groups (tribes, clans) containing cooperative individuals were more successful than those that were purely selfish and so we have evolved to be somewhat cooperative. Another possibility is Trivers' (1971) theory of 'reciprocal altruism'.

We must also note that some people are more cooperative than others. Assume we begin with a world in which everyone is a cooperator. A non-cooperative (cheating) mutant arises. The cheater will do better than the cooperators because it will get the benefits of cooperation and not pay the costs. (For example, the cheater will accept transfers today and then not repay the favour in the future.) If the non-cooperator has higher income, it will achieve higher fitness and so cheating genes will become more common. This is consistent with the argument that some traits are frequency-dependent and is very much like the hawk–dove game explained below (see section 'Frequency-dependent selection'). Two cooperators meeting each other will do better than two cheaters meeting each other. If a cooperator and a cheater meet, the cheater will do better. So, at equilibrium, there will be some cooperators and some cheaters.

For humans, however, there is another twist. Humans do not randomly associate with others. We have choices about who to interact with. Given a choice, it is always better to associate with a cooperator than with a cheater, whether one is oneself a cooperator or a cheater. It therefore pays cheaters to pretend to be cooperators. However, it also pays to be able to detect false cooperators. Thus, an arms race will evolve, with better and better deception mechanisms evolving and better and better detection mechanisms also evolving. At any given time, there will be some cheaters who are better at deception, and some who are worse, with the better cheaters being more successful (so that genes for better deception will spread). Similarly, there will be those who are better at detecting cheating, and those who are worse: and those who are better (whether themselves cheaters or cooperators) will be more successful.

Individuality versus heterogeneity

A related topic is that of heterogeneity.[4] Economists often assume that humans are homogeneous. This is of course incorrect—we differ from each other in many ways as described in the previous section. Economics has little to say about this difference, but evolutionary psychology can explain why humans differ from each other.

Gender

There are numerous differences between males and females. One of the most important from an evolutionary standpoint is risk preference. Females are more risk-averse than males (Croson and Gneezy 2009). The variance of male distributions for almost any trait is greater than the variance of female distributions. The reason is simple: the variance of number of offspring for males is much greater than for females. If a male takes a risk and wins, he can under some circumstances increase his number of descendants (and so his fitness) by a huge amount. We are all descended from males who successfully took risks. For a female, there is little to be gained in fitness by increasing risk, even if the gamble is successful. Number of offspring for a female is biologically limited and a successful risk will not increase that number by much, if any. On the other hand, an unsuccessful gamble will lead to fewer or no offspring. We are all descended from females who avoided risks. These preferences are encoded in our genes and exist for all sexually reproducing species where variance of offspring is greater for one sex.

Age

A second difference is related to age. Young males are more risk-seeking than older males (Rubin and Paul 1979). This is because a young male must accumulate enough resources or status to be able to reproduce. If a young male does not gain this level of resources, his genes will be lost. Only those who take gambles and succeed will survive to reproduce, and those are our ancestors. On the other hand, once a male has fathered offspring, then Malthusian principles become relevant and if the individual should lose a gamble his offspring might perish, thus leading to risk aversion.

Frequency-dependent selection

There are other differences between individuals. These are due to what is called 'frequency-dependent selection'. This exists when the success of some trait depends on the number of other individuals with that trait. For example, consider the classic hawk–dove game (Smith and Price 1973). The setup is that two animals meet at the site of some valuable resource, such as a piece of fruit. 'Hawks' will fight; 'doves' will give in. Then if a hawk meets a dove the hawk wins; if two doves meet, the outcome is random; and if two hawks meet there is a fight. Crucially, the average payoff for each type depends on the number of each in the population. If there are mostly doves, hawks do well since they mostly meet doves and win. If there are mostly hawks, doves do better

---

[4] This section is based on Rubin (2002).

since hawks go about injuring each other. The equilibrium (called an Evolutionarily Stable Strategy, ESS) is some percentage of each type, depending on the parameters of the model (e.g. what is the payoff? How badly are hawks injured?). The population will be made up of both hawks and doves; everyone will not be the same. There may also be intermediate strategies: pretend to be a hawk for a few minutes (where 'few' may be a variable itself) but then leave if your opponent fights back. Another possible strategy is the 'bourgeois' strategy: fight if you are first at the resource and give in if not. This implies a form of property rights.

In human societies we can think of numerous analogous games. How aggressive is someone in seeking resources? For example, how hard does one bargain in a transaction? Hawks might bargain very hard. They would then gain more if they are paired with an easy bargainer and so reach an agreement, but lose if there is a bargaining breakdown, which would occur when two hard bargainers meet. Doves, who would not be hard bargainers, would gain less from each deal but would do more deals. Another game is cooperativeness: does one cooperate when dealing with others or cheat? Again, cheaters do better when dealing with cooperators (suckers) but as the proportion of cheaters increases they do progressively worse. (An extreme form of cheater is a psychopath.) They may do well when dealing with honest citizens, but very badly when dealing with each other. (This explains why prisons are very unpleasant.)

If we think of all the games humans play, and all the possible strategies available, then it becomes clear that there will be a large number of different types of individuals in society. This is one reason why proposals to treat everyone identically are doomed. Markets, which allow for different abilities and preferences and specialization, allow for this individuality. This is one reason why markets are efficient and beneficial for humans.

Perceptions of economics

Unfortunately, humans are not very good innate economists.[5] The ancestral environment was lacking in many features that are necessary for understanding modern economics. To the extent that our brains evolved in these environments, they did not need to develop the ability to understand fundamental economic concepts. The analogy is speech and reading. Humans growing up in a natural environment automatically learn to speak, but they must be taught to read. Economics is like reading, not like speech. We can learn to understand economics, but we must be taught; it is not something that we innately understand. Consider these features of the ancestral environment:

1 The amount of trade was limited. While there was some trade, for example, in flint and other types of stone, there was not much. This was obviously true if for no other reason than there were simply not many goods to trade. Economies were simple and for most of our existence we were a mobile species, so there was little in the way of durable goods. There is another reason for the limit to trade, discussed next.

2 Division of labour was limited. There was some division of labour based on gender, and some based on age, but within a gender-age group most individuals did approximately the same thing. The reason goes back to Adam Smith's argument in *The Wealth of Nations* that 'the division of labour is limited by the extent of the market'. That is, as the size of the market (or of the economy) increases, it is possible for individuals to become more specialized. In a large

---

[5] This section is based on Rubin (2003). Rubin discusses 'folk economics'—the economic beliefs of untrained persons. See also Pinker (2002), who argues that we do not have an innate understanding of economics and that economics should be taught more commonly.

economy, for example, someone can spend all of his time making tools. In a small band, there will simply not be enough demand for tools for one person to do nothing but make tools. Our ancestors lived in small groups—probably no more than 200 individuals—and so there was not much room for specialization and its brother, division of labour. Specialization leads to trade (we specialize in teaching and writing about economics, and we trade payments received for creating that good for other goods such as food). If there is less specialization—if everyone does the same thing—there is less benefit from trade. The result is that we do not intuitively understand the benefits of trade, including international trade. This is one reason why protectionism—opposition to free international trade—is often a politically appealing notion. (Other reasons are discussed below.)

3 There were few capital goods. As mentioned above, our ancestors were mobile for most of their existence, and so would not have accumulated fixed capital. There was some capital in the form of tools, but not much, and no large-scale capital. As a result we do not have a good understanding of the value of capital. This may explain the appeal of false 'labour theories of value', such as Marxism. In addition to few capital goods, there was little technological change and little economic growth. Each individual pretty much died in the economic world in which he was born. For example, Gowlett (1992) refers to major technical change when a stone axe tradition changed over a period of a few thousand years. Thus, we are not good at understanding technological change, and we may undervalue the benefits of such change.

4 Finally, the world was pretty much a 'zero-sum' world. Since there was little trade and little benefit to innovation or capital, there was little difference in productivity between individuals. There was some difference, in that some would have been better hunters or gatherers than others, but not much. The result was that if one individual had more than another, it was probably because he had shirked in some way and exploited his fellows. With respect to access to females, the most important input to male fitness, the world was strictly zero-sum. The result is that we do not understand income inequality. We view inequality as a result of exploitation, rather than as a result of differential productivity. This can lead to inefficient policies with respect to taxation. We treat taxation as influencing income distribution, but do not realize that it also affects work effort and other inputs into the size of the economy.

This zero-sum mentality carries over to other areas of thought. For example, the number of jobs is viewed as being fixed. That means that the perception is that if one person gets a job then another person must lose one. This of course is incorrect—if one person gets a job then that person has income to spend and that income will generate another job. But this notion of the fixed number of jobs is another reason for opposition to free trade. People believe that increasing trade reduces domestic jobs, rather than understanding that increasing trade increases efficiency but has no effect on domestic jobs. Similarly, we view minimum wages as affecting earnings, but do not realize that the level of the wage also affects the number of jobs, since the number is not really fixed.

We note briefly that humans have erroneous intuitions in many areas—there are 'folk psychology', 'folk physics', 'folk chemistry', and 'folk biology'. However, in a democracy, individuals vote on economic policies, and misunderstanding of economics is socially costly. After all, if I do not understand physics there is not much harm, but if I (or most voters) have false intuitions about economics, some socially costly policies may be adopted. For example, the United States has done a decent job of reducing tariffs and other barriers to trade, but in the current recession (we write this in 2010) there is political pressure to raise these barriers again.

Moreover, there are few incentives to learn efficient decision-making principles for public policy. The vote of any one individual is virtually worthless (has no chance of being decisive), so if an individual has incorrect beliefs about public policy there is no cost to that individual. This is in contrast to ideas of efficient private decision-making (for example, with respect to the endowment effect) where correct decision-making can increase wealth (Caplan 2007). Politicians have little incentive to get things right either. If their constituents are misinformed, then the incentive of politicians seeking election is to cater to those beliefs, even if the politician knows them to be incorrect.

Conclusion

The data that behavioural and experimental economists have gathered from decisions in laboratory environments suggest that important assumptions about behaviour that economists make are inconsistent with actual behaviour. This poses a serious challenge to economists, and some have called for totally new paradigms of decision-making. In this chapter, we gave examples of how evolutionary psychology can explain anomalous behaviour including violations to rational choice, such as the status quo bias. Evolved mechanisms can also explain the curious persistence of pro-social behaviour in one-shot anonymous interactions, and the observed heterogeneity of agent types. Finally, we argue that the perception of economics is also affected by evolved mechanisms from pre-modern times. This has far-reaching policy and political implications. We strongly believe that a better understanding of our evolved behaviour and its origin can help us find ways to improve it. In particular, it is possible for individuals to learn to counter the endowment effect (and many of the other apparent irrationalities discussed in this paper). Indeed, we can study principles of optimal behaviour and so learn to improve our own decision-making. But some individuals will not undertake such learning and, in some circumstances, there is little private benefit from such learning in any case.

In a recent book, Thaler and Sunstein (2009) argued that social policy should consider these aspects of behaviour, and should lead to 'libertarian paternalism'. By libertarian paternalism, the authors mean that we should structure choices so that defaults are utility maximizing. This position is controversial (e.g. Rizzo 2009). There are some obvious examples where such paternalism is useful; for example, setting pension defaults to increase savings. From a conventional libertarian perspective, these situations are limited in number. There is, however, the danger that government officials may overreach and may themselves be subject to the same biases, and so can induce undesirable policies. Rubin (2002) has made a similar argument with respect to courts and juries. That is, the legal system itself may be subject to the same biases as individual decision-makers. If we try to apply principles derived from behavioural economics to legal institutions, juries may enforce rules that are worse, because they have little room for learning, do not bear the costs of their decisions, and are subject to the same biases.

Economists have not been hostile to evolutionary thinking. But they have not embraced it either. We believe that more explicit attention to the evolutionary bases of economic behaviour would lead to real advances in economic theory. Economics has always been based on the existence of a coherent theoretical foundation. It is in danger of losing this core and replacing it with a group of ad hoc hypotheses derived from experiments. Application of evolutionary thinking to economic behaviour can provide a core based on a firmer foundation.

References

Andreoni, J. (1995). Warm-glow versus cold-prickle: the effects of positive and negative framing on cooperation in experiments. *Quarterly Journal of Economics,* **110**, 1–21.

Brosnan, S.F., Jones, O.D., Lambeth, S.P., Mareno, M.C., Richardson, A.S., and Schapir, S.J. (2007). Endowment effects in chimpanzees. *Current Biology,* **17**, 1704–7.

Camerer, C.F. (2003). *Behavioral game theory: experiments in strategic interaction*. Princeton University Press, Princeton, NJ.

Caplan, B. (2007). *The myth of the rational voter*. Princeton University Press, Princeton, NJ.

Capra, M. and Rubin, P.H. (2011) Rationality and utility: economics and evolutionary psychology. In: Saad, G. (ed.), *Evolutionary psychology in the business sciences*. Springer, Heidelberg, Germany.

Cosmides, L. and Tooby, J. (1994). Better than rational: evolutionary psychology and the invisible hand. *American Economic Review*, **84**, 327–32.

Croson, R. and Gneezy, U. (2009). Gender differences in preferences. *Journal of Economic Literature*, **47**, 448–74.

Fehr, E. and Schmidt. K.M. (1999). A theory of fairness, competition, and cooperation. *Quarterly Journal of Economics*, **114**, 817–68.

Frank, R.H. (1988). *Passions within reason: the strategic role of the emotions*. Norton, New York.

Gigerenzer, G. (2002). *Calculated risks: how to know when numbers deceive you*. Simon and Schuster, New York.

Gigerenzer, G. and Goldstein, D.G. (1996). Reasoning the fast and frugal way: models of bounded rationality. *Psychological Review,* 103, 650–69.

Gigerenzer, G. and Selten, R. (2001). Rethinking rationality. In: G. Gigerenzer and R. Selten (eds), *Bounded rationality: the adaptive toolbox*, pp. 1–12. MIT Press, Cambridge, MA.

Gowlett, J.A.J. (1992). Tools – the Paleolithic record. In: S. Jones, R.D. Martin, and D.R. Pilbeam (eds), *The Cambridge encyclopedia of human evolution*. pp. 350–60. Cambridge University Press, Cambridge.

Guth, W., Schmittberger, R., and Schwarze, B. (1982). An experimental analysis of ultimatum bargaining. *Journal of Economic Behavior and Organization*, **3**, 367–88.

Kahneman, D. and Tversky, A. (1979). Prospect theory: an analysis of decision under risk. *Econometrica*, **47**, 263–92.

Knetsch, J.L. (1989). The endowment effect and evidence of nonreversible indifference curves. *American Economic Review*, **79**, 1277–84.

Lakshminaryanan, V., Chen, M.K., and Santos, L.R. (2008). Endowment effect in capuchin monkeys. *Philosophical Transactions of the Royal Society B*, **363**, 3837–44.

McDermott, R., Fowler, J.H., and Smirnov, O. (2008). On the evolutionary origin of prospect theory preferences. *Journal of Politics*, **70**, 335–50.

Pinker, S. (2002). *The blank slate: the modern denial of human nature*. Viking, New York.

Plott, C. and Zeiler, K. (2007). Asymmetries in exchange behavior incorrectly interpreted as evidence of prospect theory. *American Economic Review*, **97**, 1449–66.

Plott, C.R., Zeiler, K. Carbone, E., and Starmer, C. (2005). The willingness to pay–willingness to accept gap, the 'endowment effect,' subject misconceptions, and experimental procedures for eliciting valuations. *American Economic Review*, **95**, 5–20.

Rabin, M. (1993). Incorporating fairness into game theory and economics. *American Economic Review*, **83**, 1281–302.

Rizzo, M.J. (2009). Little Brother is watching you: new paternalism on the slippery slopes. *Arizona Law Review*, **51**, 685–739.

Robson, A.J. (2001). The biological basis of economic behavior. *Journal of Economic Literature*, **39**, 11–33.

Rogers, A.R. (1994). Evolution of time preference by natural selection. *American Economic Review*, **84**, 460–81.

Rubin, P.H. (2002). *Darwinian politics: the evolutionary origin of freedom*. Rutgers University Press, New Brunswick, NJ.

Rubin, P.H. (2003). Folk economics. *Southern Economic Journal*, **70**, 157–71.

Rubin, P.H. and Paul, C.W. (1979). An evolutionary model of taste for risk. *Economic Inquiry*, **17**, 585–96.

Smith, J.M. and Price, G.R. (1973). Logic of animal conflict. *Nature*, **246**, 15–18.

Thaler, R. (1980). Toward a positive theory of consumer choice. *Journal of Economic Behavior and Organization*, **1**, 39–60.

Thaler, R. (1985). Mental accounting and consumer choice. *Marketing Science*, **4**, 199–214.

Thaler, R.H. (1981). Some empirical evidence on dynamic inconsistency. *Economics Letters*, **8**, 127–33.

Thaler, R.H. (1992). *The winner's curse: paradoxes and anomalies of economic life*. Free Press, New York.

Thaler, R.H. and Sunstein, C.R. (2009). *Nudge: improving decisions about health, wealth, and happiness*. Penguin, New York.

Trivers, R.L. (1971). The evolution of reciprocal altruism. *Quarterly Review of Biology*, **46**, 35–57.

Tversky, A. and Kahneman, D. (1974). Judgment under uncertainty: heuristics and biases. *Science*, **185**, 1124–31.

Tversky, A. and Kahneman, D. (1991). Loss aversion in riskless choice: a reference-dependent model. *Quarterly Journal of Economics*, **106**, 1039–61.

Tversky, A. and Kahneman, D. (1992). Advances in prospect-theory: cumulative representation of uncertainty. *Journal of Risk and Uncertainty*, **5**, 297–323.

Chapter 3

The evolution of business and management

Nigel Nicholson

Introduction

The evolutionary paradigm is a powerful one, highly effective in predicting and providing parsimonious explanations for an almost limitless range of social phenomena, including those in the domain of business and management. This creates a challenge for a review such as this as to which aspects to focus on. My intent here is to take a broad systemic view, looking at the nature of work organizations and how they are directed and coordinated by human agents.

I shall start by introducing how co-evolution operates via multi-level selection (sometimes called group selection), without which it is not possible to make sense of the changes, successes and failures recorded in history. I shall not seek to summarize all extant neo-Darwinian work in the organizational field—it is too patchy (Nicholson and White 2006). Rather, I shall show how an evolutionary perspective can explain key aspects of one major recent business event: the so-called Credit Crunch of 2008. Then, using the evolutionary framework I shall overview the nature of work and organization and how it has changed over the ages. I shall argue that evolutionary theory contains a major gap in its treatment of the role of human agency in cultural evolution, due to its lack of attention to the self, which I shall argue is an evolved organ of special power for humans. Drawing on ideas about self-regulation from social psychology I shall seek to explain historical change in leadership and management, and the dynamics of effectiveness and failure, concluding by considering implications for future theory and practice.

The co-evolution of business

Business and management are very recent phenomena in the span of human evolutionary time, though they replicate aspects of human existence that are as old as our species (Sahlins 1972). Business can be defined as the institutions that facilitate production and economic exchange, and management is the mechanism by which agents regulate and control these processes.

To look beyond the contemporary meaning of these terms one has to consider the universal underlying factors of human life that they are representing. This is the need for all social systems to cope adaptively with: 1) widely varying and changing environmental contingencies, and 2) the relatively unchanging universals of an evolved human nature. The design of all cultures and societies has arisen to mediate between human nature and current circumstances. Periodically, social institutions fail or are transformed by revolutionary forces because of their inability to meet the adaptive challenge (Boyd and Richerson 1985). This is the process of multi-level selection—the idea, of growing importance within evolutionary theory, which recognizes that the forces of selection operate not just on the replicating unit of the gene, but on the contexts that select them—groups and societies—creating a recursive process through which local populations develop differentiated adaptive profiles (Henrich 2004; Richerson and Boyd 2005; Sober and Wilson 1998).

Co-evolution via multi-level selection, sometimes discussed under the label of Dual Inheritance Theory (McElreath and Henrich 2007), lies at the heart of our species' success story—our ability to prosper in all climates and under vastly different conditions of existence. It means that we can set the rules of the game to suit environmental conditions, allowing us to continue to work towards our ultimate goals of reproductive fitness—individually and collectively—by very different proximate routes. We establish and reinforce local norms that encourage the activation of motives and methods that fit the constraints of the time and place. These may directly affect our fitness; for example marriage laws, property rights, rules of exchange, and the like set out the conditions and criteria by which individuals will succeed and fail. In short, the processes of selection are locally tuned.

Cultural evolution is thus never arbitrary, but it can become de-tuned to the critical contingencies which it should satisfy. So it was with the Roman Empire, Communism, and the crisis in financial markets in the last decade—the last of these I shall shortly use as an illustrative case study. Co-evolution also operates at the level of business organizations, which have to satisfy the needs of their members and meet the demands of their markets (Aldrich 1999). Many go out of business through adaptive failure. Enron, a notable case study, failed dramatically because its internal selection criteria promoted people who reinforced a sub-cultural profile that was ill-tuned to its capabilities and markets (Kulik 2005). The co-evolutionary process clearly does not move smoothly in making its adjustments, but does so rather by a sequential process of adaptation–institutionalization–revolution, or sometimes, extinction—what has been called in palaeobiology 'punctuated equilibrium' (Eldredge and Gould 1972). In human institutions success is the enemy of change (Audia et al. 2000), and it is adaptation to failure that leads to development (Senge 1990).

It is a principal project of the interdisciplinary field of evolutionary psychology to analyse the nature of human nature and consider the consequences (for good or ill) of our adaptiveness to a changing world (Barrett et al. 2002; Buss 1999). Evolutionary psychologists have mainly focused on: 1) the motives that guide choices and preference; 2) the cognitive biases that shape decisions and evaluations; 3) preferred social involvements and responses; 4) communal sensitivities, 5) symbolic sensitivities; and finally, 6) individual differences.

The human imperative—explaining the Credit Crunch

Rather than reviewing specific literatures, let us use these six major themes of EP to consider a major recent social phenomenon—the so-called Credit Crunch, the financial crisis that engulfed the world in 2008—and how its origins and course tell a story about human nature.

The Credit Crunch was a classic bubble (summarized in Figure 3.1)—the latest in a long line that have afflicted markets, starting with the tulip crisis of the 14th century, and culminating in the e-commerce bubble of the 1990s. Using the six themes (all except individual differences are shown in the figure, for these have an overarching influence on the other five), let us look at the genesis of the financial crisis of 2008.

1 The motives that drive all markets are, principally, striving for status, reputation, and the rewards that go with it. All are fitness enhancing—that is, they serve the value that underpins all attributes: reproductive fitness. All the firms trading toxic assets were using simple profit and loss based direct incentives to reinforce an unrestrained maximization of sales. This was compounded by the agency problem—that the actors had no real stake in the underlying assets—hence there was no fear of loss, only an unrestrained susceptibility to the positive

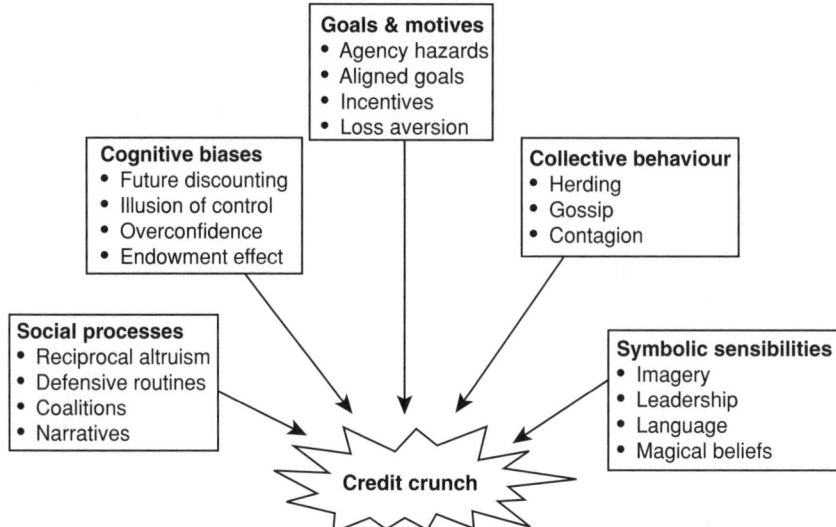

Fig. 3.1 Human nature and the Credit Crunch.

incentive effects. Yet loss aversion played a major part—the fear of being left behind in the gold rush. Loss aversion is represented in psychology by prospect theory (Kahneman and Tversky 1979), which describes an evolved form of prudence—a tendency for the utility curve to be steeper in the domain of losses than in the domain of gains, i.e. we hate to lose more than we love to win. This supported the bandwagon effect that inflates and mobilizes bubbles (De Bondt 2005)—fear of being left behind is much more tangible than the apparently remote probability of system failure. The latter disregard for distant events is called hyperbolic time discounting (Steel and König 2005): a scalar decrease in the utility of both negative and positive events with expected temporal distance; a fitness-enhancing perspective in uncertain and risky environments. In the Credit Crunch even people who foresaw an eventual crash pursued 'jam today'.

2 Both of these phenomena—loss aversion and future discounting—are at the intersection of motivation and cognition. All the cognitive biases that have been extensively documented by decision researchers play into the irrationality of bubbles. Especially relevant are illusions of control (Langer 1975), found to be a pervasive influence on behaviour in a recent study of traders in financial markets (Fenton-O'Creevy et al. 2005). It influences decision-making at all stages, not least in the belief that one can recover from the negative outcomes of risky decisions. This is a subsidiary element in the over-confidence syndrome to which decision-makers are generally susceptible. This well-documented phenomenon is, disturbingly, correlated with self-esteem (Kernis et al. 1982). In the context of human evolution such a stance has clear fitness benefits for a species endowed with more spirit than physical prowess, surviving under conditions of risk and uncertainty, but it has manifest drawbacks in the hyper-rational world of financial markets. The endowment effect—overvaluing one's assets—was also clearly relevant. This raises the interesting question of how a species endowed with the gift of reflexive cognition is capable of the various follies of bubble behaviour: obtuseness as to how emotions overcome decision-makers' reason, and how short-term values regularly obscure long-term considerations. As Trivers (2000) has explained, self-deception is an essential but dangerous tool in the human toolkit. It enables us to pursue distal fitness goals under the guise of quite

different proximate aims, helping us as actors to maintain a convincing and committed performance, undisturbed by awareness of our deeper self-serving motives.

3 Social processes are the chief source of our species' supremacy on the biosphere. It is our ability to transcend kinship networks via cooperative alliances, incorporating such processes as role differentiation, self-organization, and trading assets, that enables complex organizations to grow and markets to operate. As the pioneering researchers of the human relations school of the 1940's and 1950's recognized (Rose 1985), informal processes typically are more compelling than formal ones, and need to be congruent with the goals of an organization for it to achieve its goals. The Credit Crunch perfectly illustrates the risks of human sociality. Reciprocal altruism (Sober and Wilson 1998) bound parties into acts of dubious propriety, with coalitions acting in consort, diffusing responsibility for risky decisions—sometimes called 'groupthink' (Whyte 1998). When contrary evidence emerged, defensive routines kicked in (Argyris 1986), with buddies offering support to each other in the face of challenges to their internally consistent worldview. This was supported by the narrative processes that help us make sense of the world. Story telling is a particularly compelling human predilection, imbuing often disconnected events with the logic of human intention, often with self-serving moral overtones (Ibarra and Barbulescu 2010; Nicholson 2000). It is clearly fitness enhancing: 1) to find order and meaning in apparently unconnected events, and 2) to interpret events in ways that render self and allies blamelessly rational. The collective blindness to evidence of the bubble was sustained by narratives supporting leading actors' interests.

4 The collective behaviour that most obviously inflates bubbles is herding: a form of mimesis and a heuristic relying upon 'the wisdom of crowds' (Surowiecki 2004). This has been a primary focus of behavioural economics, where the imitation of the responses of others is framed as a mechanism for capitalizing on knowledge asymmetries (Parker and Prechter 2005). This account may be excessively rational, for human herding is also based upon simpler heuristics such as those that govern the behaviour of stampeding ungulates—the safest place is in the centre of the herd, where females and young are protected (Low et al. 2005). Following the crowd may just be safer. Yet evolutionary biologists agree that complex choices are simplified by following the majority. Communication within an informal network, especially about what other actors are doing, thinking, and feeling—what is normally considered 'gossip'—is a vital element in the informal social order of organizations. Gossip circulates through networks that are bounded by what is called 'Dunbar's number'—150—the estimated size of human network cognitive capacity (Dunbar 1992). This means that major movements are directed by sub-cultures of quite limited dimensions, whose identity and purpose is transacted through gossip. Gossip follows people's needs, not the reality they are facing, and had no value as an early warning mechanism in the Credit Crunch, though when it was well underway it served to accelerate the contagion of collective panic. Herding plus loss aversion means, in a downturn, that the bandwagon becomes a stampede.

5 Symbolic sensitivities played into the Credit Crunch in a number of ways. Cultures are replete with images and values that capture the needs and interests of their communities (Sperber 1996). Leaders sustain and create cultures—and many of the now disgraced leaders of the financial institutions at the apex of the pyramid of illusion believed that the credit boom was infinitely elastic, a classic aspect of bubble psychology (De Bondt 2005). I shall say more about the evolution of leadership later in this chapter, but here they played a major part in sustaining the magical belief that there were no limits to the growth of these risky assets. The language of credit derivatives is arcane and obscured the ephemeral underlying nature of the assets upon which they were resting.

6 Stable and largely heritable individual differences play a key part in co-evolutionary processes, mediated by sexual selection and role differentiation (Nettle 2007). In other words, at different times individuals with particular profiles may prosper. Since the most important part of the human environment is other humans, changes in the culture alter the balance of advantage and disadvantage to possessing particular characteristics. Sometimes we need more leaders, sometimes more followers; at times more warriors, at others more peace-makers (Spisak et al. 2011). The changing value of trait depending on its scarcity is called frequency dependent selection (Nettle 2006), and is a key explanatory element in understanding how cultures are formed and change. People have widely indicted the culture that arose in the financial services through the 1990s and beyond. Here was an industry virtually untouchable as the major engine of economic growth, especially in the Western world, but dominated by firms like Bear Sterns and Lehmann Brothers with internal cultures and leaders that were in thrall to the gratification of their immediate instincts (Ward 2010).

The Credit Crunch is a microcosm of many social and institutional failures, at the heart of which lie organizational behaviour processes—from the level of the actors' psychology through to the cultural context. The logic of a co-evolutionary perspective is that history is a vehicle which is driven by human impulses accommodating to changing contingencies, including many self-created ones.

For the remainder of this chapter I want to focus on two key features that have changed historically: the way we organize, and how we are led. First, though, let us look at what we can learn from the past.

A short history of organization

Human beings have always worked together and coordinated their efforts by means of authority and organization, so one can say that business and management are fundamental human processes that over time change dramatically in form. We lack direct evidence as to how our most distant ancestors lived and worked together, though much can be inferred from the fossil record and from comparative anthropology. The earliest human organization was that of semi-nomadic bands of hunter-gatherers in a savannah environment, foraging, hunting and practicing primitive forms of proto-agriculture (e.g. periodically returning to sites where edible plants have been nurtured) (Tudge 1998). The basic form has been replicated across a variety of conditions, as humans spread across the planet, from the tropics to the ice fields of the Inuit, at the end of the last Ice Age (Diamond 1991). Their model of work, organization and management, compared with those seen currently, is shown in Table 3.1.

These are very different worlds, and across the sweep of history there have been other models in between, including the highly collectivist and stratified pastoralist social organization (Nicholson 2005) plus various forms of agrarian economy. Vicissitudes of climate and the seasons govern the ways of life of these peoples—a critical input to the co-evolutionary development of social forms. The first critical discontinuity in the history of human organization was the advent of farming around fixed settlements that occurred around 10,000 years ago. For the first time individuals were able to accumulate wealth and, crucially, to pass it on between generations. Now power was not fragile, temporary, or distributed across societies but concentrated and used for the strategic self-interested purposes of those who held it. The sociopolitical forms that emerged were dynastic oligarchies, with power concentrated in the manifestation of slave states, warlords and despotic monarchs (Landes 1998). These have persisted into modern times, but have been increasingly moderated over recent millennia by the development of countervailing economic,

Table 3.1 Work, organization and management in early hunter-gatherer societies (Then) contrasted with contemporary models (Now)

Then	Now
Fuzzy boundary between work and non-work	Clearly demarcated work and leisure time
Close intertwining of consumption and production	Units of production and consumption mostly separated in time and space
Labour alongside and in cooperation with close or distant kinfolk	Collaborative endeavours often with non-kin, and often with strangers (one-shot interactions)
Labour governed by informal agreements and norms of reciprocal exchange	Labour governed by formal contracts specifying rights, obligations, and rewards
Authority fluid and shared, based on availability, expertise, interest, or experience	Authority vested in positions, often hierarchical, based on formal criteria or selection processes
Skill developed by mimesis and mentorship	Skills developed primarily by formal education and training schemes
Rewards for labour that are intrinsic or collective (food-sharing)	Rewards that are extrinsic and individual, mediated by agents and contracts

military and regional powers, establishing systems that bestow rights on the governed, by legal-judicial institutions and forms of governance.

The last 200 years has witnessed increased democratization of political and business institutions, brought about by a variety of forces, mainly education and economic diversification that have given increased power to all levels of society. Citizens become more important as agents in wealth creation and as consumers of goods and services. The response has been institutions realizing they need to move beyond 'command-and-control' and engage the motivation and cooperation of their people. The first truly complex business organizations were probably the monasteries (Kieser 1987), which combined sophisticated division of labour with elaborate management systems. The apotheosis of rational organizational systems came in the mid-20th century with so-called 'scientific management', or Taylorism after its originator. This innovation combined a number of elements to create consistent and efficient industrial production: decomposed operations, trained operators, simple rules, work measurement and incentives tied directly to production (Rose 1985).

The brilliant success of this method in sustaining efficient and consistent production has ensured that it persists in many parts of the world, though its manifest dysfunctions and costs—exposed by the human relations counter-revolution—arose directly from its failure to satisfy the needs and interests of workers. This battle between economy and psychology has been a fairly continual theme throughout the 20th century and continues into the current millennium. Evolutionary psychology not only tells about the imperative of human nature, but also about adaptability as key to our success as a species. We are able to endure privations, defer immediate wants, and practice new and difficult behaviours in order to satisfy our distal goals (Bernard et al. 2005).

This creates many of the dilemmas that characterize modern living. In order to get what we want for ourselves and our kindred, we are prepared to bend our lives, minds, and bodies out of shape. Thus does economy trump psychology. Men and women live manifestly self-destructive lives in order to keep up in the race to acquire material resources and status. Labour laws and the trade union movement grew over much of the twentieth century to check the destructive consequences of this process, and have since declined as we have co-evolved toward less monolithic forms, requiring new kinds of workers with different kinds of contracts and commitments.

These developments are uneven within as well as across regions, with highly traditional 'machine bureaucracies' persisting alongside newer ad hoc forms (Romanelli 1991). Traditional forms continue to predominate in emerging economies, while in the West we move towards creative and knowledge-based industries. Indeed, reviewing Table 3.1, one may remark that many of the newest organization designs—temporary, decentralized, and ad hoc—look in some respects like a return to ancestral models, partly to satisfy the preferences of skilled and scarce workers, and partly to meet the need for operational flexibility, innovation and high quality outputs. Small entrepreneurial family firms also resemble the 'then' model in many respects. Indeed, they owe their persistence and economic success to their ability to create cultures that satisfy human wants by embodying a more ancient and preferred kinship-based model of organization (Nicholson 2008).

Yet we must take care here of not falling into the trap of asserting what is 'natural' for humans. All human creations, including social systems and environments, are equally 'natural'—they just have different adaptive challenges and consequences. The hunter-gatherer prototype clearly satisfies a number of requirements, including a lifestyle that is consistent on most points with human wants, biases, and abilities, yet this is also a world where everyday brutality and warfare intermingle within the bounds of a highly collectivist culture (Chagnon 1997).

Self-regulation and the evolution of culture

A key to understanding co-evolutionary processes is the reflexive self, an idea that is perhaps the most challenging facing evolutionary theory at the present time. It does seem as if for 160,000 years, since the first appearance of truly modern humans, our existence was a stable and bounded hunter-gatherer lifestyle and culture, and then around 40,000 years ago a dramatic change occurred—sometimes called 'the Great Leap Forward' (Diamond 1991). At this time humans migrated to just about every corner of the planet, and started to produce a plethora of decorative and artistic artefacts.

These ranged from cave art, figurines, and the decoration of the tools and weapons of these peoples. They seem to signify a completely altered relationship with the environment, denoting such qualities as possession, veneration, and idealization. It has been speculated that this marks a truly momentous shift in human evolution, as yet not identified with any measurable anatomical shifts: the evolution of the self (Mithen 1996). For the first time, a living organism had a sense of its own agency and identity. This capacity, it has been argued, was a necessary precursor to the reading of other minds and intentions (Humphrey 1980), a process that hugely benefits complex group living.

Along with the sense of self comes awareness of mortality, ability to reflect on one's goals, to plan for future states, to defer goals, to dissemble, to manage one's moods, and a range of other highly sophisticated manipulations of self and others (Bandura 1982). The study of these processes comes under the heading of the growing sub-field of social psychology called self-regulation; mainly concerned to date with failures to self-regulate, such as procrastination, addictions, and a range of ultimately self-harming behaviours (Baumeister et al. 2007). Indeed it is via self-regulatory failure that economy trumps psychology—the lure of short-term gratification undermining longer-term fitness.

Such problems of impulse control and self-management do not arise under conditions of heavy external regulation, such as persisted for much of our evolutionary history (Leary 2004). Emotions such as regret, self-doubt, intra-psychic conflict, and chronic anxiety are rare in tribal peoples.

Under their highly regulated culture there is little scope for the more convoluted forms of agency that we now take for granted. In preliterate tribes, responsibility for many external events is attributed to supernatural external agencies, social relations are governed by custom and rules, and individual conduct is largely subordinated to direction by kin and community. There is little space left for self-obsession.

However, it is the innovation of the self that provides the engine of cultural evolution. It is true that some of our closest cousins among the great apes develop material cultures in such elements as tool use, social rituals, and expressive behaviour (McGrew 1992), with regional variations that are clearly adapted to local environmental conditions. Yet they cannot be said to be co-evolutionary since they seem not to generate any of the enduring selective processes (especially sexual selection) that marks variations in human cultures. The self is the agency of co-evolution through its inputs and how it processes them.

Let me explain. An elaborated model of the self, such as I have proposed elsewhere (Nicholson in press) incorporates the idea, first articulated by William James (1890), that we should distinguish between the self-aware executive function (the Ego, I) and various perceived selves (the Me). It is the ability to conceive of ourselves as both subject and object, using our gifts of language and autobiographical memory, in terms that are both conceptual and constructive, that facilitates much of what we do, especially in how we manage our goals and emotions.

This is the perfect vehicle for co-evolutionary processes. For example, in human societies reputation is a powerful universal value (Henrich and Gil-White 2001) but its currency changes—i.e. the basis upon which reputation is earned by actors. These criteria are readily edited and regulated, so cultures are able to change the rules of engagement by which individuals contest for personal enhancement. The changing character of status and leadership are exemplars of this process.

Status hierarchy, dominance, and gender

Leadership and management are terms—often interchangeable—that capture functions that need to be performed in social systems. Both denote the use of human agents to coordinate and direct human effort towards shared goals. In human societies collective behaviour is governed by many systems, including rules, norms, operational procedures, and routines—what have been called 'substitutes for leadership' in the literature (Howell and Dorfman 1981). Indeed, it is often viewed as desirable that agency should be removed from decision-making. Max Weber (1954 [1922]), the father of sociology, was the first to identify the capacity of bureaucracy as a species of 'legal domination' as an alternative to the capricious risks of what he termed 'charismatic' and 'traditional' domination.

Weber was not a Darwinian, though his use of the notion of domination showed recognition of one of the guiding forces for human organization. Along with all other primates, humans self-organize into dominance hierarchies, and, as among all the other great apes, these are male orders (Meder 2007). Status competition is a primary source of fitness enhancement in social animals, for both the individual and the group, ensuring that the genetically best-endowed have first access to scarce resources and breeding opportunities (Ridley 1993). Sexual selection is the mechanism driving this—males contesting for advantage and females being attracted to and selecting the prime males (Miller 2000). Dominance is thus a device for assortment, ensuring the genetic fitness of the group and its offspring are optimized.

However, dominance is fragile, and fitness is a property of the phenotype which may imperfectly express the underlying genotype. Good genes require a supportive environment and good opportunities to have the chance to develop to advantage. This makes heritability unreliable and as a consequence, dominance hierarchies in primates are characterized by fluidity and change. So it was, and remains, in hunter-gatherer groups. They are egalitarian with fluid and dispersed power-holding (Whiten 1998). Yet, as we have seen, over the course of human history what it takes to succeed has changed continually.

The criteria matter less than the outcome, which is mainly a story of undiluted benefits accruing to those of higher status (Nicholson and De Waal-Andrews 2005). The benefits of high rank have long been observed among primates and other mammals, but it has only been relatively recently recorded that within human society much the same pattern holds (Ellis 1994). Perhaps the definitive record was supplied by the so-called Whitehall epidemiological studies of thousands of members of the British Civil Service (Van Rossum et al. 2000) which blew away the myth of the stressed-out and suffering senior executive, and revealed a disturbing reverse pattern. Longevity, health, happiness, social adjustment, and of course wealth are consistently directly related to social standing. Conversely, many forms of psychological problems are 'disorders of rank'—status deprivation impairs fitness (Nesse and Williams 1994; Stevens and Price 1996).

Looking at contemporary business organizations there is a clear preference for the classical line and staff pyramid form, which combines layered hierarchy with division of labour via limited spans of management control (Jelinek 2005). This line and staff arrangement is near universal in the military, health care, industrial production, and large scale service firms. Its success owes largely but not entirely to its demonstrable efficiency, reliability, and consistency of outputs, for there are many alternative ways of organizing to produce the same results (Donaldson 1995). As we observed earlier, currently, advances in technology and the need for smarter, more flexible ways of organizing to cope with multiple sources of supply and complexity, are spawning new forms that resemble the less hierarchical forms of our hunter-gatherer ancestors (Lewin et al. 1999). In business there are several examples, mostly arrived at unconsciously, such as the highly fluid and self-organizing Gore-Tex company (Hamel 2008). One designed consciously to be consistent with the evolutionary psychology view of human nature, is the fast-growing Flight Centre travel agency (Johnson 2005)—modelled around 'family, villages and tribes' after the founder read a *Harvard Business Review* article by the author (Nicholson 1998).

Yet, one can see that when there are choices about how to organize there is a persistent revealed preference for the classical form. This design reflects hard-wired male preferences for dominance hierarchies, periodic tournaments for advancement, and highly specified roles. Organizations deploy elaborate human resource management systems of appraisals, tournament promotion systems, and performance measurement methods that appear to be multifaceted but in practice resolve to simple binary hierarchical advancement decisions. This has as its underlying logic what I have called 'the false theory' of meritocracy (Nicholson 2010a,b)—heavily disguised protocols that support the fallacious belief that the qualities of individuals can be reduced to a single merit quotient, and that this is measurable and stable.

Men understand that this is an ancient game whose rules they know well, and around the world leaders try to bias the tournaments so that their power is preserved. Christopher Boehm, an evolutionary scholar who studied primate behaviour with Jane Goodall and turned subsequently to comparative anthropology, has studied a large and diverse sample of pre-industrial tribal societies. He observed that the fluidity and democracy of all the instances he recorded were due to what he called 'reversed dominant hierarchies' (Boehm 1999). He argues that in these societies the natural tendency for males to seek to dominate others—what he calls 'upstartism'—is a threat to social stability and the achievement of goals. The tribe thus regulates power via a repertoire of forms of social pressure such as ridicule and isolation to hold back those who would dominate excessively.

The unrestrained despotism of chiefs living in fixed settlements is the result of power being unchecked, because the powerful have accrued scarce resources. The co-evolutionary argument offered earlier is that as leaders increasingly need the cooperation of the led, more democratic forms reassert themselves. As R. Hogan (2010) has observed, leadership abuses become less likely the more free followers are to defect.

My argument here is deliberately gendered, i.e. that the world of business management and organization is traditionally male-dominated and reflects male preferences in design and custom. That is changing; increasingly we are seeing not just women in the workplace but in positions where they can show how they might manage differently. Yet the so-called glass ceiling seems to be proving remarkably resistant to removal. The headline figure of around 15% of women in top corporate jobs has hardly shifted in the last two decades. Let us look at the biological origins of this and whether they are mutable.

Primatologists, observing the social life of the great apes, note that contests for supremacy are highly political, involving coalitions and alliances among competing kinship groups, with females playing a central role as power brokers (De Waal 1989; Ahmad 2009). The universal model of sex differences in mammalian kinship is that status is much more heritable for one sex, with the other sex being exogamous, i.e. dispersing from the natal home to reproduce elsewhere (Ridley 1993). For all great apes the pattern is for males' status to be transmitted intergenerationally while females are exogamous—they achieve a wider range of status outcomes by migrating at maturity from the natal home to acquire the status of their mate. For some polygynous species (humans are considered 'mildly' polygynous by various criteria), this means females are confronted with the challenge of brokering their acceptance by the new family group, especially their mate's mother and any other established wives.

This lays the basis for a quite different adaptive challenge for males and females. Males need political intelligence plus the will and ability to contest and achieve, and women need to be able to deploy social and emotional skills to negotiate entry to non-kin groups and achieve social acceptance. In our nearest primate cousins, dominance among females is based upon number of offspring and seniority (Ahmad 2009) while for males it is a mix of risk-willingness, ability to build alliances, and ability to dominate other males. Similar biases may be attributed to human sex differences in risk-taking, assertiveness, emotional sensitivity, and interpersonal skills (Geary 2010), yet, through co-evolutionary processes (contextual change) the criteria of success change over time. Men need to be able to compete with other males on whatever criteria are currently valued by the community, i.e. it is not always the tough guys who get ahead, but at other times those who are smart, entertaining, inventive, and sporting (Henrich and Gil-White 2001; Miller 2000).

One can see how these biases are expressed in contemporary life. Men are especially inventive in finding new criteria on which to play out dominance games (Miller 2000). Women who enter this world and join these contests find themselves at an immediate disadvantage on several counts: they are a minority, they lack ready access to equal networks, they are often mistrusted (by women as well as men) in terms of their ability to handle authority in ways that are acceptable to men (Eagly 2007), and they tend to show less appetite for competition (Niederle and Vesterlunde 2007). These effects are amplified by attributions. If people act on these as presumptions, women may find they are not favoured with authority, even by other women. Yet, in general there is evidence that women don't like the game as much as men, and have less relish in playing it the same way (Davey 2008).

This is strikingly illustrated in the autobiography of Carly Fiorina (2007), who rose to be become Chief Executive Officer (CEO) of Hewlett Packard. Recalling her experiences at the company ATandT, she expresses shock at the perversion of advancement by merit that she witnessed—bosses using the system to trade favours—'I'll advance your man if you advance mine'. She sighs fervently: 'I was getting my first glimpse of how prejudice can linger in an organization, and why a meritocracy is so difficult to achieve' (p. 43). Even her downfall at Hewlett Packard reveals that to the end she was an outsider, unable to read the signs and understand the games that led to her removal as CEO.

The reason for the glass ceiling is therefore not just that women are handicapped in the customary games that lead to top positions, but that these positions look very uncongenial to even the most aspiring women. The co-evolutionary argument offers hope for women in the future. It has been observed in primate colonies that when food is centrally sourced versus dispersed, contrasting forms of social organization emerge—agonistic (competitive, hierarchical) versus hedonic (cooperative, egalitarian)—mediated by the requirement in the former case to contest for resources. Pierce and White (1999, 2006) have demonstrated experimentally that this may be replicated among humans in business organization, with analogous outcomes arising from centralized versus decentralized organizational forms.

This implies that technology, monolithic supply chains and restricted markets favoured the traditional agonistic hierarchical forms, so preferred by competitive males, became the standard for corporations throughout the 20th century, but now the globalizing forces of the 21st century offer opportunities to move to more fluid, egalitarian, and networked forms, where a different set of game rules may apply. This leads to the optimistic conclusion that these developments will give more encouragement to females and non-traditional males. We have choices about how we organize and there are alternatives to the classical order. Yet one may more pessimistically conclude that the holders of traditional power will find plenty of support among their like-minded colleagues to find reasons to maintain the old order, until market forces and sharper, faster, and more cooperative competitors sweep them away.

Leadership and management—the adaptive challenge

Some writers have sought to portray leaders and managers as qualitatively different both in identity and function (Kotter 1990; Zaleznik 1992): leaders, the strategic visionaries who inspire through charisma; managers, the functionaries who execute the plans others have formulated. This argument is naïve—committing the attribution error of over-identifying persons with roles. It also reveals a romantic longing for heroic leadership to prevail over the soulless grip of the corporation (Meindl et al. 1985). Many different models of leadership have prevailed throughout human history, and can be observed today. A purely functional approach to the question might allow one to identify leadership with personal discretion, and management with the power to execute commands. It is clear that some organizations are more managerial, on the Weberian principle that bureaucracy leaves little discretion for leadership but a lot of execution for managers to discharge. Others, like family firms, are light on governance and depend upon their founders or charismatic leaders for direction (Neubauer and Lang 1998).

Yet this distinction becomes decreasingly useful in today's organization. It has been argued that the multifaceted nature of modern organizations, and the need for creative and adaptive responses to a fast changing world, requires a closing of the gap between management and leadership. Agents with managerial authority need to step up to a higher level of engagement within their organizations than has been traditional; to innovate, motivate, and integrate (Hamel 2008; Mintzberg 2009). They need increasingly to be active in how the enterprise is run, finding new solutions to non-standardized problems. In other words, as our analysis has suggested, we should not be absorbed with what are fundamentally semantic issues about management and leadership, but should consider the nature of the interplay of forces that determines the roles that agents perform in organizations.

The beginnings of a Darwinian analysis of leadership point to a need to understand the nature of the adaptive challenge facing social groups, and how leaders arise to fulfil them (Van Vugt et al. 2008). This needs to be augmented by recognition that leaders shape situations as well as administer to them. By whatever process leaders accede to leadership roles, there is a likelihood of their altering the situation they have been selected for (Nicholson 1984). It is against the risk that leaders take their organizations along paths of ultimate self-destruction that Boehm's reverse dominance operates. It is not just 'upstarts' that tribes need to beware of, but innovators.

This suggests that we need to pay attention to how leaders arise. Four routes predominate in the contemporary context: emergent, hereditary, elected, and appointed. Each reflects a different social process and co-evolutionary logic, and each has distinct advantages and disadvantages. Let us consider them.

Emergent. Emergent leaders arise only when there is a fluid and informal process of collective decision-making. Emergent leaders have, initially, a low power base—personal power, but fragile legitimacy and few resources. This may change, if the collective allows the individual to accrue resources, which of course is disbarred in reverse dominance hierarchies.

Hereditary. Once resources can be accumulated, leaders can claim legitimacy (typically by claiming divine support) and then create a dynasty. This can be an environmentally stable solution when followers are weak and unorganized and when it is desirable to centralize power and governance for purposes of warfare, defence, and major projects.

Elected. When followers also accumulate resources and some self-determination (via education and employment), the relationship requires a degree of consent by the led, and that leaders acquire legitimacy via the mandate of election. Leadership thus is provisional and conditional. This is clearly adaptive in a world where there is continuous development of the social context and a need for leaders to respond to its changing demands.

Appointed. Once one enters the 'age of institutions' (where the collective has a level of size, complexity, and stability to be able to anticipate and indeed generate the situations in which leadership is desired), then leaders can be appointed to order, and held accountable against performance criteria.

This could be seen as an ascending scale of rationality and control, but it would be mistaken to view this in any sense as progressive. Each of these has advantages and disadvantages that can only be assessed against the environmental challenge they are seeking to meet (Table 3.2; Hollander 1992). Indeed, all these forms can be found in contemporary society, and often coexist within a single organization. Emergent leadership is a property of informal workgroups, and remains a

Table 3.2 Paths to leadership: preconditions and pitfalls

Path	Preconditions	Pitfalls
Emergent	Egalitarian and fluid status relations	Unreliability; Inconsistency
Hereditary	Intergenerational accumulation and transmission of critical resources	Incompetence; Lack of commitment
Elected	Follower power and requirement for consent	Divisive; Sectional interests
Appointed	Rational order and accountability against specified criteria	Unresponsive to followers; Agency hazards

potent source of localized adjustment to immediate and changing challenges (Pescosolido 2002). Hereditary leadership is visible in family firms, and seems able to confer special performance advantages via the added value of 'family capital' to a positive business culture (Nicholson 2008). Elected leaders provide a focus for the ideals and aspirations of groups, and hence satisfy people's needs for someone accountable to shoulder the burden of their responsibilities. Appointed leaders are time limited and subject to external control, factors which afford protection against abuse of power.

The disadvantages of each are also evident. Emergent leaders lack consistency of purpose and identity and hence are unreliable agents for predictable challenges. Hereditary leaders may be congenitally unsuited to leadership roles—lacking the desire or the capability to lead. Elected leaders may be divisive by building a sectional power base, working against the interests of minorities. Appointed leaders are vulnerable to agency problems, such as lacking commitment to followers and instead being dedicated to satisfying the political interests of those who control their appointment.

Leader effectiveness—self-regulation and change

The co-evolutionary argument runs as follows: these leadership models arise, are maintained, and discarded according to how well they meet environmental challenges. The most critical element of this context is what can be loosely called 'followers' (though this term potentially paints a misleading picture of the leadership challenge, for often followers are inert, powerless, or indifferent). They may, as in many modern democracies, have quite low expectations of leaders and only be moved to take action to remove them once a threshold of dissatisfaction has been passed.

The adaptive challenge of leaders is to effect forms of coordination that satisfy the goals of the organization and its stakeholders. This has been formulated by leadership scholars as the need for setting direction, aligning systems and processes, and generating commitment from organizational members (the DAC framework of leadership: Drath et al. 2008). It is clear that a prime driver of leader effectiveness is therefore whether leaders have a correct appraisal of their leadership situation and are capable of enacting appropriate behaviours.

This raises an interesting and important question—what is an 'appropriate' behaviour? It was observed by Machiavelli that a fatal flaw of most leaders is inflexibility (Skinner 1981)—they are insufficiently versatile or rather they are too much themselves, as documented by an extensive literature on derailment (J. Hogan et al. 2010; McCall and Lombardo 1983). Yet the leadership literature is impoverished in its treatment of the leadership challenge (i.e. the leadership situation), and overly concerned with leadership types and styles, in a seeming search for the holy grail of the universal leader (Meindl et al. 1985). Contingency models, which seek to model effectiveness in terms of matching leadership tasks and styles (Vroom and Yetton 1973), and leader-member-exchange research (Graen and Uhl-Bien 1995), which similarly searches for correspondences between the behaviours of leaders and the needs of followers, both suffer from the same defect: a narrow range of inputs to either side of the equation (situation/followers versus leadership types/styles), and a simplistic and passive matching and modelling treatment of the leadership process.

The co-evolutionary argument suggests how the study of leadership can benefit from a much richer construction of the leadership situation, i.e. by considering which stakeholder groups it is critical to satisfy at any point in time, and what kinds of actions are need to master the changes the organization is facing. This is not only a challenge for leader versatility, but also to leader insight. The prior logical question this analysis raises is: how does the leader appraise the situation? Clearly leaders are dependent upon limited channels of information, and there may be various sources of filtering and distortion that impede their ability to understand with sufficient depth the situation they are in. They may be impeded in this effort by the politics of the hierarchy, which create disincentives for subordinates to tell leaders the truth. This is one of the major dysfunctions of traditional organization: how top-down control systems impede the flow of insights upward to leaders about the real problems and challenges they face.

Adding self-regulation to the co-evolutionary perspective also reveals to us processes biasing the leader's cognitions, motivations and decisions. This argues that leaders' responses will seek to balance the demands of: 1) goals; 2) perceptions of the situation; and 3) perceptions of self, with incomplete information, personal biases, and mixed motives, aided by an inborn talent for self-deception (Trivers 2000)! Elsewhere I have analysed at greater length the dynamics of this process (Nicholson in press).

The challenge for leadership effectiveness is to be able to maintain continuously an effective balance between two strategies. The first strategy is impact: leaders shaping or moulding the leadership situation, so that it suits the leader's propensities and capabilities. Many great institutions have arisen out of the ability of a leader or entrepreneur to accumulate the people and resources they need to make their dreams a reality. Many appointed leaders bring on their coat-tails small but significant teams of allies to ensure they have impact and are not overwhelmed and neutralized by their new organization's culture (Hambrick and Mason 1984; Peterson et al. 2003). This strategy has been a major source of cultural evolution throughout the ages, though it also risks failure if the leadership situations that leaders create fail to satisfy the critical environmental contingencies of the environment and key stakeholder groups (Donaldson 2001). This is a familiar cause of cultural revolution, as discussed earlier.

The second strategy is versatility: in self-regulation terms it means adapting goals, perceptions of reality, and self-concepts to become more consistent with experience (Kaplan and Kaiser 2006). This is especially important under conditions of high risk. For some leaders this is a process of self-discovery and revelation, as they reappraise their own capabilities, self-image and goals as a result of their leadership experience. There are also risks in being versatile, such as loss of purpose, surrendering one's goals to others', and being perceived as inconsistent. It is notable that research into followers' interests and desires of their leaders reveals a small set of preferences—vision, capability, integrity, and confidence are all typically highly rated (Keller 1999)—which suggests that the consistent leader may be valued more than the versatile or adaptable one, i.e. there is an implicit desire for leaders to be quasi-parental role models (Keller 2003), taking responsibility for followers and the situation.

One can see in this analysis that risks abound for leaders. To be effective over any stretch of time, a leader needs be actively regulating the elements of the system—goals, perceptions of environment, and self—through action and reflection. This has three pre-requisites. Leaders need: 1) dynamically changing strategic awareness of the correct balance of the two strategies to be deployed (as well as being able to effect a third do-nothing strategy); 2) the ability and will to be mindful of sources of bias and to correct them, whether the sources of bias are externally generated or self-engendered; and 3) the psychological and material resource to effect tactics and behaviours that monitor change and reveal what vital information may be being systemically concealed by other self-interested actors.

The self-regulation perspective is one that reflects the freedom that modern leaders have to make decisions about people, strategies and structures, coupled with a high degree of accountability for consequences. Leadership looks a tall order from this modern perspective. It is much easier when one is highly regulated by externalities and internal rules and systems (bureaucracy) or where one is immune from consequences (autocracy).

Implications and conclusion

In terms of research and theory, this review points in several directions of future need. First it argues for much more attention to the self and integrating it within evolutionary theory. Self-regulation offers a potentially powerful framework for understanding variations in adaptive response to change.

Second, an evolutionary perspective on organization recommends that research looks more closely at how groups adapt to different structures and forms, in terms of how they cooperate and exercise authority. Specifically, attention should be paid to how organizational structures may restrict the participation of women.

Third, close examination needs to be paid to the control and performance management systems that are deployed in many firms, in terms of their unintended consequences, and to consider the alternatives that might be conducive to more innovative and self-motivated performance.

Fourth, it is clear that leadership is a critical element in the co-evolution of institutions at all levels, which research should examine more closely. In particular, there would be benefits from understanding how leaders model the causal influences bearing down upon themselves, and how they seek to understand and control the effects of their actions. Self-regulation supplies a framework for doing this.

The broader implications of this analysis concern the role of business and management in the future of our society, and indeed, the planet, for business and management have become, alongside government, the chief instruments of all social change. This is not confined to large corporations and institutions, but also to the myriad small, nascent, and entrepreneurial firms that account for so much economic activity the world over. All depend upon a number of common factors: fit and healthy workers, natural resources, civil society, and a functioning economy. These conditions prevail in most parts of the world, yet threats loom on the horizon. The Credit Crunch example, discussed earlier, illustrates the vulnerability of many elements of society and economy to major risks. Especially threatening in the longer term are the challenges of climate change, energy supply, global shortages of water, and other natural resources, warfare and terrorism, plus a host of lesser but nonetheless troubling symptoms of sick societies such as crime, drug abuse, and mental illness.

The argument I have advanced here suggests that many social ills and threats are the product of what I have described as economy 'trumping' psychology—we pursue wealth and growth before peace and fulfilment. It is clear that organizational and management systems are an indispensable part of the solution to the problems of society. The most urgent is the global 'tragedy of the commons'—the process by which agents maximizing their self-interest by overgrazing their livestock on common ground bring about disaster for commonwealth, by making the common land barren when cooperative self-regulation would have ensured sufficiency for all. This is the challenge facing nation states and institutions in a world of booming population and dwindling resources.

The question this raises for theory and practice is whether we can create the leadership, social institutions, and shared cultural values that will enable us to save our environment from destruction. One notable Darwinian writer, Matt Ridley, has recently published a strongly affirmative answer to this question, arguing that human ingenuity and invention, within a framework of specialization and trade, has delivered to us miracles of plenty and accord in every area of human life, and will continue to enable us to face down the global dangers confronting us (Ridley 2010). There is much one can agree with in this upbeat analysis, yet the thesis can be challenged in its analysis of human psychology. Clearly in this, as in his previous writings, Ridley has no illusions about human instincts, but perceives that self-interest can satisfy collective interest via adaptive self-determination and exchange. Yet, the analysis I have presented here, especially around self-regulation, promotes a darker view. It suggests there are multiple ways in which we are able to deceive ourselves: self-concealment of goals, misperceptions of risk and contingency, self-serving delusions about ourselves, and overblown optimism about our capacities for self-control and remedial action. Ridley's thesis itself looks alarmingly like the latter.

Yet, my exploration of co-evolution, self-regulation, organization, and leadership in this chapter suggests that for the first time we have choices about our future identity and existence. The self-regulatory perspective reminds us of the power of insight. Now that evolutionary research is providing reliable insights into the nature of human nature—the predictable biases and drivers of our behaviour—our gift of self-consciousness enables us to take self-regulatory steps to direct human effort towards our long-term rather than our short-term interests.

As we have seen, we can organize and manage ourselves in ways that make conflict, authoritarianism, collaboration, and invention more or less likely. We can create frameworks of contracts and exchange that enable people to be more flexible, far-sighted, and cooperative in decision-making. We can use methods of recruiting, developing, and holding leaders accountable that induce styles that are more focused on the welfare of the commonwealth than on the interests of leaders and their cronies. By metrics and incentives we can change the boundaries by which we define our group identities.

All of this is possible, in theory. But can we do it? It also entails giving up many things upon which today we place a high value—some of our precious rights and freedoms, our definitions of national identity, our demands for leaders who fit stereotypic images, and our ways of making collective decisions and doing politics. It looks like a hard climb. Yet we are an ingenious and inventive species, and we will surely surprise ourselves with our adaptive capability. We may also disappoint by revealing an inability to harness our willpower and deliver through our institutions and leaders.

References

Ahmad, M.G. (2009). *The roots of leadership and primate heritage.* Working paper, INSEAD, France.

Aldrich, H. (1999). *Organizations evolving.* Sage, Thousand Oaks, CA.

Argyris, C. (1986). Reinforcing organizational defensive routines: an unintended human resources activity. *Human Resources Management,* **25**, 541–55.

Audia, P.G., Locke, E.A., and Smith, K.G. (2000). The paradox of success: an archival and laboratory study of strategic persistence following a radical environmental change. *Academy of Management Journal,* **43**, 837–53.

Bandura, A. (1982). Self-efficacy mechanisms in human agency. *American Psychologist,* **37**, 122–47.

Barrett, L., Dunbar, R., and Lycett, J. (2002). *Human evolutionary psychology.* Palgrave, Basingstoke.

Baumeister, R.F., Schmeichel, B.J., and Vohs, K.D. (2007). Self regulation and the executive function. In: A.W. Kruglanski and E.T. Higgins (eds), *Social psychology: Handbook of basic principles* (Second edition), pp. 516–38. The Guilford Press, New York.

Bernard, L.C., Mills, M., Swenson, L., and Walsh, R.P. (2005). An evolutionary theory of human motivation. *Genetic, Social and General Psychology Monographs,* **131**, 129–84.

Boehm, C. (1999). *Hierarchy in the forest: The evolution of egalitarian behavior.* Harvard University Press, Cambridge, MA.

Boyd, R. and Richerson, P.J. (1985). *Culture and the evolutionary process.* University of Chicago Press, Chicago, IL.

Buss, D.M. (1999). *Evolutionary psychology: the new science of the mind.* Allyn and Bacon, Needham Heights, MA.

Chagnon, N.A. (1997). *Yanomamo.* Wadsworth, London.

Davey, K.M. (2008). Women's accounts of organizational politics as a gendering process. *Gender, Work and Organization,* **15,** 650–71.

De Bondt, W. (2005). Bubble psychology. In: W.C. Hunter, G.G. Kaufman, and M. Pomerleano (eds), *Asset price bubbles: the implications for monetary, regulatory, and international policies,* pp. 205–16. MIT Press, Boston, MA.

De Waal, F.B.M. (1989). *Chimpanzee politics: Power and sex among apes.* The Johns Hopkins University Press, Baltimore, MD.

Diamond, J. (1991). *The rise and fall of the third chimpanzee.* Radius, London.

Donaldson, L. (1995). *American anti-management theories of organization: A critique of paradigm proliferation.* Cambridge University Press, Cambridge.

Donaldson, L. (2001). *The contingency theory of organizations.* Sage, Thousand Oaks, CA.

Drath, W.H., McCauley, C.D, Palus, C.J., Van Velsor, E., O'Connor, P.M.G., and McGuire, J.B. (2008). Direction, alignment, commitment: Towards a more integrative ontology of leadership. *The Leadership Quarterly,* **19,** 635–53.

Dunbar, R.I.M. (1992). Neocortex size as a constraint on group size in primates. *Journal of Human Evolution,* **20,** 469–93.

Eagly, A. (2007). Female leadership advantage and disadvantage: resolving the contradictions. *Psychology of Women Quarterly,* **31,** 1–12.

Eldredge, N. and Gould, S.J. (1972). Punctuated equlibria: an alternative to phyletic gradualism. In: T.J. Schopf (ed.), *Models in Paleobiology,* pp. 82–115 Freeman, Cooper, San Fransisco, CA.

Ellis, L. (1994). Social status and health in humans: the nature of the relationship and its possible causes. In: L. Ellis (ed.), *Social stratification and socioeconomic inequality,* pp. 123–44. Vol. 2: Reproductive and interpersonal aspects of dominance and status. Praeger, New York.

Fenton-O'Creevy, M., Nicholson, N., Soane, E.C., and Willman, P. (2005). *Traders: risks, decisions, and management in financial markets.* Oxford University Press, Oxford.

Fiorina, C. (2007). *Tough choices: a memoir.* Nicholas Brealey, London.

Geary, D.C. (2010). *Male, female* (Second edition). APA Press, Washington, DC.

Graen, G.B. and Uhl-Bien, M. (1995). Relationship-based approach to leadership: Development of leader-member exchange (LMX) theory of leadership over 25 years: applying a multi-level multi-domain perspective. *The Leadership Quarterly,* **6,** 219–47.

Hambrick, D.C. and Mason, P.A. (1984). Upper echelons: the organization as a reflection of its top managers. *Academy of Management Review,* **9,** 193–206.

Hamel, G. (2008). *The future of management.* Harvard University Press, Cambridge, MA.

Henrich, J. (2004). Cultural group selection, coevolutionary processes and large-scale cooperation. *Journal of Economic Behavior and Organization,* **53,** 3–35.

Henrich, J. and Gil-White, F.J. (2001). The evolution of prestige: freely conferred deference as a mechanism for enhancing the benefits of cultural transmission. *Evolution and Human Behavior,* **22,** 165–96.

Hogan, J., Hogan, R., and Kaiser, R. (2010). Management derailment: personality assessment and mitigation. In: S. Zedeck (ed.), *American Psychological Association handbook of industrial and organizational psychology* (Vol. 3), pp. 555–75. American Psychological Association, Washington, DC.

Hogan, R. (2010). *Putting leadership in an (evolutionary) context.* Presentation to SIOP annual conference, Atlanta, GA.

Hollander, E.P. (1992). The essential interdependence of leadership and followership. *Current Directions in Psychological Science,* **1,** 71–74.

Howell, J.P. and Dorfman, P.W. (1981). Substitutes for leadership: test of a construct. *Academy of Management Journal,* **24,** 714–28.

Humphrey, N. (1980). Nature's psychologists. In: V.S. Ramachandran (ed.), *Consciousness and the physical world,* pp. 55–75. Pergamon, New York.

Ibarra, H. and Barbulescu, R. (2010). Identity as narrative: prevalence, effectiveness, and consequences of narrative identity work in macro work role transitions. *Academy of Management Review*, **35**, 135–54.

James, W. (1890). *The principles of psychology* (Vol. 1). Dover, Mineola, NY.

Jelinek, M. (2005). Management: classical theory. In: N. Nicholson, P. Audia, and M. Pillutla (eds), *The Blackwell encyclopedia of management*, pp. 221–3. Blackwell, Oxford.

Johnson, M. (2005). *Family, village, tribe: the story of the Flight Centre Ltd*. Random House, Sydney.

Kahneman, D. and Tversky, A. (1979). Prospect theory: an analysis of decision under risk. *Econometrica*, **47**, 263–91.

Kaplan, R.E. and Kaiser, R.B. (2006). *The versatile leader: make the most of your strengths—without overdoing it*. Pfeiffer, San Francisco, CA.

Keller, T. (1999). Images of the familiar: individual differences and implicit leadership theories. *The Leadership Quarterly*, **10**, 589–607.

Keller, T. (2003). Parental images as a guide to leadership sensemaking: an attachment perspective on implicit leadership theories. *The Leadership Quarterly*, **14**, 141–60.

Kernis, M.H., Zuckerman, M., Cohen, and Spadafora, S. (1982). Persistence following failure: the interactive role of self-awareness and the attributional basis for negative expectancies. *Journal of Personality and Social Psychology*, **43**, 1184–91.

Kieser, A. (1987). From asceticism to administration of wealth: medieval monasteries and the pitfalls of rationalization. *Organisation Studies*, **8**, 103–23.

Kotter, J.P. (1990). *A force for change: how leadership differs from management*. The Free Press, New York.

Kulik, B.W. (2005). Agency theory, reasoning and culture at Enron: in search of a solution. *Journal of Business Ethics*, **59**, 347–60.

Landes, D.S. (1998). *The wealth and poverty of nations*. Norton, New York.

Langer, E.J. (1975). Illusion of control. *Journal of Personality and Social Psychology*, **32**, 311–28.

Leary, M.R. (2004). *The curse of the self*. Oxford University Press, New York.

Lewin, A.Y., Long, C.P., and Carroll, T.N. (1999). The coevolution of new organizational forms. *Organization Science*, **10**, 535–50.

Low, B.S., Finkbeiner, D., and Simon, C.P. (2005). Favored places in the selfish herd: trading off food and security. In: L. Booker, S. Forrest, M. Mitchell, and R. Riolo (eds), *Perspectives on adaptation in natural and artificial systems*, pp. 213–38. Oxford University Press, New York.

McCall, M.W., and Lombardo, M.M. (1983). *Off the Track: why and how successful executives get derailed*. Technical Report No. 21. Center for Creative Leadership, Greensboro, NC.

McElreath, R. and Henrich, J. (2007). Dual inheritance theory: the evolution of human cultural capacities and cultural evolution. In: R.I.M Dunbar and L. Barrett (eds), *Oxford handbook of evolutionary psychology*, pp. 555–70. Oxford University Press, Oxford.

McGrew, W.C. (1992). *Chimpanzee material culture: implications for human evolution*. Cambridge University Press, Cambridge.

Meder, A. (2007). Great ape social systems. In: W. Henke, I. Tatersall, and T. Hardt (eds), *Handbook of paleoanthropology* (Vol. 2), pp. 1235–71. Springer-Verlag, New York.

Meindl, J., Ehrlich, S.B., and Dukerich, J.M. (1985). The romance of leadership. *Administrative Science Quarterly*, **30**, 78–102.

Miller, G.F. (2000). *The Mating mind: How sexual choice shaped the evolution of human nature*. Heinemann, London.

Mintzberg, H. (2009). *Managing*. Berrett-Koehler, San Francisco, CA.

Mithen, S. (1996). *The prehistory of the mind*. Thames and Hudson, London.

Nesse, R. and Williams, G. (1994). *Why we get sick*. Times Books, New York.

Nettle, D. (2006). The evolution of personality variation in humans and other animals. *American Psychologist*, **61**, 622—31.

Nettle, D. (2007). Individual differences. In: R.I.M Dunbar, and L. Barrett (eds), *Oxford handbook of evolutionary psychology*, pp. 479–90. Oxford University Press, Oxford.

Neubauer, F and Lank, A.G. (1998). *The family business: its governance and sustainability*. Macmillan, London.

Nicholson, N. (1984). A theory of work role transitions. *Administrative Science Quarterly*, **29**, 172–91.

Nicholson, N. (1998). How hardwired is human behavior? *Harvard Business Review*, **76**(no.4, July/August), 134–47.

Nicholson, N. (2000). *Managing the human animal*. Thomson/Texere, London.

Nicholson, N. (2005). Meeting the Maasai: messages for management. *Journal of Management Inquiry*, **14**, 255–67.

Nicholson, N. (2008). Evolutionary psychology, corporate culture and family business. *Academy of Management Perspectives*, **22**, 73–84.

Nicholson, N. (2010a). The false theory of meritocracy. *Harvard Business Review* blog, 1 June.

Nicholson, N. (2010b). Gender and the future of hierarchical organization. *Harvard Business Review* blog, 7 June.

Nicholson, N. (in press). The evolved self, co-evolutionary processes and the self-regulation of leadership. In: C. Heintz, W. Callebaut, and L. Marengo (eds), *Models of man for evolutionary economics*. MIT Press, Cambridge, MA.

Nicholson, N. and De Waal-Andrews, W. (2005). Playing to win: biological imperatives, self-regulation and trade-offs in the game of career success. *Journal of Organizational Behavior*, **26**, 137–54.

Nicholson, N. and White, R. (2006). Darwinism: a new paradigm for organizational behaviour? *Journal of Organizational Behavior*, **27**, 111–20.

Niederle, M. and Vesterlande, L. (2007). Do women shy away from competition? Do men compete too much? *Quarterly Journal of Economics*, **122**, 1067–101.

Parker, W.D. and Prechter, R.R., Jr (2005). Herding: an interdisciplinary integrative review from a socionomic perspective. In: B. Kokinov (ed.), *Advances in Cognitive Economics. Proceedings of the International Conference on Cognitive Economics*, pp. 271–80. New Bulgarian University Press, Sofia.

Pescosolido, A.T. (2002). Emergent leaders as managers of group emotions. *The Leadership Quarterly*, **13**, 583–99.

Peterson, R.S., Smith, D.B., Martorana, P.V., and Owens, P.D. (2003). The impact of Chief Executive Officer personality on top management team dynamics: One mechanism by which leadership affects organizational performance. *Journal of Applied Psychology*, **88**, 795–808.

Pierce, B.D. and White, R. (1999). The evolution of social structure: why biology matters. *Academy of Management Review*, **24**, 843–53.

Pierce, B.D. and White, R. (2006). Resource context contestability and emergent social structure: an empirical investigation of an evolutionary theory. *Journal of Organizational Behavior*, **27**, 221–40.

Richerson, P.J. and Boyd, R. (2005). *Not by genes alone: how culture transformed human evolution*. University of Chicago Press, Chicago, IL.

Ridley, M. (1993). *The red queen: sex and the evolution of human nature*. Viking, London.

Ridley, M. (2010). *The rational optimist: how prosperity evolves*. Fourth Estate, London.

Romanelli, E. (1991). The evolution of new organizational forms. *Annual Review of Sociology*, **17**, 79–103.

Rose, M. (1985). *Industrial behaviour: theoretical developments since Taylor*. Penguin, Harmondsworth.

Sahlins, M.D. (1972). *Stone age economics*. Aldine, Chicago, IL.

Senge, P. (1990). *The fifth discipline*. Doubleday, New York.

Skinner, Q. (1981). *Machiavelli: A very short introduction*. Oxford University Press, Oxford.

Sober, E. and Wilson, D.S. (1998). *Unto others: the evolution and psychology of unselfish behavior.* Harvard University Press, Cambridge, MA.

Sperber, D. (1996). *Explaining culture: a naturalistic approach.* Blackwell, Oxford.

Spisak, B.R., Nicholson, N., and Van Vugt, M. (2011). Leadership in organizations: an evolutionary perspective. In: G. Saad (ed.), *Applications of evolutionary psychology in the business sciences*, pp. 165–90. Springer, Heidelberg.

Steel, P. and König, C.J. (2005). Integrating theories of motivation. *Academy of Management Review,* **31**, 889–913.

Stevens, A. and Price, J. (1996). *Evolutionary psychiatry.* Routledge, London.

Surowiecki, J. (2004). *The wisdom of crowds: Why the many are smarter than the few.* Abacus, New York.

Trivers, R.L. (2000). The elements of a scientific theory of self-deception. *Annals of New York Academy of Sciences,* **907**, 114–92.

Tudge, C. (1998). *Neanderthals, bandits and farmers.* Wiedenfeld and Nicolson, London.

Van Rossum, C.T.M, Shipley, M.J., van de Mheen, H., Grobbee, D.E., and Marmot, M.G. (2000). Employment grade differences in cause specific mortality. a 25 year follow up of civil servants from the first Whitehall study. *Journal of Epidemiology and Community Health,* **54**, 178–84.

Van Vugt, M., Hogan, R., and Kaiser, R.B. (2008). Leadership, followership, and evolution: Some lessons from the past. *American Psychologist,* **63**, 182–96.

Vroom, V.H. and Yetton, P.W. (1973). *Leadership and decision-making.* University of Pittsburg Press, Pittsburgh, PA.

Ward, V. (2010). *The Devil's casino: friendship, betrayal and the high-stakes games played inside Lehman Brothers.* Wiley, New York.

Weber, M. (1954 [1922]). *Economy and society.* Simon and Schuster, New York.

Whiten, A. (1998). The evolution of deep social mind in humans. In: M. Corballis, and S.E.G. Lea (eds), *The evolution of the hominid mind*, pp. 155–75. Oxford University Press, Oxford.

Whyte, G. (1998). Recasting Janis's groupthink model: the key role of collective efficacy in decision fiascos. *Organizational Behavior and Human Decision Processes,* **73**, 185–209.

Zaleznik, A. (1992). Managers and leaders: Are they different? *Harvard Business Review,* **72**, 126–35.

Chapter 4

The social animal within organizations

Abraham P. Buunk and Pieternel Dijkstra

The human species: a social animal

While humans are, as all other species, subject to the laws of natural selection, there are still many questions concerning the way in which humans evolved. Nevertheless, there is little doubt that humans evolved in groups. As organizations are large groups, and consist of many subgroups, evolutionary theorizing would seem very relevant to understand behaviour in organizations. It is, indeed, becoming increasingly accepted that many traits, attitudes, and behaviours relevant to the workplace do not exist simply because of social conditioning, but also because of evolved psychological mechanisms (Ilies et al. 2006). However, organizational-behavioural researchers have only recently begun to apply insights from the theory of evolution to their field (Colarelli 2003; Nicholson 1998). Applying evolutionary thinking to organizations may help understand why people in organizations behave the way they do, even if these behaviours seem counterproductive or irrational; as is the case, for instance, when people feel envious and begrudged in response to a co-worker's success or when an organization falls into financial problems due to the high costs of conspicuous consumption (i.e. luxury items that display high status).

While traditional perspectives usually provide *proximate* explanations for these phenomena, evolutionary theorizing offers *ultimate* explanations, and tries to understand human behaviour from the perspective of fitness, i.e. how behaving in particular ways may have maximized the chances of survival and reproduction. Broadly speaking, humans seem to have evolved two major needs. First, the need for a sense of belonging, the wish to be part of a group in which one feels safe and accepted, requiring the establishment of positive emotional bonds with others. The second major need—and the one on which this chapter focuses in particular—is status. The need for status is conceived here rather broadly, encompassing the striving for positive self-esteem by performing well on certain dimensions, winning competitions with others, and seeking appreciation and prestige from others (Buunk and Ybema 1997). We first discuss how the human brain seems to have evolved particularly to deal with living in large groups. We suggest that comparing oneself with others seems a basic human characteristic that may have various positive and negative consequences for individuals, as well as for organizations. Next, we focus on intrasexual competition, and discuss how this may lead not only to investing in one's career, but also to gossip, bullying, and conspicuous consumption. Finally, we discuss the role of altruistic behaviour within organizations, and link this also to intrasexual competition.

The social brain

In our evolutionary past—and that of many other species—group membership was essential to survival and reproduction as it protected against hostile environments. Groups allowed our ancestors to cope with predators, to cooperate at tasks, to find a mate, help and support each other with parenting, to share in each other's resources, and to defend oneself against hostile groups (e.g. Van Vugt and Schaller 2008). It is becoming increasingly clear that many human characteristics—like language, the ability to detect cheaters, and striving for dominance and status—evolved primarily to facilitate living in groups of varying sizes. Even more so, according to the social brain hypothesis, our large brains, and particularly the neo-cortex, developed primarily to allow humans to live in larger groups than other primates. Dunbar (1998) showed that, across primate species, there is a strong correlation between the size of the neo-cortex (relative to the rest of the brain) and the average size of the group in which the primate lives. According to Dunbar (1998), whereas primates tend to maintain their relationships with others through grooming, this is no longer feasible in the large groups in which humans traditionally lived. The growth of the neo-cortex allowed the development of language through which it is possible to obtain information—written or oral—about many other group members, and to maintain relationships with many different individuals. On the basis of the cross-species correlation, Dunbar (1993) calculated that the standard human group size should be around 150. This is the number of people with whom one can maintain personal relationships, to whom one would talk when one encounters the other in the street, and for whom one can remember the history of exchange. Many pieces of evidence support the universality of this 'magic number'. For example, it is the size of clans of hunter-gatherers, military units, church congregations, and the personal network. This suggests an explanation as to why organizations usually have to develop formal rules and procedures when their size begins to exceed around 150: as members can no longer know all others on a personal basis, informal control mechanisms no longer suffice.

It must be noted that modern organizations seem to differ from the groups in which our ancestors lived and worked. In contrast to the small ancestral groups of up to 150 people, modern organizations are often characterized by interdependence, cooperation, and competitiveness between hundreds of people. As a result, modern organizations tend to be much larger and more complex than the social environment in which our ancestors evolved. It is important to recognize this difference. Mechanisms and behaviours that evolved in the small living and working groups of our ancestors—such as envy, bullying, and physical threats—may, because of differences in social context, no longer be adaptive (Campbell 2002). It must also be noted though, that ancestral social life may have been more complex than sometimes is assumed. That is, according to Dunbar (2008), there are several layers of social life which seem to have a mathematical relation to each other. Successive layers typically number around 5, 15, 50, 150, 500, and 1500 individuals. These layers seem to reflect, respectively, the support clique of best friends, the sympathy group, the number of individuals contacted at least once a month (the band among hunter-gatherers), the social network, the mega band, and the tribal grouping. The implication of this perspective for organizations is that the optimal size of functional unities within organizations may have to correspond to the size of one of the layers. Small work teams should not compose more than four, or maximally five members; groups at the next level, probably the level at which direct, personal leadership may function, should contain no more than 15 members; the next level, and probably the basic level for an organizational unit, would be the size of the band, no more than 50 members; the following level of around 150 would be the group in which people can still know and control each other on a personal basis; and the subsequent levels of 500 and 1500 might reflect the optimal sizes of larger organizations.

Social comparison and envy

Humans have many mechanisms and adaptations to deal with the complexities of functioning in such large, multi-layered groups. One of these mechanisms is social comparison, i.e. comparing one's accomplishments and outcomes with that of fellow group members, which has likely been an unavoidable aspect of group life throughout much of human evolutionary history. The social comparison literature underlines the ubiquity of social comparison, although there are considerable individual differences in this respect (Buunk and Gibbons 2007). In modern organizations, employees may compare, for example, their performance, salary, room size, secondary benefits, or career prospects with that of others. There may be various evolutionary explanations for this tendency to compare oneself with others. First, individuals can estimate their position in the status hierarchy, and may refrain from challenging those who are clearly superior, but challenge those who can be beaten. In view of the adaptive value of adequately sizing up one's competitors, the need to compare oneself with others is phylogenetically very old, biologically very powerful, and recognizable in many species (Gilbert et al. 1995). Second, the tendency to engage in social comparisons may have evolved to monitor the fairness of the distribution of resources in the group, and to assess if others engage in reciprocity. This information may help to determine whom one can trust and whom is to be punished. Third, social comparison may have evolved because it induces individuals to undertake action to enhance their own survival and reproductive fitness. That is, seeing someone else obtaining more outcomes than oneself, or performing better than oneself, may motivate individuals to improve their own situation and teach them how to do so.

The most obvious emotion evoked by social comparisons is envy (Fischer et al. 2009). A distinction has been made between *malicious* envy and *benign* envy. Both types of envy appear to motivate individuals to decrease the status gap between themselves and others, but they do so in different ways. Benign envy leads individuals to close the gap by moving themselves up to the level of the other, whereas malicious envy leads individuals to do so by pulling the other down to one's own position (Van de Ven et al. 2009). Whereas benign envy stimulates individuals to self-promote and improve the self, for instance by observational learning and affiliation with a superior other, malicious envy encourages individuals to derogate or even damage rivals. From an evolutionary psychological point of view, both types of envy may, in our evolutionary past, have alerted individuals to fitness-relevant advantages, and may have been adaptive under specific circumstances. Benign envy may have been especially adaptive under favourable environmental conditions and in highly cooperative groups, whereas malicious envy may have been particularly adaptive under harsh environmental conditions and under high levels of competition (cf. Hill and Buss 2006, 2008; Kletner et al. 2006).

Although envy may increase the envious individual's chances of survival, it often has negative effects for an organization. In a review of studies on envy in organizations, Duffy et al. (2008) showed that workplace envy is related to lower job satisfaction, less liking for co-workers, lower group performance, higher turnover, higher absence rates, and higher social loafing. A series of experiments by Fischer et al. (2009) showed that envy may have negative effects on performance because envious individuals are less willing to share high-quality information with envied colleagues. Since information exchange is crucial for successful cooperation, group performance may suffer as a consequence. Further illustrating the potentially negative effect of envy for organizations, Gino and Pierce (2009) found that, in the visible proximity of abundant wealth or wealthy others, individuals are more likely to cheat due to feelings of envy. Indeed, employees who feel their company or employer is extremely rich may more easily engage in unethical behaviours, such as those related to financial fraud, in an attempt to close the monetary gap between themselves and their Chief Executive Officer (CEO).

Although envy is a universal human experience, there are indications that organizations can control at least some of its negative consequences. In his study, Vecchio (2000) found envy to be related to several work unit variables. As the reward system in a unit was more competitive, employees experienced more envy. In contrast, workers with more autonomy in their roles and with more considerate supervisors experienced less envy towards co-workers. Although this study was only correlational in nature, the findings do provide organizations with at least some avenues to control workplace envy and its negative consequences. That is, by uncoupling individual rewards from those of others, enhancing job autonomy, and recruiting kind and empathic managers, organizations may actively reduce levels of workplace envy. In addition, organizations may help workers enhance their feelings of self-efficacy. Research reviewed by Buunk and Ybema (1997) shows that exposure to successful others may induce positive affect, especially when workers believe they too can achieve success. This finding suggests that beliefs about self-efficacy can strengthen feelings of benign envy and reduce feelings of malicious envy. This is easy to comprehend: as workers are more convinced of their ability to become as successful as high status others, they will feel less need to derogate the target since they have access to a different strategy, i.e. working harder and becoming successful too. Organizations may therefore try to stimulate workers' self-efficacy, for instance, by self-management training and individual coaching sessions (cf. Ayres and Malouff 2007; Evers et al. 2006).

Intrasexual competition

While there is little evidence for a sex difference in levels of envy, some studies on social comparison suggest that women tend to compare themselves particularly with other women, whereas men tend to compare themselves with other men. For instance, when Buunk and Van der Laan (2002) presented women with a successful target (either male or female), women preferred to compare themselves more with the female target and saw her situation as a more likely potential future for themselves. Similar comparisons in terms of performance and pay have been noted in both men and women (Feldman and Ruble 1981; Miller 1984; Suls et al. 1979). In contrast, other studies indicate that women, more than men, view opposite-sex others as relevant for their self-evaluation (Crosby 1982). For example, a study by Steil and Hay (1997), using a sample of men and women in prestigious, male-dominated careers, showed that women were more likely to make opposite-sex comparisons than men, and that the higher a woman's income became, the more likely she was to compare her accomplishments regarding promotion, compensation, responsibility, and influence in decision-making predominantly with men (although this may, of course, be partly due to the fact that in many organizations the majority of the employees, especially in the higher echelons, are still male).

Nevertheless, it seems that, on an emotional level, comparison and competition will occur more within than between the sexes. Fierce competition over higher wages, better fringe benefits, more prestigious jobs, larger offices, and other status-related resources is the norm within organizations. From the perspective of intrasexual competition, men will tend to perceive other men as their competitors and primary standard of reference, whereas women will tend to perceive other women in the same way (e.g. Buunk et al. in press; Saad and Gill 2001). Intrasexual competition refers to rivalry with same-sex others that is, ultimately, driven by the motive to obtain and maintain access to mates. Intrasexual competition in organizations, particularly among males, has ancient evolutionary roots. Darwin (1871) already recognized the importance of intrasexual competition for the evolution of traits, such as the ornaments of male deer. According to Darwin, males compete to beat rivals with physical strength and weapons, and to impress females with features that signal health and resistance to parasites. In this context, it is important to note that males can sire offspring with a single sexual act, whereas in most species females have to invest much more than males in producing offspring (Trivers 1972). Females are therefore, as it were, a scarce resource over which males compete (Andersson 1994). Men generally compete over access to status-related resources, such as political influence and money, because these resources can be converted into reproductive opportunities, either because they are directly attractive to females or because these help conquer rival males (Sidanius and Pratto 1999; Tooby and Cosmides 1988). Indeed, research shows that, all over the world, and throughout history, status is more strongly associated with mating opportunities and reproductive success in men than in women, making men keener to compete for high status within the group (e.g. Betzig 1996; Hopcroft 2006). Although, in many organizations, there are often no women around to observe the status of males in relation to others and the direct benefits of engaging in such competition may not be immediately obvious, intrasexual competition seems to be sufficiently hard-wired for it to be evident. With the increasing influx of women in organizations, intrasexual competition among males may have become even more salient and prevalent, as the presence of women tends to make men more aware of their status, and more eager to demonstrate that they can beat other men. For example, an experiment showed that men increased their cooperation in an economic game when observed by women (Iredale et al. 2008), supposedly to impress these women.

More than in other primates, human males invest resources and parental care in their offspring. As a result, both sexes are discriminating in their choice of mates, and therefore both sexes engage in competition with same-sex conspecifics. Indeed women, like men, tend to prefer to be better-off than same-sex others (Buunk and Fisher 2009; Buunk and Ybema 1997; Hill and Buss 2006). There is evidence that women often consider other women as rivals in organizations (e.g. Jandeska and Kraimer 2005). This is nicely illustrated by two studies by Ellemers et al. (2004). In these studies, female faculty members, but not males, tended to assume that female doctoral students were less committed to a scientific career than their male counterparts. In reality, male and female doctoral students did not reliably differ in their self-reported commitment to different work aspects. This suggests that women in senior positions may actively hinder the career progression of other women. This phenomenon has been coined the 'queen bee syndrome', with the queen bee defined as 'a bitch who stings other women if her power is threatened' (Mavin 2008, p. 75).

While throughout human history men have competed primarily in the domains of status, resources, and dominance, women have tended to compete primarily in the domain of physical attractiveness (Buss 1994; Campbell 2002; Cashdan 1998; Merten 1997; Saad 2007; Saad and Peng 2006). For example, women most often nominate, perform, and rate the tactic of attracting attention to their appearance as most effective, even when no mention is made of what the competition is about (Cashdan 1998; Walters and Crawford 1994). When confronted with highly attractive rivals, women tend to 'dislike' such a rival, particular when she makes intrasexual competition salient, such as when she is conversing with a male (Baenninger et al. 1993). In addition, Hill and Buss (2006) found that women preferred the option of being less attractive in an absolute sense but more attractive than their rivals (e.g. scoring a 5 when rivals score a 3) compared with being more attractive in an absolute sense but less attractive than their rivals (e.g. scoring a 7 when rivals score a 9). There is evidence that, in organizations too, women compete with other women in terms of physical attractiveness. This is nicely illustrated by a series of studies by Luxen and Van de Vijver (2006), among both professionals and students in human resources management, that examined the effect of facial attractiveness on hiring decisions. These authors first found evidence for a mate-selection motive when the frequency of interaction between the job applicant and the participant was expected to be high. Under this condition, both men and women showed a preference to hire a highly attractive opposite-sex member over an unattractive opposite-sex other, with males showing this tendency more than women. However, with respect to same-sex candidates, a quite different pattern was found, clearly pointing to intrasexual competition among women: women were less likely to hire a highly attractive female applicant over an unattractive one, while male participants did not show this preference.

Leadership

An important implication of men's stronger tendency to compete over status is that the male status hierarchy is steeper and has a more pinched distribution: that is, we find more men at both the highest and lowest end of the status continuum, while relatively more women lie in the middle of the hierarchy. This has important implications for leadership. Many aspects of group living involving collective action—such as collective foraging and hunting, food sharing, division of labour, group defences, and communal parenting- necessitated the emergence of a group leader, i.e. a group member who takes the initiative and provides direction while others acquiesce and follow that direction (Van Vugt et al. 2008). Given their motivation to obtain status, men have been more eager to become leaders than women. Indeed, in leaderless groups, men more likely emerge as leaders than women, and male leaders are still in a clear majority at the very top of large corporations.

Of course, one may alternatively argue that the increase in numbers of women entering organizations in recent decades has been spectacular, and that it is just a matter of a few more decades before equality will be reached. However, because of their reproductive and nurturing role in our evolutionary past, women more than men value their social connectedness with other people, their relationships with family and friends, and the task of taking care of children and the home. In addition, for their reproductive success, women will be more concerned with the timing of, and care for, their offspring than men. Ample evidence supports this line of reasoning (cf. Campbell, 2002). In the Netherlands, although more young women than men receive a higher level of education (CBS, 2010), only 25% of the male workforce work part-time, compared with around 75% of women, despite facilities such as subsidized childcare and parental leave (CBS, 2009). Given that men more often drop out of school, and are, on average, becoming relatively less educated than women, women may in the future engage increasingly in gainful employment to assure adequate resources for their offspring. However, that does not necessarily imply that they will also enter the top positions in higher numbers. It may be that women, on average, will be as successful as men, but will still be less likely found in the top positions (as well as in the lowest positions). This would suggest that policies might be effectively directed towards attracting women to the organization itself, but less so to the highly demanding top positions. It may be that, given their different psychological make-up, it is unrealistic to assume that complete equality in the work place is a laudable, let alone a feasible goal.

The apparent stronger motivation of men to become a leader does not at all imply that men are *better* leaders. There are circumstances under which group members prefer a female over a male leader and in which female leaders are more effective than male leaders. Group members tend to show a strong preference for female leaders during *intra*group competition, whereas they show a strong preference for male leaders during *inter*group competition (Van Vugt and Spisak 2008). Likewise, women are preferred for 'caretaking' leader behaviours, while men are preferred for action-oriented, 'take-charge' leader behaviours (Prime et al. 2009). The preference for male and female leaders does not merely stem from gender stereotypes, but matches men and women's actual abilities as leaders (Eagly et al. 2003). The gender difference can be explained by the fact that, throughout our evolutionary history, it was more important for women to invest in creating and maintaining supportive social networks for the protection of themselves and their children. As a result, women tend to have a stronger interest in keeping the group together than men, and are more motivated and better equipped to act as intragroup peacekeepers. For men, an intergroup victory enhanced mating opportunities by gaining access to mates and prestige gains more than it did for women (Buss 1999; Tooby and Cosmides 1988). When confronted with aggressive rival groups or intergroup conflict, male group members increased levels of cooperation in order to successfully defeat the rival group. Indeed, research on traditional societies shows that tribal warfare is almost exclusively the domain of men, and that male warriors have more sexual partners and greater status within their community than other men (Chagnon 1988). Similarly, a US study on male street gangs revealed that gang members have above-average mating opportunities (Palmer and Tilley 1995). The phenomenon that men, more than women, increase intragroup cooperative behaviours during intergroup competition has been called the *male-warrior hypothesis* (Van Vugt et al. 2007a). Thus, whereas male group members compete in the main with each other, the appearance of rival groups causes a reduction in within-group intrasexual competition and men start to work together in an attempt to defeat the rival group.

It has been argued that most leaders are narcissistic individuals (Kets de Vries and Miller 2007). Brunell et al. (2008), for instance, showed that, especially in leaderless groups, narcissistic individuals (relatively self-centred individuals who are low in empathy) emerge as leaders. However, relatively altruistic individuals who are willing to sacrifice their own needs to the group are usually highly respected and admired by group members and are often nominated as leaders (Choi and Mai-Dalton 1998). From an evolutionary psychological point of view, both strategies—narcissism versus relative altruism—may be successful. Different strategies may have developed because these were adaptive under different conditions (Figueredo et al. 2005). For example, in a population composed predominantly of cooperators, there would be a niche for competitive individuals, and vice versa. Translated to modern organizational life, in an organization or unit with predominantly cooperative individuals, the competitive and narcissistic individual may be highly effective at achieving status. For organizations, this implies that people will often compete in groups to obtain leadership positions by choosing the niche that is available, and that it may be as difficult to induce all employees to become cooperative 'good citizens', as to induce all employees to be ambitious high performers.

Working hard and burnout

The drive to enhance one's status in the organization and to defeat competitors, may, especially for men, lead to high involvement in one's work. For organizations this strategy has, of course, great benefits in terms of production and profit. Nonetheless, in its extreme form, the drive to work hard may manifest itself as workaholism, i.e. the compulsion or the uncontrollable need to work incessantly, which does not necessarily enhance productivity (Oates 1971). One might view workaholism as a form of runaway selection of a trait—in this case working hard—that, while originally adaptive, is eventually not so anymore. Several studies show, for instance, that workaholics tend to suffer from ill health, and are often no more productive than their less driven co-workers (Shimazu et al. 2010). This may be related to the relative inability of workaholics to delegate work and to the high standards they set for their work (e.g. Clark et al. 2010). In addition, working hard, especially when working with people and when working is perceived as stressful, may lead to feelings of burnout, including emotional exhaustion (the feeling of depletion or draining of one's mental resources), cynicism (indifference or a distant attitude towards one's job), and a lack of professional efficacy (the tendency to evaluate one's work performance negatively), resulting in feelings of insufficiency and poor job-related self-esteem (Maslach 1993; Schaufeli and Buunk 2003). From an evolutionary perspective, stressors that involve a loss of social rank are particular potential causes of burnout (see Gilbert 2006 for a review). Although a loss of status may be controllable when one feels one can leave the situation and move to another where higher status is attainable, a loss of status becomes particularly problematic when no escape is felt to be possible. Such 'blocked' escape may induce a sense of defeat, i.e. may induce a 'giving up' state of mind, characterized by a negative self-definition ('I am a loser') and feelings of depression and exhaustion (e.g. Buunk and Brenninkmeijer 2000; Gilbert and Allan 1998; Gilbert and Miles 2000; Sloman et al. 1994). In a review, Schaufeli and Buunk (2003) found that burned-out individuals tend to feel helpless, hopeless, and powerless, and experience feelings of insufficiency, incompetence, and poor job-related self-esteem—all experiences that suggest a subjectively low status and a sense of defeat. Likewise, in a study among Spanish teachers, Buunk et al. (2007) found that a low status, a loss of status, and a sense of defeat were independent correlates of burn-out. As would be expected from the perspective of intrasexual competition, a feeling of being defeated predicted burnout among men, but not among women, in the following year. More generally, various studies in organizational contexts have documented the negative effects of subjective low status for health and well-being. For example, in a prospective study in two metal factories in the Netherlands, Geurts et al. (1994) found that a low subjective status (in terms of being worse off than others in domains such as autonomy and promotion prospects), led to a relatively high risk of taking sick leave.

Beating one's rivals: conspicuous consumption, bullying, and gossip

Because individuals see (particularly same-sex) colleagues as rivals in the quest for status-related resources, they may occasionally try to derogate them by, for example, bullying and gossip. Research in a male-dominated work setting (the Norwegian marine industry) showed that, on a weekly basis, 7% of workers reported being subjected to at least one of the following behaviours from co-workers or supervisors: ridicule and insulting teasing, verbal abuse, rumours and gossip spread about themselves, offending remarks, recurring reminders on blunders, hostility or silence when entering a conversation, or the devaluing of one's effort and work (Einarsen and Raknes 1997). As many as 22% of workers reported being subjected to one or more of these acts at least once a month. In line with the fact that competition within organizations is mainly intrasexual, a review by Schuster (1996) showed that male bullies most often victimize males, whereas females more often victimize females. Moreover, the type of attack differs. Women seem to be more spiteful, and more often talk behind others' backs, ridicule others, spread rumours, or make indirect allusions. Typical male tactics are to stop talking to someone, to permanently assign others to new tasks, and to assign tasks that violate others' self-esteem.

Men may also implicitly derogate their rivals by engaging in conspicuous consumption, i.e. by displaying luxury goods, such as expensive clothes, watches, mobile phones, and cars (e.g. Saad and Vongas 2009). By means of conspicuous consumption, men may signal to their rivals that they are superior in terms of these status-related resources, making their rivals look poor in comparison. Miller (2000) suggested that conspicuous consumption can be seen as a handicap signal. Handicap signalling refers to the evolution of an honest signal, which cannot be copied because it is very costly to produce (Zahavi and Zahavi 1997). In this way, for example, men can show off and signal to other men and potential mates that 'I can afford all this'. Lycett and Dunbar (2000) demonstrated the role of conspicuous consumption in the intrasexual competition over females. In their study, males were more inclined to conspicuously display their mobile phones as the composition of their group became more male-biased. In contrast, Hess and Hagen (in review) argued that women, particularly, tend to use gossip as a strategy for intrasexual competition. In addition, women may form coalitions with other women in order to gossip about rivals. Rucas et al. (2006) found evidence that women use gossip as a strategic tool in their study of the Tsimane of Bolivia. By derogating, for instance, other women's ability as a housekeeper, wife, and mother, these other women were seen as less desirable by men.

Like individual males, organizations as a whole may use conspicuous consumption in competition with other organizations. They may build unique and expensive offices and attract distinguished CEOs by means of high salaries and bonuses to impress clients and competitors. Both at the individual and organizational level, conspicuous consumption may also take the form of conspicuous *donation* (Van Vugt and Hardy 2010). Organizations and their leaders donate billions of dollars to charity each year. Conspicuous donation is assumed to reflect positively on the actor, revealing information about the organizations' qualities, such as generosity, reliability, and altruism. This information can be used to attract new group members or help create alliances with other groups in competition with rival groups.

Derogating a rival may also take the form of the 'sexual attribution bias' (SAB). In one study on this phenomenon, Försterling et al. (2007) found that the success of attractive same-sex others was consistently ascribed to luck and less to ability, whereas the success of attractive opposite-sex others was attributed more to ability and less to luck. Thus, both men and women view particularly attractive same-sex others as rivals, feel threatened by their success, and feel a need to 'downplay' this success. The SAB fosters a favourable assessment of the self in relation to the same-sex rival that ensures persistence of competition, reducing the rival's chances of succeeding. To remain believable, individuals usually do not derogate their rivals on all attributes. They give them credit for success in domains they regard as unimportant, but devalue their rival in domains that are perceived to be important for one's self-evaluation (Schmitt 1988).

Being bullied or gossiped about may have devastating consequences and has been found to be related to burnout, stress, and decreased job satisfaction in several segments of the job market, ranging from construction to educational and medical settings (Melia and Becerill 2007; Van Dick and Wagner 2001). Bullying and gossip have such adverse consequences because they both lead to ostracism, i.e. the perception that one is ignored or excluded by others (Williams 2001). In general, ostracism exerts immediate and detrimental effects on psychological well-being by depleting the recipient's primary needs of belonging, control, self-esteem, and meaningful existence (Williams and Zadro 2001). The fact that ostracism has such negative effects on human functioning reflects the vital importance of social groups to human survival and fitness. According to Williams (2001), social exclusion threatens people at such a basic level that it impairs their sense of meaningful existence and wellbeing. Even very small incidents of social exclusion may have serious effects, as is shown by studies in which individuals were excluded from playing a computer game with three (non-existent) individuals. Following the game, ostracized individuals reported less purpose in life and lowered well-being compared to individuals who did participate in the game or who were not experimentally excluded (e.g. Stillman et al. 2009).

Despite the potentially negative consequences of gossip, it is important to recognize that it may also have positive effects. According to Dunbar (1993), gossip permits human groups to expand in size beyond the limits imposed by physical grooming alone. It helps individuals map their social environments (Hannerz 1967). In addition, gossiping about others may strengthen alliances and group membership, create feelings of belongingness and connection, and may help individuals acquire information relevant to their fitness and survival (e.g. De Backer et al. 2007; Grosser et al. 2010). Finally, according to Wert and Salovey (2004), gossip about a superior other can lead to self-improvement. That is, as individuals learn about the achievements of others through gossip, they are motivated to compare themselves favourably with the target, which can result in a better performance (Grosser et al. 2010).

The challenge for organizations is to eliminate the negative consequences of bullying and gossip, while preserving and encouraging their positive function. A study by Kniffin and Sloan Wilson (2010) provides a possible avenue for this. These authors found that the exact effects of gossip—positive or negative—depended, at least in part, on the social context in which gossip takes place. These authors found that workplace gossip had positive consequences, such as increased cooperation, when organizational rewards are fairly allocated not to individuals but to small-scale groups that permit mutual monitoring. Under these conditions, individuals will use gossip as a tool to defend and affirm group-norms, rather than to derogate each other, and as a mechanism to cooperate rather than to compete. To profit from these benefits, however, gossip first needs to be recognized as a practice that is a natural part of organizational life that can serve socially-redeeming purposes. Kniffin and Sloan-Wilson's study also suggests that it seems wise for organizations to repartition some element of employee rewards to the level of workplace units or teams, as is for instance the case in gain sharing, when a portion of employee compensation is tied to group-level performance.

Altruism and intrasexual competition

The previous discussion suggests that workers primarily behave egotistically and competitively, which seems in contrast to the fact that, in many organizations, workers engage in Organizational Citizen Behaviours (OCB). OCB refers to behaviours that are discretionary, not directly or explicitly recognized by the formal reward system, but that promote the efficient and effective functioning of the organization (Organ 1988). Examples of OCB are helping a colleague with a problem he or she encounters in his or her work, working unpaid overtime, or organizing a farewell party for a colleague who leaves the organization. Intuitively, OCB may seem counterproductive to the actor: he or she invests costly time and energy in other workers' success and well-being, seemingly without receiving benefits in return. However from the evolutionary point of view, engaging in OCB makes perfect sense. Such behaviours can be seen as a form of *competitive altruism*: individuals and organizations compete by being generous and forgoing individual benefits (Van Vugt et al. 2007b). Competitive altruism has indeed been found to improve one's reputation and status: others often attribute charisma to those who sacrifice their own needs to those of others (De Cremer and Van Knippenberg 2004). More specifically, engaging in OCB may strengthen the actor's position in the group, because pro-social behaviours may enhance the actor's popularity as an ally, cooperation partner, and leader (Choi and Mai-Dalton 1998). Deutsch Salamon and Deutsch (2006) draw on the handicap principle to explain OCB. According to these authors, workers display positive qualities by means of OCB that otherwise would remain unobserved, such as agreeableness, conscientiousness, and reliability. Research shows that performing OCB is indeed effective as a strategy of positive self-presentation. It has, for instance, been found that employees who more often perform OCB receive more favourable performance evaluations and, consequently, more rewards (e.g. Allen and Rush 1998).

OCB may also be explained from the perspective of *reciprocal altruism* (Trivers 1972). According to this perspective, individuals help others because they expect the recipient, in the future, to reciprocate for the help received. From this point of view, individuals will only help those who have reciprocated their help in the past and those who are expected to do so in the future. In contrast, individuals will avoid helping 'cheats'. An especially important implication of this perspective is that individuals will not only avoid being exploited, but also avoid being in debt as this might lead to potential loss. Thus, both forms of a lack of reciprocity will be experienced negatively. Reciprocity can be seen as an evolutionarily rooted psychological principle: by monitoring that their relationships were governed by reciprocity, our ancestors increased their likelihood of survival (Buunk and Schaufeli 1999). Indeed, many emotions such as moral indignation, guilt, resentment, and gratitude seem to have evolved at least in part to signal the extent to which one's relationship reflect reciprocity. There is abundant evidence that, in organizations, a lack of reciprocity in one's relationships at work is accompanied by negative emotions, stress, and burnout (e.g. Buunk and Schaufeli 1999; Van Dierennonck et al. 2001).

An implication of this for organizations is that OCB may occur more or less naturally, as it, paradoxically, stems in part from competition and from the expectation that one may receive in kind. However, when employees feel systematically underbenefited in their relationships at work, this may have negative consequences for the organization, and organizations may need to pay attention to ways to restore perceptions of reciprocity. For example, Van Dierennonck et al. (1998) found that, when already burned out, workers may be helped by group-based intervention programs that aim to reduce perceptions of a lack of reciprocity by increasing the fit between the professional's goals and expectations and the actual work situation.

Conclusion

Present day organizations seem to differ in many ways from the type of groups in which humans evolved. For example, in contrast to the small groups our evolutionary ancestors lived and worked in, modern organizations are often characterized by interdependence and competitiveness between hundreds and even thousands of colleagues, by formal rules and regulations, and by an elaborate hierarchy with leaders at the top who most workers rarely or never meet. Nevertheless, in many ways, organizations today still operate in line with the adaptations humans developed for social living, and still reflect the multilayered nature of human groupings suggested by Dunbar (2008). We have tried to illuminate that, due to our evolutionary background, competition with fellow group members and with other groups seems to be a basic feature of social life, and that even altruistic behaviours may often be explained as a form of trying to outdo others. Our main focus in this chapter was on intrasexual competition, which seems to play an important role in modern organizations, often with negative consequences for the organization. For instance, intrasexual competition may cause workers to bully each other and steal from the company, may induce recruiters to reject and negatively evaluate candidates because they are perceived as rivals, and may eventually lead to burnout because workers feel they lose the battle over status. In addition, we presented evidence that the intrasexual competition mechanisms that evolved during human evolution serve men better than women when it comes to obtaining occupational success. While, for men, the features that characterize their intrasexual competition still tend to be associated with occupational success and status, intrasexual competition among human females is based more on physical attractiveness. Even in professional domains in which attractiveness is assumed to be irrelevant, women still may compete in this domain.

An evolutionary perspective does not provide unequivocal recommendations for organizational practice. It may, however, help understand why some persistent problems in organizations continue to occur. For example, by engaging in conspicuous consumption, organizations may overshoot the mark. Both organizations and workers may become trapped in a never-ending cycle: to intimidate others who show off by means of conspicuous consumption or donation, they have to come up with even more expensive or unique goods or larger donations. In times of financial crisis, workers and their organizations may not be able to continue this arms race and, when they do, end up in financial problems. Recently, in the Netherlands, a middle-sized bank (the DSB Bank) came into serious financial trouble for this exact reason. An evolutionary perspective may also explain why certain policy measures may not be effective, or may even backfire. For example, although from a societal or rational point of view positive discrimination of women or ethnic minorities may be desirable, it may clash with workers' feelings of fairness in the competitive game, and result in envy or bullying. Policy measures that demand people to resist or oppose their evolutionarily based inner drives may have a low chance of success. A policy based on an understanding of our human nature and that is aimed at preventing its pitfalls, seems more effective.

References

Allen, T.D. and Rush, M.C. (1998). The effects of organizational citizenship behavior on performance judgments: a field study and a laboratory experiment. *Journal of Applied Psychology*, **83**, 247–60.

Andersson, M.B. (1994). *Sexual selection*. Princeton University Press, Princeton, NJ.

Ayres, J. and Malouff, J.M. (2007). Problem-solving *training* to help workers increase positive affect, job satisfaction, and life satisfaction. *European Journal of Work and Organizational Psychology*, **16**, 279–94.

Baenninger, M.A., Baenninger, R., and Houle, D. (1993). Attractiveness, attentiveness, and perceived male shortage: their influence on perceptions of other females. *Ethology and Sociobiology*, **14**, 293–304.

Betzig, L. (1986). *Despotism and differential reproduction: a Darwinian view of history*. Aldine, New York.

Brunell, A.B., Gentry, W.A., Campbell, W.K., Hoffman, B.J., Kuhnert, K.W., and DeMarree, K.G. (2008). Leader emergence: the case of the narcissistic leader. *Personality and Social Psychology Bulletin*, **34**, 1663–76.

Buss, D.M. (1994). *The evolution of desire: strategies of human mating*. Basic Books, New York.

Buss, D.M. (1999). *Evolutionary psychology*. Allyn and Bacon, London.

Buunk, B.P. and Brenninkmeijer, V. (2000). Social comparison processes among depressed individuals: evidence for the evolutionary perspective on involuntary subordinate strategies? In: L. Sloman and P. Gilbert (eds), *Subordination and defeat: an evolutionary approach to mood disorders and their therapy*, pp. 147–64. Erlbaum, Mahwah, NJ.

Buunk, A.P. and Fisher, M. (2009). Individual differences in intrasexual competition. *Journal of Evolutionary Psychology*, **7**, 37–48.

Buunk, A.P. and Gibbons, F.X. (2007). Social comparison: the end of a theory and the emergence of a field. *Organizational Behavior and Human Decision Processes*, **10**, 3–21.

Buunk, A.P. and Schaufeli, W.B. (1999). Reciprocity in interpersonal relationships: an evolutionary perspective on its importance for health and well-being. *European Review of Social Psychology*, **10**, 259–91.

Buunk, A.P. and Van der Laan, V. (2002). Do women need female role models? Subjective social status and the effects of same-sex and opposite sex comparisons. *Revue Internationale de Psychologie Sociale*, **15**, 129–55.

Buunk, A.P. and Ybema, J.F. (1997). Social comparisons and occupational stress: the identification-contrast model. In: B.P. Buunk and F.X. Gibbons (eds), *Health, coping, and well-being: perspectives from social comparison theory*, pp. 359–88. Lawrence Erlbaum Associates Publishers, Mahwah, NJ.

Buunk, A.P., Peíró, J.M., Rodríguez, I., and Bravo, J.M. (2007). A loss of status and a sense of defeat: an evolutionary perspective on professional burnout. *European Journal of Personality*, **21**, 471–85.

Buunk, A.P., Dijkstra, P., Pollet, T.V., and Massar, K. (in press). Intrasexual competition within organizations. In: G. Saad (ed.), *Evolutionary psychology in the business sciences*. Springer Verlag, New York.

Central Bureau of Statistics (CBS 2009). Nederland is Europees kampioen deeltijdwerken/The Netherlands are European champions at part-time work. Retrieved from: http://www.cbs.nl/nl-NL/menu/themas/arbeid-sociale-zekerheid/publicaties/artikelen/archief/2009/2009-2821-wm.htm)

Central Bureau of Statistics (CBS 2010). Meer hoger opgeleide vrouwen, meer arbeidsdeelname/More highly educated women, higher participation in the job market. Retrieved from: http://www.cbs.nl/nl-NL/menu/themas/arbeid-sociale-zekerheid/publicaties/artikelen/archief/2010/2010-3148-wm.htm)

Campbell, A. (2002). *A mind of her own: the evolutionary psychology of women*. Oxford University Press, Oxford.

Cashdan, E. (1998). Are men more competitive than women? *British Journal of Social Psychology,* **37**, 213–29.

Chagnon, N.A. (1988). Life histories, blood revenge, and warfare in a tribal population. *Science,* **239**, 985–92.

Choi, Y. and Mai-Dalton, R.R. (1998). On the *leadership* function of self-sacrifice. *The Leadership Quarterly,* **9**, 475–501.

Clark, M. A., Lelchook, A. M., and Taylor, M. L. (2010). Beyond the big five: how narcissism, perfectionism, and dispositional affect relate to workaholism. *Personality and Individual Differences,* **48**, 786–91.

Colarelli, S.M. (2003). *No best way: an evolutionary perspective on human resource management.* Praeger Publishers/Greenwood Publishing Group, Westport, CT.

Crosby, S.M. (1982). *Relative deprivation and working women.* Yale University Press, New Haven, CT.

Darwin, C. (1871). *The descent of man and selection in relation to sex.* John Murray, London.

De Backer, C.J.S., Nelissen, M., Vyncke, P., Braeckman, J., and McAndrew, F.T. (2007). Celebrities: from teachers to friends: a test of two hypotheses on adaptiveness of celebrity gossip. Human Nature, **18**, 334–54.

De Cremer, D. and Van Knippenberg, D. (2004). Leader self-sacrifice and leadership effectiveness: the moderating role of leader self-confidence. *Organizational Behavior and Human Decision Processes,* **95**, 140–55.

Deutsch Salamon, S. and Deutsch, Y. (2006). OCB as a handicap: an evolutionary psychological perspective. *Journal of Organizational Behavior,* **27**, 185–99.

Duffy, M.K., Shaw, J.D., and Schaubroeck, J.M. (2008). Envy in organizational life. In: R.H. Smith (ed.), *Envy: theory and research,* pp. 167–89. Oxford University Press, New York.

Dunbar, R.I.M. (1993). Coevolution of neocortical size, group size and language in humans. *Behavioral and Brain Sciences,* **16**, 681–735.

Dunbar, R.I.M. (1998). The social brain hypothesis. *Evolutionary Anthropology,* **6**, 178–90.

Dunbar, R.I.M. (2008). Cognitive constraints on the structure and dynamics of social networks. *Group Dynamics,* **12**, 7–16.

Eagly, A.H., Johannesen-Schmidt, M.C., and Van Engen, M.L. (2003). *Transformational,* transactional, and laissez-faire *leadership* styles: a meta-analysis comparing *women* and *men. Psychological Bulletin,* **129**, 569–91.

Einarsen, S. and Raknes, B.I. (1997). Harassment in the workplace and the victimization of men. *Violence and Victims,* **12**, 247–63.

Ellemers, N., Van den Heuvel, H., De Gilder, D., Maass, A., and Bonvini, A. (2004). The underrepresentation of women in science: differential commitment or the queen bee syndrome? *British Journal of Social Psychology,* **43**, 315–38.

Evers, W. J. G., Brouwers, A., and Tomic, W. (2006). A quasi-experimental study on management coaching effectiveness. *Consulting Psychology Journal: Practice and Research,* **58**, 174–82.

Feldman, N.S. and Ruble, D.N. (1981). Social comparison strategies: dimensions offered and options taken. *Personality and Social Psychology Bulletin,* **7**, 11–16.

Figueredo, A.J., Sefcek, J.A., Vasquez, G., Brumbach, B.H., King, J.E., and Jacobs, W.J. (2005). Evolutionary personality psychology. In: D.M. Buss (ed.), *The handbook of evolutionary psychology,* pp. 851–77. Wiley, Hoboken, NJ.

Fischer, P., Kastenmüller, A., Frey, D., and Peus, C. (2009). Social comparison and information transmission in the work context. *Journal of Applied Social Psychology,* **39**, 42–61.

Försterling, F., Preikschas, S., and Agthe, M. (2007). Ability, luck, and looks: an evolutionary look at achievement ascriptions and the sexual attribution bias. *Journal of Personality and Social Psychology,* **92**, 775–88.

Geurts, S.A., Buunk, B.P., and Schaufeli, W.B. (1994). Health complaints, social comparisons, and absenteeism. *Work and Stress,* **8**, 220–34.

Gilbert, P. (2006). Evolution and depression: issues and implicatons. *Psychological Medicine*, **36**, 287–97.

Gilbert, P. and Allan, S. (1998). The role of defeat and entrapment (arrested flight) in depression: an exploration of an evolutionary view. *Psychological Medicine: A Journal of Research in Psychiatry and the Allied Sciences*, **28**, 585–98.

Gilbert, P. and Miles, J.N.V. (2000). Sensitivity to social put-down: its relationship to perceptions of social rank, shame, social anxiety, depression, anger and self-other blame. *Personality and Individual Differences*, **29**, 757–74.

Gilbert, P., Price, J., and Allan, S. (1995). Social comparison, social attractiveness and evolution: how might they be related? *New Ideas in Psychology*, **13**, 149–65.

Gino, F. and Pierce, L. (2009). The abundance effect: unethical behavior in the presence of wealth. *Organizational Behavior and Human Decision Processes*, **109**, 142–55.

Grosser, T.J., Lopez-Kidwell, V., and Labianca, G. (2010). A social network analysis of positive and negative gossip in organizational life. *Group and Organization Management*, **35**, 177–212.

Hannerz, U. (1967). Gossip, networks, and culture in a black American ghetto. *Ethnos*, **32**, 35–60.

Hess, N.H. and Hagen E.H. (in review). Informational warfare: coalitional gossiping as a strategy for within-group aggression. Preprint available at: http://anthro.vancouver.wsu.edu/publications/

Hill, S.E. and Buss, D.M. (2006). Envy and positional bias in the evolutionary psychology of management. *Managerial and Decision Economics*, **27**, 131–43.

Hill, S.E. and Buss, D.M. (2008). The evolutionary psychology of envy. In: R.H. Smith (ed.), *Envy: theory and research*, pp. 60–70. Oxford University Press, New York.

Hopcroft, R.L. (2006). Sex, *status*, and *reproductive success* in the contemporary United States. *Evolution and Human Behavior*, **27**, 104–20.

Ilies, R., Arvey, R.D., and Bouchard, T.J. (2006). Darwinism, behavioral genetics, and organizational behavior: a review and agenda for future research. *Journal of Organizational Behavior*, **27**, 121–41.

Iredale, W., Van Vugt., M., and Dunbar, R.I.M. (2008). Showing off in humans: male generosity as mate signal. *Evolutionary Psychology*, **6**, 386–92.

Jandeska, K.E. and Kraimer, M.L. (2005). Women's perceptions of organizational culture, work attitudes, and role-modeling behaviors. *Journal of Managerial Issues*, **17**, 461–78.

Kets de Vries, M.F.R. and Miller, D. (2007). Narcissism and leadership: an object relations perspective. *Human Relations*, **38**, 583–601.

Kletner, D., Haift, J., and Shiota, M.N. (2006). Social functions and the evolution of emotions. In: M.A. Schaller, J.A. Simpson, and D.T. Kenrick (eds), *Evolution and social psychology*, p. 115–42. Psychology Press, New York.

Kniffin, K, M. and Sloan Wilson, D. (2010). Evolutionary perspectives on workplace gossip: why and how gossip can serve groups. *Group and Organization Management*, **35**, 150–76.

Luxen, M.F. and Van de Vijver, F.J.R. (2006). Facial attractiveness, sexual selection, and personnel selection: when evolved preferences matter. *Journal of Organizational Behavior*, **27**, 241–55.

Lycett, J. and Dunbar, R.I.M. (2000). Mobile phones as lekking devices among human males. *Human Nature*, **11**, 93–104.

Maslach, C. (1993). Burnout: a multidimensional perspective. In: W.B. Schaufeli, C. Maslach, and T. Marek (eds), *Professional burnout: recent developments in theory and research*, pp. 19–32. Taylor and Francis, Washington, DC.

Mavin, S. (2008). Queen Bees, wannabees, and afraid to bees: no more 'best enemies' for women in management? *British Journal of Management*, **19**, s75-84.

Melia, J. L. and Becerril, M. (2007). Psychosocial sources of stress and burnout in the construction sector: a structural equation model. *Psicothema*, **19**, 679–86.

Merten, D.E. (1997). The meaning of meanness: popularity, competition, and conflict among junior high school girls. *Sociology of Education*, **70**, 175–91.

Miller, C.T. (1984). Self-schemas, gender, and social comparison: a clarification of the related attributes hypothesis. *Journal of Personality and Social Psychology*, **46**, 1222–9.

Miller, G.F. (2000). *The mating mind*. Heinemann, London.

Nicholson, N. (1998). Seven deadly syndromes of management and organization: the view from evolutionary psychology. *Managerial and Decision Economics*, **19**, 411–26.

Oates, W. (1971). *Confessions of a workaholic: the facts about work addiction*. World, NewYork.

Organ, D.W. (1988). *Organizational citizenship behavior—the good soldier syndrome* (1st edition). Lexington Books, Lexington, MA.

Palmer, C.T. and Tilley, C.F. (1995). Sexual access to females as a motivation for joining gangs: an evolutionary approach. *Journal of Sex Research*, **32**, 213–17.

Prime, J.L., Carter, N.M., and Welbourne, T.M. (2009). Women "take care", men "take charge": managers' stereotypic perceptions of women and men leaders. *Psychologist-Manager Journal*, **12**, 25–49.

Rucas, S.L., Gurven, M., Kaplan, H., Winking, J., Gangestad, S., and Crespo, M. (2006). Female intrasexual competition and reputational effects on attractiveness among the Tsimane of Bolivia. *Evolution and Human Behavior*, **27**, 40–52.

Saad, G. (2007). A multitude of environments for a consilient Darwinian meta-theory of personality: the environment of evolutionary adaptedness, local niches, the ontogenetic environment and situational contexts. *European Journal of Personality*, **21**, 624–6.

Saad, G. and Gill, T. (2001). Sex differences in the ultimatum game: an evolutionary psychology perspective. *Journal of Bioeconomics*, **3**, 171–93.

Saad, G. and Peng, A. (2006). Applying Darwinian principles in designing effective intervention strategies: the case of sun tanning. *Psychology and Marketing*, **23**, 617–38.

Saad, G. and Vongas, J. G. (2009). The effect of conspicuous consumption on men's testosterone levels. *Organizational Behavior and Human Decision Processes*, **110**, 80–92.

Schaufeli, W. and Buunk, B.P. (2003). Burnout: an overview of 25 years of research and theorizing. In: M.J. Schabracq, J.A.M. Winnubst, and C.L. Cooper (eds), *Handbook of work and health psychology*, pp. 383–425. Wiley, Chichester.

Schmitt, B.D. (1988). Social comparison in romantic jealousy. *Personality and Social Psychology Bulletin*, **14**, 374–87.

Schuster, B. (1996). Rejection, exclusion, and harassment at work and in schools. An integration of results from research on mobbing, bullying, and peer rejection. *European Psychologist*, **1**, 293–317.

Shimazu, A., Schaufeli, W.B., and Toon, T.W. (2010). How does workaholism affect worker health and performance? The mediating role of coping. *International Journal of Behavioral Medicine*, **17**, 154–60.

Sidanius, J. and Pratto F. (1999). *Social dominance*. Cambridge University Press, Cambridge.

Sloman, L, Price, J., Gilbert, P., and Gardner, R. (1994). Adaptive function of depression: psychotherapeutic implications. *American Journal of Psychotherapy*, **48**, 401–16.

Steil, J.M. and Hay, J. L. (1997). Social comparison in the workplace: a study of 60 dual-career couples. *Personality and Social Psychology Bulletin*, **23**, 427–38.

Stillman, T.F., Baumeister, R.F., Lambert, N.M., Crescioni, W.A., DeWall, N.C., and Fincham, F. (2009). Alone and without purpose: life loses meaning following social exclusion. *Journal of Experimental Social Psychology*, **45**, 686–94.

Suls, J., Gaes, G., and Gastorf, J.W. (1979). Evaluating a sex-related ability: comparison with same-, opposite-, and combined-sex norms. *Journal of Research in Personality*, **13**, 294–304.

Tooby, J. and Cosmides, L. (1988). *The evolution of war and its cognitive foundations* (Institute for Evolutionary Studies Tech. Rep. No. 88–1). Institute for Evolutionary Studies, Palo Alto, CA.

Trivers R.L. (1972). Parental investment and sexual selection. In: B. Campbell (ed.), *Sexual selection and the descent of man, 1871–1971*, pp. 136–79. Aldine, Chicago, IL.

Van de Ven, N., Zeelenberg, M., and Pieters, R. (2009). Leveling up and down: the experiences of benign and malicious envy. *Emotion*, **9**, 419–29.

Van Dick, R. and Wagner, U. (2001). Stress and strain in teaching: a structural equation approach. *British Journal of Educational Psychology*, **71**, 243–59.

Van Dierennonck, D., Schaufeli, W. B., and Buunk, A. P. (1998). The evaluation of an individual burnout intervention program: the role of inequity and social support. *Journal of Applied Psychology*, **83**, 392–407.

Van Dierennonck, D., Schaufeli, W. B., and Buunk, A. P. (2001). The evaluation of burnout and inequity among human service professionals: a longitudinal study. *Journal of Occupational Health Psychology*, **6**, 43–52.

Van Vugt, M. and Hardy, C.L. (2010). Cooperation for reputation: wasteful contributions as costly signals in public goods. *Group Processes and Intergroup Relations*, **13**, 101–111.

Van Vugt, M. and Schaller, M. (2008). Evolutionary approaches to group dynamics: an introduction. *Group Dynamics: Theory, Research,* and Practice, 12, 1–6.

Van Vugt, M. and Spisak, B.R. (2008). Sex differences in the emergence of leadership during competitions within and between groups. *Psychological Science*, **19**, 854–58.

Van Vugt, M., De Cremer, D., and Janssen, D.P. (2007a). Gender differences in cooperation and competition: the male-warrior hypothesis. *Psychological Science*, **18**, 19–23.

Van Vugt, M., Roberts, G., and Hardy, C. (2007b). Competitive altruism: development of reputation-based cooperation in groups. In: R.I.M. Dunbar and L. Barrett (eds), *Handbook of Evolutionary Psychology*, pp. 531–40. Oxford University Press, Oxford.

Van Vugt, M., Hogan, R., and Kaiser, R. B. (2008). Leadership, followership, and evolution: some lessons from the past. *American Psychologist*, **63**, 182–96.

Vecchio, R.P. (2000). Negative emotion in the workplace: employee jealousy and envy. *International Journal of Stress Management*, **7**, 161–79.

Walters, S. and Crawford, C.B. (1994). The importance of mate attraction for intrasexual competition in men and women. *Ethology and Sociobiology*, **15**, 5–30.

Wert, S.R. and Salovey, P. (2004). A social comparison account of gossip. *Review of General Psychology*, **8**, 122–37.

Williams, K. D. (2001). *Ostracism: the power of silence*. Guilford Press, New York.

Williams, K. D. and Zadro, L. (2001). Ostracism: on being ignored, excluded and rejected. In: M.R. Leary (ed.), *Interpersonal rejection*, pp. 21–53. Oxford University Press, New York.

Zahavi, A. and Zahavi, A. (1997). *The handicap principle: a missing piece of Darwin's puzzle.* Oxford University Press, Oxford.

Section 2

Family

Chapter 5

The evolved child: adapted to family life

David F. Bjorklund and Patrick Douglas Sellers II

Introduction

Application of an evolutionary perspective to the success or failure of any organism typically leads to a focus on adulthood. Indeed, only after the age of fertility is reached can one actively engage in behaviours directly associated with the superordinate goal of passing on one's genes: mate selection, procreation, and caring for offspring (and perhaps grandoffspring). Explicit survival behaviours operating in adulthood rightfully provide the blood pumping through the veins of an evolutionarily informed discussion. However, the journey to adulthood begins many years earlier, in fact at conception, and before organisms can engage in the all-important behaviours of adulthood, they must survive the prenatal period, infancy, and the juvenile period. Moreover, in many instances, success in adulthood will be dependent on achievements made during earlier stages of life. We argue that no evolutionary theory of psychology can be complete without a serious examination of adaptations of early development (see Bjorklund and Pellegrini 2002; Konner 2010). Along these lines, Konner (2010) has boldly stated that 'life *is* development' (p. 741) and, modifying Theodosius Dobzhansky's famous line, 'nothing in childhood makes sense except in the light of evolution' (p. 749).

A major contention of *evolutionary developmental psychology* (e.g. Bjorklund and Hernández Blasi 2005; Bjorklund and Pellegrini 2000, 2002; Burgess and McDonald 2005; Ellis and Bjorklund 2005; Geary and Bjorklund 2000) is that natural selection has operated on all phases of the lifespan, but not necessarily equally. Selection will have its greatest effects on early stages of development: being born, developing to sexual maturity, and finding a mate and rearing offspring to reproductive age. For example, although more than 95% of children born in developed countries today can expect to live to adolescence, the rate is closer to 50% in traditional cultures and was likely as low or lower for our hunter-gatherer ancestors, making infancy and childhood the crucible for natural selection for humans (Volk and Atkinson 2008). Any benefits that foster development through these stages will be favored, even if they have deleterious effects later in life.

Although 'development matters' in evolutionary explication for any species, it is especially important for a slow-developing animal such as *Homo sapiens*. In fact, humans' leisurely sojourn to adulthood is accompanied by increased risk of death before reproducing compared to faster-developing animals. In evolutionary biology, when a feature has substantial costs, there must also be substantial offsetting benefits, otherwise it would have been eliminated by natural selection. There have been many proposals for the benefits of an extended developmental period. Contemporary theory emphasizes that the complexity of human social interactions required that children have a long time in order to learn the ways of their community. The *social-brain hypothesis* holds that humans evolved superior intelligence not so much to deal with predators or prey or their physical environment, but in order to cooperate and compete with conspecifics (e.g. Alexander 1989; Dunbar 1992, 2010; see papers in Kappeler and Silk 2010), and that humans' advanced form of social cognition was the result of the confluence of an extended juvenile period, a large brain, and increased social complexity (e.g. Bjorklund and Bering 2003). Consistent with this, Joffe (1997) investigated the relationships between brain size, social complexity, and length of the juvenile period across 27 primate species and reported positive linear relationships among the three variables. That is, among primates, the size of neocortical brain tissue not associated with the visual system was positively related to the length of the juvenile period and the size of the social group.

Childhood is ripe with areas for possible application for an evolutionary psychological perspective, and research from this perspective abounds, ranging from cognitive processes of attention and deontic reasoning to social forces and pressures governing cooperation and competition (see chapters in Burgess and McDonald (2005) and Ellis and Bjorklund (2005) for some examples). Our focus in this chapter is on children's reactions to, and developmental consequences of family life from an evolutionary perspective. As mammals, human infants are highly dependent on maternal care for the first years of their life, and as a slow-developing species that does not reach maturity until well into the teen years, the family, as well as other social agents, plays a critical role in shaping children's behaviour and preparing them for adulthood. Before examining children's adaptations to family life, we first examine briefly adaptations of infancy and childhood and then discuss some aspects of children's dispositions toward social life.

Adaptations of infancy and childhood

In evolutionary biology and psychology, adaptations refer to reliably developing features that solved some recurrent problem of survival or reproduction in a species' phylogeny. Although in evolutionary psychology adaptations usually refer to behaviours or cognitions of adults, natural selection has operated as thoroughly on the phenotypes and genotypes of infants and children as on those of adults, and we describe three such classes of adaptations: deferred, ontogenetic, and facultative, or conditional.

Deferred adaptations are those that prepare a child for adulthood (Hernández Blasi and Bjorklund 2003). Deferred adaptations are most likely to evolve when ecological conditions remain relatively stable over time. Some sex differences in social behaviours or cognition exemplify such adaptations. For instance, differences in play styles between boys and girls may prepare them for the life they will lead (or would have led in ancient environments) as adults. Boys engage in more rough-and-tumble play than girls, which some have proposed prepares boys for fighting as adults (see Geary 2009; Pellegrini and Smith 1998). Sex differences in symbolic, or fantasy, play are also observed, with the play of girls being more focused on relationships (e.g. playing house, school), whereas the fantasy play of boys tends to emphasize dominance themes (see Pellegrini 2011; Pellegrini and Bjorklund 2004). This is consistent with roles of men and women in traditional cultures, and likely for our ancestors, with women's relationships being more intimate, whereas men's are based more on status. Similarly, there are robust sex differences in interest in infants, with girls from all cultures showing more interest in nurturing babies than boys (see Hrdy 1999; Maestripieri and Pelka 2002). As with most mammals, females are primarily responsible for the care of infants, and practice caring for babies during childhood and adolescence would prepare girls for the role that they will likely play (or would have played in the environment of evolutionary adaptedness) in adulthood. Both sex differences in play styles, particularly rough-and-tumble play and interest in infants, are also found in other mammals (e.g. see Maestripieri and Roney 2006; Pellegrini 2011), suggesting that they are not a quirk of socialization but evolved adaptations—ones that have little consequence for children's current functioning, but rather prepare them for the future. An evolutionary developmental perspective does not propose that such sex differences are inevitable or genetically determined, however. Rather they are evolved biases that interact with children's early environments to produce mature behaviour.

In contrast to deferred adaptations, *ontogenetic adaptations* (Bjorklund 1997; Oppenheim 1981) refer to features of infancy and childhood that serve to adapt children to their immediate environment and not to prepare them for a future one. Obvious physiological examples include some structures and functions associated with fetal development. For example, the yolk sack in birds and the placenta and umbilical cord in mammals serve vital adaptive functions at a specific time in development, keeping the animal alive while in the egg or uterus, but have no further function and are discarded after birth. Similarly, the sucking reflex functions in the first months of life, but disappears when infants can exert better control of their own oral behaviour.

A number of cognitive or behavioural characteristics have been nominated as candidates for ontogenetic adaptations. For example, young infants' limited perceptual abilities might protect them from overstimulation and competition between developing senses (e.g. Turkewitz and Kenny 1982); preschool children's tendencies to overestimate their physical and cognitive abilities might bolster their self-esteem, causing them to persist at tasks that children with more realistic self-evaluations would cease (e.g. Shin et al. 2007); and newborns' tendency to imitate the facial gestures of adults (e.g. Meltzoff and Moore 1977; Nagy and Molnar 2004) might serve to enhance social interaction between an infant and its mother at a time when infants have little intentional control over their social responding (e.g. Bjorklund 1987; Byrne 2005; Legerstee 1991).

The immature appearance of infants and young children, and the nurturing such visages prompts in adults, may be another example of an ontogenetic adaptation. Lorenz (1943) observed that the immature features of juvenile animals trigger caregiving behaviours in many species, including humans. In humans, infants have heads that are proportionally larger than their bodies, foreheads that are large in relation to the rest of the face, large eyes, round cheeks, flat noses, and short limbs, all of which adults find 'cute' (e.g. Alley 1981, 1983). Although these immature features of infants and young children may qualify as ontogenetic adaptations, adults must recognize these features and respond accordingly if they are to provide any benefit to the child. There is evidence that the preference for 'babyness' is first seen between the ages of 12 and 14 years of age in girls and a couple of years later in boys (Fullard and Reiling 1976), suggesting that responding to 'babyness' may have evolved to prepare adolescents for parenthood. Recent evidence suggests that some forms of immature cognition as typically expressed by preschoolers may also positively bias adults toward young children (Bjorklund et al. 2010a).

Especially important for the current chapter are *conditional* adaptations (Boyce and Ellis 2005; Konner 2010). Konner (2010) defines facultative adaptations as 'reproductive advantages gained through different evolved developmental paths in different external conditions detected in early life' (p. 746). Similarly, Boyce and Ellis (2005) define *conditional adaptations* as 'evolved mechanisms that detect and respond to specific features of childhood environments—features that have proven reliable over evolutionary time in predicting the nature of the social and physical world into which children will mature—and entrain developmental pathways that reliably matched those features during a species' natural selective history' (p. 290). The concept of facultative adaptations implies a high degree of plasticity, with children being able to adjust their course of development dependent upon current environmental conditions. However, plasticity is not unlimited but is constrained both by a child's environment and by his or her biology. The central idea of facultative adaptations is that one's current environment is the best predictor of later environments. Children are sensitive to certain environmental conditions at specific times in ontogeny and adjust aspects of their development accordingly, essentially in anticipation of what environmental conditions will likely be in the years ahead.

As an example of a facultative adaptation, consider patterns of weight gain as a function of nutrition during the fetal period. Somewhat ironically, pregnant women who experience poor nutrition, although giving birth to lighter infants, have children who are more likely to be overweight. Such children develop 'thrifty phenotypes', producing higher levels of the appetite-regulating hormone leptin and storing more fat than children with better prenatal diets. Such a response to poor fetal nutrition would have been adaptive to our ancestors, when prenatal malnutrition was predictive of a later scarcity of calories (Gluckman and Hanson 2005). In modern environments, however, complete with grocery stores and fast-food restaurants, such an adaptation may lead to obesity. Gluckman and Hanson (2005) refer to fetuses responding to current conditions (in this case poor nutrition) not for immediate benefit but in anticipation of later advantage after birth as *predictive adaptive responses*.

The most investigated example of a facultative psychological adaptation is one proposed by Belsky et al. (1991) to explain how aspects of children's early environments bias them, particularly girls, toward developing a reproductive strategy that emphasizes either mating opportunities over investment in children or the reverse. Belsky and his colleagues proposed that 'a principal evolutionary function of early experience—the first 5 to 7 years—is to induce in the child an understanding of the availability and predictability of resources (broadly defined) in the environment, of the trustworthiness of others, and of the enduringness of close interpersonal relationships, all of which will affect how the developing person apportions reproductive effort' (1991, p. 650). Children who grow up in homes with adequate resources and predictable and supportive parents learn that other people are reliable. In comparison, children who grow up in homes with limited resources, social stress, father absence, and anxious and undependable attachment relationships learn a very different lesson. Compared to children growing up in more supportive homes, these latter children reach puberty earlier (at least girls), become sexually active earlier, and have more children but invest less in them. Although this can be seen as a maladaptive outcome in modern society, such a pattern reflects a way to enhance inclusive fitness in less-than-optimal environments—environments that were surely frequently experienced by many of our ancestors. This pattern of accelerated pubertal development has been found in numerous studies, with age of attainment of puberty for girls being associated with socioemotional stress (e.g. Ellis et al. 1999; Graber et al. 1995), maternal depression (Ellis and Graber 2000), harsh parenting styles (Belsky et al. 2010), father absence (e.g. Quinlan 2003; Tither and Ellis 2008), and increased sexual activity and adolescent pregnancy (e.g. Belsky et al. 2010; Ellis et al. 2003), among other stress-related factors (see Belsky 2007; Del Giudice 2009; Ellis 2004 for reviews). One study reported increased interest in infants as well as earlier age of menarche for girls, consistent with the proposition that girls are becoming 'prepared' for early reproduction and parenting (Maestripieri et al. 2004).

The point we wish to make is that natural selection has produced physical and psychological adaptations associated with infancy and childhood, in addition to adaptations in adults with which evolutionary psychologists are more familiar. Some adaptations function to adjust infants and children to their current ecological niche (ontogenetic adaptations), whereas others serve to prepare them for life as adults (deferred adaptations), with some directing children down one of several alternative developmental paths, anticipating that future environments will be similar to current ones (facultative adaptations).

An orientation toward social life

As we commented earlier, humans are a social species. Judith Harris (1995) proposed that humans share with other social apes four dispositions toward social life: 1) group affiliation and in-group favoritism; 2) fear of, and/or hostility toward, strangers; 3) within-group status seeking; and

4) the seeking and establishment of close dyadic relationships. Humans, however, are the most social of apes, because they display not only an ability to read and anticipate the *behaviour* of group members, but also an ability to read *minds*, if imperfectly. This is referred to as *theory of mind*, a person's concepts of how people conceptualize mental activity and how they attribute intention to and predict the behaviour of others. Although chimpanzees (*Pan troglodytes*) and other great apes may possess a sophisticated theory of behaviour, they seem to lack, or are deficient in, thinking about another's thoughts and intentions, and as a result are limited in their ability to cooperate, compete, or, in general, socialize with one another (see papers in Kappeler and Silk 2010).

According to some theorists, children are born with *skeletal competencies* (Geary 2005) or *core knowledge* (Spelke and Kinzler 2007) about important aspects of the world, among them social relations. Infants are biased to attend to faces (or face-like stimuli) shortly after birth (e.g. Batki et al. 2000), and are especially sensitive to eye gaze (e.g. Farroni et al. 2002; Gava et al. 2008). By 3 or 4 months of age they process upright faces more efficiently than upside-down faces (e.g. de Haan et al. 1998). This is initially extended to any type of faces (e.g. human or monkey), but by 9 months is limited to faces of their own species (Pascalis et al. 2002). Beginning around 9 months, infants also display the first evidence of *shared attention*, in which two people attend to a common third object, each one being aware of what the other is seeing (Carpenter et al. 1998; Tomasello and Carpenter 2007). Shared attention reflects that infants view other people as *intentional agents*, an understanding that other peoples' behaviour is based on their goals and intentions (see Tomasello 1999; Tomasello et al. 2007), a first step in theory of mind. Shared attention seems to be unique to humans (e.g. Tomasello and Carpenter 2005; but see Leavens et al. 2005, who argue that pointing behaviour in chimpanzees reflects a form of shared attention) and sets the stage for more advanced forms of social cognition that separate *Homo sapiens* from the rest of the great apes. For example, beginning around 4 years of age, children are able to solve *false-belief tasks*, understanding that someone can believe something that contradicts reality. For instance, they understand that a child who saw a cookie hidden in one location (e.g. a jar) will continue to believe that it is still there, even though, unbeknownst to the child, it was moved to a different location (e.g. a cabinet) (see, e.g. Perner et al. 1987). No chimpanzee has been able to solve non-verbal versions of false-belief tasks (e.g. Krachun et al. 2009). Thus, although human children may possess the same dispositions toward social life as do chimpanzees and bonobos, they do so with substantially more sophisticated cognitive systems (see Bjorklund et al. 2010b; Tomasello and Moll 2010).

Returning to some of Harris's dispositions for social life, children show the beginnings of in-group favouritism in infancy and a bit later, out-group hostility. For example, by 6 months of age, infants are better able to discriminate between faces from their own race versus those from other races (e.g. Kelly et al. 2007, 2009). This is not an indication of implicit racism in infancy, but rather reflects the role of familiarity in shaping infants' perceptual abilities. Beginning in the preschool years and increasingly into the school years, children voluntarily segregate themselves into play groups by sex (when there are enough boys and girls in a peer group to make this possible), an early indication of children's in-group favouritism and out-group discrimination (Patterson and Bigler 2006; Powlishta 1995). Consistent with this interpretation, in one study 8- to 10-year-old children watched videos of unfamiliar boys and girls and rated them on a variety of dimensions. Much like adults in similar situations, both the boys and the girls rated the same-sex children more positively than opposite-sex children (Powlishta 1995; see also Egan and Perry 2001).

Although sex seems to be the first in-group/out-group distinction that children make, in the United States white children show a pro-white/anti-black bias by 6 years of age, and perhaps younger (Aboud 2003; Bigler and Liben 1993), which typically begins to decline by age 7 and disappears in most children by 12 years (see Aboud 1988; Apfelbaum et al. 2008). In a similar vein, preschool and first-grade children tend to view peers who play with out-group members less positively than peers who play only with other in-group children, with this effect decreasing or disappearing by 9 years of age (Castelli et al. 2007). Research has suggested that children's initial out-group discrimination reflects a strong favouritism for the in-group rather than overt hostility for the out-group (Brewer 1999; Pfeifer et al. 2007). Consistent with this, in research by Aboud (2003), white Canadian 4- to 7-year-old children showed in-group favouritism earlier, which was stronger than out-group discrimination toward black children. Research using versions of the *Implicit Association Test* (Greenwald et al. 1998), in which the speed with which people make decisions about words or concepts (for example, 'good' and 'bad') associated with different social groups (for example, white versus black people), have been conducted with children and, as for adults, reveals an implicit (unconscious) in-group favouritism. For example, using a child-friendly Implicit Association Test, Baron and Banji (2006) reported that 6- and 10-year-old white children and adults displayed faster reaction times to white faces paired with positive attributes and to black faces paired with negative attributes, reflecting a positive same-race bias. A similar same-race bias has been reported in Japanese children (Dunham et al. 2006), suggesting that this effect is universal and emerges early in development (see Dunham et al. 2008 for a review).

Children's in-group favouritism is seen not only in laboratory studies, but also in 'real life'. For example, in a classic study by Muzafer Sherif and his colleagues (1961), two groups of unacquainted 10- to 11-year-old boys spent several weeks at a summer camp. Dominance hierarchies were established early in each group (reflecting another of Harris's dispositions toward social life, within-group status seeking), as was group solidarity. After establishing group identities (each group adopted names: Rattlers and Eagles), the groups were introduced to one another and engaged in some friendly competitive activities (e.g. ball games, tug-of-war). Within-group conflict arose when a group lost a competition, often resulting in changes in leadership. Over the course of competition, however, between-group hostility increased, with the groups verbally insulting each other, raiding each other's camps, and engaging in physical aggression. Counter to the expectations of the counsellors, hostility did not decrease when the groups engaged in some non-competitive activities (e.g. meals, movie night), although it did toward the end of the summer after the counsellors arranged a series of events that required the cooperation of members of both groups (e.g. finding the source of a disturbance in the water supply).

Children's propensity for in-group favouritism and out-group hostility can help explain the prevalence and adaptive nature of youth gangs. Youth gangs are found mostly in low-income neighbourhoods, claiming (and defending) ownership of a specific territory, frequently for the purpose of selling drugs. Gangs also have rivalries with other gangs, with many gang-related homicides being members of one gang killing members of another, often as revenge for earlier deaths or insults (Howell and Egley 2005). Although gangs surely represent a 'social evil', they may reflect adaptive choices for some children growing up in impoverished and dangerous neighbourhoods. Gangs can be viewed as a special form of peer group, playing into children's bias to belong to a group. They consist of people 'like me' with well-defined rules by which the group behaves. Gangs are more than a collection of delinquent individuals, but people who share similar ethnicity, socioeconomic background, and who have undergone 'rites of passages' to solidify their membership. Gangs afford a way of achieving status in the local neighbourhood, of demonstrating loyalty to friends, possibly family (especially if older siblings are already gang members), and the larger youth community. Importantly, because low-income neighbourhoods are often hotbeds of violence, they can afford protection against rival gangs. To a large degree, gangs are popular with some youth from low-income neighbourhoods because there are few other options for them to succeed. Academic performance among low-income children and adolescents is often poor, conventional jobs in the local neighbourhood often scarce, and gang membership can afford such adolescents opportunities for status, prestige, and financial success through criminal activities that they otherwise could not obtain.

Although social relations with peers become increasingly important as children grow up, as in all mammals, children's first social relationship is with their mothers. Moreover, humans are among a small set of social species that raise their offspring in families—both biparental families, where both mother and father provide some support, and extended families, with several genetically related adults typically helping to tend the young (Emlen 1995). The family is critical for rearing children to adulthood, and natural selection has shaped the behaviours and cognitions of children, their parents, and perhaps others who sometimes care for the welfare of young dependent members of our species.

Family matters

Investment among families

Faced with a long period of vulnerability prior to reproductive viability, an individual's success must be predicated on an active and meaningful support system to aid in survival. Providing a solution to this problem in the human species is family, both immediate and extended. Families provide protection, advice, guidance, food, and shelter, all in an attempt to foster the development of their children and relatives. However, not all family members treat children equally. Extended and even immediate family members vary in their willingness to assist children, and most importantly for evolutionary psychology, they do so in a predictable manner.

Two theories most relevant to an evolutionary developmental investigation of the family are William Hamilton's (1964) inclusive fitness theory and Robert Trivers' (1972) parental investment theory. Inclusive fitness theory holds that an individual's willingness to help another is directly related to their degree of relatedness. Parents share 50% of their genes with their children, but only 25% with their nieces and nephews. Therefore, all other things being equal, individuals should be more apt to sacrifice for their children than for their nieces and nephews. Children are the best way to pass on genes, but members of an extended family are enough alike to warrant altruism. First cousins, who share 12.5% of their genes, ought to cooperate more than second cousins or unrelated peers.

While compelling and certainly true, Hamilton overlooked critical sex differences that significantly influence family members' willingness to invest in children. Parental investment refers to any efforts undertaken by a parent or caretaker for the betterment, safety, or well-being of a child that could have been used to directly benefit the parent. Types of investment include behaviours as varied as emotional concern, provision of food and shelter, and monetary aid. According to Trivers (1972), animals partition their effort between mating and parenting, with the females of most species, humans included, having greater obligatory investment in their offspring (e.g. internal gestation, nursing) than men. In addition, there is the issue of paternity uncertainty. A woman undoubtedly knows when she is related to a child; men, however, can never have 100% paternity certainty. Risks of cuckoldry—spending time and resources raising another man's offspring—are ever present for males. Men protect against this possibility by investing reliably less than women in children, even those who are their true genetic offspring. However, men, unlike 95% of male mammals (Clutton-Brock 1991), do provide care for their children after birth. This was necessitated by the demands of a slow-developing, dependent offspring, who required more than a mother's care to reach adulthood.

Humans are among a small number of mammals that engage in *cooperative breeding* (Hrdy 2007, 2009), with people other than the mother providing care for their offspring. In most human cultures, fathers provide some care and resources for their mates and children, which increases the likelihood that their children will survive to adulthood. However, in most cultures the most common *alloparents* (people other than the biological mother who provide care for children) are the mother's female kin, including sisters, older daughters, and especially, the mother's mother. In addition to genetic relatives, today and surely over evolutionary history, people who are not biologically related to children have also been responsible, in part, for children's care, in particular step-parents. Thus, over our species' history, children have had to depend on a variety of people for their care, each with his or her own inclusive self-interest in seeing that the child reaches reproductive age. A variety of factors other than genetic similarity contribute to whether and how much a person invests in a particular child, which influences that child's likelihood of success. In the following sections we briefly outline some of these factors, separately for mothers, fathers, grandparents, and step-parents.

Maternal investment

Despite receiving assistance in child rearing from many sources, human mothers are undoubtedly the primary caregivers the world over (Hrdy 1999, 2007). They have 100% genetic certainty that the child is theirs, and, until technology provided a substitute for mother's milk, an infant without a nursing mother was unlikely to survive. Women begin to prepare for their role as mother and caretaker from a young age, with girls' interest in infants far exceeding that of boys during childhood and adolescence (Maestripieri and Pelka 2002). However, as much as we may like to view mothers as willing to sacrifice anything for their children, the natural-selective calculus tells a different story. Maternal investment is far from universal in kind and amount, varying widely given differences in resource availability, child viability, and numerous maternal variables (Gardiner and Bjorklund 2007; Keller and Chasiotis 2007). Ancestral women who could allocate their resources 'strategically' among their offspring—investing differentially in each of their children depending on the children's likelihood of reaching adulthood in a given ecological niche—were women who passed on more of their genes than women who invested indiscriminately among their children.

In general, how much a woman invests in each of her children depends on the availability of resources and her assessment of how a particular child can make use of those investments to reach adulthood (Bugental and Beaulieu 2003). Among the factors that mothers consider (albeit unconsciously) in making investment decisions are a child's health, a child's age, her own age, and the availability of resources to support herself, the child, and other children she may have either currently or in the future.

A child's health is probably the clearest indicator of potential reproductive fitness. Investing substantially in a sickly child may represent a great cost with little likely benefit, particularly for ancestral women, when resources were scarce and the likelihood of a child living to adulthood was no greater than 50% (Volk and Atkinson 2008). The heartiness of an infant's cry may be a sign of health and thus a cue a mother could use to guide investment. For example, DeVries (1984) reported that among a group of Masai pastoralists, only six of 13 infants born during a time of famine survived, and most of these were 'fussy' babies who cried a lot. DeVries proposed that infants who cried a lot received more maternal attention and were more likely to be soothed by their mothers and thus fed. Infants with heartier cries may also have been healthier, with their cries being an honest cue to fitness. In other research in the United States, mothers of premature and extremely low birthweight twins displayed more positive behaviour toward the healthier twin when the infants were 8 months of age (Mann 1992). Mann proposed the *healthy baby hypothesis*, in which mothers differentially invest in the healthier of two infants, based on appearance and behaviour (see also Sameroff and Suomi 1996). At the extremes of reduced maternal investment is abuse and filicide, both of which are more likely to occur for less-healthy children. For example, children with intellectual impairment or congenital defects are two to ten times more likely to be abused sometime during childhood than developmentally typical children (see Daly and Wilson 1981). With respect to filicide, in many traditional societies, infanticide is sanctioned if a child is sickly or if nurturing the newborn would imperil the survival of older children (see Daly and Wilson 1988).

The age of both the child and the mother are also variables that influence a woman's investment decisions. Young children, years away from evolutionary payoff, require more maternal investment, whereas older children are at least comparably more self-sufficient and may survive periods of lowered maternal care, making older children, all other things being equal, more worthy of investment when resources are scarce (Daly and Wilson 1988; Keller and Chasiotis 2007). For example, the probability of being killed by a genetic parent drops sharply after the first year of life (see Daly and Wilson 1988). Maternal age also impacts on the distribution of care. Younger women have more childbearing years ahead of them than older women and are thus more selective in resource allocation than older mothers (see Keller and Chasiotis 2007). As a result, they are less apt to invest highly in a high-risk infant than are older women (Beaulieu and Bugental 2008) and are more apt to abuse and neglect their children than older women (Lee and George 1999).

Using myriad variables, mothers assess constraints placed on their caregiving capabilities and decide, typically without conscious awareness, how to parcel out these resources for maximum reproductive fitness. It is important to note that maternal investment decisions of this nature carry no moral or emotional preferential implications; they simply reflect our species' history of living in uncertain environments with finite resources where successful parenting often meant differentially providing resources to children.

As we mentioned earlier, however, mothers are not the sole provider of parenting and investment in children. Hrdy's (2007) cooperative breeding hypothesis promotes the vital importance of alloparents in rearing children, both today and for our ancestors. Raising a child with the help of family in a larger social context affords mothers support and aid of all kinds. Here we examine briefly the roles different alloparents play in child rearing, how and why they invest, and the direct consequences of alloparent investment for children.

Paternal investment

In contrast to the comparably unflinching role of women in caring for children, men in all cultures devote considerably less time to their offspring. Even under conditions of government-mandated paternity leave, men take advantage less reliably than women and contribute significantly less to childrearing (Lammi-Taskula 2008). Nonetheless, as we mentioned earlier, men do far exceed the effort into childcare exhibited by most mammalian fathers. Immature birth of human infants, relative to other primates, and subsequent lengthy childhoods, required greater paternal investment when compared to other male mammals, as mothers found themselves weighed down by a slow-developing child and in need of assistance (Gardiner and Bjorklund 2007). Thus, one reason for human fathers' substantially greater investment in childrearing than for most other mammal fathers is that it was in their best interest, in terms of fitness, to do so. The likelihood of their children surviving and attaining higher status is related to the amount they invest (at least under some ecological conditions, see Geary 2007).

Another driving force behind paternal investment is the relatively low risk of cuckoldry in humans. While rates of cuckoldry vary from 0.8% to 30% between specific studies and populations, the median rate determined by meta-analysis is 3.8% (Bellis et al. 2005). Paternal certainty should reliably influence investment, with those fathers most certain of their genetic relatedness to the child willing to invest to a greater extent. Just as with mothers, numerous factors influence a father's childcare behaviours including hormone levels, marital satisfaction, and environmental differences in resource availability (see Geary 2007).

Levels of a father's emotional involvement and resource provision predicts increased social competence, higher academic achievement, lower rates of delinquent behaviour among adolescents, and overall well-being (Cabrera et al. 2000). Paternal investment in children also predicts reproductive fitness variables for those children as they age. The presence of a biological father or father figure facilitates an early age of initiation into manhood among members of an Australian Aboriginal tribe as well as early parenthood (Scelza 2010). Interestingly, the presence of the biological father in the household after age 7 delays the age of first menarche for girls, which delays first sexual intercourse (Neberich et al. 2010). This may result in a different reproductive strategy for girls, who will establish stable adult relations and invest substantially in a few children, compared to girls from father-absent homes, who reach puberty earlier, establish less stable adult relationships, and invest less in the (often many) children they have (see Belsky et al. 1991; Del Giudice 2009; Ellis 2004 and discussion earlier in this chapter).

Interestingly, within species in which fathers invest, amount of paternal investment is related to male longevity. In addition to humans, males of several primate species invest significantly in their offspring, including Goeldi's monkeys, siamangs, owl monkeys, and titi monkeys. In an examination of sex differences in survival rates, Allman (1999) found that the survival rates of males relative to females increase as a function of the degree of paternal care: the greater the paternal investment, the longer the expected male lifespan. Allman (1999; Allman and Hasenstaub 1999; Allman et al. 1998) proposed that male caregivers are less likely to take risks than non-caregivers. If the survival of offspring is highly dependent on paternal investment, males who avoid risk would be favoured by natural selection. This is similar to the argument made by Campbell (1999) for the generally less-risky dispositions of females. Conversely, if paternal investment is not crucial for offspring survival, natural selection would favour males who take risks in order to increase status and win mating opportunities (Daly and Wilson 1988), even if it results in a shorter lifespan. Paternal investment is thus facultative, high when survival of offspring is dependent upon male care and low when offspring are likely to survive with little or no postnatal investment from fathers (Geary 2007).

Grandparental investment

Maternal grandmothers are thought to be the most important alloparent, providing care more dependably and reliably than even fathers (Euler and Michalski 2007). This comes as no surprise from the point of inclusive fitness theory, as maternal grandmothers are also unequivocally certain of their genetic relatedness to the child. Assisting their daughters in childrearing affords women the opportunity to positively impact their inclusive fitness post-fertility. Sear and Mace (2008) reviewed 45 studies to determine the influence that specific alloparents have on child survival in natural-fertility populations. They reported that aid from maternal grandmothers significantly increased rates of survival for children. Using historical data from German births in the 18th and 19th centuries, Beise and Voland (2002) reported that maternal grandmothers significantly increased the probability of survival for children. Lack of a maternal grandmother increased the probability of death by as much as 60% for children between the ages of 6 and 12 months.

Not only is the presence of a maternal grandmother a protective factor, the more obvious prediction, her absence is itself a risk factor for premature death.

Amount of investment from a child's other three grandparents is predictable based on the degree of their genetic relatedness to the child. After maternal grandmothers, most care is given by maternal grandfathers, then paternal grandmothers, and lastly paternal grandfathers. This descending order of investment is perhaps the most robust finding in the grandparental investment literature (see Coall and Hertwig 2010) and follows the pattern of decreasing certainty of genetic relatedness.

Step-parents

Step-parental investment is a curious case. The contributing step-parent is completely aware that the child in whom he or she is investing is not a genetic relative, seemingly making any investment in the child a total evolutionary loss. Accordingly, step-parental investment, specifically stepfather investment due to the tendency for children to live with their biological mother, is generally lower in quantity and quality (see Anderson et al. 1999; Marlowe 1999). Resident stepfathers have significantly fewer interactions with their partner's children, spend less time with the children, and spend less money on education expenses than do resident genetic fathers (Anderson et al. 1999). However, Anderson and his colleagues (1999) reported that resident stepfathers invest more than non-resident genetic fathers, the very definition of cuckoldry, begging the question of why step-parental investment occurs at all. Stepfather investment in a partner's children is viewed primarily as a mating strategy—an attempt to win over the mother and thus access to future mating opportunities (Anderson et al. 1999; Berger et al. 2008). Demonstrating one's willingness to invest in a woman's children increases the likelihood that he will be selected as a mate.

Unfortunately, a well-documented effect of step-parenting is increased rates of child abuse, unintentional death, and even murder. Daly and Wilson (1996) found higher rates of filicide among stepfamilies than genetic families. Differences existed not just in the number of murders, but in their nature as well. Stepfathers were more likely to brutally murder a child, and the murder was almost never accompanied by suicide (0.015%). However, genetic fathers who murdered children did so in less brutal ways and often (28%) in the context of suicide. Children of a stepfamily are 40 times more likely to suffer from abuse, more likely to be apprehended by police as runaways, and more likely to be arrested for criminal offences, than children in genetic families (Daly and Wilson 1985). Children in stepfamilies are also at higher risk for all types of unintentional death, most markedly drowning (Tooley et al. 2006). Many step-parents simply find it difficult to develop and maintain a close connection and emotional bond with stepchildren, leading to lower overall investment and, often, tragic outcomes.

Despite the dark side of step-parents, it is noteworthy that most step-parents love and care for their stepchildren, creating a positive two-parent environment in which children can grow and flourish. According to Gardiner and Bjorklund (2007): 'we perhaps should be impressed with how flexible human behaviour can be and with the vast majority of step parents who provide loving care to their stepchildren' (p. 349).

Parental investment and healthy psychological development

It is unquestionable that parents (and other caretakers) are crucial to children's survival. However, how good must parents be in order to produce psychologically healthy children? Sandra Scarr (1992, 1993) proposed that children have evolved to tolerate a wide range of parenting behaviours and that individual differences in parenting style have little effect on children's psychological development. While child-rearing practices vary greatly around the world and among families, children across cultures develop into productive and reproductive members of their societies. According to Scarr, unless the family environment falls well outside the normal range, 'good enough' parents will raise children who develop equally as well as the children of 'superparents'.

These assertions are based on an extension of Scarr's genotype → environment theory (Scarr 1992, 1993; Scarr and McCartney 1983), which contends that children seek environments consistent with their genotypes. Experiences in these environments shape children's personalities and intellects, but one's genes ultimately drive such experiences. These genotype → environment effects increase as children increasingly choose environments in which to interact, while their parents' non-genetic influences decrease.

Scarr's theory sustained immediate and substantial criticism (e.g. Baumrind 1993; Jackson 1993). For example, critics argued that although children reared in a wide range of environments grow into reproductive members of their species, the particular mating strategies they employ are consequences of parent–child interactions and not simply of children's genes. Additionally, if children do not have the opportunities to develop in environments to which they are drawn by their genetically influenced tendencies, their development will be influenced to a greater degree by their parents than if they did have such opportunities. Moreover, there is an extensive literature relating parenting styles to psychological characteristics of children (see Bornstein 2006; Sternberg et al. 2006). Although there are multiple influences on children's developing psyches, including peers (e.g. Harris 1995), the media (Comstock and Scharrer 2006), and other cultural institutions such as schools (Epstein and Sanders 2002), parents clearly matter.

Scarr (1993) agreed that in some environments 'good enough' parents may be insufficient to produce children who live up to the standards of their culture, although the effects of inadequate parenting can be mediated through education. The primary difference between Scarr and her critics is the level of focus: Scarr is concerned with how individual children develop into functioning members of their species, whereas her critics focus on the generation of individual differences among children within a given culture, some of which are associated with greater economic success and psychological well-being than others. Consistent with both viewpoints is the existence of 'resilient' children, who develop normally despite being reared in impoverished and high-risk environments (e.g. Masten and Coatsworth 1998).

Recent research has shown that Scarr was likely right, in that children are able to survive and become reproductive members of their community in a wide range of environments. And she was also likely right in suggesting that children's genotypes greatly influence the choices they make in forming adult personalities. However, rather than (or perhaps in addition to) children's genotypes causing them to select environments consistent with their inherited temperaments, children's genotypes are sensitive to early-life conditions, most the result of family interactions, which set them on courses that, on average, were associated with inclusive fitness for our ancestors. These are a form of conditional adaptations, discussed earlier in this chapter. Moreover, individual differences exist in children's ability to adapt to a range of environments, resulting in what are often some counterintuitive outcomes.

Differential susceptibility to experiences

Environmental influence on development is certainly an area abundant with a variety of research, theory, and debate. However, most work in the field examines potential negative effects of adverse environmental influence, such as deleterious influences of abuse, neglect, and familial conflict on children's emotional, social, and intellectual competence. The current dominant approach to evaluating gene–environment interactions in the development of psychopathology is the *diathesis-stress* model, which contends that certain individuals are predisposed to develop psychopathological conditions when exposed to stressful environments. Predisposition is the presence of genetic or socioemotional variables resulting in an individual who is more likely to develop any number of clinical psychopathologies when faced with negative environmental stressors. The predisposed individual can be thought of as possessing a weakened psychological immune system. Understanding variables that contribute to the development and progression of illness, in an effort to identify at-risk individuals and successfully intervene, is a critical arena for research.

Counter to the diathesis-stress model is the evolutionary developmental perspective that recognizes that stressful environments have been part of human experience throughout our evolutionary history. When children encounter stressful environments, this does not so much *disturb* their development as *direct* it toward social strategies that are adaptive under stressful conditions, even though these adaptations may well be harmful in terms of the long-term welfare of the individual (see Ellis et al. 2005, 2011). In particular, natural selection has resulted in children (and presumably adults) who are *differentially susceptible to experiences*. Some children are highly sensitive to both positive and negative early environments, showing the best and worst developmental outcomes depending on the level of support and stress they experience. In contrast, other children are little influenced by variation in early experience and will develop in a species-typical way under most rearing conditions.

This was first illustrated in a study by Boyce et al. (1995), who reported a curvilinear relationship between stress reactivity and early environment for respiratory illness outcomes. High-reactive children had higher rates of respiratory illness in high-adversity conditions (about 3.5 infections per 3-month period), but had the lowest rates of illness when living in low-adversity conditions (slightly over two infections per 3-month period). In contrast, low-reactive children had equivalent rates of illness across different levels of environmental adversity (slightly fewer than three infections per 3-month period). Recently, Obradović and colleagues (2010) extended this reasoning to school readiness and social adjustment and found that highly reactive children took advantage of positive environments, showing the lowest levels of externalizing behaviour and the highest levels of pro-social behaviour and school engagement, but were negatively influenced by adverse environments, all relative to low-reactive children.

The central idea behind differential susceptibility to experience is that susceptible individuals are influenced by their environment 'for better *and* for worse', while environment plays a less significant role for other individuals (Belsky 2007; Boyce and Ellis, 2005; Ellis et al. 2011). A helpful analogy to clarify this relationship is comparing the differences between these types of children to the differences between dandelions and orchids (Boyce and Ellis 2005). Dandelions are resilient to environmental influence, growing and thriving almost anywhere, much to the dismay of gardeners. The end product, a fully-grown dandelion, varies little with respect to the hospitability of its environment. Orchids on the other hand are renowned for their beauty, but also for their high maintenance and difficult nature. Given a nurturing and supportive environment, orchids will flourish, but placed under adverse environmental conditions they will perish.

Research in this area has confirmed the existence of a distinction between 'types' of individuals, those defined by susceptibility and those defined by resilience to environmental influence. For example, Belsky and his colleagues (1998) reported that parenting influence predicted externalizing problems and inhibition in 3-year-olds, but only for those children who exhibit high negativity. In this example, infant negativity acted as a susceptibility factor moderating the relationship between parenting and behavioural outcomes. Recent findings from Pluess and Belsky (2010) provide longitudinal data extending the effects of differential susceptibility to rearing influence well into childhood. Data on parenting behaviour, non-maternal parenting, child behaviour, and academic achievement were collected on children who were followed from the age of 1 month to 11 years. Parenting behaviour, both maternal and non-maternal, predicted child academic and social outcomes in middle childhood, but this relationship was more pronounced for those children measured as temperamentally difficult as infants. Among difficult children, deleterious parenting was associated with negative outcomes, but positive parenting was associated with positive outcomes.

Research has consistently found that infants with difficult temperaments, and fearful and anxious children, are more influenced by parental behaviour for a host of dependent measures, including attachment, externalizing behaviour, IQ, and self-control (e.g. Blair 2002; Crockenberg 1981; Denham et al. 2000; Feldman et al. 1999; Kochanska et al. 2007; Stright et al. 2008; see also Belsky and Pluess 2009). This evidence however, is correlational in nature. Direct testing of the differential susceptibility hypothesis through experimentation would provide a more stringent test of its principles and thus stronger evidence for its existence.

Although experimental studies of differential susceptibility to experience are difficult to do with humans, they can be more easily done with other species. For example, Boyce and his colleagues (1998) examined biological reactivity to novel experiences in a group of rhesus monkeys living in a wooded enclosure of approximately 2 hectares at the National Institute of Mental Health Primate Center. In this low-stress environment, high-reactive monkeys had the lowest rates of injuries. However, when the group was moved into a 93m^2 building, the increased living stress resulted in these same high-reactive monkeys having the highest rates of injuries. In comparison, injury rates of monkeys rated as low reactivity did not vary between the high- and low-stress environments.

Although they are difficult to perform and somewhat limited in the conclusions that can be drawn from them, experimental assessments of the differential susceptibility hypothesis have been done with humans. For instance, Velderman and colleagues (2006) subjected first-time mothers to one of two parenting intervention programmes or a control condition. Parenting intervention consisted of maternal sensitivity training over four sessions when the child was between 7 and 10 months old. The goal of sensitivity training was to teach first-time mothers to be responsive and sensitive to the needs and behaviour of their children. Intervention produced a higher instance of positive infant attachment security, however this effect was more pronounced for those children classified as highly reactive. Highly reactive children, defined as scoring at or above the 80th percentile on a measure of temperament, benefited most from their mother's participation in the intervention. In other research, Blair (2002) followed low birthweight premature infants from the age of 12 months to 3 years as their mothers participated in the Infant Health and Development Program (IHDP), which was designed to give low-income families with premature infants access to training, education, and social services in a comprehensive manner. Children classified as high in negative emotionality received the most benefit from the intervention compared to children lower in negative emotionality. Benefits included decreased occurrence of IQ lower than 75, clinical behaviour problems, and clinically at-risk internalizing and externalizing scores at age 3. These experiments show differential susceptibility to rearing environment through direct experimental manipulation, at least for one extreme, as highly reactive children were specially poised to take advantage of positive influence from maternal training and education.

Moderating variables from both the correlational and experimental data are clearly theoretical relatives and point to the existence of a higher-order conceptualization of these clustering moderator variables. High negative reactivity and difficult temperament may evidence a heightened sensitivity to all environmental stimuli, both positive and negative (Pluess and Belsky 2010) that results in greater impact on development. Greater susceptibility and sensitivity of this nature is necessarily neurobiological in nature, resulting from increased nervous system activation (see Ellis et al. 2011).

Currently there are two competing evolutionary developmental models to explain the growing evidence of differential susceptibility to environment influence—the *biological sensitivity to context* model of Ellis and Boyce (Boyce and Ellis 2005; Ellis et al. 2005; Ellis and Boyce 2008) and Belsky's differential susceptibility theory (Belsky 2005; Belsky and Pluess 2009) (see Ellis et al. 2011 for a review of both theories). Both theories are grounded in the understanding that natural selection would favour variation in phenotypic expression of environmental influence. However, they differ in how natural selection would produce such variation.

Advocates of the biological sensitivity to context model propose that children's stress reactivity is tuned to their early environment as a predictive measure for the future. Given a dangerous, stressful, uncertain environment, or a particularly supportive early environment, children may find it advantageous to develop a heightened reactivity to stressors in order to survive and/or take advantage of opportunities. What results is an acute sensitivity to environmental stressors, both positive and negative, through a quick-to-fire stress response. High reactivity of the stress response system is produced along a curvilinear U-shaped relationship with early environmental conditions. Children exposed to particularly nurturing or especially adverse early environments both develop up-regulated biological responses to stressors (Boyce and Ellis 2005). These would be the 'orchid' children described earlier. Thus, the human stress response system functions not only as a preparation for responding to environmental situations of peril—fight or flight; it also serves as a conditional (or facultative) adaption, as we discussed earlier in this chapter, to facilitate resource attainment from the surrounding environment when individuals serve to benefit most from responsiveness (see Ellis et al. 2011). The stress response system of children who find themselves in a positive, nurturing, and supportive early environment will up-regulate in an attempt to take advantage of the resource-rich environment. Children in an unsupportive, hostile early environment will necessarily be faced with high levels of resource competition from peers and family, in addition to outright threats to survival, for which an overactive stress response system is beneficial. Thus, while 'orchid' children may be especially apt to display behaviour problems associated with adverse environments, such dispositions, although deleterious in the long run, may serve to adapt children to a stressful contemporary environment, as well as to prepare them to cope in such environments in the future.

While the biological sensitivity to context model argues for environmentally induced plasticity, Belsky's differential susceptibility theory argues that there are genetic differences in children's differential susceptibility to experiences and posits that parents hedge their evolutionary bets by producing offspring who vary in susceptibility, producing some children who are highly responsive to environmental influences ('orchid' children) and others who are not ('dandelion' children). By producing two types of children, parents effectively insulate against the possibility of drastic environmental changes. Individuals of any given generation do not vary their own responses to the environment, but rather it is their children's variation that has fitness implications. Direct benefits of this variation are derived from the simple fact that, while often stable over time, the resource make-up of an environment is subject to change. Highly susceptible children are the evolutionary gamblers in the equation who stand to benefit most from a supportive environment but suffer under an adverse one. (Although such suffering in terms of increased levels of internalizing and externalizing behaviours may help such children adapt to a less-than-optimal environment.) Less-susceptible children relinquish the possibility to extract maximum resources from a highly supportive environment, but are apt to adjust well in 'average' environments, which are most likely to be encountered. Boyce and Ellis (2005) estimated that most children (about 85%) are 'dandelions', whereas only a minority (15%) are 'orchids'.

At this time, it is not necessary to choose between Ellis and Boyce's biological sensitivity to context model and Belsky's differential susceptibility theory. Both are rooted in evolutionary developmental theory and extend the diathesis-stress model that currently dominates studies of psychopathology. Moreover, one theory being right does not mean the other is wrong, as surely both behavioural-physiological plasticity and genetic differences play important parts in ontogeny and, following evolutionary developmental theory, played a significant role in phylogeny. In the words of Belsky and Pluess (2009), '[Differential susceptibility] emphasize[s] the role of nature in shaping individual differences in plasticity, without excluding a role for nurture; [Biological sensitivity to context] has emphasized nurture, without excluding nature' (p. 886).

Conclusion

Childhood can be a perilous time, and this was likely especially true for our forechildren who needed to survive without the modern comforts and advancements that make surviving childhood commonplace in the technological world. Children's slow trek to adulthood meant that they were highly dependent on the kindness of others if they were to survive and thrive. Mammalian parents, of course, are predisposed by natural selection to care for their young (at least mothers, and to a lesser extent other family members). But natural selection has not ignored infants and children, moulding them in such a way as to increase the likelihood that they would become reproductive members of their species. Adaptations prepare children for the challenges of adulthood, help them solve recurrent problems of childhood, and allow for plasticity in choosing the most advantageous developmental pathways given environmental constraints. Humans are the most social of creatures and a disposition toward social life, particularly life in a family, was critical in our evolution. Therefore, many of the adaptations of infancy and childhood serve to orient children to social others. But not all social situations are created equal, and natural selection has prepared children (or at least some children) to be sensitive to their surroundings and respond in ways that will be adaptive to their inclusive fitness, even if seemingly maladaptive by contemporary standards (e.g. internalizing and externalizing behaviour). An evolutionary perspective provides insights into the study of children, their development, and family life, and it also provides new ways of looking at some of the pathologies and difficulties of childhood, which may result in new solutions to old problems.

References

Aboud, F.E. (1988). *Children and prejudice*. Blackwell, Cambridge, MA.

Aboud, F.E. (2003). The formation of in-group favoritism and out-group prejudice in young children: are they distinct? *Developmental Psychology*, **39**, 48–60.

Alexander, R.D. (1989). Evolution of the human psyche. In: P. Mellers and C. Stringer (eds), *The human revolution: behavioural and biological perspectives on the origins of modern humans*, pp. 455–513. Princeton University Press, Princeton, NJ.

Alley, T.R. (1981). Head shape and the perception of cuteness. *Developmental Psychology*, **17**, 650–4.

Alley, T.R. (1983). Growth-produced changes in body shape and size as determinants of perceived age and adult caregiving. *Child Development*, **54**, 241–8.

Allman, J.M. (1999). *Evolving brains*. Scientific American Library, New York.

Allman, J. and Hasenstaub, A. (1999). Brains, maturation times, and parenting. *Neurobiology of Aging*, **20**, 447–54.

Allman, J., Rosin, A., Kumar, R., and Hasenstaub, A. (1998). Parenting and survival in anthropoid primates: caretakers live longer. *Proceedings of the National Academy of Sciences of the USA*, **95**, 6866–69.

Anderson, K.G., Kaplan, H., Lam, D., and Lancaster, J. (1999). Paternal care by genetic fathers and stepfathers II: reports by Xhosa High School students. *Evolution and Human Behavior*, **20**, 433–51.

Apfekbaum, E. P., Pauker, K., Ambaby, N., Norton, M. I., and Sommers, S. R. (2008). Learning (not) to talk about race: when older children underperform in social categorization. *Developmental Psychology*, **44**, 1513–18.

Baron, A. S. and Banaji, M. R. (2006). The development of implicit attitudes. *Psychological Science*, **17**, 53–8.

Batki, A., Baron-Cohen, S., Wheelwright, S., Connellan, J., and Ahluwalia, J. (2000). Is there an innate gaze module? Evidence from human neonates. *Infant Behavior and Development*, **23**, 223–9.

Baumrind, D. (1993). The average expectable environment is not good enough: a response to Scarr. *Child Development*, **64**, 1299–317.

Beaulieu, D.A. and Bugental, D.B. (2008). Contingent parental investment: An evolutionary framework for understanding early interaction between mothers and children. *Evolution and Human Behavior*, 29, 249–55.

Beise, J. and Voland, E. (2002). A mutlilevel event history analysis of the effects of grandmother on child mortality in a historical German population. *Demographic Research*, **7**, 469–98.

Bellis, M.A., Hughes, K., Hughes, S., and Ashton, J.R. (2005). Measuring paternal discrepancy and its public health consequence. *Journal of Epidemiology*, **59**, 749–54.

Belsky, J. (2005). Differential susceptibility to rearing influences: an evolutionary hypothesis and some evidence. In: B. Ellis and D. Bjorklund (eds), *Origins of the social mind: evolutionary psychology and child development*, pp. 139–63. Guildford, New York.

Belsky, J. (2007). Experience in childhood and the development of reproductive strategies. *Acta Psychologica Sinica*, **39**, 454–68.

Belsky, J. and Pluess, M. (2009). Beyond diathesis stress: differential susceptibility to environmental influences. *Psychological Bulletin*, **135**, 885–908.

Belsky, J., Steinberg, L., and Draper, P. (1991). Childhood experience, interpersonal development, and reproductive strategy: an evolutionary theory of socialization. *Child Development*, **62**, 647–70.

Belsky, J., Hsieh, K., and Crnic, K. (1998). Mothering, fathering, and infant negativity as antecedents of boy's externalizing problems and inhibition at age 3: differential susceptibility to rearing influence? *Development and Psychopathology*, **14**, 311–32.

Belsky, J., Steinberg, L., Houts, R.M., and Halpern-Felsher, B.L. (2010). The development of reproductive strategy in females: early maternal harshness → earlier menarche → increased sexual risk taking. *Developmental Psychology*, **46**, 120–8.

Berger, L.M., Carlson, M.J., Bzostek, S.H., and Osborne, C. (2008). Parenting practices of resident fathers: the role of marital and biological ties. *Journal of Marriage and Family*, **70**, 625–39.

Bigler, R.S. and Liben, L.S. (1993). A cognitive-developmental approach to racial stereotyping and reconstructive memory in Euro-American children. *Child Development*, **64**, 1507–19.

Bjorklund, D.F. (1987). A note on neonatal imitation. *Developmental Review*, **7**, 86–92.

Bjorklund, D.F. (1997). The role of immaturity in human development. *Psychological Bulletin*, **122**, 153–69.

Bjorklund, D.F. and Bering, J.M. (2003). Big brains, slow development, and social complexity: the developmental and evolutionary origins of social cognition. In: M. Brüne, H. Ribbert, and W. Schiefenhövel (eds), *The social brain: evolutionary aspects of development and pathology*, pp. 133–51. Wiley, New York.

Bjorklund, D.F. and Hernández Blasi, C. (2005). Evolutionary developmental psychology. In: D. Buss (ed.), *Evolutionary psychology handbook*, pp. 828–50. Wiley, New York.

Bjorklund, D.F. and Pellegrini, A.D. (2000). Child development and evolutionary psychology. *Child Development*, **71**, 1687–798.

Bjorklund, D.F. and Pellegrini, A.D. (2002). *The origins of human nature: evolutionary developmental psychology.* American Psychological Association, Washington, DC.

Bjorklund, D.F., Hernández Blasi, C., and Periss, V. (2010a). Lorenz revisited: the adaptive nature of children's supernatural thinking. *Human Nature,* 21, 371–92.

Bjorklund, D.F., Causey, K., and Periss, V. (2010b). The evolution and development of human social cognition. In: P. Kappeler and J. Silk (eds), *Mind the gap: tracing the origins of human universals,* pp. 351–71. Springer Verlag, Berlin.

Blair, C. (2002). Early intervention for low birth weight, preterm infants: the role of negative emotionality in the specification of effects. *Development and Psychopathology,* 14, 311–32.

Bornstein, M.H. (2006). Parenting science and practice. In: W. Damon and R.M. Lerner (gen. eds), *Handbook of child psychology (6th edition),* K.A. Renninger and I.E. Sigel (vol. eds), Vol. 4, Child psychology in practice, pp. 893–949. Wiley, New York.

Boyce, W.T. and Ellis, B.J. (2005). Biological sensitivity to context: I. An evolutionary-developmental theory of the origins and functions of stress reactivity. *Development and Psychopathology,* 17, 271–301.

Boyce, W.T., Chesney, M., Alkon–Leonard, A., *et al.* (1995). Psychobiologic reactivity to stress and childhood respiratory illnesses: results of two prospective studies. *Psychosomatic Medicine,* 57, 411–22.

Boyce, W.T., O'Neill–Wagner, P., Price, C. S., Haines, M., and Suomi, S.J. (1998). Crowding stress and violent injuries among behaviorally inhibited rhesus macaques. *Health Psychology,* 17, 285–9.

Brewer, M.B. (1999). The psychology of prejudice: in-group love or out-group hate? *Journal of Social Issues,* 55, 429–44.

Bugental, D.B. and Beaulieu, D.A. (2003). A bio-social-cognitive approach to understanding and promoting the outcomes of children with medical and physical disorders. In: R.V. Kail (ed.), *Advances in child development and behavior,* Vol. 31, pp. 329–61. Elsevier, New York.

Burgess, R. and MacDonald, K. (eds) (2005). *Evolutionary perspectives on human development.* Sage Publications, Thousand Oaks, CA.

Byrne, R.W. (2005). Social cognition: imitation, imitation, imitation. *Current Biology,* 15, R489–R500.

Cabrera, N., Tamis-LeMonda, C.S., Bradley, R.H., Hofferth, S., and Lamb, M.E. (2000). Fatherhood in the twenty-first century. *Child Development,* 71, 127–36.

Campbell, A. (1999). Staying alive: evolution, culture, and women's intrasexual aggression. *Behavioral and Brain Sciences,* 22, 203–14.

Carpenter, M., Nagell, K., and Tomasello, M. (1998). Social cognition, joint attention, and communicative competence from 9 to 15 months of age. *Monographs of the Society for Research in Child Development,* 63, 1–143.

Castelli, L., De Amicis, L., and Sherman, S J. (2007). The loyal member effect: on the preference for ingroup members who engage in exclusive relations with the ingroup. *Developmental Psychology,* 43, 1347–59.

Clutton-Brock, T. H. (1991). *The evolution of parental care.* Princeton University Press, Princeton, NJ.

Coall, D.A. and Hertwig, R. (2010). Grandparental Investment: past, present, and future. *Behavioral and Brain Sciences,* 39, 1–59.

Comstock, G. and Scharrer, E. (2006). Media and popular culture. In: W. Damon and R.M. Lerner (gen. eds), *Handbook of Child Psychology (6th edition),* K.A. Renninger and I.E. Sigel (vol. eds), Vol. 4, Child psychology in practice, pp. 817–63. Wiley, New York.

Crockenburg, S.B. (1981). Infant irritability, mother responsiveness, and social support influences on the security of infant-mother attachment. *Child Development,* 52, 857–65.

Daly, M. and Wilson, M. (1981). Abuse and neglect of children in evolutionary perspective. In: R.D. Alexander and D.W. Tinkle (eds), *Natural selection and social behavior,* pp. 405–416. Chiron, New York.

Daly, M. and Wilson, M. (1985). Child abuse and other risks of not living with both parents. *Ethology and Sociobiology,* 6, 197–210.

Daly, M. and Wilson, M. (1988). *Homicide.* Aldine, New York.

Daly, M. and Wilson, M. (1996). Violence against stepchildren. *Current Directions in Psychological Science*, **5**, 77–81.

de Haan, M., Oliver, A., and Johnson, M.H. (1998). Electrophysiological correlates of face processing by adults and 6-month-old infants. *Journal of Cognitive Neural Science* (Annual Meeting Supplement), 36.

Del Giudice, M. (2009). Sex, attachment, and the development of reproductive strategies. *Behavioral and Brain Sciences*, **32**, 1–21.

Denham, S.A., Workman, E., Cole, P.M., Weissbrod, C., Kendziora, K T., and Zahn-Waxler, C. (2000). Prediction of externalizing behavior problems from early to middle childhood: the role of parental socialization and emotion expression. *Development and Psychopathology*, **12**, 23–45.

DeVries, M.W. (1984). Temperament and infant mortality among the Masai of East Africa. *American Journal of Psychiatry*, **141**, 1189–94.

Dunbar, R.I.M. (1992). Neocortex size as a constraint on group size in primates. *Journal of Human Evolution*, **20**, 469–93.

Dunbar, R.I.M. (2010). Brain and behaviour in primate evolution. In: P.M. Kappeler and J.B. Silk (eds), *Mind the gap: tracing the origins of human universals*, pp. 315–30. Springer, New York.

Dunham, Y., Baron, A.S., and Banaji, M.R. (2006). From American city to Japanese village: a cross-cultural investigation of implicit race attitudes. *Child Development*, **77**, 1268–81.

Dunham, Y., Baron, A.S., and Banaji, M.R. (2008). The development of implicit intergroup cognition. *Trends in Cognitive Science*, **12**, 248–53.

Egan, S.K. and Perry, D.G. (2001). Gender identity: a multidimensional analysis with implications for psychosocial adjustment. *Developmental Psychology*, **37**, 451–63.

Ellis, B.J. (2004). Timing of pubertal maturation in girls: an integrated life history approach. *Psychological Bulletin*, **130**, 920–58.

Ellis, B.J. and Bjorklund, D.F. (eds) (2005). *Origins of the social mind: evolutionary psychology and child development*. Guilford, New York.

Ellis, B.J. and Boyce, W.T. (2008). Biological sensitivity to context. *Current Directions in Psychological Science*, **17**, 183–7.

Ellis, B.J., McFadyen-Ketchum, S., Dodge, K.A., Pettit, G. S., and Bates, J.E. (1999). Quality of early family relationships and individual differences in the timing of pubertal maturation in girls: a longitudinal test of an evolutionary model. *Journal of Personality and Social Psychology*, **77**, 387–401.

Ellis, B.J., Bates, J.E., Dodge, K.A., *et al.* (2003). Does father absence place daughters at special risk for early sexual activity and teenage pregnancy? *Child Development*, **74**, 801–21.

Ellis, B.J., Essex, M.J., and Boyce, W.T. (2005). Biological sensitivity to context: II. Empirical explorations of an evolutionary–developmental theory. *Development and Psychopathology*, **17**, 303–28.

Ellis, B.J., Boyce, W.T., Belsky, J., Bakermans-Kranenburg, M.J., and van Ijzendoorn, M.H. (2011). Differential susceptibility to the environment: an evolutionary-neurodevelopmental theory. *Development and Psychopathology*, **32**, 7–28.

Emlen, S.T. (1995). An evolutionary theory of the family. *Proceedings of the National Academy of Sciences of the USA*, **92**, 8092–9.

Epstein, J.L. and Sanders, M.G. (2002). Family, school, and community partnerships. In: M.H. Bornstein (ed.), *Handbook of parenting (2nd edition), Vol. 5: practical issues in parenting*, pp. 407–37. Erlbaum, Mahwah, NJ.

Euler, H.A. and Michalski, R.L. (2007). Grandparental and extended kin relationships.
In: C.S. Salmon and T.K. Shackelford (eds), *Family relationships: an evolutionary perspective*, pp. 39–68. Oxford University Press, New York.

Farroni, T., Csibra, G., Simion, F., and Johnson, M.H. (2002). Eye contact detection in humans from birth. *Proceedings of the National Academy of Sciences of the USA*, **99**, 9602–5.

Feldman, R., Greenbaum, C.W., and Yirmiya, N. (1999). Mother-infant affect synchrony as an antecedent of the emergence of self-control. *Developmental Psychology*, **35**, 223–31.

Fullard, W. and Reiling, A.M. (1976). An investigation of Lorenz's 'babyness'. *Child Development*, **47**, 1191–93.

Gardiner, A. and Bjorklund, D.F. (2007). All in the family: an evolutionary development perspective. In: C.S. Salmon and T.K. Shackelford (eds), *Family relationships: an evolutionary perspective*, pp. 39–68. Oxford University Press, New York.

Gava, L., Valenza, E., Turati, C., and de Schonen, S. (2008). Effect of partial occlusion on newborns' face preference and recognition. *Developmental Science*, **11**, 563—74.

Geary, D.C. (2005). *The origin of mind: evolution of brain, cognition, and general intelligence.* American Psychological Association, Washington, DC.

Geary, D.C. (2007). Evolution of fatherhood. In C. Solomon and T. K. Shackelford (eds), *Family relationships: evolutionary perspectives*, pp. 1–99. Oxford University Press, New York.

Geary, D.C. (2009). *Male, female: the evolution of human sex differences* (2nd edition). American Psychological Association, Washington, DC.

Geary, D.C. and Bjorklund, D.F. (2000). Evolutionary developmental psychology. *Child Development*, **71**, 57–65.

Gluckman, P. and Hanson, M. (2005). *The fetal matrix: evolution, development, and disease.* Cambridge University Press, Cambridge, UK.

Graber, J.A., Brooks-Gun, J., and Warren, M.P. (1995). The antecedents of menarchael age: heredity, family environment and stressful life events. *Child Development*, **66**, 346–59.

Greenwald, A.G., McGhee, D.E., and Schwartz, J.K.L. (1998). Measuring individual differences in implicit cognition: the implicit association test. *Journal of Personality and Social Psychology*, **74**, 1464–80.

Hamilton, W.D. (1964). The genetical evolution of social behavior. *Journal of Theoretical Biology*, **7**, 17–52.

Harris, J.R. (1995). Where is the child's environment? A group socialization theory of development. *Psychological Review*, **102**, 458–89.

Hernández Blasi, C. and Bjorklund, D.F. (2003). Evolutionary developmental psychology: a new tool for better understanding human ontogeny. *Human Development*, **46**, 259–81.

Howell, J.C. and Egley, Jr, A. (2005). Moving risk factors into developmental theories of gang membership. *Youth Violence and Juvenile Justice*, **3**, 334–54.

Hrdy, S. B. (1999). *Mother Nature: a history of mothers, infants, and natural selection.* Pantheon Books, New York.

Hrdy, S.B. (2007). Evolutionary context of human development: the cooperative breeding model. In: C.S. Salmon and T.K. Shackelford (eds), *Family relationships: an evolutionary perspective*, pp. 39–68. Oxford University Press, New York.

Hrdy, S.B. (2009). *Mothers and others: the evolutionary origins of mutual understanding.* Belknap Press, Cambridge, MA.

Jackson, J.F. (1993). Human behavioral genetics: Scarr's theory, and her views on intervention: a critical review and commentary on their implications for African American children. *Child Development*, **64**, 1318–32.

Joffe, T.H. (1997). Social pressures have selected for an extended juvenile period in primates. *Journal of Human Evolution*, **32**, 593–605.

Kappeler, P. and Silk, J. (eds) (2010). *Mind the gap: tracing the origins of human universals.* Springer Verlag, Berlin.

Keller, H. and Chasiotis, A. (2007). Maternal investment. In: C.S. Salmon and T.K. Shackelford (eds), *Family relationships: an evolutionary perspective*, pp. 39–68. Oxford University Press, New York.

Kelly, D.J., Quinn, P.C., Slater, A.M., Lee, K., Ge, L., and Pascalis, O. (2007). The other-race effect develops during infancy. *Psychological Science*, **18**, 1084–9.

Kelly, D.J., Liu, S., Lee, K., *et al.* (2009). Development of the other-race effect in infancy: evidence toward universality? *Journal of Experimental Child Psychology*, **104**, 105–14.

Kochanska, G., Askan, N., and Joy, M.E. (2007). Children's fearfulness as a moderator of parenting in early socialization: two longitudinal studies. *Developmental Psychology*, **43**, 222–37.

Konner, M. (2010). *The evolution of childhood: relationships emotions, mind*. Belknap Press, Cambridge, MA.

Krachun, C., Carpenter, M., Call, J., and Tomasello, M. (2009). A competitive nonverbal false belief task for children and apes. *Developmental Science*, **12**, 521–35.

Lammi-Taskula, J. (2008). Doing fatherhood: understanding the gendered use of parental leave in Finland. *Fatherhood*, **6**, 133–48.

Leavens, D.A., Hopkins, W.D., and Bard, K.A. (2005). Understanding the point of chimpanzee pointing. epigenesis and ecological validity. *Current Directions in Psychological Science*, **14**, 185–9.

Lee, B.J. and George, R.M. (1999). Poverty, early childbearing and child maltreatment: a multinomial analysis. *Children and Youth Services Review*, **21**, 755–80.

Legerstee, M. (1991). The role of person and object in eliciting early imitation. *Journal of Experimental Child Psychology*, **51**, 423–33.

Lorenz, K.Z. (1943). Die angeboren Formen moglicher Erfahrung [The innate forms of possible experience]. *Zeitschrift fur Tierpsychologie*, **5**, 233–409.

Maestripieri, D. and Pelka, S. (2002). Sex differences in interest in infants across the lifespan: a biological adaptation for parenting? *Human Nature*, **13**, 327–44.

Maestripieri, D. and Roney, J.R. (2006). Evolutionary developmental psychology: contributions from comparative research with nonhuman primates. *Developmental Review*, **26**, 120–37.

Maestripieri, D., Roney, J.R., DeBias, N., Durante, K.M., and Spaepen, G.M. (2004). Father absence, menarche and interest in infants among adolescent girls. *Developmental Science*, **7**, 560–6.

Mann, J. (1992). Nurture or negligence: maternal psychology and behavioral preference among preterm twins. In: J. Barkow, L. Cosmides, and J. Tooby (eds), *The adapted mind: evolutionary psychology and the generation of culture*, pp. 367–90. Oxford University Press, New York.

Marlowe, F. (1999). Showoffs or providers? The parenting effort of Hazda men. *Evolution and Human Behavior*, **20**, 391–404.

Masten, A.S. and Coatsworth, J.D. (1998). The development of competence in favorable and unfavorable environments. *American Psychologist*, **53**, 205–20.

Meltzoff, A.N. and Moore, M.K. (1977). Imitation of facial and manual gestures by human neonates. *Science*, **198**, 75–8.

Nagy, E. and Molnar, P. (2004). Homo imitans or homo provocans? Human imprinting model of neonatal imitation. *Infant Behavior and Development*, **27**, 54–63.

Neberich, W., Penke, L., Lehnart, J., and Asendorpf, J.B. (2010). Family of origin, age at menarche, and reproductive strategies: a test of four evolutionary-developmental models. *European Journal of Developmental Psychology*, **7**, 153–77.

Obradovič, J., Bush, N.R., Stamperdahl, J., Adler, N.E., and Boyce, W.T. (2010). Biological sensitivity to context: the interactive effects of stress reactivity and family adversity on socioemotional behavior and school readiness. *Child Development*, **81**, 270–89.

Oppenheim, R.W. (1981). Ontogenetic adaptations and retrogressive processes in the development of the nervous system and behavior. In: K.J. Connolly and H.F.R. Prechtl (eds), *Maturation and development: biological and psychological perspectives*, pp. 73–108. International Medical Publications, Philadelphia, PA.

Pascalis, O., de Haan, M., and Nelson, C.A. (2002). Is face processing species-specific during the first year of life? *Science*, **296**, 1321–3.

Patterson, M.M. and Bigler, R.S. (2006). Preschool children's attention to environmental messages about groups: social categorization and the origins of intergroup bias. *Child Development*, **77**, 847–60.

Pellegrini, A.D. (2011). Play. In: P. Zelazo (ed.), *Oxford handbook of developmental psychology*, Oxford University Press, New York.

Pellegrini, A.D. and Bjorklund, D.F. (2004). The ontogeny and phylogeny of children's object and fantasy play. *Human Nature*, **15**, 23–43.

Pellegrini, A.D. and Smith, P.K. (1998). Physical activity play: the nature and function of neglected aspect of play. *Child Development*, **69**, 577–98.

Perner, J., Leekam, S.R., and Wimmer, H. (1987). Three-year-olds' difficulty with false belief: the case for a conceptual deficit. *British Journal of Developmental Psychology*, **5**, 125–37.

Pfeifer, J.H., Rubble, D.N., Bachman, M.A., Alvarez, J.M. Cameron, J.A., and Fuligni, A.J. (2007). Social identities and intergroup bias in immigrant and nonimmigrant children. *Developmental Psychology*, **43**, 496–507.

Pluess, M. and Belsky, J. (2010). Differential susceptibility to parenting and quality child care. *Developmental Psychology*, **46**, 379–90.

Powlishta, K.K. (1995). Intergroup processes in childhood: social categorization and sex role development. *Developmental Psychology*, **31**, 781–8.

Quinlan, R.J. (2003). Father absence, parental care, and female reproductive development. *Evolution and Human Behavior*, **24**, 376–90.

Sameroff, A.J. and Suomi, S.J. (1996). Primates and persons: a comparative developmental understanding of social organization. In: R.B. Cairns, G.H. Elder, Jr, and E.J. Costello (eds), *Developmental science*, pp. 97–120. Cambridge University Press, New York.

Scarr, S. (1992). Developmental theories for the 1990s: development and individual differences. *Child Development*, **63**, 1–19.

Scarr, S. (1993). Biological and cultural diversity: the legacy of Darwin for development. *Child Development*, **64**, 1333–53.

Scarr, S. and McCartney, K. (1983). How people make their own environments: a theory of genotype → environment effects. *Child Development*, **54**, 424–35.

Scelza, B.A. (2010). Fathers' presence speeds the social and reproductive careers of sons. *Current Anthropology*, **51**, 295–303.

Sear, R. and Mace, R. (2008). Who keeps children alive? A review of the effects of kin on child survival. *Evolution and Human Behavior*, **29**, 1–18.

Sherif, M.H., Harvey, O.J., White, B.J., Hood, W.R., and Sherif C.W. (1961). *Inter-group conflict and cooperation: the Robbers Cave experiment*. University of Oklahoma Press, Norman, OK.

Shin, H-E., Bjorklund, D.F., and Beck, E.F. (2007). The adaptive nature of children's overestimation in a strategic memory task. *Cognitive Development*, **22**, 197–212.

Spelke, E.S. and Kinzler, K.D. (2007). Core knowledge. *Developmental Science*, **10**, 89–96.

Sternberg, K.J., Baradaran, L.P., Abbott, C.B., Lamb, M.E., and Guterman, E. (2006). Type of violence, age, and gender differences in the effects of family violence on children's behavior problems: a mega-analysis. *Developmental Review*, **26**, 89–112.

Stright, A.D., Gallagher, K.C., and Kelly, K. (2008). Infant temperament moderates relations between maternal parenting in early childhood and children's adjustment in first grade. *Child Development*, **79**, 186–200.

Tither, J.M. and Ellis, B.J. (2008). Impact of fathers on daughters' age at menarche: a genetically and environmentally controlled sibling study. *Developmental Psychology*, **44**, 1409–20.

Tomasello, M. (1999). *The cultural origins of human cognition*. Harvard University Press, Cambridge, MA.

Tomasello, M. and Carpenter, M. (2005). The emergence of social cognition in three young chimpanzees. *Monographs of the Society for Research in Child Development*, **70**, 123–32.

Tomasello, M. and Carpenter, M. (2007). Shared intentionality. *Developmental Science*, **10**, 121–5.

Tomasello, M. and Moll, H. (2010). The gap is social: human shared intentionality and culture. In: P. Kappeler and J. Silk (eds), *Mind the gap: tracing the origins of human universals*, pp. 331–50. Springer, New York.

Tomasello, M., Carpenter, M., and Liszkowski, U. (2007). A new look at infant pointing. *Child Development*, **78**, 705–22.

Tooley, G.A., Karakis, M., Stokes, M., and Ozanne-Smith, J. (2006). Generalising the Cinderella Effect to unintentional childhood fatalities. *Evolution and Human Behavior*, **27**, 224–30.

Trivers, R.L. (1972). Parental investment and sexual selection. In: B. Campbell (ed.), *Sexual selection and the descent of man 1871–1971*, pp. 136–79. Aldine Publishing, Chicago, IL.

Turkewitz, G. and Kenny, P. (1982). Limitations on input as a basis for neural organization and perceptual development: a preliminary theoretical statement. *Developmental Psychobiology*, **15**, 357–68.

Velderman, M.K., Bakermans-Kranenburg, M.J., Juffer, F., and van IJzendoorn, M.H. (2006). Effects of attachment-based interventions on maternal sensitivity and infant attachment: differential susceptibility of highly reactive infants. *Journal of Family Psychology*, **20**, 266–74.

Volk, T. and Atkinson, J. (2008). Is child death the crucible of human evolution? *Journal of Social, Cultural, and Evolutionary Psychology*, **2**, 247–60.

Chapter 6

Application of evolutionary psychology to academic learning

David C. Geary

Introduction

Schooling is at the interface between culture and evolved biases in children's learning because in modern societies school is one of the primary contexts in which culturally-important knowledge and skills are transmitted across generations. An evolutionarily-informed study of this interface is critical for better understanding how to prepare people to live and work in these societies and, through this preparation, for maintaining the continued success of such societies. These are particularly critical issues because of the vast and constantly increasing store of knowledge (e.g. in books, digitally) generated in modern economies and because the gap between these forms of knowledge and the knowledge and competencies needed for living in traditional societies is widening. My proposal has been that there is a corresponding gap between what children find easy to learn and are motivated to learn based on human evolutionary history, and what they need to learn to be successful adults in the modern world (Geary 1995, 2007, 2008, in press-a).

Unfortunately, the distinction between evolved biases in children's learning and motivation to learn, and their ability and motivation to learn evolutionarily novel cultural knowledge and skills, has not been made in educational research or practice, and in fact, the application of insights from evolutionary theory has not frequently occurred at all in the field of education, despite increasing acceptance among psychological scientists (e.g. Buss 2005). In this chapter, I provide a brief overview of evolutionary educational psychology and begin with an outline of the evolved cognitive, developmental, and motivational foundations for learning in evolutionarily-novel contexts. I close with several illustrations of how this framework can be used to better understand children's motivation and learning in modern schools.

Evolution of the human mind

Over the past 4 million years, and especially over the past several hundred thousand years, hominid (our bipedal ancestors) brain size has tripled (Holloway et al. 2004; McHenry 1994; Tobias 1987). The mechanisms that drove these evolutionary changes and a likely increase in our ability to learn throughout the lifespan are debated. The proposals range from the ability to anticipate and make accommodation for climatic fluctuations (Ash and Gallup 2007; Kanazawa 2004) to learning complex hunting skills (Kaplan et al. 2000), and to the demands of living in large, dynamic social groups (Alexander 1989; Bailey and Geary 2009; Dunbar 1998; Flinn et al. 2005; Geary 2005; Humphrey 1976). Despite the different foci, at the core of all of these proposals is the ability to anticipate changing conditions and to generate and mentally rehearse potential behavioural responses. I have proposed that the mechanisms that resulted in our ability to anticipate and cope with change during our evolutionary history are also the mechanisms that now allow us to create culture and to accumulate a wealth of evolutionarily novel knowledge (Geary 2005). The corresponding cognitive systems support children's ability to learn evolutionarily novel information (e.g. linear algebra) and skills (e.g. reading) in school, and the evolved motivational components influence their motivation, or lack thereof, to engage in school learning.

Learning in school

All hypotheses regarding the evolution of the human mind focus on our ability to cope with conditions that were not entirely predictable from one situation to the next and had the potential to influence survival or reproductive prospects during our evolutionary history. As an example, there are aspects of social life—marriage, investment in children, competition with others—that are found in all human cultures and presumably during human evolution (Brown 1991), but the specifics differ from one group to the next, from one person to the next, and across time. This is true in other primates as well, but their social behaviour is much more stereotypical. In Hamadryas baboons (*Papio hamadryas*), for instance, the behaviour of subordinate individuals to dominant ones follows the same script, more or less, across dyads (see Parker 2004). The social behaviour of chimpanzees (*Pan troglodytes*), and other great apes, is scripted to some extent but it is also more varied and flexible in comparison to that of monkeys, requiring some degree of anticipation and planning. Parker argued that the latter requires a social apprenticeship whereby the scripts for some of the more complex features of social life are elaborated upon during development, with the aid of older kin. This pattern is taken to the extreme in humans, making scripted 'hardwired' responses to social relationships maladaptive, especially if social cooperation and competition drove hominid brain evolution and are thus where people focus their mental efforts. For instance, people who always respond in the same way are very predictable and thus easily outmanoeuvred by others.

I proposed in *Origin of Mind: Evolution of Brain, Cognition, and General Intelligence* (Geary 2005) that the mechanisms that allow people to cope with novelty and change include general fluid intelligence and the underlying ability to focus attention and inhibit irrelevant information from entering into working memory. These mechanisms use information generated from the folk-modular systems described below. The simultaneous representation of different pieces of information in working memory appears to link the associated brain systems. In this way, evolved modular systems can be linked in novel ways (Sporns et al. 2000), creating evolutionarily novel abilities. The linking of language systems to visual object-naming systems may contribute to our ability to read, as an example (Geary 2007). Many modular systems are also internally modifiable in response to experience, especially early experience (Geary and Huffman 2002). In other words, even evolved modular systems can respond to a range of information, within constraints (Sperber 1994).

Motivation to learn in school

Understanding the selection pressure or pressures that drove the evolution of the human brain and mind is important, because it will inform us about the types of information that had to be processed to cope with attendant survival or reproductive demands. Differences in the content of these pressures, as in processing human faces as contrasted with the biological movement of hunted species, create advantages to modularized folk systems, as detailed below. The selection pressures also tell us what people are motivated to think about and better understand. If anticipating and adapting to climatic change was a key evolutionary pressure, then humans are predicted to be biased to attend to corresponding information (e.g. cloud patterns) and engage in activities during development that facilitates the ability to predict climate change. If climatic variation was a selective pressure that drove the evolution of the human mind and brain, then children should be biased to learn about weather patterns. This knowledge cannot be 'hardwired' because weather is too variable and thus a motivational bias to learn how to predict this variation should be part of children's natural development. If learning the nuances of social dynamics drove brain and cognitive evolution, then learning and motivational biases are predicted to be organized around common social relationships.

Whether the selection pressures were climatic, hunting, or social competition, the associated motivational biases are for content that differs from much of that taught in modern schools. The implication is that children are not inherently motivated to master much of the knowledge or skills needed to succeed in the modern world. Of course, new topics of any type will likely engender some curiosity and engagement, but this is not likely to result in the sustained engagement needed to master fractions, linear algebra, ancient history, or most other topics covered with modern schooling. There are also individual differences in the dimension of personality, openness, or culture, that when combined with high fluid intelligence will result in some people who enjoy and excel at academic learning (Geary 1995, 2005), but at the same time, we must keep the goal of universal education in mind: if the goal is to educate all children such that they can function successfully in the modern world, then we cannot rely on any inherent curiosity or motivation to learn to drive this learning.

Modular domains of mind

Information that is stable across generations, correlated with survival or reproductive prospects, and partly heritable creates the conditions needed for the evolution of brain and cognitive systems that quickly identify the corresponding information and result in adaptive (not optimal, but good enough) behavioural responses. Simon (1956) termed these ecology-behaviour links bounded rationality, Gigerenzer and colleagues as fast and efficient heuristics (e.g. Gigerenzer et al. 1999), Gelman (1990) as the skeletal structure of evolved domains, Geary and Huffman (2002) as soft modularity, and Timberlake (1994) pointed out that these include the unconditioned stimulus-unconditioned response component of classical conditioning. The combination of advantages conferred by quickly identifying and responding to evolutionarily important information results in high potential for the evolution of modularized cognitive systems in the human mind (Pinker 1997; Tooby and Cosmides 1995).

A taxonomy of these evolved, *biologically primary* or core domains of human cognition is shown in Figure 6.1 (Geary 2005). These domains coalesce around folk psychology, folk biology, and folk physics (Atran 1998; Baron-Cohen 1995; Geary 1995; Leslie et al. 2004; Mithen 1996; Spelke 2000; Wellman and Gelman 1992). I have suggested these folk systems are composed of a constellation of 'soft' modular mechanisms. They are modular in that they have evolved to process specific forms of information, and they are soft because they are sensitive to, and change in response to, variation in the corresponding information patterns. A bias to orient to the human face and to automatically process key pieces of facial information (e.g. eyes) is an example of a modular primary ability (Kanwisher et al. 1997). If the ability to discriminate one person from the next affords a social advantage over those who cannot make these discriminations, then evolution should produce a soft module, one with some degree of plasticity, but within constraints. Plasticity results in a capacity for the face-processing system to respond to experience with different people such that the system codes and stores information that allows people to discriminate one person from the next.

I have argued that the function of these folk competencies is to focus behaviour on attempts to achieve access to and control of the social, biological, and physical resources that tended to enhance survival or reproductive prospects during human evolution (Geary 2005).

Achieving control is not guaranteed, and if it occurs it is typically without conscious awareness of the corresponding evolved function. Peer relationships are supported by folk-psychological competencies, and much of the corresponding social behaviour is not explicitly guided by a motivation to control the behaviour of friends. Children's peer relationships allow them to learn the nuances of social dynamics and to come to understand how they can influence other children and how others in the social group try to influence them. In traditional societies and during much of human evolution, these peers will become the adults with whom they will cooperate and compete for mates and other key resources. Early peer relationships become the social-reproductive milieu of adulthood, and thus early learning can have evolutionary outcomes. In other words, whether or not people are aware of the evolved function, friendships are social resources that can enhance survival and reproductive prospects under the types of conditions found in traditional societies and presumably throughout human evolution.

Folk systems

Psychology

The folk-psychological systems at the top of Figure 6.1 represent three sets of modules that process information related to the self, other individuals, and group dynamics, respectively. The first includes awareness of the self as a social being and awareness of one's relationships with other people (e.g. Harter 2006). Self awareness is tied to the ability to mentally project the self backwards in time to recall episodes that are of personal importance, and to project oneself forwards in time to create a self-centred mental simulation of potential future states (Suddendorf and Corballis 1997; Tulving 2002). Individual-level modules process the information needed for one-on-one social dynamics and to maintain social relationships (Bugental 2000). Group-level modules enable individuals to parse their social world into categories of kin, members of favoured in-groups, and members of disfavoured out-groups.

People also have the unique ability to form in-groups on the basis of ideology, such as nation (Alexander 1989). These ideologies include moral edicts regarding the treatment of in-group members, and mechanisms for their enforcement (Haidt 2007). They support the formation of large-scale cooperative communities, provide stability across generations, and support the cross-generational accumulation of cultural knowledge. These biases evolved because they allow the formation of large cooperative in-groups that are better able to compete with other groups to control ecologies and social politics.

These ideologies are the foundation for the emergence of modern societies. The formation of these societies supports the division of labour needed for the creation of vast amounts of evolutionarily novel information currently available in the modern world and supports the development of schools. Most adults are freed from foraging and hunting demands and some specialize in generating new knowledge (e.g. scientists, poets). The formation of institutions that specialize in the creation of new knowledge, such as universities, has resulted in the exponential increase in novel knowledge over the past several thousand years, and especially during the 20th century. This trend is only going to increase and will result in an even greater importance for children's schooling during the 21st century.

Biology and physics

People living in traditional societies use the local ecology to support their survival and reproductive needs. The associated activities are supported by, among other things, the folk biological and folk physical modules shown in Figure 6.1 (Geary 2005; Geary and Huffman 2002). The folk biological systems support the categorizing of flora and fauna in the local ecology, especially species

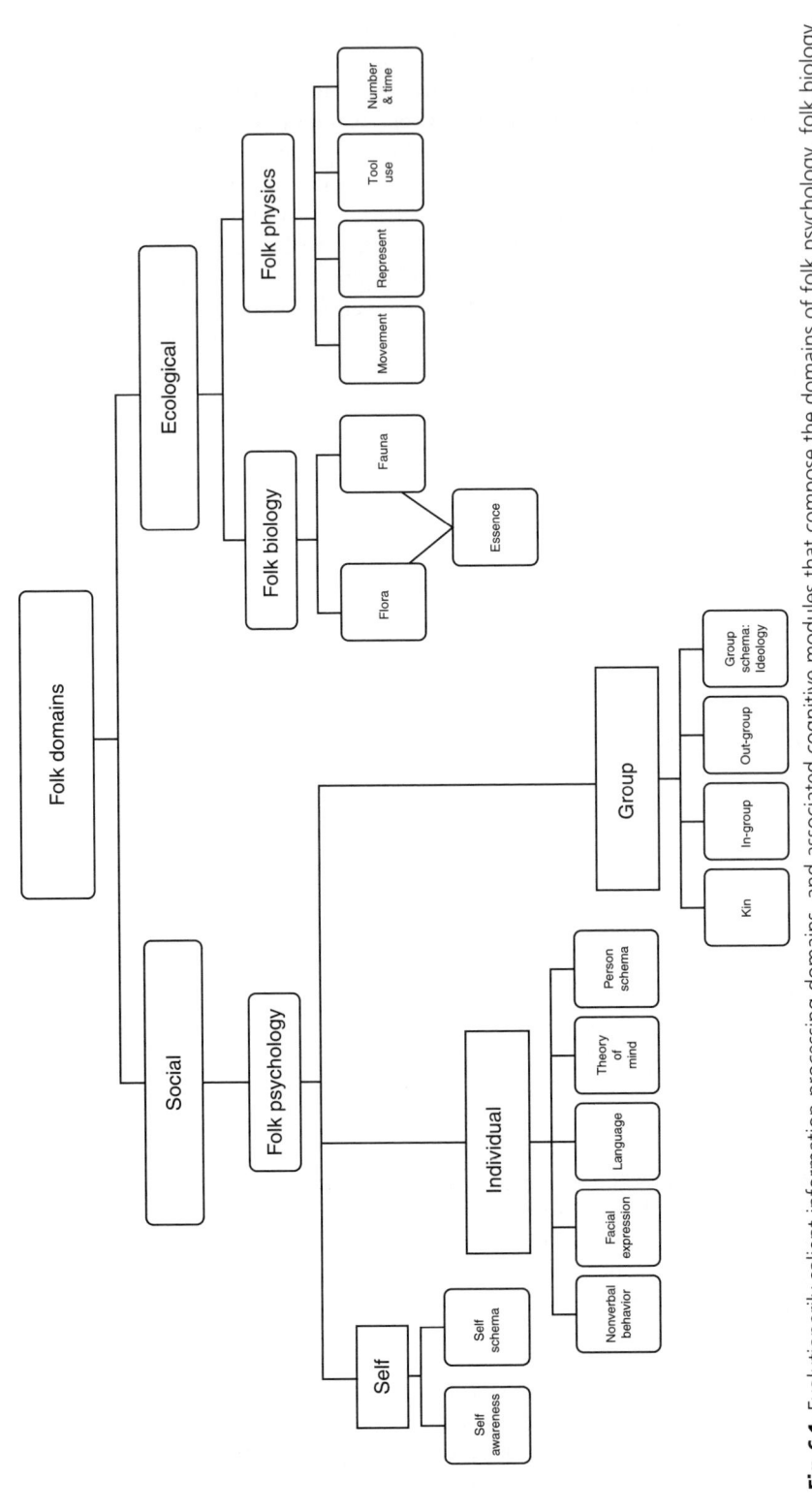

Fig. 6.1 Evolutionarily salient information-processing domains, and associated cognitive modules that compose the domains of folk psychology, folk biology, and folk physics.

used as food, medicines, or in social rituals (Atran 1998; Berlin et al. 1973; Malt 1995; Medin and Atran 2004; New et al. 2007). One result of the categorization of physical features and behaviours of other species (as well as our own) is a belief that these features and behaviours result from an underlying essence that makes all members of the same species the same. Although this essentialism results in an underestimation of the variation within species, it is pragmatic in terms of generating heuristic-based decisions regarding the likely behaviour of these species that in turn facilitates hunting and other activities involved in securing and using them as food and medicine (Givón 2005).

Physical modules are for guiding movement in three-dimensional physical space, mentally representing this space (e.g. demarcating the in-group's territory), and for using physical materials (e.g. stones, metals) for making tools (Pinker 1997; Shepard 1994). The associated abilities support a host of evolutionarily significant activities, such as hunting, foraging, and the use of tools as weapons. There is also strong evidence for interrelated neural and cognitive systems that represent both number and time (Gallistel and Gelman 1992; Meck and Church 1983).

Heuristics

The behavioural features of folk domains can be described as 'rules of thumb' (Gigerenzer et al. 1999). The corresponding information is processed implicitly and the behavioural component is more or less automatically executed (Simon 1956). Returning to face processing, the pattern generated by the shape of the eyes and nose provides information on the sex of the individual, whereas the pattern generated by the configuration of the mouth provides information about the individual's emotional state (Schyns et al. 2002). These patterns are automatically and implicitly processed by the receiver, who in turn expresses corresponding emotional and other social signals (e.g. smile). The receiver may also make implicit decisions regarding the interaction, but these do not need to be explicitly represented in working memory and made available to conscious awareness. These quick, rule-of-thumb decisions can be based on automatically generated feelings and other social information. Negative feelings, such as fear elicited by an angry expression, may prompt withdrawal; and positive feelings, such as happiness generated by a smile, a continuance of the interaction (Damasio 2003).

Folk heuristics can also include explicit inferential and attributional biases. People often make attributions about the cause of their failures and often attribute such failures to bad luck or other people's biases. An evolved tendency to make attributions of this type has the benefit of maintaining effort and control-related behavioural strategies in the face of inevitable failures (Heckhausen and Schultz 1995). Similar attributional biases have been identified in the areas of folk biology and folk physics (Atran 1998; Clement 1982).

These biases provide good enough explanations for day-to-day living and self-serving explanations for social and other phenomena. Critically, an evolved usefulness for everyday living is not the same as scientific accuracy. Indeed, descriptions of psychological, physical, and biological phenomena are often correct (Wellman and Gelman 1992), but many of the explicit explanations regarding the causes of these phenomena are objectively and scientifically inaccurate. The differences in children's folk understanding and the scientifically more accurate understanding that they are expected to learn in school create a substantial educational hurdle.

Evolution and cognitive development

Cognitive development

Cognitive development occurs in all cultures and is distinguished from academic development that occurs in societies with modern schools. The former is the experience-driven adaptation of biologically primary modular competencies to the nuances of the local social, biological, and physical ecologies (Geary and Bjorklund 2000). As noted, modular systems are predicted to be open to experiential modification to the extent that sensitivity to variation within these domains has been of survival or reproductive significance. Broadly, prenatal brain organization provides the skeletal structure that comprises neural and perceptual modules that guide attention to, and the processing of, stable forms of information (e.g. the general shape of the human face) in the folk domains shown in Figure 6.1 (see also Gelman 1990). In keeping with this proposal, studies of infants' attentional biases and preschool children's nascent and implicit knowledge are often organized around these three folk domains, that is, folk psychology, folk biology, and folk physics (Gelman 2003; Mandler 1992; Wellman and Gelman 1992).

The result is biases in postnatal attentional, affective, and information-processing capacities, and in self-initiated behavioural engagement of the environment (Bjorklund 2007; Bjorklund and Pellegrini 2002; Scarr 1992). Behavioural engagement generates evolutionarily-expectant experiences, that is, experiences that provide the social and ecological feedback needed to adjust modular systems to within-domain variation (Greenough et al. 1987). These behavioural biases are expressed as common juvenile activities, such as social play and exploration of the ecology. Experience-expectant processes result in the modification of plastic features of the modular systems, resulting in the ability to identify and respond to variation (e.g. to discriminate one individual from another) within folk domains, and the ability to create the forms of category shown in Figure 6.1, such as in-groups/out-groups or flora/fauna.

Folk domains

Psychology

Human infants are inherently biased to attend to human faces, movement patterns, and speech, reflecting the initial and inherent organizational and motivational structure of folk psychological modules (Freedman 1974). These biases reflect the evolutionary significance of social relationships and recreate the microconditions (e.g. parent–child interactions) associated with the evolution of the corresponding modules (Caporael 1997). Attention to and processing of this information provides exposure to the within-category variation needed to adapt modular architecture to variation in parental faces, behaviour, and so forth (Gelman and Williams 1998). One illustrative result is that infants are able to discriminate the voice of their parents from the voice of other adults with only minimal exposure. When human fetuses (~ 38 weeks' gestation) are exposed *in utero* to human voices, their heart-rate patterns suggest they are sensitive to and learn the voice patterns of their mother, and discriminate her voice from that of other women (Kisilevsky et al. 2003).

Biology and physics

The complexity of hunting and foraging demands varies across ecologies and creates a situation that should select for plasticity in the folk biological and physical systems. Children's implicit folk biological knowledge and inherent interest in living things reflect a motivation to engage in activities that automatically create taxonomies of local flora and fauna and result in the accrual of an extensive knowledge base of these species. In traditional societies, these experiences include assisting with foraging and play hunting (Blurton Jones et al. 1997; Bock 2005). Anthropological research indicates that it often takes many years of engaging in these forms of play and early work to master the skills and knowledge needed for successful hunting and foraging in many of these societies (Kaplan et al. 2000).

Learning about the physical world is a complex endeavour for humans and requires an extended developmental period, in comparison with the more rapid learning that occurs in species that occupy a more narrow range of physical ecologies (Gallistel 2000). The importance of early experience in this domain is illustrated by development of the ability to mentally form map-like representations of the large-scale environment. The initial ability to form these representations emerges by the age of 3 years (DeLoache et al. 1991), improves gradually through adolescence, and often requires extensive exploration and exposure to the local environment to perfect (Matthews 1992). Matthews' research shows that children automatically attend to geometric features of the environment and landmarks within this environment and at a later time can generate a cognitive representation of landmarks and their geometric relations. Children's skill at generating these representations increases with repeated explorations of the physical environment. Chen and Siegler's (2000) finding that 18 month-olds have an implicit understanding of how to use simple tools, and with experience learn to use these tools in increasingly effective ways, suggests that similar processes occur for tool use (see also Gredlein and Bjorklund 2005).

Academic development

Our extraordinary ability to create evolutionarily novel—biologically secondary—knowledge and skills has resulted in the many benefits provided by modern societies, but comes at a cost: during the last several thousand years the cross-generational accumulation of cultural knowledge (e.g. through books) has occurred at such a rapid pace (Richerson and Boyd 2005), that the cognitive and motivational biases that facilitate the modification of folk abilities during children's natural activities do not have evolved counterparts to facilitate the learning of secondary abilities. A thorough discussion is provided elsewhere (Geary 2007, 2008, in press-a), but I highlight a few key aspects in the following sections.

Motivation to learn

Research on children's early attentional biases and activity preferences should be placed within an evolutionary perspective, as a fuller understanding of these preferences and how they are expressed in school settings has the potential to significantly improve our understanding of children's motivation (or not) to learn biologically secondary material.

A core prediction is that children's evolved motivational biases will be focused on learning in folk domains and that they will prefer to engage in this learning through play and exploration. Stated differently, children's attention is predicted to be easily captured by the types of information and dynamics, such as peer relationships, that were correlated with survival and reproductive prospects during our evolutionary history, and biased to engage in activities that allow them to learn in these areas. Children are also predicted to show a preference for the activities that promote the cross-generational transfer of knowledge in traditional societies. In these societies, the corresponding activities involve stories to convey morals (i.e. cultural rules for social behaviour) and other themes relevant to day-to-day living, and apprenticeships whereby culturally important skills (e.g. hunting, tool making) are learned through observation of, or direct instruction by, more skilled individuals (Brown 1991). The specific content of these activities is centred on features of social living or the ecology that children will need to learn before assuming adult responsibilities. In other words, there are universal mechanisms that support the learning of culture-specific information (e.g. observational learning; Bandura 1986), in addition to the attentional, motivational, and cognitive mechanisms described in the 'Evolution of the human mind' section. The combination results in human universals, such as face processing and language, as well as many cultural particulars that are variations on these themes.

From this perspective, it is not surprising that many school children value achievement in sports–ritualized practice of organized in-group/out-group competition (Geary et al. 2003)—more than achievement in core academic areas (Eccles et al. 1993). It is also not surprising that many students report in-school activities to be a significant source of negative affect (Larson and Asmussen 1991). Csikszentmihalyi and Hunter (2003) found that the lowest levels of happiness were experienced by children and adolescents while they were doing homework, listening to lectures, and doing mathematics, and the highest levels were experienced when they were talking with friends. For adolescents, the weekend is the highlight of their week because they can socialize with their peers (Larson and Richards 1998).

A preference for engagement in peer relationships will not promote mastery of any biologically secondary competency taught in school, but it is consistent with the prediction of an evolved bias for children to self-organize social activities during development; it is necessary to learn about one's specific peer group and how to manage and influence dynamics in this group. Schooling is not, however, at odds with all evolved learning and motivational biases. This is because a long developmental period is predicted to have co-evolved with an interest in and ability to transfer culturally important information across generations (Bjorklund 2007; Flinn 1997; Richerson and Boyd 2005). As noted earlier, a species-typical curiosity about, and an ability to learn evolutionarily novel information, especially when first introduced, is predicted, but so are substantive individual differences in the motivation and ability to learn this information. The gist is if there were not a gap between the secondary knowledge needed to function well in modern societies and evolved motivational and learning biases, then the motivational dispositions, interests, and abilities of the creative-productive individuals who developed this secondary knowledge (e.g. Murray 2003) would be mundane and easily acquired without schooling. This is not the case (Ericsson et al. 1993).

If our goal is universal education that encompasses a variety of evolutionarily novel academic domains (e.g. mathematics) and skills (e.g. phonetic decoding as related to reading), then we cannot assume that an inherent curiosity or motivation to learn will be sufficient for most children and adolescents. Children's and adolescents' explicit valuation of academic learning, the perceived utility of academic skills, and the centrality of these areas to their overall self-esteem is predicted to be highly dependent on social-cultural valuation of academic competencies, such as explicit rewards for academic achievement (e.g. honour rolls) and valuation of cultural innovators (e.g. Thomas Edison). In contrast, the child's and adolescent's valuation and perceived efficacy of their physical traits or social relationships are implicit features of their evolved folk psychology and will manifest with or without cultural supports.

Learning in school

Biologically secondary learning is the acquisition of culturally important information and skills using the evolved mechanisms that enable people to cope with novelty and change within their lifetimes, and that enable the cross-generational transfer of cultural knowledge. Again, the details are provided elsewhere (Geary 2005, 2007), but I illustrate the process below.

Writing is the primary means through which secondary knowledge has accumulated over the past several millennia and reading this material is the primary means of transmitting this knowledge across generations. From an evolutionary perspective, writing initially emerged from the motivation of people to communicate with and influence other people, and the desire to read emerged from the benefits of learning from others. In other words, the cognitive and motivational systems that support reading and writing are embedded in the folk psychological systems shown in Figure 6.1, especially language (e.g. Mann 1984; Rozin 1976). I illustrate this by discussing how learning how to read and reading comprehension may be linked to several folk psychological domains.

There is considerable evidence that children's reading acquisition is dependent on the language system (Bradley and Bryant 1983; Hindson et al. 2005; Wagner and Torgesen 1987). Core early components include phonemic awareness—explicit awareness of distinct language sounds—and the ability to decode unfamiliar written words into basic sounds. Decoding requires an *explicit* representation of the sound (e.g. *ba, da, ka*) in phonemic working memory and the association of this sound and blends of sounds with corresponding visual patterns (Bradley and Bryant 1983). The ease of learning basic word-decoding skills in first grade is predicted by the fidelity of children's phonological processing systems (e.g. skill at discriminating language sounds) in kindergarten (Wagner et al. 1994). Children who show a strong explicit awareness of basic language sounds easily learn to associate these sounds with the symbol system of the written language.

Unlike natural language learning, the majority of children acquire these competencies most effectively with systematic, organized, and teacher-directed explicit instruction on phoneme identification, blending, and word decoding (e.g. Hindson et al. 2005). Skilled reading also requires text comprehension which is dependent on several component skills, such as locating main themes and distinguishing relevant from less relevant passages. As with more basic reading skills, many children require explicit instruction in the use of these strategies to aid in text comprehension (Connor et al. 2004).

From the lens of evolution, text comprehension will be dependent in part on theory of mind and other folk psychological domains, at least for genre that involve human relationships (Geary 2010). Most of these stories involve the recreation of social relationships, more complex patterns of social dynamics, and even elaborate person schema knowledge for main characters. The theme of many of the most popular genre involves the dynamics of mating relationships and competition for mates. Once people learn how to read, they engage in this biologically secondary activity because it allows for the representation of evolutionarily salient themes, particularly the mental representation and rehearsal of social dynamics. In addition, some people are predicted to be interested in reading about mechanical things and biological phenomena, reflecting interests associated with folk-physical and -biological systems.

Conclusion

Evolution has shaped the creation of cultural ideologies and rules for social behaviour. These provide the structure for the formation and stability of large cooperative groups (Baumeister 2005; Richerson and Boyd 2005). Children and adults have evolved corresponding learning and motivational mechanisms that support the cross-generational transfer of knowledge that is useful in their culture. In traditional societies, the corresponding mechanisms include child-initiated play, observational learning, and adults' use of stories and apprenticeships to teach cultural knowledge, but not typically explicit, adult-directed teaching. These mechanisms may no longer be sufficient for preparing children for adulthood in the modern world, because of the vast amount of cultural knowledge that has accumulated during the past several thousand years and because of the sometimes substantial differences between this knowledge and evolved folk knowledge and biases. In fact, in many cases, evolved biases lead to inferenccs and beliefs about the social and natural world that are scientifically inaccurate—science itself is a secondary activity that emerged only recently (Geary in press-b).

Schools emerged in modern societies to address these and other limitations of folk systems and to formalize the cross-generational transfer of knowledge. In other words, schools are the central interface between evolution and culture—they are the venues in which children's evolved biases in learning and motivation intersect with the need to learn the secondary abilities and knowledge needed to be successful in modern societies. This perspective has the potential to answer many questions in instruction and learning that are not otherwise fully understandable, such as why most children need explicit instruction to learn word decoding and text comprehension but do not need instruction to produce and understand natural language.

Among many implications, effective instruction in secondary academic domains will be dependent on the same attentional control and working memory systems that evolved to cope with variation and novelty within lifetimes (see Geary 2005, 2007), and leads to the hypothesis that many children will need to have any associated problem-solving steps explicitly organized by instructional materials and extensively practised for long-term retention (e.g. Sweller 2004). Unlike the adaptation of primary systems (e.g. language, folk biology) to the nuances of the local social group and ecology, learning in these domains is not privileged by inherent attentional, cognitive, and motivational systems. Because of this, teachers must provide the structure and organization to secondary learning that has been provided to primary learning by our evolutionary history. With respect to motivation, children's natural curiosity and desire to learn cannot be assumed to be sufficient to support the long and effortful learning needed to master secondary domains, such as algebra.

References

Alexander, R.D. (1989). Evolution of the human psyche. In: P. Mellars and C. Stringer (eds), *The human revolution: behavioural and biological perspectives on the origins of modern humans,* pp. 455–513. Princeton University Press, Princeton, NJ.

Ash, J. and Gallup, G.G. Jr (2007). Paleoclimatic variation and brain expansion during human evolution. *Human Nature,* **18**, 109–24.

Atran, S. (1998). Folk biology and the anthropology of science: cognitive universals and cultural particulars. *Behavioral and Brain Sciences,* **21**, 547–609.

Bailey, D.H and Geary, D.C. (2009). Hominid brain evolution: testing climatic, ecological, and social competition models. *Human Nature,* **20**, 67–79.

Bandura, A. (1986). *Social foundations of thought and action: a social cognitive theory.* Prentice-Hall, Englewood Cliffs, NJ.

Baron-Cohen, S. (1995). *Mindblindness: an essay on autism and theory of mind.* MIT Press/Bradford Books, Cambridge, MA.

Baumeister, R.F. (2005). *The cultural animal: human nature, meaning, and social life.* Oxford University Press, New York.

Berlin, B., Breedlove, D.E., and Raven, P.H. (1973). General principles of classification and nomenclature in folk biology. *American Anthropologist,* **75**, 214–42.

Bjorklund, D.F. (2007). *Why youth is not wasted on the young: immaturity in human development.* Blackwell Publishing, Malden, MA.

Bjorklund, D. F. and Pellegrini, A.D. (2002). *The origins of human nature: evolutionary developmental psychology.* American Psychological Association, Washington, DC.

Blurton Jones, N.G., Hawkes, K., and O'Connell, J.F. (1997). Why do Hadza children forage? In: N.L. Segal, G.E. Weisfeld, and C.C. Weisfeld (eds), *Uniting psychology and biology: integrative perspectives on human development,* pp. 279–313. American Psychological Association, Washington, DC.

Bock, J. (2005). What makes a competent adult forager? In: B.S. Hewlett and M.E. Lamb (eds), *Hunter-gatherer childhoods: evolutionary, developmental and cultural perspectives,* pp. 109–28. Transaction Publishers, New Brunswick, NJ.

Bradley, L. and Bryant, P.E. (1983). Categorizing sounds and learning to read—A causal connection. *Nature,* **301**, 419–21.

Brown, D.E. (1991). *Human universals.* Temple University Press, Philadelphia, PA.

Bugental, D.B. (2000). Acquisition of the algorithms of social life: a domain-based approach. *Psychological Bulletin,* **126**, 187–219.

Buss, D.M. (ed.) (2005). *The evolutionary psychology handbook.* Wiley and Sons Hoboken, NJ.

Caporael, L.R. (1997). The evolution of truly social cognition: the core configurations model. *Personality and Social Psychology Review,* **1**, 276–98.

Chen, Z. and Siegler, R.S. (2000). Across the great divide: bridging the gap between understanding toddlers' and older children's thinking. *Monographs of the Society for Research in Child Development,* **65** (2), 1–96.

Clement, J. (1982). Students' preconceptions in introductory mechanics. *American Journal of Physics,* **50**, 66–71.

Connor, C.M., Morrison, F.J., and Petrella, J.N. (2004). Effective reading comprehension instruction: examining child X instruction interactions. *Journal of Educational Psychology,* **96**, 682–98.

Csikszentmihalyi, M. and Hunter, J. (2003). Happiness in everyday life: the uses of experience sampling. *Journal of Happiness Studies,* **4**, 185–99.

Damasio, A. (2003). *Looking for Spinoza: joy, sorrow, and the feeling brain.* Harcourt, Inc., Orlando, FL.

DeLoache, J.S., Kolstad, D.V., and Anderson, K.N. (1991). Physical similarity and young children's understanding of scale models. *Child Development,* **62**, 111–26.

Dunbar, R.I.M. (1998). The social brain hypothesis. *Evolutionary Anthropology,* **6**, 178–90.

Eccles, J., Wigfield, A., Harold, R.D., and Blumenfeld, P. (1993). Age and gender differences in children's self- and task perceptions during elementary school. *Child Development,* **64**, 830–47.

Ericsson, K.A., Krampe, R.T., and Tesch-Römer, C. (1993). The role of deliberate practice in the acquisition of expert performance. *Psychological Review,* **100**, 363–406.

Flinn, M.V. (1997). Culture and the evolution of social learning. *Evolution and Human Behavior,* **18**, 23–67.

Flinn, M.V., Geary, D.C., and Ward, C.V. (2005). Ecological dominance, social competition, and coalitionary arms races: why humans evolved extraordinary intelligence. *Evolution and Human Behavior,* **26**, 10–46.

Freedman, D.G. (1974). *Human infancy: an evolutionary perspective.* John Wiley and Sons, New York.

Gallistel, C.R. (2000). The replacement of general-purpose learning models with adaptively specialized learning modules. In: M.S. Gazzaniga (Editor-in-chief), *The new cognitive neurosciences (second edition),* pp. 1179–91. Bradford Books/MIT Press, Cambridge, MA.

Gallistel, C.R. and Gelman, R. (1992). Preverbal and verbal counting and computation. *Cognition,* **44**, 43–74.

Geary, D.C. (1995). Reflections of evolution and culture in children's cognition: implications for mathematical development and instruction. *American Psychologist,* **50**, 24–37.

Geary, D.C. (2005). *The origin of mind: evolution of brain, cognition, and general intelligence.* American Psychological Association, Washington, DC.

Geary, D.C. (2007). Educating the evolved mind: conceptual foundations for an evolutionary educational psychology. In: J.S. Carlson and J.R. Levin (eds), *Educating the evolved mind, Vol. 2, Psychological perspectives on contemporary educational issues,* pp. 1–99. Information Age, Greenwich, CT.

Geary, D.C. (2008). An evolutionarily informed education science. *Educational Psychologist,* **43**, 279–95.

Geary, D.C. (2010). *Male, female: the evolution of human sex differences* (Second Edition). American Psychological Association, Washington, DC.

Geary, D.C. (in press-a). Evolutionary educational psychology. In: K.R. Harris, S. Graham, and T. Urdan, (eds), *Educational psychology handbook.* American Psychological Association, Washington, DC.

Geary, D.C. (in press-b). The evolved mind and scientific discovery. In: J. Shrager and S.M. Carver (eds), *From child to scientist: mechanisms of learning and development.* American Psychological Association, Washington, DC.

Geary, D.C. and Bjorklund, D.F. (2000). Evolutionary developmental psychology. *Child Development,* **71**, 57–65.

Geary, D.C. and Huffman, K.J. (2002). Brain and cognitive evolution: forms of modularity and functions of mind. *Psychological Bulletin,* **128**, 667–98.

Geary, D.C., Byrd-Craven, J., Hoard, M.K., Vigil, J., and Numtee, C. (2003). Evolution and development of boys' social behavior. *Developmental Review,* **23**, 444–70.

Gelman, R. (1990). First principles organize attention to and learning about relevant data: number and animate-inanimate distinction as examples. *Cognitive Science,* **14**, 79–106.

Gelman, S.A. (2003). *The essential child: origins of essentialism in everyday thought.* Oxford University Press, New York.

Gelman, R. and Williams, E.M. (1998). Enabling constraints for cognitive development and learning: domain-specificity and epigenesis. In: D. Kuhl and R.S. Siegler (volume eds), *Cognition, perception, and language, Volume 2,* pp. 575–630; W. Damon (general ed.), *Handbook of child psychology* (Fifth Edition). John Wiley and Sons, New York.

Gigerenzer, G., Todd, P.M., and ABC Research Group (eds) (1999). *Simple heuristics that make us smart.* Oxford University Press, New York.

Givón, T. (2005). *Context as other minds: the pragmatics of sociality, cognition, and communication.* John Benjamins Publishing Company, Philadelphia, PA.

Gredlein, J.M. and Bjorklund, D.F. (2005). Sex differences in young children's use of tools in a problem-solving task. *Human Nature,* **16**, 211–32.

Greenough, W.T., Black, J.E., and Wallace, C.S. (1987). Experience and brain development. *Child Development,* **58**, 539–59.

Haidt, J. (2007). The new synthesis in moral psychology. *Science,* **316**, 998–1002.

Harter, S. (2006). The self. In: N. Eisenberg (volume ed.), *Social, emotional, and personality development, Vol 3,* pp. 505–70. W. Damon and R.M. Lerner (general eds), *Handbook of child psychology* (Sixth Edition). John Wiley and Sons, New York.

Heckhausen, J. and Schulz, R. (1995). A life-span theory of control. *Psychological Review,* **102**, 284–304.

Hindson, B., Byrne, B., Shankweiler, D., Fielding-Barnsley, R., Newman, C., and Hine, D.W. (2005). Assessment and early instruction of preschool children at risk for reading disability. *Journal of Educational Psychology,* **97**, 687–704.

Holloway, R.L., Broadfield, D.C., and Yuan, M.S. (2004). *The human fossil record, Volume three: brain endocasts – the paleoneurological record.* John Wiley and Sons, Hoboken, NJ.

Humphrey, N.K. (1976). The social function of intellect. In: P.P.G. Bateson and R.A. Hinde (eds), *Growing points in ethology,* pp. 303–17. Cambridge University Press, New York.

Kanazawa, S. (2004). General intelligence as a domain-specific adaptation. *Psychological Review,* **111**, 512–23.

Kanwisher, N., McDermott, J., and Chun, M.M. (1997). The fusiform face area: a module in human extrastriate cortex specialized for face perception. *Journal of Neuroscience,* **17**, 4302–11.

Kaplan, H., Hill, K., Lancaster, J., and Hurtado, A.M. (2000). A theory of human life history evolution: diet, intelligence, and longevity. *Evolutionary Anthropology,* **9**, 156–85.

Kisilevsky, B.S., Hains, S.M.J., Lee, K., *et al.* (2003). Effects of experience on fetal voice recognition. *Psychological Science,* **14**, 220–4.

Larson, R. and Asmussen, L. (1991). Anger, worry, and hurt in early adolescence: an enlarging world of negative emotions. In: M.E. Colten and S. Gore (eds), *Adolescent stress: causes and consequences,* pp. 21–41. Aldine de Gruyter, New York.

Larson, R. and Richards, M. (1998). Waiting for the weekend: Friday and Saturday night as the emotional climax of the week. *New Directions for Child and Adolescent Development,* **82**, 37–51.

Leslie, A.M., Friedman, O., and German, T.P. (2004). Core mechanisms in 'theory of mind'. *Trends in Cognitive Science,* **8**, 528–33.

Malt, B.C. (1995). Category coherence in cross-cultural perspective. *Cognitive Psychology,* **29**, 85–148.

Mandler, J.M. (1992). How to build a baby: II. Conceptual primitives. *Psychological Review,* **99**, 587–604.

Mann, V.A. (1984). Reading skill and language skill. *Developmental Review,* **4**, 1–15.

Matthews, M.H. (1992). *Making sense of place: children's understanding of large-scale environments.* Barnes and Noble Books, Savage, MD.

McHenry, H.M. (1994). Tempo and mode in human evolution. *Proceedings of the National Academy of Sciences of the USA,* **91**, 6780–6.

Meck, W.H. and Church, R.M. (1983). A mode control model of counting and timing processes. *Journal of Experimental Psychology: Animal Behavior Processes,* **9**, 320–34.

Medin, D.L. and Atran, S. (2004). The native mind: biological categorization and reasoning in development and across cultures. *Psychological Review,* **111**, 960–83.

Mithen, S. (1996). *The prehistory of the mind: the cognitive origins of art and science.* Thames and Hudson, Inc., New York.

Murray, C. (2003). *Human accomplishment: the pursuit of excellence in the Arts and Sciences, 800 B.C. to 1950.* HarperCollins, New York.

New, J., Cosmides, L., and Tooby, J. (2007). Category-specific attention for animals reflects ancestral priorities, not expertise. *Proceedings of the National Academy of Sciences of the USA,* **104**, 16598–603.

Parker, S.T. (2004). The cognitive complexity of social organization and socialization in wild baboons and chimpanzees: guided participation, socializing interactions, and event representation. In: A.E. Russon and D.R. Begun (eds), *The evolution of thought: evolutionary origins of great ape intelligence,* pp. 45–60. Cambridge University Press, Cambridge.

Pinker, S. (1997). *How the mind works.* W.W. Norton and Co, New York.

Richerson, P.J. and Boyd, R. (2005). *Not by genes alone: how culture transformed human evolution.* University of Chicago Press, Chicago, IL.

Rozin, P. (1976). The evolution of intelligence and access to the cognitive unconscious. In: J.M. Sprague and A.N. Epstein (eds), *Progress in psychobiology and physiological psychology, Vol. 6,* pp. 245–80. Academic Press, New York.

Scarr, S. (1992). Developmental theories of the 1990s: developmental and individual differences. *Child Development,* **63**, 1–19.

Schyns, P.G., Bonnar, L., and Gosselin, F. (2002). Show me the features! Understanding recognition from the use of visual information. *Psychological Science,* **13**, 402–9.

Shepard, R.N. (1994). Perceptual-cognitive universals as reflections of the world. *Psychonomic Bulletin and Review,* **1**, 2–28.

Simon, H.A. (1956). Rational choice and the structure of the environment. *Psychological Review,* **63**, 129–38.

Spelke, E.S. (2000). Core knowledge. *American Psychologist,* **55**, 1233–43.

Sperber, D. (1994). The modularity of thought and the epidemiology of representations. In: L.A. Hirschfeld and S.A. Gelman (eds), *Mapping the mind: domain specificity in cognition and culture,* pp. 39–67. Cambridge University Press, New York.

Sporns, O., Tononi, G., and Edelman, G.M. (2000). Connectivity and complexity: the relationship between neuroanatomy and brain dynamics. *Neural Networks,* **13**, 909–22.

Suddendorf, T. and Corballis, M.C. (1997). Mental time travel and the evolution of the human mind. *Genetic, Social, and General Psychology Monographs,* **123**, 133–67.

Sweller, J. (2004). Instructional design consequences of an analogy between evolution by natural selection and human cognitive architecture. *Instructional Science,* **32**, 9–31.

Timberlake, W. (1994). Behavior systems, associationism, and Pavlovian conditioning. *Psychonomic Bulletin and Review,* **1**, 405–20.

Tobias, P.V. (1987). The brain of *Homo habilis*: a new level of organization in cerebral evolution. *Journal of Human Evolution,* **16**, 741–61.

Tooby, J. and Cosmides, L. (1995). Mapping the evolved functional organization of mind and brain. In: M.S. Gazzaniga (ed.), *The cognitive neurosciences,* pp. 1185–97. Bradford Books/MIT Press, Cambridge, MA.

Tulving, E. (2002). Episodic memory: from mind to brain. *Annual Review of Psychology,* **53**, 1–25.

Wagner, R. K. and Torgesen, J.K. (1987). The nature of phonological processing and its causal role in the acquisition of reading skills. *Psychological Bulletin,* **101**, 192–212.

Wagner, R.K., Torgesen, J.K., and Rashotte, C.A. (1994). Development of reading-related phonological processing abilities: new evidence of bidirectional causality from a latent variable longitudinal study. *Developmental Psychology,* **30**, 73–87.

Wellman, H.M. and Gelman, S.A. (1992). Cognitive development: foundational theories of core domains. *Annual Review of Psychology,* **43**, 337–75.

Chapter 7

Serial monogamy and clandestine adultery: evolution and consequences of the dual human reproductive strategy

Helen E. Fisher

Introduction

Considerable data suggest that *Homo sapiens* has evolved a dual reproductive strategy: life-long and/or serial monogamy in conjunction with clandestine adultery (Fisher 1992). This paper explores the underlying biochemical and genetic mechanisms likely to contribute to this flexible, yet specific human reproductive system, and explores some of the implications of this dual human reproductive strategy for contemporary partnerships.

Neurobiology of human attachment

Pair-bonding is a hallmark of humanity. Data from the Demographic Yearbooks of the United Nations on 97 societies canvassed in the 1980s indicate that approximately 93% of women and 92% of men in that decade married by age 49 (Fisher 1989, 1992). Worldwide marriage rates have declined somewhat since then; but today 85–90% of men and women in the United States are projected to marry (Cherlin 2009). Cross-culturally, most individuals wed one person at a time: monogamy. Polygyny is permitted in 84% of human societies; but in the vast majority of these cultures, only 5–10% of men actually have several wives simultaneously (van den Berghe 1979; Frayser 1985). Because polygyny in humans is regularly associated with rank and wealth, monogamy may have been even more prevalent in pre-horticultural, unstratified societies (Daly and Wilson 1983). Human monogamy is not, however, always life-long. Nearly half of all marriages in the United States end in divorce. By age 35, 10% of American women have had more than one husband (Cherlin 2009). Data collected from the Demographic Yearbooks of the United Nations on 58 societies from 1947–1989, as well as a host of ethnographic studies, indicate that divorce and remarriage are also common cross-culturally (Fisher 1989, 1992).

   This human disposition for pair-bonding appears to have a biological basis. The contemporary psychological investigation of human attachment began with Bowlby (1969, 1973) and Ainsworth et al. (1978) who proposed that, to promote the survival of the young, primates have evolved an innate attachment system designed to motivate infants to seek comfort and safety from their primary caregiver, generally mother. Since then, extensive psychological research has been done on the behaviours and feelings associated with this attachment system in adults (Fraley and Shaver 2000); researchers have proposed that this biologically-based attachment system remains active throughout the life course, serving as a foundation for attachment between pair-bonded spouses for the purpose of raising offspring (Hazan and Shaver 1987; Hazan and Diamond 2000). Hatfield (1988, p. 191) refers to the human feelings associated with these attachment behaviours as companionate love, which she defines as 'a feeling of happy togetherness with someone whose life has become deeply entwined with yours'.

The human penchant to form pair-bonds is rare among mammals; only 3% of mammalian species form pair-bonds to rear their young. In contrast, pair-bonding is common in birds: some 90% of more than 8000 species practise pair-bonding to rear their young. Moreover, in all avian and mammalian species where monogamy is the primary reproductive strategy, it is associated with a particular suite of behaviours, including mutual territory defence and/or nest building, mutual feeding and grooming, maintenance of close proximity, affiliative behaviours, and shared parental chores. The ethological literature commonly regards this constellation of pair-bonding behaviours as a behavioural syndrome that evolved primarily to motivate mating partners to sustain an affiliative connection for long enough to complete species-specific parental duties.

The most informative biological research on pair-bonding in mammals has been conducted on prairie voles (*Microtus ochrogaster*). Individuals mate soon after puberty and maintain a monogamous relationship throughout life, raising a series of litters as a team. Some of the neural underpinnings of these pair-bonding behaviours have been established. When prairie voles engage in sex, copulation triggers the activity of oxytocin (OT) in the nucleus accumbens in females and arginine vasopressin (AV) in the ventral pallidum in males, which then facilitates dopamine release in these reward regions and motivates females and males to prefer a particular mating partner, initiate pair-bonding, and express attachment behaviours (Carter 1992; Lim and Young 2004; Lim et al. 2004).

These data are corroborated in other species. Promiscuous white-footed mice and promiscuous rhesus monkeys do not form pair-bonds or express attachment behaviours for a specific mate, and the males of these species do not express the same distribution of V1a receptors in the ventral pallidum (Wang et al. 1997; Young et al. 1997; Bester-Meredith et al. 1999; Young 1999). Moreover, when the genetic variant in the vasopressin system associated with pair-bonding in male prairie voles is transgenically inserted into the ventral pallidum of male meadow voles (an asocial promiscuous species), vasopressin receptors were up-regulated (Pitkow et al. 2001; Lim and Young 2004); these males also began to fixate on a particular female and to mate exclusively with her, even when other females were available (Lim et al. 2004). When this gene was inserted into non-monogamous male mice, these creatures also began to exhibit attachment behaviours (Young et al. 1999).

A number of groups have reported that the basic human motivations and emotions arise from distinct systems of neural activity and that these brain systems derive from mammalian precursors (Davidson 1994; Panksepp 1998), so it appears parsimonious to suggest that the underlying physiology associated with human monogamy is similar to that of other mammalian pair-bonding species. Moreover, activity in the ventral pallidum has been linked with longer-term pair-bonding in humans (Aron et al. 2005; Acevedo et al. 2008, 2011; Fisher et al. 2010a). Furthermore, although the AVPR1A gene among *Homo sapiens* is not homologous to the one found in prairie voles, humans do have similar alleles in this genetic region (Walum et al. 2008), suggesting that a related biological system plays a role in human monogamy.

Infidelity

Monogamy is only part of the human reproductive strategy. Infidelity is also widespread (Fisher 1992; Buunk and Dijkstra, 2006; Tsapelas et al. 2010). The National Opinion Research Center in Chicago reports that some 25% of American men and 15% of American women philander at some point during marriage (Laumann et al. 1994). Other studies of American married couples indicate that 20–40% of heterosexual married men and 20–25% of heterosexual married women have an extramarital affair during their lifetime (e.g. Greeley, 1994; Laumann et al. 1994; Tafoya and Spitzberg 2007). Still other researchers indicate that some 30–50% of American married men and women are adulterous (Gangestad and Thornhill 1997). When polled, approximately 2–4% of American men and women had had extramarital sex in the past year (Forste and Tanfer 1996).

The *Oxford English Dictionary* defines adultery as sexual intercourse by a married person with someone other than one's spouse. However, current researchers have broadened this definition to include romantic infidelity (romantic exchanges with no sexual involvement), sexual infidelity (sexual exchange with no romantic involvement), and sexual and romantic involvement (Glass and Wright 1992). When considering these varieties of adultery, statistics vary. In a meta-analysis of 12 studies of infidelity among American married couples, Thompson (1983) reported that 31% of men and 16% of women had had a sexual affair that entailed no emotional involvement; 13% of men and 21% of women had been romantically but not sexually involved with someone other than their spouse; and 20% of men and women had engaged in an affair that included both a sexual and emotional connection.

Currently, 70% of American dating couples report an incidence of infidelity in their partnership (Allen and Baucom 2006). Furthermore, in a recent survey of single American men and women, 60% of men and 53% of women admitted to 'mate poaching', trying to woo an individual away from a committed relationship with another to begin a relationship with them instead (Schmitt and Buss 2001). Mate poaching is also common in 53 other cultures (Schmitt et al. 2004).

Infidelity was also widespread in former decades. Reports in the 1920s indicated that 28% of American men and 24% of women were adulterous at some point after wedding (Lawrence 1989). In the late 1940s and early 1950s, approximately 33% of men and 26% of women in an American sample were adulterous (Kinsey et al. 1948, 1953), and in the 1970s, some 41% of men and 25% of women reported infidelity (Hunt 1974). Infidelity was also common among the classical Greeks and Romans, in pre-industrial Europeans, in historical Japanese, Chinese, and Hindus, among the traditional Inuit of the arctic, Kuikuru of the jungles of Brazil, Kofyar of Nigeria, Turu of Tanzania, and in many other tribal societies (Fisher 1992). Extra-pair copulations also occur frequently in every other society for which data are available (Frayser 1985). Human testes size suggests that adultery by both sexes was also common in hominin prehistory (Short 1977; Møller 1988).

Extra-pair copulations (EPCs) are prevalent in over 100 species of monogamous birds and several mammalian species examined (Wittenberger and Tilson 1980; Mock and Fujioka 1990; Westneat et al. 1990). Only 10% of some 180 species of monogamous songbirds are sexually faithful to their mating partners; the rest engage in EPCs. Among swift foxes (*Vulpes velox*) over 59% of a female's offspring were not genetically related to the male with whom she was pair-bonded (Kitchen et al. 2006); EPCs are also common among gibbons (*Hylobates lar*: Reichard 1995).

In fact, infidelity is so widespread and persistent in monogamous avian and mammalian species, including humans, that scientists now refer to monogamous species as practising 'social monogamy', in which partners display the array of social and reproductive behaviours associated with monogamy, regardless of sexual fidelity.

Myriad psychological, sociological, and economic variables play a role in the frequency and expression of infidelity (Tsapelas et al. 2010). But recent genetic studies suggest that biology plays a role. Walum et al. (2008) investigated 552 couples who were either married or had been co-habiting for at least 5 years. Men carrying the 334 allele in a specific region of the vasopressin system scored significantly lower on the Partner Bonding Scale, indicating lower feelings of spousal attachment. Moreover, their scores were dose-dependent: those homozygous for this allele (carrying two copies) showed the lowest scores for feeling of attachment, followed by those carrying only one allele and then by those carrying no copies. Men carrying the 334 allele also experienced more marital crises (including threat of divorce) during the past year, with homozygotes being approximately twice as likely to have had a marital crisis than those with either one or no copies of the allele. Men with at least one copy were also significantly more likely to be involved in a partnership without being married. Last, the spouses of men with at least one copy scored significantly lower on questionnaires measuring marital satisfaction.

This study did not measure infidelity directly; instead it measured several factors likely to contribute to infidelity. Nevertheless, among prairie voles, polymorphisms in a similar gene in the vasopressin system contribute to the *variability* in the strength of the monogamous pair-bond (Hammock and Young 2005), including the degree to which individuals express sexual fidelity (Ophir et al. 2008). Moreover, in a more direct study of infidelity in a sample of 181 young adult humans, Garcia et al. (2010) found an association between specific alleles in the dopamine system (DRD47R+) and frequency of both uncommitted sexual intercourse (one-night stands) and sexual infidelity.

Another biological system may contribute to infidelity. In the now classic 'sweaty t-shirt' experiment, women sniffed the t-shirts of several anonymous men and selected those they felt were the sexiest. They selected the shirts of men with different alleles (from themselves) in a specific part of the immune system, the major histocompatibility complex (MHC) (Wedekind et al. 1995). Women in established relationships show a stronger preference for odours of MHC-dissimilar men than do single women, and amongst women in relationships, those with highest preference for dissimilarity tend to be more likely to fantasize about extra-pair partners (Roberts et al. 2008). Indeed, women in romantic partnerships with men sharing relatively high numbers of MHC alleles are also more likely to fantasize about, and engage in extra-pair sex, compared to women sharing relatively few MHC alleles with their partner (Garver-Apgar et al. 2006; for a review of MHC effects in human mate choice, see Havlicek and Roberts 2009).

Brain architecture may also contribute to infidelity, due to the connections between three distinct yet interrelated brain systems that evolved for mating, reproduction, and parenting: the sex drive, romantic attraction, and attachment (Fisher 1998). In mammals, the sex drive is associated primarily with the oestrogens and androgens; however in humans, only the androgens, particularly testosterone, are central to sexual desire in both men and women (Sherwin 1994; Van Goozen et al. 1997). Studies using functional magnetic resonance imaging (fMRI) indicate that specific networks of brain activation are associated with the sex drive, among the regions involved are the hypothalamus (Arnow et al. 2002; Karama et al. 2002) and amygdala (Karama et al. 2002). Romantic attraction (also known as romantic love, obsessive love, passionate love, or being in love) is primarily associated with elevated dopamine activity in reward pathways of the brain (Bartels and Zeki 2000, 2004; Aron et al. 2005; Acevedo et al. 2008, 2011; Fisher et al. 2005, 2010a). Finally, as discussed above, attachment in humans and other mammals is associated primarily with oxytocin and vasopressin activity in the nucleus accumbens and ventral pallidum, respectively (Lim and Young 2004; Lim et al. 2004; Fisher et al. 2010a; Acevedo et al. 2011).

These three interrelated but distinct neural systems interact with one another and many other brain systems in myriad flexible, combinatorial patterns to provide the range of cognitions, emotions, motivations, and behaviours necessary to orchestrate our complex human reproductive strategy (Fisher et al. 2002). Nevertheless, they are not always directly connected, making it possible for one to express deep feelings of attachment for one individual, *while* one feels intense romantic attraction toward another, *while* one also feels the sex drive for more extra-dyadic partners (Fisher 2004). The relative biological independence of these three neural systems for mating and reproduction enable *Homo sapiens* to engage opportunistically in social monogamy and clandestine adultery simultaneously (Fisher 2004).

Evolution of monogamy

Phylogenetic origin

Monogamy could have evolved at any point in hominin evolution. However two lines of evidence suggest that the neural circuitry for human pair-bonding evolved with the basal radiation of the hominin stock (Fisher 1992), most likely in tandem with the hominin adaptation to the woodland/savannah eco-niche some time prior to 4 million years (my) before present (BP). *Ardipithecus ramidus*, currently dated at 4.4my BP, displays traits associated with reduced sexual dimorphism, so Lovejoy (2009) suggests that human monogamy had evolved by this time. Anthropologists have also re-measured *Australopithicus afarensis* fossils for skeletal size, reporting that by 3.5my BP hominins exhibited roughly the same degree of sexual dimorphism in several traits as found today, and have proposed that these hominins were 'principally monogamous' (Reno et al. 2003, p. 1073).

The emergence of facultative bipedalism may have been a primary contributing factor to the evolution of the neural circuitry for hominin monogamy (Fisher 1992). While foraging and scavenging in the woodland/savannah eco-niche, bipedal ardipithecine females were most likely obliged to carry infants in their arms instead of on their backs, as quadrupedal female apes do, thus needing the protection and provisioning of a mate while they transported nursing young. Meanwhile, ardipithecine males may have had considerable difficulty protecting and providing for a harem of females in this open woodland/savannah eco-niche. But a male could defend and provision a single female with her infant as they walked near one another, within the vicinity of the larger community.

So the exigencies of facultative bipedalism in conjunction with hominin expansion into the woodland/savannah eco-niche may have pushed ardipithecines over the 'monogamy threshold', selecting for the brain chemistry and architecture for pair-bonding and associated attachment behaviours (Fisher 1992, 2004).

Evolution of serial monogamy

Contemporary cross-cultural patterns of divorce suggest that *serial* social monogamy may have also evolved as part of the suite of traits associated with hominin adaptation to the expanding woodland/savannah eco-niche prior to 4my BP. Data on 58 human societies taken from the Demographic Yearbooks of the United Nations between 1947 and 1989 indicate three worldwide divorce patterns. Divorce occurs most frequently among couples with one dependent child, among couples at the height of their reproductive and parenting years (ages 25–29), and among couples married a modal duration of 4 years (Fisher 1989, 1992). Because 4 years is the common duration of birth spacing in hunting/gathering societies, and because many monogamous avian and mammalian species form pair-bonds that last only long enough to rear the young through infancy, this cross-cultural modal divorce peak may represent the remains of an ancestral hominin reproductive strategy to remain pair-bonded at least long enough to raise a single child through infancy (Fisher 1992).

Children in hunting/gathering societies characteristically join a multi-age playgroup soon after being weaned, becoming the responsibility of older siblings and other relatives in the band. So in ancestral environments, the ecological pressure on couples to remain pair-bonded *after* offspring weaning would have been substantially reduced, unless the couple conceived another child. Moreover, ancestral hominins that practised serial social monogamy, in association with offspring weaning, would have created disproportionately more genetic variety in their lineages, an adaptive phenomenon (Fisher 1992).

Evolution of clandestine adultery

Infidelity often involves considerable time and metabolic energy. It also involves risk; adultery can lead to diseases, unwanted pregnancy, and many adverse social consequences, including losing one's home, spouse, children, job, community, and/or health. Yet, despite near universal disapproval of infidelity, it occurs with regularity. Most curious, regardless of the many correlations between relationship dissatisfaction and adultery (see Tsapelas et al. 2010), Glass and Wright (1985) report that among Americans who engage in infidelity, 56% of men and 34% of women rate their marriage as 'happy' or 'very happy'. It seems likely that infidelity is a core aspect of our primary human reproductive strategy, and that it evolved in tandem with hominin serial social monogamy for adaptive purposes, because: 1) philandering is prevalent worldwide; 2) it is associated with a wide range of psychological and sociological factors; 3) it is correlated with several biological underpinnings discussed above; 4) promiscuity is the primary reproductive strategy among our closest primate relatives (bonobos and chimpanzees); and 5) because infidelity occurs even in 'happy' and 'very happy' marriages today.

Many hypotheses have been proposed regarding the selective value of infidelity (see Buss 1994). Among these, it has been suggested that in the ancestral woodland/savannah eco-niche, philandering males and females would have disproportionately reproduced, as well as reaped the reproductive benefits of genetically more varied offspring (Fisher 1992). Unfaithful females may have also garnered economic resources from extra-dyadic liaisons, as well as parenting support if their primary partner died or deserted them (Fisher 1992). Hence ancestral clandestine infidelity (in conjunction with serial and/or life-long social monogamy) may have had reproductive payoffs for both males and females, selecting for the biological underpinnings of infidelity in both sexes today.

Along with the evolution of serial/lifelong social monogamy and clandestine adultery, several other neural systems may have evolved. Three are considered next: the brain system associated with feelings of intense romantic attraction to a *specific* individual; four proposed temperament dimensions that may have begun to play a guiding role in mate choice; and the brain networks associated with rejection in love. Each of these neural systems most likely served other purposes among our hominoid forebears prior to the hominin radiation into the woodland/savannah eco-niche. Nevertheless, these neural systems may have taken on new functions with the evolution of the hominin dual reproductive strategy.

Evolution of romantic attraction

Human romantic love is a cross-cultural phenomenon (Jankowiak and Fischer 1992). In a survey of 166 societies, Jankowiak and Fischer (1992) found evidence of romantic love in 147 of them. (No negative evidence was found; in the 19 remaining cultures, anthropologists had failed to ask the appropriate questions, cases of ethnographic oversight.) Jankowiak and Fischer (1992) concluded that romantic love constitutes a 'human universal... or near universal'.

Romantic attraction is associated with a specific suite of psychological, behavioural, and physiological traits (Tennov 1979; Hatfield and Sprecher 1986; Hatfield et al. 1988; Harris 1995; Fisher 1998; Gonzaga et al. 2001). Romantic love generally begins as an individual starts to regard another individual as special and unique. The lover focuses attention on the beloved, aggrandizing the beloved's worthy traits and overlooking or minimizing their flaws. The lover expresses increased energy, ecstasy when the love affair is going well, and mood swings into despair during times of adversity. Adversity and social barriers tend to heighten romantic passion, 'frustration attraction' (Fisher 2004). The lover often suffers 'separation anxiety' when apart from the beloved, and experiences a host of sympathetic nervous system reactions when with the beloved, including sweating, stammering, butterflies in the stomach, and/or a pounding heart. Lovers are emotionally dependent; they change their priorities and daily habits to remain in contact with and/or to impress the beloved. Smitten humans also exhibit increased empathy for the beloved; many are willingness to sacrifice, even die for this 'special' other. The lover also expresses sexual desire for the beloved, as well as intense sexual possessiveness: 'mate guarding'. Yet the lover's craving for emotional union with the beloved tends to supersede craving for sexual union. Most characteristic, the lover thinks obsessively about the beloved: 'intrusive thinking'.

Several neuroimaging studies of romantic love indicate the physiological underpinnings of this near-universal human experience (Bartels and Zeki 2000, 2004; Fisher et al. 2003, 2010a; Aron et al. 2005; Ortigue et al. 2007; Acevedo et al. 2008). It is predominantly associated with increased activity in several regions of the reward system, most likely mediated by increased dopamine release (Aron et al. 2005; Fisher et al. 2005, 2010a; Acevedo et al. 2011). Activity in the ventral tegmental area (VTA), a central region of the brain's reward system associated with pleasure, general arousal, focused attention, and motivation to pursue and acquire rewards (Delgado et al. 2000; Schultz 2000; Elliot et al. 2003) is central to the experience.

Considerable data suggest that the human brain system for romantic attraction arose from mammalian antecedents. Like humans, all birds and mammals exhibit mate preferences, focusing courtship energy on favoured conspecifics (Fisher 2004). This phenomenon is so common that the ethological literature regularly uses several terms to describe it, including 'female choice', 'mate preference', 'individual preference', 'favouritism', 'sexual choice', 'female choice', 'selective proceptivity' (Andersson 1994), and 'courtship attraction' (Fisher 2004). Furthermore, most of the basic traits associated with human romantic love are also characteristic of mammalian courtship attraction, including increased energy, focused attention, obsessive following, affiliative gestures, possessive mate-guarding, goal-oriented behaviours, and motivation to win a preferred mating partner (Fisher et al. 2002; Fisher 2004).

The biological underpinnings of human romantic attraction and mammalian courtship attraction are also similar. When a female laboratory-maintained prairie vole is mated with a male, she forms a distinct preference for him associated with a 50% increase of dopamine in the nucleus accumbens (Gingrich et al. 2000), a central region of the reward system. When a dopamine antagonist is injected into the nucleus accumbens, the female no longer prefers this partner; and when a female is injected with a dopamine agonist, she begins to prefer the conspecific who is present at the time of the infusion, even if she has not mated with this male (Wang et al. 1999; Gingrich et al. 2000). An increase in central dopamine is also associated with courtship attraction in female sheep (Fabre-Nys et al. 1998). In male rats, too, increased striatal dopamine release has been shown in response to the presence of a receptive female rat (Robinson et al. 2002; Montague et al. 2004). In most species, however, courtship attraction is brief (lasting only minutes, hours, days or weeks), while in humans, intense, early-stage romantic love can last 12–18 months (Marazziti et al. 1999) or longer (Acevedo et al. 2011). Because human romantic attraction shares many behavioural and biological characteristics with mammalian courtship attraction, it is likely that human romantic love is a developed form of this mammalian courtship biobehavioural mechanism (Fisher 1998, 2004).

Evolution of mate choice

The evolution of romantic love, serial social monogamy, and clandestine adultery may have also stimulated new patterns of mate choice. Mate selection is governed by myriad cultural and biological factors. Men and women tend to be more attracted to individuals who have similar attitudes and values (Krueger and Caspi 1993; Shaikh and Suresh 1994), and those with a similar level of education and intelligence, religious and political views, social goals, sense of humour, financial stability, social and communication skills, as well as those from a similar socioeconomic and ethnic background (e.g. Byrne et al. 1986; Rushton 1989; Cappella and Palmer 1990; Laumann et al. 1994; Pines 1999; Buston and Emlen 2003). Freud (1965 [1905]) (and many others) have proposed that one's parents play a primary role in one's romantic choices, and Harris (1999) proposes that individuals choose a partner who reflects the values, interests, ideals, and goals of the friends they knew during their formative years. Timing also plays a role (Hatfield 1988), as does proximity (Pines 1999) and many aspects of physical attractiveness (reviewed in Gangestad and Scheyd 2005; Roberts and Little 2008).

Temperament may also play a role (Fisher 2009). Personality is composed of two basic types of traits (Cloninger et al. 1994): those that an individual acquires (dimensions of character) and those with biological underpinnings (dimensions of temperament). Many traits of temperament are heritable, relatively stable across the life course, and linked to specific gene pathways and/or hormone or neurotransmitter systems. Moreover, although many neural systems orchestrate human survival and reproduction, only four brain systems are regularly associated with human cognition, feelings, motivations, and behaviours: the dopamine, serotonin, testosterone, and oestrogen/oxytocin systems (Fisher 2009; Fisher et al. 2010b,c,d). A literature review indicates that each of these four neural systems is associated with a distinct *constellation* of related biobehavioural traits (temperament dimensions or behaviour syndromes), as summarized in Table 7.1.

To study the possible role of these four broad temperament dimensions in mate choice, a questionnaire was first designed to measure the traits associated with each of these brain systems. This measure comprised 56 questions: 14 to measure an individual's expression of each scale (i.e. the dopamine, serotonin, testosterone, and oestrogen/oxytocin scales). Data were collected, and the questions modified, on a regular basis between 2006–2007, using the Internet dating site 'Chemistry.com' (a division of 'Match.com'), until reliability was obtained in a US sample of 39,913 anonymous men and women. Participants completed demographic information, the questionnaire, and 9 validity questions, with the goal of finding a romantic partner. Respondents ranged in age from 18–88 years ($M=37.0$; $SD=12.6$); 56.4% were female ($N=22,521$); 89.6% ($N=35,759$) were seeking an opposite-sex partner. All individuals expressed all four temperament dimensions, though they varied in the degree to which they expressed them.

The finalized measure (with the Cronbach's alpha consistency coefficient ranging from 0.78–0.80) was named the Fisher, Rich, Island Neurochemical Questionnaire (FRI-NQ) (Fisher et al. 2010b,c,d). It was then placed on an international dating site (Match.com) in 39 other countries and data to measure reliability were collected on 15,000 individuals in each of five of these translations: German, French, Spanish, English (Australian sample), and Swedish. After this questionnaire had achieved adequate reliability and had correlated positively with 9 validity measures (Fisher 2009; Fisher et al. 2010a,b,c,d), it was employed to investigate the initial attraction phase of mate choice. A random sample of anonymous participants from Chemistry.com was examined, consisting of 28,128 heterosexual adults (17,776 men; 10,352 women) who had just had an initial meeting with a potential partner.

Men and women who predominantly expressed the constellation of biobehavioural traits associated with the proposed dopamine and serotonin scales were significantly more likely to choose to meet individuals who predominantly expressed the same temperament dimension. In contrast, individuals who predominantly expressed traits associated with the proposed testosterone scale were more likely to choose to meet their opposite (those who predominantly expressed traits

Table 7.1 Biological and behavioural traits associated with the dopamine, serotonin, testosterone, and oestrogen/oxytocin neural systems

Neural system	Trait	Source
Dopamine	Exploratory behaviour, thrill and adventure seeking, boredom susceptibility, disinhibition	Cloninger et al. 1991, 1994; Espejo 1997; Zuckerman 2005
	Mania and hypersocial behaviour	Depue and Collins 1999
	Enthusiasm	Zuckerman 2005
	Lack of introspection	Gerbing et al. 1987; Cloninger et al. 1991; Ebstein et al. 1996
	Energy, motivation and achievement striving	Depue and Collins 1999; Wacker et al. 2006
	Abstract intellectual exploration	DeYoung et al. 2002
	Cognitive flexibility	Ashby et al. 1999
	Plasticity	DeYoung et al. 2005
	Curiosity	Zuckerman 2005
	Idea generation and verbal and non-linguistic creativity	Flaherty 1995; Reuter et al. 2006
Serotonin	Reduced anxiety, higher hypomania and extraversion scores, higher sociability	Golimbet et al. 2004
	Harm avoidance	Golimbet et al. 2004; Parks et al. 2006
	Positive mood	Opbroek et al. 2002; Flory et al. 2004
	Religiosity	Borg et al. 2003
	Observing social norms	Golimbet et al. 2004
	Conformity, adhering to plans, methods, habits	DeYoung et al. 2002
	Orderliness	DeYoung and Gray 2005
	Conscientiousness	Manuck et al. 1998
	Managerial skills	Knutson et al. 1998
	Persistence and patience	Davidge et al. 2004
	Precision and interest in details	Cloninger et al. 1991
	Concrete thinking	Zuckerman 2005
	Self-control	Manuck et al. 2000
	Sustained attention	Zuckerman 2005
	Reduced novelty seeking	Serretti et al. 2006
	Figural and numeric creativity	Reuter et al. 2006
Testosterone	Attention to detail, intensified focus; restricted interests	Baron-Cohen et al. 2005; Knickmeyer et al. 2005
	Emotional containment	Dabbs 1997
	Emotional flooding, especially rage	Manning 2002
	Sensitivity to social dominance, drive for rank	Mazur et al. 1997
	Aggressiveness	Manning 2002
	Reduced social sensitivity	Baron-Cohen et al. 2005
	Poorer language skills	Manning 2002

Table 7.1 (continued) Biological and behavioural traits associated with the dopamine, serotonin, testosterone, and oestrogen/oxytocin neural systems

Neural system	Trait	Source
	Less empathy	Knickmeyer et al. 2006
	Confidence, forthrightness, boldness	Nyborg 1994
	Spatial and mathematical acuity	Nyborg 1994; Manning 2002
Oestrogen and oxytocin	Verbal fluency, language skills	Manning 2002; Baron-Cohen et al. 2005; Knickmeyer et al. 2005
	Empathy, nurturing, pro-social skills	Kendrick 2000; Taylor et al. 2000; Baron-Cohen 2002
	Drive to maintain social attachments	Skuse et al. 1997; Carter 1992; Baron-Cohen 2002
	Agreeableness, cooperation	Baron-Cohen 2002
	Theory of mind	Baron-Cohen 2002; Domes et al. 2007
	Generosity, trust	Kosfeld 2005; Zak et al. 2007
	Heightened memory for emotional experiences	Canli et al. 2002
	Contextual thinking	Fisher 1999; Dabbs and Dabbs 2000; Baron-Cohen et al. 2005
	Imagination	Fisher 2009
	Mental flexibility	Skuse et al. 1997

associated with the proposed oestrogen/oxytocin dimension), and vice versa (Fisher 2009; Fisher et al. 2010d).

It has been hypothesized that variations in human personality stem from their reproductive advantages in the shifting ecological and social environment of prehistory (Buss 1991; MacDonald 1995). Unions between individuals predominantly expressing testosterone and those predominantly expressing oestrogen and oxytocin may have increased their fecundity by pooling suites of complementary temperament traits; while mates who were both expressive of the proposed serotonin system may have capitalized on some very effective parenting traits, including loyalty, calm, and caution. Mates who were both expressive of the proposed dopamine system (and who thus were equally novelty-seeking, curious, and creative) may have engaged in more extra-pair copulations and serial partnerships, thereby producing disproportionate genetic variation in their lineages.

In summary, my proposal so far is that the human predisposition to seek partners with specific biochemical profiles evolved in conjunction with the evolution of monogamy to facilitate more effective mate choices in the expanding woodland/savannah environment of southern and eastern Africa prior to 4my BP (Fisher 2009).

Evolution of romantic rejection

The evolution of serial/long-term social monogamy, coupled with clandestine adultery, most likely elevated the trauma of rejection in love. To study the neural systems associated with romantic rejection, ten female and five male college-age heterosexuals were studied using fMRI; all had recently been rejected by their partners but reported that they were still intensely in love (Fisher et al. 2010a). The average length of time since the initial rejection and the participants' enrolment in the study was 63 days. All scored high on the Passionate Love Scale (Hatfield and Sprecher 1986), a self-report questionnaire that measures the intensity of romantic feelings. All participants said that they spent more than 85% of their waking hours thinking of the person who rejected them and yearned for their abandoning partner to return to the relationship. Brain activations coupled with romantic rejection included activity in the ventral tegmental area (associated with feelings of intense romantic love), the nucleus accumbens and orbitofrontal/prefrontal cortex (associated with craving and addiction), the insular and anterior cingulate cortex (associated with physical pain and distress), and the ventral pallidum (associated with feelings of attachment). Thus, rejected individuals are experiencing extreme feelings of romantic passion, intense craving, and severe physical and mental distress (Fisher et al. 2010a).

Rejected individuals also often experience 'abandonment rage' (Meloy 2001). The primary rage system is closely connected to centres in the prefrontal cortex that anticipate rewards (Panksepp 1998), producing a response to unfulfilled expectations known as 'frustration-aggression'. Abandonment rage stresses the heart, raises blood pressure, and suppresses the immune system (Dozier 2002). Romantic rejection can also stimulate feelings of resignation, despair, lethargy, despondency, and depression (Panksepp 1998; Najib et al. 2004); some broken-hearted lovers die from heart attacks or strokes caused by their depression (Rosenthal 2002).

Few men or women avoid the suffering of rejection at some point over the life course. In one American college community, 93% of both sexes reported that they had been spurned by someone they passionately loved and 95% reported they had rejected someone who was deeply in love with them (Baumeister et al. 1993). Moreover, rejected individuals most likely suffer for good evolutionary reasons: they have wasted precious courtship time and metabolic energy, and their social alliances and reproductive future have been jeopardized. So today's rejected individuals are most likely experiencing the effects of a strong survival system that evolved to provide them with the energy and motivation to renew or sustain a foundering partnership crucial to reproduction in ancestral environments (Fisher 2004).

Conclusion

Critics of evolutionary psychology fail to find the profound value of this budding discipline. For example, people have long regarded romantic love as part of the supernatural, or as an invention of the Troubadours in 12th-century France, or as the result of childhood exposure and cultural experiences. On the contrary, romantic love engages primary regions of the brain's primitive 'reward system' associated with focus, energy, craving, and intense motivation to win and/or sustain a partnership; it most likely emerged from mammalian antecedents during hominin evolution to enable our forebears to focus their mating energy on a single mate and initiate a pair-bond essential to their reproductive and genetic survival. If the medical and legal communities were to understand that romantic love is an evolved drive (Fisher 2004) that can lead to severe social and personal consequences, they might develop new procedures for dealing with the negative aspects of this powerful neural mechanism.

Indeed, it would be appropriate to treat romantic rejection as a biologically-based addictive state. Because romantic love is associated with focused attention, euphoria, craving, obsession, compulsion, distortion of reality, personality changes, emotional and physical dependence, inappropriate and dangerous behaviours, tolerance, withdrawal symptoms, relapse, and loss of self-control, several psychologists have regarded romantic love as an addiction (Peele 1975; Tennov 1979; Hunter et al. 1981; Halpern 1982; Carnes 1983; Schaef 1989; Griffin-Shelley 1991; Mellody et al. 1992). Data from neural imaging confirm this: romantic rejection activates three basic brain regions associated with craving and addiction (Fisher et al. 2010a).

Researchers and therapists might also design their therapies differently if they were to acknowledge the varying ways that the sexes process rejection (Baumeister et al. 1993; Buss 1994; Hatfield and Rapson 1996). Men are two to three times more likely to commit suicide after being rejected (Hatfield and Rapson 1996); and men are more likely to stalk a rejecting partner, as well as batter or kill her (Meloy 2001). Rejected women report more severe feelings of depression (Mearns 1991) and more chronic strain (Nolen-Hoeksema et al. 1999). Women are also more likely to talk about their trauma, sometimes inadvertently re-traumatizing themselves (Hatfield and Rapson 1996). It is also important to recognize that feelings of romantic love after rejection recede with time. Activity in the right ventral putamen/pallidum (the brain region associated with attachment) is negatively correlated with time since rejection (Fisher et al. 2010a). Furthermore, areas associated with reappraising difficult emotional situations and assessing one's gains and losses, are activated after rejection, suggesting that rejected individuals are trying to understand and learn from their difficult situation.

Knowledge of evolutionary psychology could also help professionals (and many others) understand the underlying evolutionary predispositions that lead to unstable partnerships (for further discussion on this issue, see Roberts et al. 2010). The brain mechanisms associated with serial monogamy and clandestine adultery surely contribute to many contemporary and cross-cultural patterns of philandering and divorce, as well as the high incidence of sexual jealousy, partner stalking, spousal abuse, love homicide, love suicide, and clinical depression (Meloy and Fisher 2005).

Last, I feel it is essential that the medical and legal communities begin to embrace the possible consequences of contemporary antidepressant usage (Fisher and Thomson 2007). Over 100 million prescriptions for antidepressants are written annually in the United States; most are for SSRIs (selective serotonin reuptake inhibitors) that elevate serotonin at the synapse. It is well known that these drugs cause sexual dysfunction in as many as 73% of patients. But these drugs may also adversely affect the neural systems that evolved to enable people to assess potential mates, prefer and choose specific partners, feel extended romantic passion, and sustain feelings of attachment during a long-term relationship (Fisher 1999; Fisher and Thomson 2007). The number of neural mechanisms associated with mate selection, romantic love, and long-term partnership stability are unknown, and many operate outside of conscious awareness. If professionals prescribing these medications for *long-term use* were aware of their potential effects on conscious and unconscious neural systems associated human reproduction, they might consider informing patients of their potential side effects, and undertaking far more expansive studies of these effects (Fisher and Thomson 2007).

References

Acevedo, B., Aron, A., Fisher, H.E., and Brown, L.L. (2008). *Neural correlates of long-term pair-bonding in a sample of intensely in-love humans.* Poster presented at the Annual Meeting of the Society for Neuroscience, Washington, DC.

Acevedo, B., Aron, A., Fisher, H.E., and Brown, L.L. (2011). Neural correlates of long term intense romantic love. *Social Cognitive and Affective Neuroscience*, First published online: January 5, 2011.

Ainsworth, M., Behar, M., Waters, E., and Wall, S. (1978). *Patterns of attachment: a psychological study of the strange situation.* Erlbaum, Hillsdale, NJ.

Allen, E.S. and Baucom, D.H. (2006). Dating, marital, and hypothetical extradyadic involvements: how do they compare? *The Journal of Sex Research*, **43**, 307–317.

Andersson, M. (1994). *Sexual selection.* Princeton University Press, Princeton, NJ.

Arnow, B.A., Desmond, J.E., Banner, L.L., et al. (2002). Brain activation and sexual arousal in healthy, heterosexual males. *Brain*, **125**, 1014–23.

Aron, A., Fisher, H.E., Mashek, D.J., Strong, G., Li, H.F., and Brown, L.L. (2005). Reward, motivation, and emotion systems associated with early-stage intense romantic love: an fMRI study. *Journal of Neurophysiology*, **94**, 327–37.

Ashby, F.G., Isen, A.M., and Turken, A.U. (1999). A neuropsychological theory of positive affect and its influence on cognition. *Psychological Review*, **106**, 529–50.

Baron-Cohen, S. (2002). The extreme male brain theory of autism. *Trends in Cognitive Sciences*, **6**, 248–54.

Baron-Cohen, S, Knickmeyer, R.C., and Belmonte, M.K. (2005). Sex differences in the brain: implications of explaining autism. *Science*, **310**, 819–23.

Bartels, A. and Zeki, S. (2000). The neural basis of romantic love. *Neuroreport*, **11**, 3829–34.

Bartels, A. and Zeki, S. (2004). The neural correlates of maternal and romantic love. *NeuroImage*, **21**, 1155–66.

Baumeister, R.F., Wotman, S.R., and Stillwell, A.M. (1993). Unrequited love: on heartbreak, anger, guilt, scriptlessness and humiliation. *Journal of Personality and Social Psychology*, **64**, 377–94.

Bester-Meredith, J.K., Young, L J., and Marler, C.A. (1999). Species differences in paternal behavior and aggression in *Peromyscus* and their associations with vasopressin immunoreactivity and receptors. *Hormones and Behavior*, **36**, 25–38.

Borg, J., Andree, B.A., Soderstrom, H., and Farde, L. (2003). The serotonin system and spiritual experiences. *American Journal of Psychiatry*, **160**, 1965–69.

Bowlby, J. (1969). *Attachment and loss: attachment (Vol 1)*. Basic Books, New York.

Bowlby, J. (1973). *Atttachment and loss: separation (Vol 2)*. Basic Books, New York.

Buss, D.M. (1991). Evolutionary personality psychology. *Annual Review of Psychology*, **42**, 459–91.

Buss, D.M. (1994). *The evolution of desire: strategies of human mating*. Basic Books, New York.

Buston, P.M. and Emlen, S.T. (2003). Cognitive processes underlying human mate choice: the relationship between self-perception and mate preference in Western society. *Proceedings of the National Academy of Sciences of the USA*, **100**, 8805–10.

Buunk, A.P. and Dijkstra, P. (2006). Temptation and threat: extra-dyadic relations and jealousy. In: A.L. Vangelisti and D. Perlman (eds), *The Cambridge handbook of personal relationships*, pp. 533–55. Cambridge University Press, New York.

Byrne, D., Gerald, L.C., and Smeaton, G. (1986). The attraction hypothesis: do similar attitudes affect anything? *Journal of Personality and Social Psychology*, **51**, 1167–70.

Canli, T., Desmond, J.E., Zhoa, Z. and Gabrieli, J.D.E. (2002). Sex differences in the neural basis of emotional memories. *Proceedings of the National Academy of Sciences USA*, **99**, 10789–10794.

Cappella, J.N. and Palmer, M.T. (1990). Attitude similarity, relational history, and attraction: the mediating effects of kinesic and vocal behaviors. *Communication Monographs*, **57**, 161–83.

Carnes, P. (1983). *Out of the shadows: understanding sexual addiction*. CompCare, Minneapolis, MN.

Carter, C.S. (1992). Oxytocin and sexual behavior. *Neuroscience and Biobehavioral Reviews*, **1**, 131–44.

Cherlin, A.J. (2009). *The marriage-go-round: the state of marriage and the family in America today*. Alfred A. Knopf, New York.

Cloninger, C.R., Przybeck, T.R. and Svrakic, D.M. (1991). The tridimensional personality questionnaire: U.S. normative data. *Psychological Reports*, **69**, 1047–57.

Cloninger, C.R., Przybeck, T.R. Svrakic, D.M. and Wetzel, R.D. (1994). *The temperament and character inventory (TCI): a guide to its development and use*. Washington University, Center for Psychobiology of Personality, St Louis, MO.

Dabbs, J.M. (1997). Testosterone, smiling and facial appearance. *Journal of Nonverbal Behavior*, **12**, 45–55.

Dabbs, J.M. and Dabbs, M.G. (2000). *Heroes, rogues and lovers: testosterone and behavior*. McGraw-Hill, New York.

Daly, M. and Wilson, M. (1983). *Sex, evolution and behavior* (2nd ed). Willard Grant, Boston, MA.

Davidge, K.M., Atkinson, L., Douglas, L., et al. (2004). Association of the serotonin transporter and 5HT1K[beta] receptor genes with extreme persistent and pervasive aggressive behaviour in children. *Psychiatric Genetics*, **14**, 143–46.

Davidson, R.J. (1994). Complexities in the search for emotion-specific physiology. In: P. Ekman and R.J. Davidson (eds), *The nature of emotion: fundamental questions*, pp. 237–42. Oxford University Press, New York.

Delgado, M.R., Nystrom, L.E., Fissel, C., Noll, D.C., and Fiez, J.A. (2000). Tracking the hemodynamic responses to reward and punishment in the striatum. *Journal of Neurophysiology*, **84**, 3072–77.

Depue, R.A. and Collins, P.F. (1999). Neurobiology of the structure of personality: dopamine, acilitation of incentive motivation, and extraversion. *Behavioral and Brain Sciences*, **22**, 491–569.

DeYoung, C.D., Peterson, J.B., and Higgins, D.M. (2002). Higher-order factors of the Big Five predict conformity: are there neuroses of health? *Personality and Individual Differences*, **33**, 533–52.

DeYoung, C.D. Peterson, J.B., and Higgins, D.M. (2005). Sources of openness/intellect: cognitive and neuropsychological correlates of the fifth factor of personality. *Journal of Personality*, **73**, 825–58.

Domes, G. Heinrichs, M., Michel, A., Berger, C., and Herpertz, S.C. (2007). Oxytocin improves 'mind-reading' in humans. *Biological Psychiatry*, **61**, 731–33.

Dozier, R.W. (2002). *Why we hate: understanding, curbing, and eliminating hate in ourselves and our world.* Contemporary Books, Chicago, IL.

Ebstein, R.P., Novick, O., Umansky, R., et al. (1996). Dopamine D4 (D4DR) exon III polymorphism associated with the human personality trait of novelty seeking. *Nature Genetics*, **12**, 78–80.

Elliott, R., Newman, J.L., Longe, O.A., and Deakin, J.F.W. (2003). Differential response patterns in the striatum and orbitofrontal cortex to financial reward in humans: a parametric functional magnetic resonance imaging study. *Journal of Neuroscience*, **23**, 303–307.

Espejo, E.F. (1997). Selective dopamine depletion within the medial prefrontal cortex induces anxiogenic-like effects in rats placed on the elevated plus maze. *Brain Research*, **762**, 281–84.

Fabre-Nys, C. (1998). Steroid control of monoamines in relation to sexual behaviour. *Reviews of Reproduction*, **3**, 31–41.

Fisher, H.E. (1989). Evolution of human serial pair-bonding. *American Journal of Physical Anthropology*, **78**, 331–54.

Fisher, H.E. (1992). *Anatomy of love*. WW Norton, New York.

Fisher, H.E. (1998). Lust, attraction, and attachment in mammalian reproduction. *Human Nature*, **9**, 23–52.

Fisher, H.E. (1999). *The first sex: the natural talents of women and how they are changing the world.* Random House, New York.

Fisher, H.E. (2004). *Why we love*. Henry Holt, New York.

Fisher, H. (2009). *Why him? Why her?* Holt and Co., New York.

Fisher, H.E., Aron, A., Mashek, D., Strong, G., Li, H., and Brown, L.L. (2002). Defining the brain systems of lust, romantic attraction and attachment. *Archives of Sexual Behavior*, **31**, 413–419.

Fisher, H.E., Aron, A., Mashek, D., Strong, G., Li, H., and Brown, L.L. (2003). *Early stage intense romantic love activates cortical-basal-ganglia reward/motivation, emotion and attention systems: An fMRI study of a dynamic network that varies with relationship length, passion intensity and gender.* Poster presented at the Annual Meeting of the Society for Neuroscience, New Orleans.

Fisher, H.E., Brown, L.L., Aron, A., Strong, G., and Mashek, D. (2010a). Reward, addiction, and emotion regulation systems associated with rejection in love. *Journal of Neurophysiology*, **104**, 51–60.

Fisher, H.E., Rich, J., Island, H.D., and Marchalik, D. (2010b). The second to fourth digit ratio: a measure of two hormonally-based temperament dimensions. *Personality and Individual Differences*, **49**, 773–77.

Fisher, H.E., Rich, J., Island, H.D., Marchalik, D., Silver, L., and Zava, D. (2010c). *Four primary temperament dimensions.* Poster presented at the Annual Meeting of the American Psychological Association, San Diego.

Fisher, H.E., Rich, J., Island, H.D., Marchalik, D., Silver, L., and Zava, D. (2010d). *Four primary temperament dimensions in the process of mate choice*. Poster presented at the Annual Meeting of the American Psychological Association, San Diego.

Fisher, H.E. and Thomson, J.A., Jr (2007). Lust, romance, attachment: do the side-effects of serotonin-enhancing antidepressants jeopardize romantic love, marriage and fertility? In: S.M. Platek, J.P. Keenan, and T.K. Shackelford (eds), *Evolutionary cognitive neuroscience*, pp. 245–83. MIT Press, Cambridge, MA.

Flaherty, A.W., and Graybiel, A.M. (1995). Motor and somatosensory corticostriatal projection magnifications in the squirrel monkey. *Journal of Neurophysiology*, **74**, 2638–48.

Flory, J.D., Manuck, S.B., Matthews, K.A. and Muldoon, M.F. (2004). Serotonergic function in the central nervous system is associated with daily ratings of positive mood. *Psychiatry Research,* **129**, 11–19.

Forste, R. and Tanfer, K. (1996). Sexual exclusivity among dating, cohabiting, and married women. *Journal of Marriage and the Family*, **58**, 33–47.

Fraley, R.C. and Shaver, P.R. (2000). Adult romantic attachment: theoretical developments, emerging controversies, and unanswered questions. *Review of General Psychology*, **4**, 132–54.

Frayser, S. (1985). *Varieties of sexual experience: an anthropological perspective of human sexuality*. HRAF Press, New Haven, CT.

Freud, S. (1965 [1905]). *Three essays on the theory of sexuality*. Basic Books, New York.

Gangestad, S.W. and Scheyd, G.J. (2005). The evolution of human physical attractiveness. *Annual Review of Anthropology*, **34**, 523–48.

Gangestad, S.W. and Thornhill, R. (1997). The evolutionary psychology of extra-pair sex: the role of fluctuating asymmetry. *Evolution and Human Behavior*, **18**, 69–88.

Garcia, J.R, MacKillop, J., Aller, E.L., Merriwether, A.M., Wilson, D.S., and Lum, J.K. (2010). Associations between dopamine D4 receptor gene variation with both infidelity and sexual promiscuity. *PLoS ONE*, **5**, e14162.

Garver-Apgar, C.E., Gangestad, S.W., Thornhill, R., Miller, R.D., and Olp, J.J. (2006). Major histocompatibility complex alleles, sexual responsivity, and unfaithfulness in romantic couples. *Psychological Science*, **17**, 830–35.

Gerbing, D.W., Ahadi, S.A., and Patton, J.H. (1987). Toward a conceptualization of impulsivity: components across the behavioral and self-report domains. *Multivariate and Behavioral Research*, **22**, 357–79.

Gingrich, B., Liu, Y., Cascio, C., Wang, Z., and Insel, T.R. (2000). D2 receptors in the nucleus accumbens are important for social attachment in female prairie voles (*Microtus ochrogaster*). *Behavioral Neuroscience*, **114**, 173–83.

Glass, S. and Wright, T. (1985). Sex differences in type of extramarital involvement and marital dissatisfaction. *Sex Roles*, **12**, 1101–20.

Glass, S. and Wright, T. (1992). Justifications for extramarital relationships: the association between attitudes, behaviors, and gender. *Journal of Sex Research*, **29,** 361–87.

Golimbet, V.E., Alfimova, M.V., and Mityushina, N.G. (2004). Polymorphism of the serotonin 2A receptor gene (5HTR2A) and personality traits. *Molecular Biology*, **38**, 337–44.

Gonzaga G.C., Keltner, D., Londahl, E.A., and Smith, M.D. (2001). Love and the commitment problem in romantic relations and friendship. *Journal of Personality and Social Psychology*, **81**, 247–62.

Greeley, A. (1994). Marital infidelity. *Society,* **31**, 9–13.

Griffin-Shelley, E. (1991). *Sex and love: addiction, treatment and recovery.* Praeger, New York.

Halpern, H.M. (1982). *How to break your addiction to a person.* McGraw-Hill, New York.

Hammock, E.A.D. and Young, L.J. (2005). Microsatellite instability generates diversity in brain and sociobehavioral traits. *Science*, **308**, 1630–34.

Harris, H. (1995) Rethinking heterosexual relationships in Polynesia: a case study of Mangaia, Cook Island. In: W. Jankowiak (ed.), *Romantic passion: a universal experience?*, pp. 95–127. Columbia University Press, New York.

Harris, J.R. (1999). *The nurture assumption: why children turn out the way they do*. Touchstone, New York.

Hatfield, E. (1988). Passionate and companionate love. In: R.J. Sternberg, and M.L. Barnes (eds), *The psychology of love*, pp. 191–217. Yale University Press, New Haven, CT.

Hatfield, E. and Rapson, R.L. (1996). *Love and sex: cross-cultural perspectives*. Allyn and Bacon, Boston, MA.

Hatfield, E., Schmitz, E., Cornelius, J., and Rapson, R.L. (1988). Passionate love: how early does it begin? *Journal of Psychology and Human Sexuality*, **1**, 35–51.

Hatfield, E. and Sprecher, S. (1986). Measuring passionate love in intimate relations. *Journal of Adolescence*, **9**, 383–410.

Havlicek, J. and Roberts, S.C. (2009). The MHC and human mate choice: a review. *Psychoneuroendocrinology*, **34**, 497–512.

Hazan, C. and Diamond, L.M. (2000). The place of attachment in human mating. *Review of General Psychology*, **4**, 186–204.

Hazan, C. and Shaver, P.R. (1987). Romantic love conceptualized as an attachment process. *Journal of Personality and Social Psychology*, **52**, 511–24.

Hunt, M. (1974). *Sexual behavior in the 1970s*. Playboy Press, Chicago, IL.

Hunter, M.S., Nitschke, C., and Hogan, L. (1981). A scale to measure love addiction. *Psychological Reports*, **48**, 582.

Jankowiak, W.R. and Fischer, E.F. (1992). A cross-cultural perspective on romantic love. *Ethnology*, **31**, 149–55.

Karama, S., Lecours, A.R., Leroux, J.M., et al. (2002). Areas of brain activation in males and females during viewing of erotic film excerpts. *Human Brain Mapping*, **16**, 1–13.

Kendrick, K.M. (2000). Oxytocin, motherhood and bonding. *Experimental Physiology*, **85**, 111S–24S.

Kinsey, A.C., Pomeroy, W.B., and Martin, C.E. (1948). *Sexual behavior in the human male*. W.B. Saunders, Philadelphia, PA.

Kinsey, A.C., Pomeroy, W.B., Martin, C.E., and Gebhard, P. (1953). *Sexual behavior in the human female*. W.B. Saunders, Philadelphia, PA.

Kitchen, A.M., Gese, E.M., Walts, L.P., Karki, S.M., and Schauster, E.R. (2006). Multiple breeding strategies in the swift fox, *Vulpes velox*. *Animal Behaviour*, **71**, 1029–38.

Knickmeyer, R., Baron-Cohen, S., Raggatt, P., and Taylor, K. (2005). Foetal testosterone, social relationships and restricted interests in children. *Journal of Child Psychology and Psychiatry*, **46**, 198–210.

Knickmeyer, R., Baron-Cohen, S., Raggatt, P., Taylor, K., and Hackett, G. (2006). Fetal testosterone and empathy. *Hormones and Behavior*, **49**, 282–92.

Knutson, B., Wolkowitz, O.M., Cole, S.W., et al. (1998). Selective Alteration of personality and social behavior by serotonergic intervention. *American Journal of Psychiatry*, **155**, 373–78.

Kosfeld, M., Heinrichs, M., Zak, P.J., Fischbacher, U., and Fehr, E. (2005). Oxytocin increases trust in humans. *Nature*, **435**, 673–76.

Krueger, R.F. and Caspi, A. (1993). Personality, arousal, and pleasure: a test of competing models of interpersonal attraction. *Personality and Individual Differences*, **14**, 105–111.

Laumann, E.O., Gagnon, J.H., Michael, R.T., and Michaels, S. (1994). *The social organization of sexuality: sexual practices in the United States*. University of Chicago Press, Chicago, IL.

Lawrence, R.J. (1989). *The poisoning of Eros: sexual values in conflict*. Augustine Moore Press, New York.

Lim, M.M., Murphy, A.Z., and Young, L.J. (2004). Ventral striatopallidal oxytocin and vasopressin V1a receptors in the monogamous prairie vole *(Microtus ochrogaster)*. *Journal of Comparative Neurology*, **468**, 555–70.

Lim, M.M. and Young, L.J. (2004). Vasopressin-dependent neural circuits underlying pair bond formation in the monogamous prairie vole. *Neuroscience*, **125**, 35–45.

Lovejoy, O.C. (2009). Reexamining human origins in light of *Ardipithecus ramidus*. *Science*, **326**, 74–78.

MacDonald, K. (1995). Evolution, the five-factor model, and levels of personality. *Journal of Personality*, **63**, 525–67.

Manning, J.T. (2002). *Digit ratio: a pointer to fertility, behavior, and health*. Rutgers University Press, New Brunswick, NJ.

Manuck, S.B., Flory, J.D., Ferrell, R.E., Mann, J.J., and Muldoon, M.F. (2000). A regulatory polymorphism of the monoamine oxidase-A gene may be associated with variability in aggression, impulsivity, and central nervous system serotonergic responsivity. *Psychiatry Research*, **95**, 9–23.

Manuck, S.B., Flory, J.D., McCaffery, J.M., Matthews, K.A., Mann, J.J., and Muldoon, M.F. (1998). Aggression, impulsivity, and central nervous system serotonergic responsivity in a nonpatient sample. *Neuropsychopharmacology*, **19**, 287–99.

Marazziti, D., Akiskal, H.S., Rossi, A., and Cassano, G.B. (1999). Alteration of the platelet serotonin transporter in romantic love. *Psychological Medicine*, **29**, 741–45.

Mazur, A., Susman, E.J., and Edelbrock, S. (1997). Sex differences in testosterone response to a video game contest. *Evolution and Human Behavior*, **18**, 317–26.

Mearns, J. (1991). Coping with a breakup: negative mood regulation expectancies and depression following the end of a romantic relationship. *Journal of Personality and Social Psychology*, **60**, 327–34.

Meloy, J.R. (2001). When stalkers become violent: the threat to public figures and private lives. *Psychiatric Annals*, **33**, 658–65.

Meloy, J.R. and Fisher, H.E. (2005). A neurobiological theory of stalking. *Journal of Forensic Sciences*, **50**, 1472–80.

Mellody, P., Miller, A.W., and Miller, J.K. (1992). *Facing love addiction*. Harper San Francisco, New York.

Mock, D.W. and Fujioka, M. (1990). Monogamy and long-term bonding in vertebrates. *Trends in Ecology and Evolution*, **5**, 39–43.

Møller, A.P. (1988). Ejaculate quality, testes size and sperm competition in primates. *Journal of Human Evolution*, **17**, 479–88.

Montague P.R., McClure, S.M., Baldwin, P.R., et al. (2004). Dynamic gain control of dopamine delivery in freely moving animals. *Journal of Neuroscience*, **24**, 1754–59.

Najib, A., Lorberbaum, J.P., Kose, S., Bohning, D.E., and George, M.S. (2004). Regional brain activity in women grieving a romantic relationship breakup. *American Journal of Psychiatry*, **161**, 2245–56.

Nolen-Hoeksema, S., Larson, J., and Grayson, C. (1999). Explaining the gender difference in depressive symptoms. *Journal of Personality and Social Psychology*, **77**, 1061–72.

Nyborg, H. (1994). *Hormones, sex and society: the science of physiology*. Praeger, Westport, CT.

Opbroek, A., Delgado, P.L., Laukes, C., et al. (2002). Emotional blunting associated with SSRI-induced sexual dysfunction. Do SSRIs inhibit emotional responses? *International Journal of Neuropsychopharmacology*, **5**, 415–416.

Ophir, A.G., Wolff, J.O., and Phelps, S.M. (2008). Variation in the neural V1aR predicts sexual fidelity and space use among male prairie voles in semi-natural settings. *Proceedings of the National Academy of Sciences of the USA*, **105**, 1249–54.

Ortigue. S., Bianchi-Demicheli, F., Hamilton, A.F., and Grafton, S.T. (2007). The neural basis of love as a subliminal prime: an event-related functional magnetic resonance imaging study. *Journal of Cognitive Neuroscience*, **19**, 1218–30.

Panksepp, J. (1998). *Affective neuroscience: the foundations of human and animal emotions*. Oxford University Press, New York.

Parks. C.L., Robinson, P.S., Sibille, E., Shenk, T., and Toth, M. (1998). Increased anxiety of mice lacking the sertonin 1A receptor. *Proceedings of the National Academy of Sciences USA*, **95**, 10734–39.

Peele, S. (1975). *Love and addiction*. Taplinger, New York.

Pines, A.M. (1999). *Falling in love: why we choose the lovers we choose*. Routledge, New York.

Pitkow, L.J., Sharer, C.A., Ren, X., Insel, T.R., Terwilliger, E.F., and Young, L.J. (2001). Facilitation of affiliation and pair-bond formation by vasopressin receptor gene transfer into the ventral forebrain of a monogamous vole. *Journal of Neuroscience*, **21**, 7392–96.

Reichard, U. (1995). Extra-pair copulation in the monogamous gibbon (Hylobates lar). *Ethology*, **100**, 99–112.

Reno, P.L., Meindl, R.S., McCollum, M.A., and Lovejoy, C.O. (2003). Sexual dimorphism in *Australopithecus afarensis* was similar to that of modern humans. *Proceedings of the National Academy of Sciences of the USA*, **100**, 9404–9.

Reuter, M., Roth, S., Holve, K. and Hennig, J. (2006). Identification of first candidate genes for creativity: a pilot study. *Brain Research*, **1069**, 190–7.

Roberts, S.C., Gosling, L.M., Carter, V., and Petrie, M. (2008). MHC-correlated odour preferences in humans and the use of oral contraceptives. *Proceedings of the Royal Society B*, **275**, 2715–22.

Roberts, S.C., and Little, A.C. (2008). Good genes, complementary genes and human mate choice. *Genetica*, **132**, 309–21.

Roberts, S.C., Miner, E.M., and Shackelford, T.K. (2010). The future of an applied evolutionary psychology for human partnerships. *Review of General Psychology*, **14**, 318–29.

Robinson, D.L., Heien, M.L., and Wightman, R.M. (2002). Frequency of dopamine concentration transients increases in dorsal and ventral striatum of male rats during introduction of conspecifics. *Journal of Neuroscience*, **22**, 10477–86.

Rosenthal, N.E. (2002). *The emotional revolution: how the new science of feelings can transform your life.* Citadel Press, New York.

Rushton, J.P. (1989). Epigenesis and social preference. *Behavioral and Brain Sciences*, **12**, 31–2.

Schaef, A.W. (1989). *Escape from intimacy: the pseudo-relationship addictions.* Harper and Row, New York.

Schmitt, D.P., Alcalay, L., Allik, J., et al. (2004). Patterns and universals of mate poaching across 53 nations: the effects of sex, culture, and personality on romantically attracting another person's partner. *Journal of Personality and Social Psychology*, **86**, 560–84.

Schmitt, D.P. and Buss, D.M. (2001). Human mate poaching: tactics and temptations for infiltrating existing mateships. *Journal of Personality and Social Psychology*, **80**, 894–917.

Schultz, W. (2000). Multiple reward signals in the brain. *Nature Reviews. Neuroscience*, **1**, 199–207.

Serretti, A., Mandelli, L., Lorenzi, C., et al. (2006). Temperament and character in mood disorders: influence of DRD4, SERTPR, TPH and MAO-A polymorphisms. *Neuropsychobiology*, **53**, 9–16.

Shaikh, T. and Suresh, K. (1994). Attitudinal similarity and affiliation needs as determinants of interpersonal attraction. *Journal of Social Psychology*, **134**, 257–9.

Short, R.V. (1977). Sexual selection and descent of man. In: J.H. Calaby, and C. Tyndale-Biscoe (eds), *Reproduction and evolution*, pp. 3–19. Australian Academy of Science, Canberra.

Sherwin, B.B. (1994). Sex hormones and psychological functioning in postmenopausal women. *Experimental Gerontology*, **29**, 423–30.

Skuse, D.H., James, R.S., Bishop, D.V.M., et al. (1997). Evidence from Turner's syndrome of an imprinted X-linked locus affecting cognitive function. *Nature*, **387**, 705–8.

Tafoya, M.A. and Spitzberg, B.H. (2007). The dark side of infidelity: its nature, prevalence, and communicative functions. In: B.H. Spitzberg, and W.R. Cupach (eds), *The dark side of interpersonal communication* (2nd Ed.), pp. 201–42. Lawrence Erlbaum Associates, Mahwah, NJ.

Taylor, S.E., Klein, L.C., Lewis, B.P., Gruenewald, T.L., Gurung, R.A.R. and Updegraff, J.A. (2000). Biobehavioral responses to stress in females: tend-and-befriend, not fight–or–flight. *Psychological Review*, **107**, 441–29.

Tennov, D. (1979). *Love and limerence: the experience of being in love.* Stein and Day, New York.

Thompson, A.P. (1983). Extramarital sex: A review of the research literature. *Journal of Sex Research*, **19**, 1–22.

Tsapelas, I., Fisher, H.E., and Aron, A. (2010). Infidelity: who, when, why. In: W.R. Cupach, and B.H. Spitzberg (eds), *The dark side of close relationships II*, pp. 175–96. Routledge, New York.

Van den Berghe, P.L. (1979). *Human family systems: an evolutionary view*. Greenwood Press, Westport, CT.

Van Goozen, S., Wiegant, V.M., Endert, E., Helmond F.A., and Van de Poll, N.E. (1997). Psychoendocrinological assessment of the menstrual cycle: the relationship between hormones, sexuality, and mood. *Archives of Sexual Behavior*, **26**, 359–82.

Wacker, J., Chavanon, M.L. and Stemmler, G. (2006). Investigating the dopaminergic basis of extraversion in humans: a multilevel approach. *Journal of Personality and Social Psychology,* **91**, 171–87.

Walum, H., Westberg, L., Henningsson, S., et al. (2008). Genetic variation in the vasopressin receptor 1a gene (AVPR1A) associates with pair-bonding behavior in humans. *Proceedings of the National Academy of Sciences of the USA*, **10537**, 14153–56.

Wang, Z., Toloczko, D., Young, L.J., Moody, K., Newman, J.D., and Insel, T.R. (1997). Vasopressin in the forebrain of common marmosets (*Calithrix jacchus*): studies with *in situ* hybridization, immunocytochemistry and receptor autoradiography. *Brain Research*, **768**, 147–56.

Wang, Z., Yu, G., Cascio, C., Liu, Y., Gingrich, B., and Insel, T.R. (1999). Dopamine D2 receptor-mediated regulation of partner preferences in female prairie voles (*Microtus ochrogaster*): a mechanism for pair bonding? *Behavioral Neuroscience*, **113**, 602–611.

Wedekind, C., Seebeck, T., Bettens, F., and Paepke, A.J. (1995). MHC-dependent mate preferences in humans. *Proceedings of the Royal Society B*, **260**, 245–49.

Westneat, D.F., Sherman, P.W., and Morton, M.L. (1990). The ecology and evolution of extra-pair copulations in birds. In: D.M. Power (ed.), *Current ornithology* (Vol. 7), pp. 331–70. Plenum Press, New York.

Wittenberger, J.F. and Tilson, R.L. (1980). The evolution of monogamy: hypotheses and evidence. *Annual Review of Ecology and Systematics*, **11**, 197–232.

Young, L.J. (1999). Oxytocin and vasopressin receptors and species-typical social behaviors. *Hormones and Behavior*, **36**, 212–21.

Young, L.J., Nilsen, R., Waymire, K.G., MacGregor, G.R., and Insel, T.R. (1999). Increased affiliative response to vasopressin in mice expressing the V1a receptor from a monogamous vole. *Nature*, **400**, 766–8.

Young, L.J., Winslow, J.T., Nilsen, R., and Insel, T.R. (1997). Species differences in V1a receptor gene expression in monogamous and nonmonogamous voles: behavioral consequences. *Behavioral Neuroscience*, **111**, 599–605.

Zak, P.J., Stanton, A.A., and Ahmadi, S. (2007). Oxytocin increases generosity in humans. *PLoS One*, **2**, 54–71.

Zuckerman, M. (2005). *Psychobiology of personality* (2nd edition). Cambridge University Press, New York.

Section 3

Society

Chapter 8

The evolutionary psychology of mass politics

Michael Bang Petersen

Introduction

One of the extraordinary developments in human evolution is the advent of modern mass society. In terms of social complexity, the societies of today are unrivalled. Each member has his own interests and needs and, at the same time, his fate is linked to that of millions of other members he does not know and will never meet—each with their own interests. In this intricate social web, politics has a central role. It is the primary means through which solutions to collective problems are devised, implemented and sustained. In that process, a key element is the distribution of the costs and benefits entailed by each solution—that is, the determination of whose interests are taken care of and whose interests are disregarded (Easton 1953).

Given this, it is not surprising that politics is high on the media agenda. Whenever an individual turns on the television, picks up a newspaper, or browses the Internet, it is likely that he or she will be confronted with the developments in national or international politics. In this chapter, I will investigate how lay individuals process these developments and form opinions on them using evolved decision-making mechanisms. As will be seen, ancestral life in small hunter/gatherer groups has selected for a suite of mechanisms that are engaged by mass politics, but at the same time these mechanisms evolved to function within a radically different context. By implication, whenever a mass political problem engages the evolved mind, it seems to be psychologically reduced to a problem of small-scale interaction. This is the key message of the chapter.

The political animal

As group-living animals, a range of social problems such as sharing decisions, formation of collective action, punishment of free-riders, management of intergroup relations, and hierarchy formation would directly have impinged on the survival and reproduction of our ancestors (Buss 2005). The evidence supporting this assertion is overwhelming. Anthropologists have, for example, carefully compiled a list of human universals (i.e. traits that are present in all known human cultures). This list features inherently collective traits such collective identities, conflict, conflict mediation, cooperation, ethnocentrism, government, group living, law, leaders, property, sanctions for crime, and trade (Pinker 2002). Furthermore, the fossil record provides evidence that many of these activities are not only universal but have deep evolutionary roots. Hence, archaeological findings document that our ancestors have cared for the crippled and injured at least since 1.77 million years ago (Hublin 2009), have had elements of social organization at least since 750,000 years ago (Alperson-Afil et al. 2009), have hunted cooperatively and shared meat within groups at least since 400,000 years ago (Stiner et al. 2009), and have used weapons to engage in conspecific aggression since at least the Middle Palaeolithic (Walker 2001). Comparisons with other primate species push these activities even further back into the evolutionary past. Organized between-group aggression has, for example, frequently been observed among non-human primates, most notably our closest relative, the chimpanzee. If this last observation is taken as evidence that already the last common ancestor between humans and chimpanzees engaged in warfare, this would imply that war has been with the human lineage for between 5 and 6 million years (Wrangham and Peterson 1996). Similarly, a number of non-human primates display sophisticated capacities for engaging in cooperative endeavours, sanctioning exploitive behaviour, and forming hierarchies (de Waal 1996) and, hence, these and related behaviours could have been present in the last common ancestor shared by these species and modern humans.

To survive and reproduce over evolutionary history, our ancestors needed to solve the problems posed by social life. Importantly, from a problem-solving perspective, social life is not a monolith but is rather constituted by a multitude of separate problem domains, each posing its own sets of challenges and requiring its own set of solutions. Choosing a competent leader requires, for example, the solution of different problems compared with choosing a valuable social partner from among equals. Similarly, inflicting the optimal amount of punishment on an exploitive individual requires the solution of different problems than choosing whether to punish at all (Petersen et al. 2010). A key insight in the cognitive sciences is that such distinct domains of problems are optimally solved by distinct mental procedures that each take the particularities of the specific problem into account (Tooby and Cosmides 1992). Given that processes of natural selection drive biological design towards greater optimization, it is very likely that the adaptive problems inherent in ancestral social life have selected for a multitude of domain-specific decisions rules, each tailored to a particular class of problems. In line with this, research in evolutionary psychology has uncovered the existence of domain-specific mental programmes specifically geared to solve problems relating to the detection of cheaters in situations of social exchange (Cosmides and Tooby 2005), tracking alliances (Kurzban et al. 2001), estimating others' fighting ability (Sell et al. 2008), and many other issues (for an overview of research, see Buss 2005).

These insights from evolutionary psychology are important for our understanding of political behaviour because ancestral problems related to social living carry structural similarities to modern political problems such as welfare, tax payments, criminal sanctions, immigration, warfare, race relations, and redistribution (Alford and Hibbing 2004; Schreiber 2006; Petersen 2009). Both sets of problems are, essentially, about distributions of costs and benefits within and between groups. By implication it is plausible that, during normal development, the human mind builds a toolbox of mental programs directly applicable to the issues of mass politics. This toolbox should facilitate the formation of political attitudes and the execution of political decisions. In this sense, research in evolutionary psychology seems to be aligned with Aristotle's famous dictum: humans are political animals.

To the evolved mind, all politics is local

Our social decision-making apparatus was designed by natural selection to produce decisions that would constitute adaptive 'best bets' in the particular environment in which it evolved (Tooby and Cosmides 1992). In this respect, it is notable and important that modern politics is played out in large-scale nation states comprising millions of individuals, because large-scale societies are extremely recent evolutionary phenomena. States first emerged in the world around 5000 before present (BP) and, while state technology rapidly diffused to some parts of the world, states in other parts of the world were only formed within the last centuries (Petersen and Skaaning 2010). For millions of years prior to that, human evolved as hunters and gatherers in small-scale groups with between 25 and 200 individuals (Kelly 1995). Most parts of our species-typical social decision-making apparatus are, in other words, designed to be operative within the causal structure of a small-scale social environment rather than a large-scale mass society. This implies, first, that the input this apparatus is designed to extract and process from decision-making contexts would be cues causally relevant for adaptive choice in small-scale interaction and, second, that the cognitive and motivational output this apparatus produces is that which would solve a given problem in a small-scale context.

We have no reason to believe that modern individuals should be particularly aware of these built-in assumptions and, hence, they may not be able to correct for them. Rather, prior research in other fields suggests that evolved decision-rules operate in an automated fashion. This has important consequences. Price (2008), for example, argues that because of their automated nature, evolved decision-rules are switched on by any cues that mimic ancestrally recurrent cues even when these are present in a novel situation and, upon triggering, cause individuals to misapprehend the novel situation as a related evolutionarily recurrent one (see also Hagen and Hammerstein 2006). For example, in the domain of popular culture, one obvious example is pornography which is psychologically represented as if mating opportunities were present, thereby triggering sexual arousal (e.g. Saad and Gill 2000). Similarly, research shows that people have difficulties in distinguishing between their real friends and people they see on television in the sense that their satisfaction with their friendships is paradoxically influenced by both (Kanazawa 2002). Other research show that misapprehension even occurs in direct interaction with others. Seemingly, modern individuals process anonymous one-shot interactions of mass society as if they were of the iterated kind to which we have, most likely, adapted to (e.g. Haley and Fessler 2005; Hagen and Hammerstein 2006; Price 2008). For example, decisions about whether to cooperate in fully anonymous one-shot economic games are influenced by subtle and rationally irrelevant cues indicating surveillance, such as the presence of stylized eyes (e.g. Haley and Fessler 2005; see also Chapter 10, this volume).

These arguments boils down to a single assertion with wide-ranging consequences: to the extent that individuals' attitudes on modern political issues are formed using evolved decision-making mechanisms, these attitudes will implicitly assume that mass politics are in fact played out in a local small-scale setting. In other words, to paraphrase an often heard expression about politics: to the evolved mind, all politics is local. The information that people find intuitively relevant when producing political choices will be information of relevance in small-scale social environments; and the political solutions that people will find intuitively correct will be solutions that work within such environments. This effect, it must be stressed, does not arise because modern individuals are not consciously aware that the problems of state-centred politics are different from the problems they encounter in local settings with, for example, their wife, friends, neighbours, and colleagues. Rather, it is because the automatic operations of evolved decision-rules swamp the opinion formation process of individuals and make the differences seem irrelevant and uninteresting.

Misapprehension and specific political issues

From this perspective of misapprehension, I will review the adaptive problems associated with a range of evolutionarily recurrent social situations, the computational mechanisms these problems have selected for, and how these mechanisms structure modern political attitudes on a number of issues. This discussion is not intended to be an exhaustive review of studies tackling mass politics from an evolutionary perspective, but an issue-by-issue illustration of how citizens think about mass politics based on factors that would be ancestrally relevant but seem irrelevant in a modern context.

Ancestral help-giving and the politics of social welfare

The current consensus suggests that sharing evolved as a strategy to help our ancestors buffer variance in food supply (Kaplan and Gurven 2005). At the same time, however, such help-giving is an evolutionarily risky strategy. As is well-known from the free-rider problem, help-giving is only adaptive to the extent it is reciprocal (i.e. if those who receive also give: Trivers 1971; Cosmides and Tooby 2005). The challenge, in this respect, is that adaptive help-giving requires one to avoid *cheaters*: individuals that strategically receive more than they give. In line with this, detailed experiments conducted over 30 years have demonstrated that individuals have specialized abilities designed to detect cheating in contexts of social exchange (Cosmides and Tooby 2005). In discriminating between cheaters and non-cheaters, our minds have seemingly been designed to pay particular attention to cues of effort (i.e. the displayed motivation to accrue resources: Cosmides and Tooby 2005). On the one hand, we decline to share with needy individuals whose needs reflect a lack of effort but, on the other, we feel morally obliged to share with individuals whose needs reflect bad luck. As demonstrated by a number of studies of help-giving in everyday situations (e.g. whether or not to lend exam notes to a fellow student), these effort cues regulate behavioural response particularly through two emotional systems: anger and compassion (for a review, see Weiner 1995). Lack of effort in alleviating one's own need triggers anger; effort triggers compassion towards the needy. (Petersen in press; Petersen et al. in press)

From a political science perspective, the important observation is that provisioning of benefits to the poor through redistribution of income and social welfare programmes mimics in many ways a situation of help-giving. Hence, the misapprehension hypothesis implies that when social welfare is framed in terms of effort cues, evolved mechanisms designed for ancestral help-giving should swamp the opinion formation processes of modern individuals and cause them to think about social welfare in terms of small-scale help-giving (Petersen in press; Petersen et al. in press).

In line with this, studies have found that one of the strongest predictors of social expenditure across countries are national differences in beliefs about whether recipients of social welfare are lazy or not (e.g. Fong et al. 2006). Another study has demonstrated that these beliefs shape opinions through specific emotional reactions (Petersen et al. in press). Consistent with the above-mentioned studies of emotions in small-scale help-giving, anger is triggered towards lazy welfare recipients, while compassion is triggered towards unlucky ones. In line with the misapprehension perspective, this suggests that it is indeed the same psychological mechanisms that are used to form judgments in small- and large-scale situations.

Direct evidence for the role of misapprehension has recently been produced by using a psychological method designed to unobtrusively measure how people categorize information (Petersen in press). The protocol is based on the fact that subjects more readily confuse individuals they categorize together. Subjects are presented with a number of individuals and then asked to recall information about these individuals. By analysing the patterns of recall errors (who is confused with whom), the researcher is able to measure how subjects spontaneously process the targets in the presentation phase. In the experiment focusing on social welfare and effort cues, subjects were exposed to one of two conditions. In the first, subjects were presented with a number of individuals receiving social welfare and a number receiving everyday types of help from a friend (a small-scale interaction). In the other condition, cues to the effort of the specific recipient in alleviating his own need were added to the descriptions. The analyses demonstrated that, in the absence of effort cues, citizens think of social welfare provisioning as very different from everyday help-giving between friends. Small-scale situations were confused with other small-scale situations but not those involving state provisioning of help. When directly provided with effort cues, however, citizens spontaneously picked up on these cues and came to process situations of welfare provisioning and everyday situations of help-giving using the same categories. Cheaters were confused with cheaters, and non-cheaters were confused with non-cheaters. In fact, as predicted from the misapprehension hypothesis, citizens stopped thinking of situations involving state-sponsored social welfare and situations involving everyday help-giving as different. For example, subjects failed to remember whether a lazy individual received help from the state or whether he received help from a friend. The subjects, in other words, implicitly treated the two situations as identical, despite their critical differences in terms of scale.

These sets of studies provide strong support for the idea that social welfare issues are misapprehended as small-scale problems. More specifically, misapprehension is activated in the domain of social welfare when evolved mechanisms are triggered by the availability of evolutionarily relevant cues about effort. Under these circumstances, individuals literally stop thinking about mass politics and small-scale interaction as different. Modern individuals, to a significant extent, judge anonymous social welfare recipients as if they were directly interacting with them.

Ancestral resource conflicts and the politics of redistribution

Forming opinions about whether to help the poor does not just entail judging whether the needy are lazy or unlucky, and then giving accordingly. Redistribution—and politics in general—also entails a significant element of conflict. Giving more to one group of people implies that another group gets less, or even has to relinquish some of the resources they already have. Redistribution of wealth is an example in point. Redistribution benefits individuals with lower socio-economic status (SES), while decreasing redistribution benefits individuals with higher SES. The latter will have their resources taken, and the former will be given those resources.

Importantly, such resource conflicts have been evolutionarily recurrent (Sell et al. 2009) and we should expect our evolved bargaining psychology to be designed to manage such conflicts within the context of ancestral social life. Here, one important factor for the outcome of conflicts would have been the contestants' fighting ability in the form of their physical strength (e.g. Maynard Smith and Parker 1976; Sell et al. 2009). Stronger males would have been more successful in both claiming resources from others (when they did not have resources) and defending their resources from others (when they did). For females, however, strength would have been much less important. Men are, on average, 90% stronger than females and women would rarely have found themselves decisively above average in a social group (Lassek and Gaulin 2009). From these selection pressures, studies have proposed and demonstrated that the human mind contains mental mechanisms designed to assess physical strength in self and others, and, in the case of males, use these assessments to regulate their disposition to assert their interests (for evidence, see Sell et al. 2008, 2009, 2010).

To the extent that this evolved bargaining psychology influences modern political attitudes on issues of redistribution, the physical strength of males should regulate the degree to which they assert their SES-dependent interests in relation to redistribution. The stronger males are, the more they should feel comfortable with asserting their economic interests. Specifically, Petersen et al. (in prep. - a) predicted that in males low in SES, physical strength should correlate positively with support for redistribution. For males high in SES, physical strength should correlate negatively with support for redistribution. To test this theory cross-culturally, data on SES, redistribution attitudes and upper-body strength was collected in three countries: the US, Argentina, and Denmark. As predicted, the analyses showed that physical strength regulated the degree to which males in all three countries asserted their economic self-interest. Among those with low SES, strong males were significantly more likely to support redistribution than weak males. For those with high SES, we found the opposite pattern: strong males were significantly more likely to *oppose* redistribution than weak males. For females, strength had no effect on their support for redistribution.

With this effect of physical strength, yet another piece of evidence supports the basic argument of this chapter: individuals think about mass politics in terms of ancestral small-scale interaction—whether or not this is rational from a modern-day perspective. Hence, males react as if disputes over national policies were a matter of direct physical confrontation among small numbers of individuals, rather than abstract electoral dynamics among millions.

Ancestral counter-exploitation and the politics of criminal justice

At their core, adaptive problems of whether to transfer resources to others concern the structure and depth of within-group cooperation. Another set of adaptive problems relates to how to react when individuals seek to circumvent cooperative life altogether. As in other species, the social world of our ancestors contained individuals who were poised to impose costs on others if such acts were self-beneficial (e.g. Daly and Wilson 1988). Theft, violence, murder, and rape are all examples of such exploitative acts. Several researchers have argued that exploitation as an evolutionarily recurrent phenomenon has selected for a range of computational mechanisms designed to effectively identify and counter exploiters (Robinson et al. 2007; Petersen et al. 2010).

Punishment is one obvious counter-strategy against exploitation: the infliction of costs on another. Accounts of revenge as a motivation are cross-culturally and historically ubiquitous (e.g. Jacoby 1983; Daly and Wilson 1988) and a range of non-human animals display punitive reactions against exploiters or antagonists (Clutton-Brock and Parker 1995; de Waal 1996), suggesting that this form of interaction has existed within our evolutionary lineage for tens of millions of years. Punishment, however, should be allocated with care. Punishment is not only costly, but can also trigger retaliation. At the computational level, our ancestors have therefore needed to solve the quantitative problem of how much punishment to inflict (Petersen et al. 2010).

As demonstrated by a large criminological literature, this quantitative decision about how much to react is driven by perceptions of the seriousness of exploiters' actions (cf. Darley and Pittman 2003). Increasing seriousness is reflected in increasingly severe sanctions. This perception of the seriousness of the act is predominantly formed by integrating information about the amount of harm inflicted through the act (Stylianou 2003). From an evolutionary perspective of small-scale groups, the role of a crime's seriousness makes sense. Gauging the harm done provides an estimate of how little the perpetrator values the victim and how much this valuation needs to be up-regulated before future exploitation becomes unlikely (Petersen et al. 2010). Consistent with this evolutionary interpretation, criminological studies have demonstrated a high level of cross-cultural consistency in how harmful people perceive different crimes to be (cf. Robinson et al. 2007). Furthermore, studies have demonstrated that evolutionarily novel factors that would be rational to consider from a modern crime-prevention perspective—such as how easy it is for the police to detect a crime or the public character of the crime—do not seem to influence how severely individuals want to punish different crimes (Carlsmith et al. 2002).

Adaptations for countering exploitation in ancestral groups shape modern criminal justice attitudes even more profoundly. Hence, it is important to notice that in an evolutionary perspective, social relations exist because they create reproductive benefits. Because of possible future benefits, it will, under some circumstances, be adaptive for the exploited party to avoid a breakdown of a cooperative relationship—even in the face of occurrences of exploitation. Given this selection pressure, it is likely that reparation is an evolved strategy (de Waal 1996; Petersen et al. 2010). In deciding how to react against exploiters, our ancestors have, in other words, also confronted the problem of whether to punish or whether to react with reconciliation instead. Evolutionary analysis suggests that the key decision element in this regard is the decision-maker's estimated net lifetime future value of maintaining interactions with the exploiter—in other words, the exploiter's *association value* (Petersen et al. 2010). Cues of high association values include lack of intentionality, lack of prior exploitive behaviour, past cooperative behaviour, remorse, resourcefulness, sexual attractiveness, and kinship. Low association values should trigger punitive sentiments, while high association values should trigger conciliatory motives.

In a recent study, it was demonstrated how these mechanisms for gauging the association value of exploiters shape attitudes to modern criminal justice (Petersen et al. in review). Subjects were presented with a series of highly different types of crime. As predicted, for all crimes, subjects' preferences for rehabilitation over punishment were regulated by their perceptions of the criminals' association value. Perceptions of the seriousness of the crime, as judged by the subjects, regulated consecutive decisions about how much to punish or how much effort to put into reparative endeavour. Furthermore, in line with the notion that different adaptive problems are solved by different mechanisms, the study demonstrated that perceptions of association value are computed based on cues distinct from those used to compute crime seriousness. Hence, cues to the offender's past criminal history, as well his status as an in-group or out-group member, had large effects on the computation of a criminal's association value, but little or no effect on computations of the crime's seriousness.

Today, there is little chance that an individual victim's personal welfare will be affected by whether the state punishes or rehabilitates a specific criminal, or even fails to react at all. Nevertheless, our intuitions seem to reflect a strategic social calculus that operates as if crimes occurred in an intimate social setting where we ourselves required punitive protection or could harvest the social value of a repaired relationship. In this way, lay individuals intuitively fit the sanction to the criminal, not just the crime; in contrast to modern criminal justice institutions, individuals allow for and accept unequal treatments for equal crimes, depending on how valuable the specific criminal is perceived to be.

Ancestral hierarchies and political leaders

Current evidence suggests that ancestral social groups were relatively egalitarian (Boehm 2000). Still, some sort of hierarchies would most likely have existed and anthropological descriptions of both hunter-gatherer societies and non-human primates contain numerous examples of individuals and groups with elevated positions (e.g. de Waal 2000; Sidanius and Pratto 2001). Given these observations, we should expect our evolved minds to be designed to think in terms of hierarchies and to evaluate leaders. As argued in this section, the underlying mechanisms seem to structure how modern citizens cast their votes.

Ancestrally, the negotiation of hierarchies would often have been a collective endeavour. Studies of chimpanzees, for example, show how ambitious individuals form coalitions with others to move themselves into higher positions (de Waal 2000). Among humans, anthropologists have similarly argued that coalitions from below are formed to overthrow exploitive leaders (Boehm 2000). In these collective negotiations, for each individual, it would be important to provide support for the specific person that would convey most benefits to him or her. Leaders would, for example, have fulfilled a number of roles beneficial to subordinates in general, such as mitigating conflicts between group members (de Waal 1996) and organizing collective action (Tooby et al. 2006). For every person in a group, it would be preferable to have a leader who was competent at fulfilling these roles. At the same time, leaders would be in a position to provide benefits for themselves and their allies at the expense of others (Hibbing and Alford 2005).

Studies in political science have demonstrated that competence and integrity ratings of modern politicians are indeed highly important for their electoral success (e.g. McCurley and Mondak 1995). What is important from the misapprehension perspective is that these ratings are seemingly formed based on automatic processing of a range of subtle cues. Hence, studies have documented that subjects are remarkably good at predicting the actual electoral success of political candidates simply by looking at their faces (Little et al. 2007) for as little as 100ms (Willis and Todorov 2006).

More detailed studies provide support for the evolutionary underpinnings of such effects. In ancestral small-scale groups, it would have been adaptive to move kin members into positions of power. One recurrent cue of kinship would be facial similarity. To test whether facial similarity influenced vote choice, Bailenson et al. (2009) morphed the faces of subjects into the face of political candidates and had the subjects rate the faces. As expected, as the blend of the two faces contained more of the subjects' facial features, the subjects' provided more support for the candidate.

Another important study investigated the effects of facial masculinity of political candidates on vote choice. From an evolutionary perspective, facial masculinity is important because it seemingly tracks physical strength and behavioural aggressiveness (Sell et al. 2008, 2009). Ancestrally, these characteristics would be important in a leader in times of external conflict, as such a leader would be better prepared to organize and take part in aggressive collective behaviour. In times of peace, however, such traits would have been less preferable as they could indicate exploitative tendencies. By implication, it is possible that our competence-detection mechanisms include context-sensitive components that gauge the masculinity of potential leaders and place positive or negative weight on this depending on the situation facing the group. In a study, Little et al. (2007) demonstrated that such an inference system reveals itself in modern voting decisions. In one condition, subjects were asked to choose between a masculine and a feminine computer-generated male candidate while imagining that the country was engaged in war. In the other condition, subjects were asked to imagine that the country was at peace. Consistent with a misapprehension hypothesis, under conditions of war, subjects (of both sexes) preferred the masculine candidate. Under conditions of peace, subjects on average preferred the feminine candidate. While it seems irrational to think that the strength of the president would influence how effective the modern military would be in times of war, this study demonstrated that voters react as if it were so.

Ancestral coalitional conflict and political groups

The anthropological literature suggests that the overarching structure of ancestral social life would have been constituted by relatively stable groups within which smaller hunter-gatherer bands would have undergone continuous fission and fusion (Kelly 1995). The near universal tendencies of non-human primates to live in social groups (with orang-utans as a notable exception) suggest that this way of life goes back millions of years and, given the importance of reciprocity in the evolution of cooperation, the evolutionary necessity of relatively stable groups is apparent. In the four previous subsections, the focus has been on adaptive problems that relate to distribution of costs and benefits within such groups. Yet, a large number of adaptive problems relate to the interaction *between* groups. As we will see, the selection for mental mechanisms, in response to these problems, shape how modern individuals reason about political groups and their interactions at all levels including racial groups, nation-states, and political parties.

Race constitutes a politically important line of division, not least in the US. According to Kurzban et al. (2001), one important reason for this is that the cues surrounding modern racial groups mimic evolutionarily recurrent coalitional cues. Most importantly, people experience racially segregated societies where patterns of cooperation run along racial lines. Given this, Kurzban et al. (2001) predicted that racial factions are tagged and treated by the mind as coalitions. While a large strand of research shows that racial cues are massively attended to by subjects, the coalitional account of Kurzban et al. implied that subjects could easily be prompted to disregard racial cues when alternative and stronger cues to coalitional affiliation was available. To test this theory, they made use of the memory confusion protocol described earlier in this section. Specifically, they presented subjects with a social situation in which two groups were arguing. Each of these groups contained black and white persons and, hence, race was not predictive of the group membership of any individual. As predicted, in the recall phase, subjects stopped confusing people along racial lines and, instead, confused them along the appropriate group lines. People, in other words, do not have dedicated systems for representing race, but given the available cues, race is tracked by our coalitional psychology.

Ancestrally, systems of reciprocity and social exchange would be dense within-coalitions, and sparse between-coalitions. Given that racial groups are tracked by our coalitional psychology, we should, therefore, expect less willingness to engage in country-wide cooperation in racially diverse countries. In line with this, Alesina and Glaeser (2005) have shown that racial and ethnic diversity has a huge effect on the level of redistribution in a country. Similarly, within the US, states with less diversity redistribute to a greater extent than more diverse states. In support of an evolutionary interpretation of these effects of racial and ethnic diversity, experiments by Habyarimana et al. (2007) demonstrate that a member of an ethnic group only treats other co-members more favourably under conditions of non-anonymity (i.e. when the co-member knows his or her identity). As argued by Yamagishi and Mifune (2008), this suggests that in-group-favouritism in cooperation is driven by systems designed for reciprocal exchange as such as system would exactly downregulate favouritism when there was no possibility for the co-member to return the favour (as in the case of anonymity).

Groups do not only delimit patterns of cooperation. Over the course of human evolution, groups would have competed against each other with regard to, among other things, tangible resources, territory, mates, and social status (Tooby et al. 2006). In this regard, sex differences become important. Among chimpanzees, males but not females engage in deadly intergroup conflict, in part to ensure access to reproductive females. Among humans, the anthropological record similarly shows that small-scale warfare is largely carried out by men (Wrangham and Peterson 1996; see also Chapter 14, this volume). The evolutionary background for this sex difference relates to differences in parental investment. In humans, like other mammals, females invest significantly more in offspring than males, paying the costs of both pregnancy and lactation (see Chapter 9, this volume). Females thus become a limiting resource for male reproduction. Consequently, males invest heavily in competition (both direct and indirect), adopting a suite of risky strategies designed to increase their number of matings (Daly and Wilson 1988).

Given this, we should expect that males exhibiting features that correlate with large gains (in terms of increased reproduction) from group-conflict would be those most supportive of engaging in conflict. For example, young males who have just reached sexual maturity would be more in need of reproductive opportunities compared to older males in established relationships. To test how this selection pressure structures modern-day decisions about war, Mesquida and Wiener (1999) investigated armed conflicts occurring after 1980. They found that countries with many young males are particularly like to engage in deadly warfare. Importantly, this relationship is robust to the inclusion of a large number of control variables relating to, for example, socioeconomic development and democratization.

While the need for reproductive opportunities would be present among most young males, some males would be more successful in gaining these through warfare. In particular,

Sell et al. (2009) argue that it would be more adaptive for physically strong males to engage in group aggression. Ancestral hunter/gatherer bands would most likely not contain more than a dozen adult males and, hence, the fighting ability of each and every individual male would have influenced the probability that the group as a whole prevailed. In line with this, physical strength shapes modern attitudes about war: stronger males are more supportive of using war as a means of solving international disputes. As predicted, this relationship does not hold for females who engage less in physical aggression.

Our evolved coalitional machinery should be activated in all contexts containing cues that ancestrally would have disclosed the existence of group-based cooperation and conflict. In modern politics, one such context is conflict between political parties. Political parties are characterized by high levels of within-party cooperation and between-party conflict. By implication then, some of the above mechanisms should also take effect in modern party politics. In one study, affiliations with a political party were found to be psychologically represented as coalitional affiliations—at least for males. Stanton et al. (2009) followed party supporters on the night of the 2007 US presidential elections, measuring their testosterone levels. Studies of humans and other animals have demonstrated that male testosterone levels fall in response to losing status, presumably to motivate a de-escalation of conflict. The same hormonal pattern was observed among supporters of the Republican party immediately after Barack Obama, the Democratic candidate, was announced as the winner. Having voted for the losing party felt just like losing a direct status competition.

Another study focused on the notion that, if political affiliation is conceptualized as a coalitional identity, political attitudes can come to serve as coalitional badges (Petersen et al. in prep.-b). That is, individuals can come to have certain attitudes to signal that they are loyal partisans. If parties change position on an issue, voters usually do not shift allegiance to another party but rather change their position too (e.g. Goren 2005; Zaller 1992). In support of an evolutionary interpretation of these effects, Petersen et al. (in prep.-b) identified sex differences in the way citizens' attitudes are used as coalitional badges. Subjects were primed with demeaning statements about a party attributed to a spokesman from another party. After priming, they were asked about their opinion on a policy proposal that was either attributed to the attacked party or the attacking party (keeping the content of the proposal constant) and, hence, were provided with an opportunity for signalling coalitional loyalty. As expected, males reacted in a highly coalitional manner to the primes. When their preferred party was attacked in the statement, they became more willing to support policy when attributed to their party, and less willing to support it when it was attributed to the other party. Females, however, reacted in a conciliatory fashion. When their preferred party was attacked by another party, they become *more* willing to accept the policy proposal by this other party, as if to appease the attacker. Female appeasement strategies have been observed among females in small-scale conflict situations (Salter et al. 2005) and make evolutionary sense in this context given the lack of female physical strength. In modern party politics, however, there seems to be less, if any, rational reason to back off in this manner.

The anonymity of mass politics and processing implications

The above examples illustrate situations where modern humans misapprehend political issues as ancestral adaptive problems. From this, one obvious expectation would be that people intuitively find politics interesting and engaging. After all, the argument implies that politics is implicitly represented as about matters of survival and reproduction. While there is certainly truth to this, it is, however, not the whole truth. The human political animal is, in fact, one with a paradoxically low interest in day-to-day affairs of modern politics. Since the first studies of public opinion and voting behaviour, political scientists have repeatedly documented how little citizens know, think, and care about politics (e.g. Converse 1964; Zaller 1992). Specifically, researchers have demonstrated that citizens do not know basic facts concerning their political system, such as the number of seats in the legislature or the identity of those holding specific office (Carpini and Keeter 1996). Furthermore, a number of studies have demonstrated that a large number of people find it difficult to decide on political issues. On issues such as abortion and criminal justice, the public are ambivalent and most see merits in the arguments of both sides (Craig and Martinez 2005). Finally, a large number of studies have revealed that citizens have difficulties in sticking to their principles. Even though individuals in general support freedom of speech for everybody, they are often quite willing to limit the freedoms of specific groups (McClosky and Brill 1983). Similarly, individuals who support harsh punishments for crimes can be quite lenient towards specific criminals (Roberts 1992).

Any attempt to apply evolutionary psychology to the study of modern political attitudes and behaviour needs to reconcile the conclusion from evolutionary biology that humans are political animals with these observations from political science. If humans were endowed with a natural sense of politics, would citizens not think of mass politics as an interesting and intuitive affair?

Processing without cues

One reason that this turns out not to be the case is that some modern issues bear little relationship to recurrent adaptive problems and, hence, do not activate natural intuitions. Technological advances in modern society have spawned a range of new issues (e.g. regulation of the macroeconomy). From the perspective of evolutionary psychology, such novel issues that do not fit evolved decision-mechanisms should indeed appear uninteresting and difficult to grasp. Yet, the technological advances of modern societies are insufficient to explain the lack of political attention among modern citizens. In addition, one needs—again—to focus on the contextual differences between modern mass politics and ancestral social interaction. Hence, the key to understanding how a political animal can be quite politically inattentive lies in the fact that our social decision-making mechanisms evolved in information-dense situations characterized by repeated face-to-face interactions. By implication, to disclose and solve adaptive problems of social life, our decision-making mechanisms are designed to rely on intimate social cues such as facial expressions, knowledge of past interactions, and so on.

In the context of mass politics, these myriad subtle cues will often be lacking. Given that modern politics is played out in large-scale nation states comprising millions of individuals, the majority of citizens do not know and will never meet each other, and most often, citizens are required to form opinions and impressions of anonymous strangers. Modern politics, in other words, is characterized by information scarcity and our decision-making apparatus is unable to extract the detailed information necessary for its execution.

This problem is attenuated because the vastness of modern political systems imposes a certain structure on modern political debates. While ancestral social interaction revolved around the specific other, modern politics is concerned with establishing general rules that apply across specific cases (Petersen 2009). This affects the way in which politicians discuss political solutions in public debates. Specifically, the problems confronting citizens in these debates are not whether a specific criminal should be punished, whether a specific welfare recipient is deserving, or whether a specific immigrant should be allowed entry. Rather, debates turn on how to treat all criminals, all welfare recipients, and all immigrants. In other words, political debates ask citizens to form *general* political opinions rather than opinions on specific cases and, thereby, make judgements about whole social categories at once.

This explains why many citizens find core political issues such as redistribution, criminal justice and immigration both interesting and important, while at the same time being highly ambivalent about adoption of specific policy. General debates about criminal justice will, for example, arouse counter-exploitation circuitry but at the same time the lack of cues will hinder the production of unambiguous motivational output. In order for this and other social decision-making circuitries to execute a decision in the context of mass politics, they need to rely much more heavily on internally-generated cues than in other contexts. It is, in other words, necessary for modern individuals themselves to provide the vivid social cues that would have been directly available in intimate social interaction. By recruiting more domain-general processes such as stereotyping and imagination, individuals need to piece together a vivid imagery from the cues available from, for example, news media (see Chapter 22, this volume), elite communications, personal experience, and lose individual genetic predispositions (Petersen 2009; on the role of genetics in political attitudes, see Alford et al. 2005). As the decision-maker cannot see the criminals, for example, he has to imagine them. In that sense, making political judgements is like judging the characters in a book.

There is some evidence for this proposed process. Studies have demonstrated that individuals scoring high on the personality trait Openness to Experience, which is linked to imaginative abilities, are more interested in, and participate more in politics (Mondak et al. 2010). Other evidence comes from investigations of what modulates the level of ambivalence individuals feel towards political issues. Traditional political science has argued that characteristics like education and political knowledge are of fundamental importance (e.g. Zaller 1992). It turns out, however, that the degree to which individuals' stereotypes fit the input conditions of relevant evolved inference systems is much more important. People who have a fixed belief that all criminals have low (or high) association value also have a remarkably unambivalent attitude about general criminal justice issues (Petersen 2009). Those with more mixed beliefs, in contrast, are highly ambivalent and struggle to choose between punishment and rehabilitation as the main goals of the criminal justice system. Hence, no matter whether opinion formation is driven by externally or internally generated cues, the key to understanding how these cues impinge on opinion formation is the degree of fit between the cues and the input conditions of our evolved decision-making systems.

Circumventing imagination

It is processing necessities, rather than anything else, that prompts the recruitment of higher-order reasoning processes in the formation of political attitudes. Relying on vivid, externally provided cues—if available—is not only swifter and easier but also, in a certain sense, unavoidable given that the inference mechanisms these cues would activate work in an automated fashion. Internally-generated imagery is, in other words, best thought of as a cognitive surrogate for direct cues. Because of this, the decision-making context facing modern individuals when forming political attitudes will, from time to time, circumvent imaginative processes. In contexts containing vivid cues, individuals will begin diverging from their general opinions. From some traditional political science perspectives, this would look like a lack of principle rooted in the mindlessness of citizens (Converse 1964). From an evolutionary perspective, however, it is a simple consequence of the fact that when the political decision-making context begins to resemble, either by chance or political design, a more ecologically-valid context, the facultative (cue-sensitive) nature of our reasoning system reveals itself.

One decision-making context that mimics ancestrally valid contexts is when citizens are invited to judge specific individuals: for example, when news media report on the specifics of a crime case. In this case, citizens are directly provided with a range of cues that should turn off the need for engaging in higher-order reasoning. Research demonstrates that this is in fact what happens. While citizens' beliefs about criminals in general shape their general opinions (see 'Processing without cues' section), these beliefs are simply not predictive of their opinions on specific cases when these cases contain ancestrally relevant cues (Petersen 2009). Hence, while individuals disagree in the abstract about issues relating to crime, punishment, and rehabilitation, they can be quite in agreement in a specific situation.

As further evidence of this argument, other research demonstrates that citizens' reliance on higher-order reasoning process is a continuous function of the degree to which the specific case contains clear cues with ancestral relevance (Petersen et al. 2011). The more these cues fit evolved inference mechanisms, the more citizens disregard their priors. A further study demonstrated that this effect is to a large extent driven by emotional reactions (Petersen et al. in press). Consistent with the arguments concerning adaptations for social exchange and the politics of social welfare (see 'Ancestral help-giving and the politics of social welfare' section), it has been demonstrated that conservatives' and liberals' opinions converged in the face of specific welfare recipients because both sides responded with equal amounts of anger and compassion to lazy and unlucky recipients, respectively.

Another related decision-making context reflects that politicians often deliberately exploit the way individuals' reason, to seek to gain support for their policies. A common trick in political communication is to link a general policy with specific individuals; for example, reminding people of a specific welfare recipient when discussing welfare policies. By framing policies in this way, politicians seemingly hope to shut off higher-reasoning processes, prompt citizens to process the same cues and, thereby, induce the same attitudes in voters across the political spectrum. Research demonstrates that such episodic frames (as they are known) are effective in shaping attitudes to both the specific individual used to frame the policy and the policy itself (Iyengar 1991). More detailed studies have compared the mediators of such episodic frames with the effects of thematic frames that only provide abstract information about issues (Aarøe 2011). As expected from an evolutionary perspective, the key to episodic framing seems to be their capacity to activate specific emotions in the audience.

Conclusions

Over the course of human evolutionary history, our species has confronted many social problems with consequences for survival and reproduction. As demonstrated by a vast number of studies in evolutionary psychology, these adaptive problems have selected for an elaborate social psychology. At the core of mass politics, the problems confronting modern individuals carry a range of similarities to these ancestral problems of social life. Ancestrally, as today, questions such as 'How do we divide our resources?', 'How should we treat those who break the rules?', 'Who are our enemies and how should we deal with them?', are in need of answers.

Our evolved decision-making mechanisms are designed to be triggered by ancestrally relevant cues and to motivate ancestrally relevant solutions. When analysing how these mechanisms shape the political attitudes and behaviours of modern individuals, the key is to carefully analyse the cues available in the context of a given political issue and their degree of fit with the input conditions of evolved mechanisms. If cues are available that, ancestrally, would have disclosed the existence of a specific adaptive problem, we should expect the related mechanisms to activate and cause citizens to process the political issues as if it were in fact that same adaptive problem. Given the small-scale nature of ancestral social environments, this effectively causes citizens to misapprehend mass political issues as small-scale social problems.

In determining cue availability, the analysis of mass politics is particularly challenging. In small-scale interaction, cues are directly present. In mass politics, in contrast, cues are most often either indirectly provided by news media and the political elite—and often so with a particular agenda—or internally generated by individuals' own higher-order reasoning processes. In order to understand which mechanisms are triggered in decision-making situations, one needs in other words to analyse the context of the situation closely. Whenever externally-provided and evolutionarily recurrent cues are provided, we should expect these to be given priority in terms of processing and, hence, to prompt individuals to calibrate their political attitudes and behaviours closely to these cues. Whenever cues are sparse, we should expect them to fall back on their own internally-generated images of the mass society.

References

Aarøe, L. (2011). Investigating frame strength: the case of episodic and thematic frames. *Political Communication*, 28, 207–26.

Alesina, A. and Glaeser, E.L. (2005). *Fighting poverty in the US and Europe*. Oxford University Press, Oxford.

Alford, J., Funk C., and Hibbing, J. (2005). Are political orientations genetically transmitted? *American Political Science Review*, **99**, 153–67.

Alford, J. and Hibbing, J. (2004). The origins of politics: an evolutionary theory of political behavior. *Perspectives on Politics*, **2**, 707–23.

Alperson-Afil, N., Sharon, G., Kislev, M., *et al.* (2009). Spatial organization of hominin activities at Gesher Benot Ya'aqov, Israel. *Science*, **326**, 1677–80.

Bailenson, J.N., Iyengar, S., Yee, N., and Collins, N.A. (2009). Facial similarity between voters and candidates causes influence. *Public Opinion Quarterly*, **72**, 935–61.

Boehm, C. (2000). Conflict and the evolution of social control. *Journal of Consciousness Studies*, **7**, 79–101.

Buss, D.M. (ed). (2005). *The handbook of evolutionary psychology*. John Wiley and Sons, Inc., Hoboken, NJ.

Carlsmith, K.M., Darley, J.M., and Robinson, P.H. (2002). Why do we punish? Deterrence and just deserts as motives for punishment. *Journal of Personality and Social Psychology*, **83**, 284–99.

Carpini M.D. and Keeter, S. (1996). *What Americans know about politics and why it matters*. Yale University Press, New Haven, CT.

Clutton-Brock, T.H. and Parker, G.A. (1995). Punishment in animal societies. *Nature*, **373**, 209–16.

Converse, P.E. (1964). The nature of belief systems in mass publics. In: D.E. Apter (ed.), *Ideology and discontent*, pp. 206–261. The Free Press, New York.

Cosmides, L. and Tooby, J. (2005). Neurocognitive adaptations designed for social exchange. In: D.M. Buss (ed.), *The handbook of evolutionary psychology*, pp. 584–627. John Wiley and Sons, Inc, Hoboken, NJ.

Craig, S.C. and Martinez, M.D. (eds). (2005). *Ambivalence and the structure of public opinion*. Palgrave Macmillan, New York.

Daly, M. and Wilson, M. (1988). *Homicide*. Aldine, Hawthorne, NY.

Darley, J.M. and Pittman, T.S. (2003). The psychology of compensatory and retributive justice. *Personality and Social Psychology Review*, **7**, 324–6.

de Waal, F. (1996). *Good natured*. Harvard University Press, Cambridge, MA.

de Waal, F. (2000). *Chimpanzee politics*. John Hopkins University Press, Baltimore, MD.

Easton, D. (1953). *The political system: an inquiry into the state of political science*, Alfred A. Knopf, New York.

Fong, C., Bowles, S., and Gintis, H. (2006). Strong reciprocity and the welfare state.
In: S. Kolm, and J. Mercier Ythier (eds), *Handbook on the economics of giving, reciprocity and altruism*, pp. 1439–64. Elsevier, Amsterdam.

Goren, P. (2005). Party identification and core political values. *American Journal of Political Science*, **49**, 881–96.

Habyarimana, J., Humphreys, M., Posner, D.N., and Weinstein, J. M. (2007). Why does ethnic diversity undermine public goods provision? *American Political Science Review*, **101**, 709–25.

Hagen, E.H. and Hammerstein, P. (2006). Game theory and human evolution: a critique of some recent interpretations of experimental games. *Theoretical Population Biology*, **69**, 339–48.

Haley, K.J. and Fessler, D. (2005). Nobody's watching? Subtle cues affect generosity in an anonymous dictator game. *Evolution and Human Behavior*, **26**, 245–56.

Hibbing, J. and Alford, J. (2004). Accepting authoritative decisions: humans as wary cooperators. *American Journal of Political Science*, **48**, 62–76.

Hublin, J-J. (2009). The prehistory of compassion. *Proceedings of the National Academy of Sciences of the USA*, **106**, 6429–30.

Iyengar, S. (1991). *Is Anyone Responsible?* The University of Chicago Press, Chicago, IL.

Jacoby, S. (1983). *Wild justice: the evolution of revenge.* Harper and Row, New York.

Kanazawa, S. (2002). Bowling with our imaginary friends. *Evolution and Human Behavior*, **23**, 167–71.

Kaplan, H. and Gurven, M. (2005). In: H. Gintis, S. Bowles, R. Boyd, and R. Fehr (eds), *Moral sentiments and material interests,* pp. 75–113. MIT Press, New York.

Kelly, R. (1995). *The foraging spectrum.* Smithsonian Institution Press, Washington, DC.

Kurzban, R., Tooby, J., and Cosmides, L. (2001). Can race be erased? Coalitional computation and social categorization. *Proceedings of the National Academy of Sciences of the USA*, **98**, 15387–92.

Lassek, W.D. and Gaulin, S. (2009). Costs and benefits of fat-free muscle mass in men: relationship to mating success, dietary requirements, and native immunity. *Evolution and Human Behavior*, **30**, 322–8.

Little, A.C., Burriss, R.P, Jones, B.C, and Roberts, S.C. (2007). Facial appearance affects voting decisions. *Evolution and Human Behavior*, **28**, 18–27.

Maynard Smith, J. and Parker, G.A. (1976). The logic of asymmetric contests. *Animal Behaviour*, **24**, 159–75.

Mesquida, C.G. and Wiener, N.I. (1999). Male age composition and severity of conflicts. *Politics and the Life Sciences*, **18**, 181–9.

McClosky, H. and Brill, A. (1983). *Dimensions of tolerance.* Russell Sage Foundation, New York.

McCurley, C. and Mondak, J.J. (1995). Inspected by #1184063113: The influence of incumbents' competence and integrity in U.S. House elections. *American Journal of Political Science*, **39**, 864–85.

Mondak, J.J., Hibbing, M.V., Canache, D., Seligson, M.A., and Anderson, M.R. (2010). Personality and civic engagement: an integrative framework for the study of trait effects on political behavior. *American Political Science Review*, **104**, 85–110.

Petersen, M.B. (2009). Public opinion and evolved heuristics: the role of category-based inference. *Journal of Cognition and Culture*, **9**, 367–89.

Petersen, M.B. (in press). Social welfare as small-scale help: on the evolved roots of the deservingness heuristic *American Journal of Political Science.*

Petersen, M.B. and Skaaning, S.E. (2010). Ultimate causes of state formation: the significance of biogeography, diffusion, and Neolithic revolutions. *Historical Social Research*, **35**, 200–26.

Petersen, M.B., Sell, A., Tooby, J., and Cosmides, L. (2010). Evolutionary psychology and criminal justice: a recalibrational theory of punishment and reconciliation. In: H. Høgh-Olesen (ed.), *Human morality and sociality: evolutionary and comparative perspectives*, pp. 72–131. Palgrave Macmillan, Hampshire.

Petersen, M.B., Slothuus, R., Stubager, R., and Togeby, L. (2011). Deservingness versus values in public opinion on welfare: the automaticity of the deservingness heuristic. *European Journal of Political Research*, **50**, 24–52.

Petersen, M.B., Sznycer, D., Sell, A., Cosmides, L. and Tooby, J. (in prep.-a). The ancestral logic of politics: upper body strength regulates the assertion of self-interest over income redistribution.

Petersen, M.B., Sznycer, D., Cosmides, L., and Tooby, J. (in press). Who deserves help? Evolutionary psychology, social emotions, and public opinion on welfare. *Political Psychology.*

Petersen, M.B., Delton, A., Robertson, T., Cosmides, L., and Tooby, J. (in prep.-b). *Politics in the evolved mind: Males and females reason differently about political parties.*

Petersen, M.B., Sell, A., Tooby, J. Cosmides, L. (in review). To punish or repair? Evolutionary psychology and lay intuitions about modern criminal justice.

Pinker, S. (2002). *The blank slate*. Penguin Books, London.

Price M.E. (2008). The resurrection of group selection as a theory of human cooperation. A review of Foundations of human sociality: economic experiments and ethnographic evidence from fifteen small-scale societies, edited by J. Henrich, R. Boyd, S. Bowles, C. Camerer, E. Fehr, and H. Gintis, & Moral sentiments and material interests: the foundations of cooperation in economic life, edited by H. Gintis, S. Bowles, R. Boyd and E. Fehr. *Social Justice Research*, **21**, 228–40.

Roberts, J.V. (1992). Public opinion, crime, and criminal justice. *Crime and Justice*, **16**, 99–180.

Robinson, P.H., Kurzban, R., and Jones, O. (2007). The origins of shared intuitions of justice. *Vanderbilt Law Review*, **60**, 1633–85.

Saad, G. and Gill, T. (2000). Applications of evolutionary psychology in marketing. *Psychology and Marketing*, **17**, 1005–34.

Salter, F., Grammer, K., and Rikowski, A. (2005). Sex differences in negotiating with powerful males. *Human Nature*, **16**, 306–21.

Schreiber, D. (2006). Political cognition as social cognition: are we all political sophisticates? In: W.R. Neuman, G.E. Marcus, A.N. Crigler, and M. Mackuen (eds), *The affect effect*, pp. 48–70. University of Chicago Press, Chicago, IL.

Sell, A., Bryant, G.A., Cosmides, L., et al. (2010). Adaptations in humans for assessing physical strength from the voice. *Proceedings of the Royal Society B*, **277**, 3509–18.

Sell, A., Cosmides, L., Tooby, J., Sznycer, D., von Rueden, C., and Gurven, M. (2008). Human adaptations for the visual assessment of strength and fighting ability from the body and face. *Proceedings of the Royal Society B*, **276**, 575–84.

Sell, A., Tooby, J., and Cosmides, L. (2009). Formidability and the logic of human anger. *Proceedings of the National Academy of Sciences of the USA*, **106**, 15073–8.

Sidanius, J. and Pratto, F. (2001). *Social dominance: an intergroup theory of social hierarchy and oppression*. Cambridge University Press, Cambridge.

Stanton, S.J., Beehner, J.C., Saini, E.K., Kuhn, C.M., and LaBar, K.S. (2009). Dominance, politics, and physiology: voters' testosterone changes on the night of the 2008 United States presidential election. *PloS One*, **4**, 1–6.

Stiner, M., Barkai, R., and Gopher, A. (2009). Cooperative hunting and meat sharing 400–200 kya at Qesem Cave, Israel. *Proceedings of the National Academy of Sciences of the USA*, **106**, 13207–12.

Stylianou, S. (2003). Measuring crime seriousness perceptions: what have we learned and what else do we want to know. *Journal of Criminal Justice*, **31**, 37–56.

Tooby, J. and Cosmides, L. (1992). The psychological foundations of culture. In: J.H. Barkow, L. Cosmides, and J. Tooby (eds), *The adapted mind*, pp. 19–135. Oxford University Press, Oxford.

Tooby, J., Cosmides, L., and Price, M. (2006). Cognitive adaptations for n-person exchange: the evolutionary roots of organizational behavior. *Managerial and Decision Economics*, **27**, 103–29.

Trivers, R. (1971). The evolution of reciprocal altruism. *Quarterly Review of Biology*, **46**, 35–57.

Walker, P. (2001). A bioarchaeological perspective on the history of violence. *Annual Review of Anthropology*, **30**, 573–96.

Weiner, B. (1995). *Judgments of responsibility: a foundation for a theory of social conduct*. The Guilford Press, New York and London.

Willis, J. and Todorov, A. (2006). First impressions: making up your mind after 100 ms exposure to a face. *Psychological Science*, **17**, 592–8.

Wrangham, R. and Peterson, D. (1996). *Demonic males: apes and the origins of human violence*. Houghton Mifflin, New York.

Yamagishi, T. and Mifune, N. (2008). Does shared group membership promote altruism? *Rationality and Society*, **20**, 5–30.

Zaller, J.R. (1992). *The nature and origins of mass opinion*. Cambridge University Press, New York.

Chapter 9

Gender equity issues in evolutionary perspective

Bobbi S. Low

Introduction

Overview

'Gender equity' and 'gender equality' are relatively recent terms; the concept is really a 20th-century, Western idea.[1] Typically today we focus on three major areas of concern regarding gender equity: job opportunity and pay equity, equity in political impacts from voting to office-holding, and equity in the autonomy of reproductive decisions. But in our evolutionary and historical past, many more basic divisions of labour existed, and in many, monetary 'pay' was not a known currency. Thus, important though these terms are today, people in a fair proportion of the world's known societies would have had, until very recently, no clue of their meaning—and a few still would have no idea. Why? Our modern nation states are truly novel environments in evolutionary terms, from developing nations to the most developed. They are extremely large, with heterogeneous populations, have formal rules where many traditional societies had, well, tradition; and men's and women's roles have, at the least, considerable overlap today, in contrast to roles in traditional societies.

Here I examine gendered patterns, first in traditional pre-modern and historical societies. There I look for patterns: are some divisions of labour (e.g. who hunts large animals) underlain by ecological constraints? What patterns are driven by the differences in resource payoffs for male versus female mammals? (For that, of course, is where we have our beginnings.) What patterns are culturally determined—and are any of those driven by male-female conflicts of interests? Then I ask: where do such issues stand today, in our evolutionarily novel world? How does gender equity vary around the world? What are its correlates?

How our evolutionary past matters

Evolved sex biases

Some biases in gendered behaviour are even older than ancient human societies. As mammals, we begin with evolved sex biases. For example, most primates are polygynous, whether they simply form consort pairs briefly and then separate, or live in single-male or multi-male groups. Bound by providing milk as the offspring's post-partum nutrition, females tend to spend their time differently from males. In sum, males tend to spend mating effort, while females spend

---

[1] In the United States, 'equity' is a branch of jurisprudence in common law; the term is used in many other situations, always in the context of *fairness*. 'Equality' has the additional meaning of *desirability for all* (or some defined subset) of individuals to be treated in the same way.

offspring-specific parental effort such as nursing (review in Low 2000a). As a result, in most non-human primates, the female and her offspring are the ecological/economic unit; males and females do not spend a lot of time together. In some species (e.g. baboons) a particular male may offer friendship and protection, usually to a female with whom he has mated (e.g. Smuts 1985).

The importance of resources for the two sexes

In biological terms, getting and dispensing resources are the core issues of reproduction. It takes resources to grow, to achieve whatever kind of success (status, resources) that makes one a desirable mate, to have and rear offspring, and to provide for their future. The first two we call somatic and mating effort, and the last two are parental effort. The conundrum is that what makes an individual successful in one endeavour makes that individual less successful at the other activity. Seeking and finding mates, fighting, or performing expensive displays to impress them: such activities are antithetical to getting food safely and feeding and protecting an offspring.

This biological reality generates different returns-for-effort of mating and parental activities (e.g. Low 2000a). These different return curves mean that, under most conditions and in most species, males and females will be most successful in reproductive terms by doing different activities. In males, for example, striving for status (including fighting and things like growing antlers) typically entails what economists call a 'fixed cost'—a large expenditure is required even to begin competing for mates. Note, too, that this expenditure is not specific to any particular offspring. In contrast, when a mother nurses her baby, that expenditure is specific to this particular offspring, and must be spent all over again for another offspring.

The result is that men and women, like other male and female mammals, *ceteris paribus*, will spend their time and energy differently. Thus, a division of labour between the sexes or genders does not, in and of itself, suggest inequity.[2] In traditional societies, a woman's value in the mate market is her reproductive value—the number of daughters she is (statistically) likely to have in the rest of her life. A man's value is his resource (or power) value (Low 2000a).

Most traditional societies are, like most mammals, polygynous (Murdock 1957, 1967, 1981; Murdock and White 1969); men who can compete successfully are allowed more than one wife at a time. Monogamy and polyandry (137/849 and 4/849 societies, respectively, in Murdock's *Ethnographic Atlas*, 1967) are so rare that statistical testing of hypotheses is often impossible.[3] Men do typically live with wives in most societies, although women may have separate residences in some.

More important for the question of women's access to resources and influence is the fact that even socially monogamous groups typically are *genetically* polygynous (Reichard and Boesch 2003). This means that the payoff for controlling resources is higher for men than for women; there are great implications for male behaviour in general, and especially for male parental care. It is no surprise that inheritance in polygynous (and some socially monogamous) societies is typically biased toward sons (Hartung 1982), for sons can marry more wives and have more grandchildren the wealthier they are, while daughters are limited by the biological constraints of being female mammals. This is one of many biases that only make sense once one understands the

---

[2] As a biologist, I tend to think of 'sex' as the term for genetic 'males' and 'females'. Indeed, among most mammals, including primates, the genetic composition (e.g. XX, XY) predicts behaviour. I will restrict my use of 'gender' to socially-shaped roles, as in humans: not only 'typical' male–female roles, but also Zuni berdaches, for example.

[3] Polyandry, in particular, is associated with resources that lose value when divided, combined with resource constraints. Women typically marry brothers; it provides a way to keep resources in the family line.

evolutionary background.[4] Another finding that should not surprise us when viewed through an evolutionary lens is that for both men and women, one thing never changes: in all sorts of societies, traditional and modern, based on all sorts of subsistence types, parents everywhere try to be sure their children get more than others (Borgerhoff Mulder et al. 2009; Shenk et al. 2009).

It is clear that in traditional societies, men's and women's activities and roles differ, and that getting substantial amounts of resources affects men's lineage success more, and more directly, than it affects women's lineage success. The result—and the reason the concept of gender equity may never have arisen for women in a wide variety of traditional societies—is that men generally act in the more 'public' sphere and compete for bonanza resources, while women are more involved in the home-and-kin-group arena, and deal primarily with resources related to children's growth and health.

The correlates of women's power in traditional societies

What *does* influence women's larger-scale power in traditional societies? Some early ideas—more women's 'voice' in small-scale societies (Leacock and Lee 1982) or in protein-stressed societies (Divale and Harris 1976)—have not tested out. Whyte (1978, 1979) suggested that there was no single measure of women's power or voice, and his and following studies reinforce this finding. There are ecological (subsistence type, predictability) and cultural (contribution to subsistence, residence pattern) correlates of women's abilities to control and inherit various kinds of resources (Low 1989, 2007).

Most women today have free mate choice and are not bound by arranged marriages, but from traditional societies through to at least the Middle Ages, arranged marriages were common. Roughly three-quarters of the societies in the *Ethnographic Atlas* are bride price societies (men literally buy their wives), and most of these are polygynous—which means that getting *many* resources was really important for men. There is evidence that the kinds of marriage preferred in a society are influenced by who controls what kind of resources.

Thus, in contrast to many modern societies, the older generation in most traditional societies makes most of the decisions about the marriages of younger family members (and it is mainly men, talking and thinking about resources and familial alliances). Further, in many societies the preferred marriages are among first cousins, which can concentrate resources within a lineage.[5]

There are four kinds of first-cousin marriages, and their distribution among traditional societies is related both to resource control and to potential inbreeding avoidance (review in Flinn and Low 1986). 'Cross-cousin' marriages are MBD, mother's brother's daughter, and FZD, father's sister's daughter. The two 'parallel-cousin' marriage types are FBD, father's brother's daughter, and MZD, mother's sister's daughter. In the West today, we seldom distinguish these niceties (and we tend to discourage first-cousin marriages), but in traditional societies they had a great impact on resource access and control. In particular, FBD strengthens reciprocity among male paternal kin; it is associated with patrilocal residence and strong male control of resources. MZD, because alliances among women are largely ineffective in getting and holding large-scale resources, is virtually unknown.

---

[4] For example, among the Kipsigis, Borgerhoff Mulder (1988, 1989) found that the bride price reflected a woman's reproductive value (the number of daughters she was likely to have in the rest of her life, given the prevailing age-specific fertility and mortality schedules).

[5] In the US, 25 states prohibit first-cousin marriage, 20 allow it, and six allow it under some conditions (e.g. partner too old to bear children). South Carolina prohibits double-cousin marriage (http://www.ncsl.org/default.aspx?tabid=4266).

What about women's direct control of resources? Women's ability to control the fruits of men's labour increases as women contribute to the subsistence base; women also have the most control in matrilocal societies, intermediate control in bilocal and neolocal societies, and the least resource control (not surprisingly) in patrilocal societies (Low 1990a). Within polygynous societies, independent of marital residence, the greater the degree of polygyny (the more wives, on average), the lower is women's ability to control fruits of labour or inherit property (Low 1990a). Nonetheless, women are not without allies in polygynous and patrilocal societies (Yanca and Low 2004).

Women's ability to inherit economically valuable property has numerous associations (Low 1990a). Polygynous societies, the most common arrangement, typically favour sons as inheritors. Yet within polygynous societies, as the degree of polygyny increases, women's ability to control the fruits of men's labour decreases and their ability to control the fruits of their own labour increases—women can function as independent economic units, just as in other primates. Not only resource control, but inheritance, is affected by marriage: the greater the degree of polygyny, the less likely are women to be able to inherit property, for sons in polygynous societies are better able than daughters to turn resources into grandchildren (Low 2000a, chapter 6). Women are most likely to inherit property in matrilocal societies (including matrilocal polygynous groups), where they live among their own kin, and least likely to inherit property in patrilocal societies, living among their husband's kin.

Across cultures, the more women contribute to the subsistence base, the more control they have over various resources. In Western society as well, there is evidence that an increase in professional working women has resulted in greater economic independence for them. In some agricultural societies (notably West African), women may hold their own land and market their crops independently of their husbands. In societies based largely on irrigation agriculture, pastoral activities, and horticulture, men tend to control more resources. As the importance of hunting and fishing in subsistence grows, so does women's ability to inherit; this is true also for the intensity of agriculture (irrigation or non-irrigation). So there appear to be some ecological correlates of resource control patterns.

Gendered work patterns

Murdock and Provost's (1973) study of the gender allocation of 50 common work activities done by men and women in 188 traditional societies reflects individually and socially efficient responses to ecological pressures. The tasks found to be 'entirely male' included: hunting large aquatic and terrestrial fauna, lumbering, ore smelting, fowling, etc. There was surprisingly little variation around the world: in only seven of 1215 cases (society × activity) was one of these activities reported as predominantly or exclusively female. These cases are truly scattered, and often something of a misnomer. For example, 'mining' is coded as a female activity among the Fur of Sudan; what Fur women actually do is collect dust containing iron ore for sale to smiths.

Activities done almost entirely by women included spinning, laundering, food preparation, and cooking (all of which can be done with small children close by), as well as food and fuel gathering, and dairy production (but not herding). There were no technological activities that were exclusively female in many societies; in fact, men participated, sometimes equally, in many of the 'mostly female' activities.

When simpler technology is replaced by more complex machines in daily activities across cultures (e.g. the plough in horticulture), men become more involved. And when an occupational specialty begins to involve commoditization, profit, and a larger market—even if women otherwise are the principals—men tend to take over. Examples include male potters among the Aztecs, Babylonians, Romans, Hebrews, and Ganda; male weavers among the Burusho and Punjabi; male mat makers among the Aztecs, Babylonians, and Javanese. As Murdock and Provost noted: 'Even the most feminine tasks... cooking and the preparation of vegetal foods, tend to be assumed by specialized male bakers, chefs, and millers in the more complex civilizations of Europe and Asia'—and our own. There can be flexibility: consider the building of shelter or homes. In sedentary peoples, this tends to be men's work, but in nomadic societies, when people arrive at a new site, women tend to put up the shelter while men hunt and, if necessary, organize defences. The gendered division of labour is pragmatic rather than ideological.

Women, war, and politics

Warfare in traditional societies was almost an entirely male endeavour. Note that war is not sex-specific in some other primates. When two groups of vervet monkeys dispute territorial boundaries, for example, males fight males and females fight females. But in humans, as in many primates (e.g. chimpanzees), fighting is typically a male endeavour to get mates (e.g. Manson and Wrangham 1991). Inter-group conflict—warfare—is, in traditional societies, largely about stealing women, or the resources (horses, cattle) used as bride price to buy women[6] (see review in Low 2000a, chapters 4, 5, and 13).

Similarly, large-scale politics tended to be a male bastion, although in many societies women have or had informal influence (above; also review in Low 2000a, chapter 12). The reproductive benefits of high political office to men were usually direct (e.g. there might be rules that a chief could have a certain number of wives, while other men could have only one). So it is no surprise that in cross-cultural studies of political leadership, about 70% of societies have only male political leaders, and even when there are women political leaders, they are never as numerous, and seldom coded as being as powerful, as men (Whyte 1978, 1979).

In most traditional societies, women's influence and striving for resources are focused on household issues; women are less prominent in community or larger-scale affairs than men. In only 10 of the 93 societies studied by Whyte were women able to hold formal leadership positions; in one (the Cree) the influence was informal but the people considered it important (Low 1992). In 35/93 societies women were able to participate openly, if informally, in community decisions. Women could hold leadership positions *within the kin group* in a number of societies; for example among the Ashanti and the Ibo in Africa, among the Truk in the Pacific, the Haida in North America, and the Cubeo in South America, women are coded as having greater influence than men in the kingroup. In another set of cultures, men and women are coded as having equal kinship influence (review in Low 2000a, chapter 12). The few societies in which women hold or held 'large-scale' political power show no pattern: they are on several continents; some are settled, some migratory; some are patrilocal, some matrilocal; some are highly polygynous; and some have subsistence bases that are not usually associated with women's control of resources (e.g. animal husbandry).

The strongest correlate of women's engagement in large-scale politics is that of descent rules. Whyte found that 9/93 societies have some reason to report women as holding broad political power. Five of these nine are matrilineal or double descent. Why? These are the conditions in which a woman's power can be used to assist her son's (though not her own) reproductive success. Of the 18 matrilineal and double-descent societies in Whyte's sample, five have women political office holders (statistically, we would expect two); of the 56 other descent systems such as patrilineal, four have women in political power (we would expect seven).

---

[6] Although in most nation states today, women choose husbands freely, in most traditional societies, marriage arrangements concerned family alliances, and the bride and groom had relatively little to say. Typically older men were involved. Many arrangements prevailed, but bride price and bride service by the groom were common.

Attitudes and behaviour regarding sex and marriage

The arena in which men clearly have the 'upper hand' is that of sex and marriage. In sexual matters, egalitarianism is rare among traditional societies. How individual women may fare varies greatly (as it does among modern nation states), and women could have considerable allies and influence even in some polygynous societies (e.g. Yanca and Low 2004).

Societal attitudes vary, for example, about sexual behaviour, and whether rules apply equally to men and women. Today there is considerable concern over clitoridectomies in a number of developing nations. These were, in many traditional societies, a coming-of-age ritual—and in many societies, both sexes underwent sometimes-bloody rituals. Circumcision for men was typical, and sub-incision (slitting open the underside of the penis) occurred in a number of societies. Originally, both sexes were likely to undergo painful and dangerous rituals. It would be interesting to know the extent of sex bias in current practices.

Braude and Greene (1976) found differences in attitudes in traditional societies that mirror today's, and which relate to the different costs to men versus women of unplanned pregnancies. Men have frequent pre-marital sex in 64/107, or 59.8% of societies in the Standard Cross-Cultural Sample; women in 56/114 (49.1%). In 50/116 (43.1%) societies, extramarital sex is allowed for the husband, but not the wife. In 26 more societies (22.4%), extramarital sex is condemned for both sexes, but women are punished more severely (e.g. Braude and Greene note: husband is scolded, but wife is divorced).

Rape is accepted, ignored, ridiculed, or given a token fine or punishment in 55% of societies for which there are codes; and rape is coded as common in 41.2% of societies. The obvious theoretical background for this bias is that of the huge difference in how important paternal confidence is, as compared to maternal confidence: women know their own children from birth, but a husband must fear wifely deception in terms of the investment 'fathers' make in children. So men watch women; they confine and control women; they hire watchers; and they punish any discovered female transgression (e.g. see Low 2000a).

Thus, sexual equality in sexual matters is decidedly uncommon across traditional societies. Men typically control women. To put it bluntly, women have reproductive value to men, and men seek to control that value. Women are stolen, bought, and traded. In evolutionary terms the array of male strategies, from high parental care to absence and abuse, is related to the potentially high cost of paternal care and the uncertainty of paternity in any particular case. Men use an array of strategies to ensure only they impregnate their wives; and when they cannot do so (as in societies in which men are gone much of the time, or men and women sleep apart), investment in a woman's children comes from her brother(s), not her husband (e.g. Alexander 1979).

The bottom line from what we know about traditional societies is that division of labour is extensive in them, although, except for sexual matters, this outcome appears to be largely a matter of efficiency of different types of work. There are societies in which women's lives are relatively independent of those of men, but few or none of the arrangements look anything like the pattern of women's lives in Western developed nations today.

How the modern world differs

Convergence of gender roles

In the world of traditional societies, then, some clear divisions of labour made ecological and social—perhaps even psychological—sense. Men tended to specialize as resource-seekers (often on the grand scale—fabulous resource returns usually meant improved reproductive success). Women sought more modest resources (honey, plant foods), and prepared and distributed food; all the while, they were raising children. Such divisions make ecological sense. Despite Maxine

Hong Kingston's wonderful fable at the beginning of *The Woman Warrior*, fighting large opponents while pregnant or nursing is ineffective, perhaps lethal, as is hunting large game. Before modern medicine, if a woman died, her young children were likely to die also. Thus, sexual divisions of labour like those above have arisen from ordinary natural selection. In other areas, men clearly held the upper hand, and women were subservient.

All of these facts may seem a bit quaint to us today, because men's and women's work converge; and because gender roles in modern life are evolutionarily novel, and far more culturally than ecologically defined. This opens the door for expanding women's roles, and even for ideologically-driven biases in treatment of men and women.[7]

Smaller families facilitate converging gender roles

Modern conditions seem to be related to demographic transitions in Europe in the 19th century and in much of the world today: in these trends family sizes fell (or are falling) from about ten to about two. Face it: a mother of ten children has little time, typically, to devote to things like a well-paying job. Smaller family sizes mean two things: first, the *kind* of investment (face-time socialization, excellent schooling made possible by a good job) that is most effective changes, and second, fewer children means a woman can devote more time to garnering resources so that her children can afford the best of everything.

How did this start? Beginning in Europe in the 19th century, and continuing around the world today (Borgerhoff Mulder 1998; Low 2000b), life has become ever more competitive, and other sorts of work have replaced hunting and gathering—work that requires considerable training and practice, even new kinds of work for children that may be specialized in ways the parents have not mastered—and work that may be equally well done by a man or a woman. In many countries today, men's value to women is no longer solely or primarily resource value, and women's value to men is no longer solely or primarily reproductive value.

To raise children to adulthood, to see them married and having their own children, has come to require ever more monetary (as opposed to time) investment by parents (see review in Low 2000a). Gender- or sex-specific behavioural investment (e.g. nursing versus hunting) is not important, but resources are.

The question becomes important: as children become increasingly expensive, how many children can one afford to raise? So women might come to have value as resource producers, not simply as reproductive devices.

However, the equalizing impact of demographic transitions on women's roles was slow in coming. Throughout much of the 19th century, even with the impacts of the demographic transition, the wealthiest men in Europe married younger women and sired more children than poorer men, and women did not hold large-scale resources. Even in socially monogamous and explicitly egalitarian 19th-century Sweden, this was true: wealthy men married younger women than did than poor men (and had roughly 1.5 more children). After the death of a spouse, wealthy men remarried more frequently, and had more children, than poor men (e.g. see Low 1990b, 1991; Low and Clarke 1991a,b, 1992). Women did not inherit land.

---

[7] As noted recently in Maureen Dowd's column in the *New York Times*, many young women today shun the label 'feminist' because to them it represents a harsh and biased approach. As I note later, there were conscious decisions by leading feminists to focus on gains in the work world, and they have made significant progress.

How gender-equal are modern nations?

In the modern developed nations, making more expensive offspring means making fewer offspring, unless one has ever more resources (like the wealthy men, above). As a result of demographic changes, then, in many societies men's and women's values to each other have shifted. When family sizes approximate replacement value (two children), women may be valued more as resource-acquirers, and less as baby-makers. Fertility is late and low, but *per capita* investment in each child is high. A man's income may be less important today, because it is not the sole income, than in earlier times. Men's and women's work converge. In the modern world, 'gender equity' can have real meaning. Today, issues of gender equity may involve complex assessments of occupation, political power, and many issues of control.

Of course, women do not experience the same conditions in all modern nations. To explore the variety and covariates of gender equity/inequity in modern nation states can be daunting: over and above biological reflections of how healthy a society is (life expectancy, infant and child mortality, age at first birth) measures such as wealth and wealth disparities (e.g. the Gini coefficient), cultural history and values, religion, and the state and expense of health care, can all affect how different men's and women's lives are: whether they hold the same jobs, receive the same pay, and more. Relevant modern data exist at the national level for many countries; these will not capture within-country variance, but can help us understand broad patterns.

Materials and methods

Measures

The United Nations Human Development Report has been published since 1990, with different emphases in each wave. The 2007/2008 document[8] reports a large number of measures of human health, knowledge, standards of living, and various disparities (e.g. wealth, gender).

The basic measure, the *Human Development Index* (HDI), attempts to measure achievements of each country in three basic dimensions: long and healthy life (life expectancy at birth), knowledge (adult literacy rate and gross enrolment ratio), and decent standard of living (gross domestic product (GDP)/capita). Two measures focus on gender differences (Human Development Report 2007/2008, Technical Note 1). The *Gender-related Development Index* (GDI) measures achievement in the same basic capabilities as the HDI, but examines inequality in achievement between women and men. The GDI falls when the achievement levels of both women and men in a country are low, or when the disparity between their achievements is high. The greater the gender disparity in basic capabilities, the lower a country's GDI compared with its HDI. Thus, the GDI is simply the HDI discounted, or adjusted downwards, for gender inequality. This measure is important in the least well-off countries, and to some extent, in countries with religious restrictions on women's behaviour. The GDI does not itself reflect equity; HDI and GDI are almost identical, with an R^2 of 0.999. One divides GDI/HDI to obtain what I call here the *Gender Ratio*.

The *Gender Empowerment Measure* (GEM) focuses on women's opportunities; it evaluates women's standing in political and economic forums. It examines the extent to which women and men are able to participate in economic and political life, and take part in decision-making. It measures something different from the *Gender Ratio*, although the two measures are loosely correlated ($R^2 = 0.345$; d.f. $= 1,68$; $\beta = 0.587$) Informal influence is difficult to measure in this context and this measure varies considerably in high-HDI countries.

---

[8] See http://hdr.undp.org/en/reports/global/hdr2007–2008/

The World Economic Forum (WEF) also began in 2006 to assess gender differences. They have produced somewhat comparable measures to those of the United Nations Development Programme (UNDP) but because the measures are derived somewhat differently, they correlate but are not easily substituted in analyses. The economic and political measures are separate in the WEF data, but combined in the *GEM* of the UN data. The UN measures (overall) cover 177 nations, while the WEF data cover 128 countries. The *Gender Ratio* and the variable from WEF (here named *TotGend*) have an adjusted R^2 of 0.602. GEM, which in the UNDP data attempts to measure women's economic and political empowerment, shows an R^2 of 0.425 with the WEF *EconGap* measure, and an R^2 of 0.445 with the WEF *PolitGap* measure. Thus, while the measures are correlated they cannot substitute for one another; here I use the UN measures.

It is difficult to assess cultural influences on measures of women's performance and opportunities, compared to men's. I explore three measures here, two from the World Values Survey (Inglehart and Welzel 2005), and a crude assessment of whether any major religion obtains. The World Values Survey was designed to measure major areas of human concern. More than 70% of the trans-national variance is captured by two dimensions: Tradition/Secular-rational, and Survival/Self-expression.[9]

Although it is only one cultural measure, when one or another religion predominates, that religion may well reflect deep attitudes (e.g. Inglehart 2004). The Central Intelligence Agency (CIA) Factbook summarizes within-country religion for almost all of the countries in the UN database (a few Caribbean islands are missing). Here, I used an estimate of 55% or more to signify a single predominant religion. If two varieties of Christianity predominated (e.g. 40% Protestant, 50% Roman Catholic, 10% other), I termed the mixture 'Judeo Christian'. For mixtures such as 40% Muslim, 50% Christian, I termed the country 'Mixed'. Finally, China officially has no official religion, and the CIA data reflected religious identity for less than 5% of the population; it is designated 'None'. While Russia had no specified state religion, a large percentage of the population was reported as having some version of Christianity in practice; I termed it Judeo Christian.

Statistical analyses

Previous work (Low et al. 2008) has confirmed a suite of variables that tend to co-vary across less-developed nations; some variables appear to be relatively biologically driven (e.g. life expectancy at birth, age at first birth) and some essentially entirely culturally driven (female secondary enrolment). Here, all likely variables were first regressed pairwise with the dependent variables *Gender Equity* and (separately) the *GEM*. Then stepwise and backward stepwise regressions were performed, separating the analyses for 'high', 'moderate', and 'low' Human Development level, because so much depends on a country's wealth and general level of development. In all cases, the R^2 reported is the adjusted R^2, and the β is the standardized β. A variety of descriptive and distributional measures were calculated as well, in part to determine whether (and which) additional analyses should be done.

Results

How women fare in basic living compared to men: the gender ratio

The overall HDI, in which high scores reflect 'a long and healthy life, knowledge, and a decent standard of living' is closely tracked, as noted above, by the GDI. The *Gender Ratio*, the direct

---

[9] E.g. see www.worldvaluessurvey.org/wvs/articles/folder_published/article_base_54

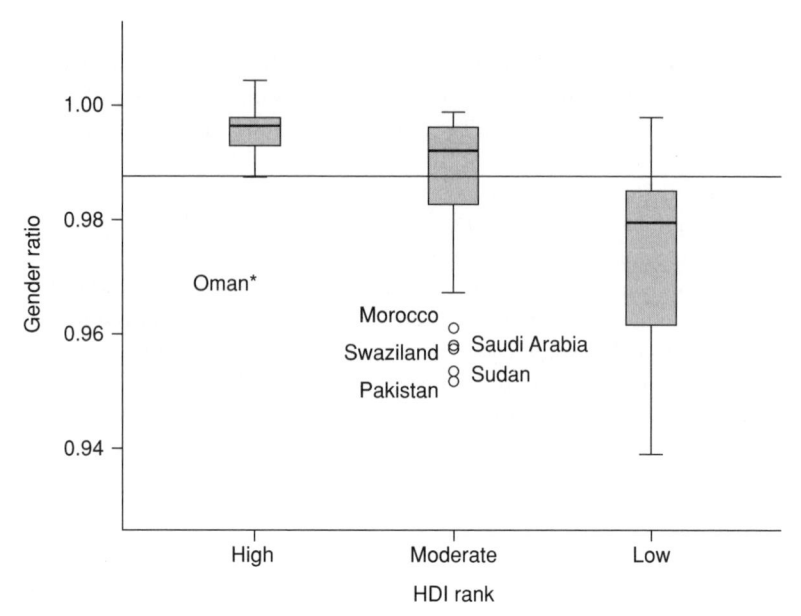

Fig. 9.1 The *Gender Ratio*, which reflects overall gender equality or inequality in the United Nations Development Programme database, differs across the ranks of Human Development Index (HDI). See text for further information.

comparison of women's lives compared to men's, averages 0.988 ± 0.014. The *Gender Ratio* differs across the best-off, moderate, and worst-off countries (Figure 9.1; ANOVA: $F = 32.04$; d.f. = 2, 133; $p <0.000001$). The variance of the *Gender Ratio* also differs across the three HDI levels (Levene statistic = 25.73; $p <0.00001$). In post hoc comparisons, the *Gender Ratio* in each level differed significantly from both other levels at $p <0.002$ or greater. All 'HDI-1' countries except Oman showed a *Gender Ratio* above the overall average. The average *Gender Ratio* of the HDI-2 countries was above the overall average, but a very long 'tail' of the distribution (Figure 9.1) included Morocco, Saudi Arabia, Swaziland, Sudan, and Pakistan. The worst-off HDI-3 countries, not surprisingly, had, overall, relatively low *Gender Ratios*.

What influences or correlates with the *Gender Ratio*? The variables that looked promising from pairwise tests were *Education Index*, *Female Employment*, *Female Secondary School Enrolment*, and (marginally) *Health Expenditures per capita*. Using stepwise regression, in the 52 high-HDI countries, the overall level of education (the *Education Index* in the HDI data) and the level of *Female Employment* were most significant ($R^2 = 0.512$; d.f. = 2,49; p <0.00001; $\beta_{educ} = 0.475$, p <0.001; $\beta_{femempl} = 0.322$, $p = 0.018$). Both female secondary school enrolment and health expenditures per capita (which are important in other tests) were deleted; I suspect they are above some threshold in the HDI-1 countries of this analysis.

Similarly, in 58 moderate-HDI countries, the *Education Index* and *Female Employment* were most significant ($R^2 = 0.723$; d.f. = 2,56; p <0.00001; $\beta_{educ} = 0.500$, p <0.001; $\beta_{femempl} = 0.626$, $p = 0.018$). The impact of female employment was stronger than in HDI-1 countries.

In the 24 low-HDI countries for which there are data, the *Education Index* and *Female Employment* again were most significant in stepwise regression ($R^2 = 0.499$; d.f. = 2,22; $p <0.0001$; $\beta_{educ} = 0.546$, $p <0.001$; $\beta_{femempl} = 0.559$; $p = 0.018$). Partial plots are, not surprisingly, more scattered than for the other two analyses. This is partly a result of the smaller sample size (data are more often missing for many variables in the worst-off societies), but also may reflect the fact that the array of workable arrangements is narrower than the possible non-effective arrangements. Tolstoy (*Anna Karenina*) seems to be right: happy families are all alike, while every unhappy family is unhappy in its own way.

Despite different dynamics, the *Education Index* and *Female Employment* were the most significant predictor variables in all three country levels. From a policy point of view, this should be encouraging; raising educational levels and opportunities, though not always easy, is likely to be politically popular—and it is not specifically women's education that matters here, simply the overall literacy rate of the population.

Both because these variables might be important for policy, and because the particular procedure in regression can influence results, I also used backward stepwise regression. It produced different results for the HDI-2 countries. For the HDI-1 ($R^2 = 0.526$), and HDI-3 ($R^2 = 0.443$) countries, results were similar to stepwise results: the important independent influences were *Female Employment* and the *Education Index*. But for the HDI-2 countries, backward stepwise regression suggested a different model ($R^2 = 0.695$; $\beta_{fsecenrol} = 0.509$, $p < 0.001$; $\beta_{femempl} = 0.762$; $p < 0.00001$), in which *Female Secondary Enrolment* and *Female Employment* were important, but overall *Educational Index* was not. Again, I suspect that this is the group of countries in which there is considerable variation in female secondary enrolment compared to the HDI-1 (high) and HDI-3 countries. From a policy point of view, then, for the 'middle' HDI countries, attention not only to general literacy but specifically to female secondary school enrolment may be important.

Women's economic and political opportunities: the gender empowerment measure

The *GEM*, which reflects women's economic and political opportunities, similarly varies (Figure 9.2) with the HDI ranking. The variables that looked promising from pairwise tests were

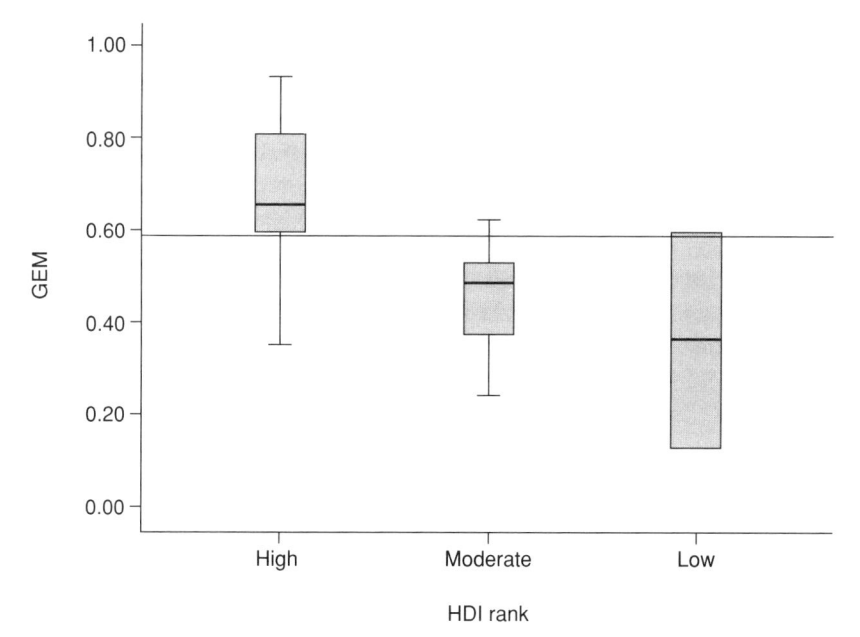

Fig. 9.2 The *GEM*, which reflects women's abilities to be effective in economic and political spheres, also differs by Human Development rank.

Education Index, *Female Employment*, *GDP per capita*, and *Health Expenditures per capita*. In the 45 high-HDI countries for which there are data, *Health Expenditures per capita* and *Education Index* were most significant ($R^2 = 0.675$; d.f. = 2,43; $p < 0.00001$; $\beta_{HealthExp} = 0.540$, $p = 0.00001$, $\beta_{EducaIndex} = 0.389$, $p = 0.001$). In 26 moderate-HDI countries, only *Female Employment* proved significant in stepwise regression and the fit was quite loose ($R^2 = 0.335$, d.f. = 1,25; $p = 0.001$; $\beta = 0.600$).

The bottom line: in high HDI countries, putting effort into getting and keeping children in school, and working toward more effective healthcare, should have the most impact. In the mid-HDI countries, getting more women in paying jobs is a strong correlate of their ability to move up. Tests could not be completed for the lowest-ranked countries; only one country (Tanzania) had relevant data.

World values

The World Values data on attitudes, like most variables, differ across the HDI 1, 2, and 3 groups. But along a Traditional-Rational scale and along a survival-self expression scale, these did not show patterns of interest with the *Gender Ratio*. Regressing Traditional-Rational with *Gender Ratio* yielded an adjusted R^2 of 0.297 ($\beta = 0.558$); with Survival-Self expression, $R^2 = 0.067$ and $\beta = 0.293$. So we gain little or no explanatory power from introducing these variables.

Religions and women's status

To the best of my knowledge, no major religion today proclaims inferiority of either gender. But there may remain some biases in practice like those discussed above for traditional societies (men are scolded, women are divorced, etc.). Blood feuds and vendettas, more common when the rule of law is weak, can result in death of either sex, and can promote such relatively sex-specific acts as rape.

The *Gender Ratio* varies with major religion (Figure 9.3). Any religion may differ in its liturgy versus its practice. Practice more than liturgy may reflect broad societal attitudes, rather than anything specific to the religion itself. The mean *Gender Ratio* is above the overall mean (higher gender equity) for the following religious designations: None (China), general Judeo-Christian, Roman Catholic, and Buddhist (Figure 9.3). Note, however, the long low-value 'tail' on the Judeo-Christian distribution. These countries have a predominance of mixed forms of Christianity, but their *Gender Ratios* are below the overall mean (low gender equity): Equatorial Guinea, Guatemala, Botswana, Democratic Republic of Congo, and Swaziland. Except for Guatemala, all of these are Christianized African countries, which have an underlayment of local traditional religions onto which Christianity is grafted.

The two Hindu nations represented, India and Nepal, have low *Gender Ratios*.[10] Both the 'Mixed' (Muslim and Christian) and the Muslim countries show considerable variation.

Discussion

The UNDP data show that considerable variation in women's status exists around the world and that a small number of variables appear to matter in thinking about raising women's access to resources, jobs, and power.

---

[10] It is of interest, also, I think, that when one asks about religion and age at first birth, the Hindu nations have much earlier AFB than any other designation.

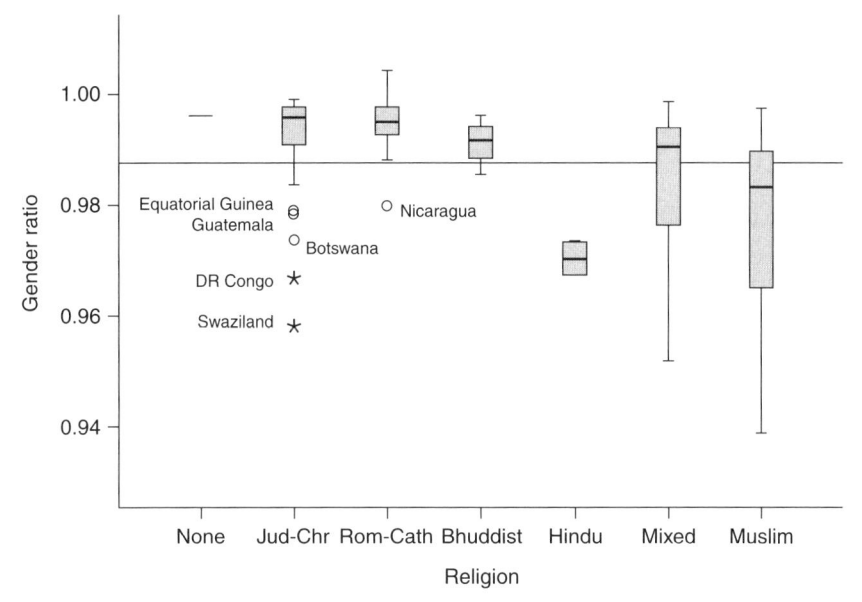

Fig. 9.3 The *Gender Ratio*, reflecting women's overall 'status' compared to men's, shows some differences with major religious patterns of nations. (Jud-Chr, Judeo Christian; Rom-Cath, Roman Catholic). See text for further interpretation.

Cross-culturally, especially in least-developed nations, women's employment may have little impact on women's resource control and independence. Women's work in less-developed nations may be proto-industrial or unpaid; the type of work is typically something compatible with home life and childcare. In developed nations, women tend to hold a range of jobs and professions, and work is seldom done at home. As women's and men's jobs and professions converge and work migrates out of the home, the conflict between work (getting resources) and childcare (dispersing resources) is exacerbated. This is not a new conflict (Hurtado et al. 1992): even in traditional hunter-gatherers like the Ache, nursing women can gather less than other women; however, working in a workplace with others, rather than at home, seems to restrict options further. The result is a clear trade-off between work advancement and family responsibilities.

These results barely scratch the surface of gender equity issues. For example, a form of gender inequality that was somewhat hidden until relatively recently is that of sex-preferential infanticide. Infanticide certainly occurred in traditional and historical societies; however, most infanticide or abandonment centred on unwanted newborns, or newborns with deformities (e.g. Daly and Wilson 1988). Sex-preferential infanticide was not uncommon in stratified societies, in which women could marry up but men could not; daughters were a cost to high-status parents (e.g. Dickemann 1979, Hrdy 2000). But sex-preferential infanticide in traditional societies never reached the level that exists today in some parts of the world (principally India, Japan, China, and Korea; e.g. Hudson and den Boyer 2004; Jha et al. 2006). Today, there are serious demographic consequences of female infanticide: for example, young men have difficulty finding brides.

What about the often-broken link between attitudes and actual behaviour? Recently, the Pew Research Center (2010) conducted a survey of attitudes related to gender equity in 22 countries; its results on attitudes are consistent with the patterns found here. On the general question of whether women should have equal rights with men, support ranged from 45% in Kenya to 99%

in France and Spain (US = 97%). On the question of whether women should be able to work outside the home, agreement ranged from 61% (Egypt) to 97% (US, Britain, France, Germany, Spain). There are religious dividing lines that echo the data here: within Lebanon, for example, Sunni Muslims (75%) and Christians (73%) agree that women should be able to work outside the home; among Lebanese Shia, the proportion is 63%.

In 19 of 22 countries in the Pew survey, respondents felt a marriage was more satisfying when both husband and wife have jobs. In contrast, in China, Pakistan, and Nigeria, views of marriage have become more traditional since an earlier survey in 2002. As with the question of outside work for women, responses were more mixed in Muslim countries than others. In Jordan, Egypt, Pakistan, and Turkey (as well as non-Muslim South Korea), more women than men favoured egalitarian marriage. The issue of women workers 'taking away' men's jobs in tough economic times led majorities in nine countries (India, Pakistan, Nigeria, Egypt, Indonesia, China, Jordan, Turkey, South Korea) to suggest men should have job preferences in those circumstances. Interestingly, in 18/22 countries, people strongly disagreed with the proposition that university education is more important for sons than for daughters; this was another issue on which women's responses differed from men's in a number of countries.

There are two lessons here. First, although attitudes do not reliably predict behaviour in many contexts, it is encouraging that egalitarianism appears to have considerable attitudinal support. Second, when men's and women's opinions differ, women are, not surprisingly, the gender that both has fewer opportunities, and seeks more egalitarian conditions.

Those of us in western democracies might be tempted to feel that our opportunities are equal—but there can be hidden costs (Smock and Noonan 2005). In the US, we might point to the fact that we have three women Supreme Court justices (of a total of nine) as reflecting real equity. By many measures, we are among the most gender-equal societies in existence. But anecdotes don't tell the whole story; there are two complications that suggest hidden complexities. First, the 'glass ceiling' appears to be real, and to limit most women; there are only 15 Fortune 500 companies with a female chief executive. And, overall, income of full-time women workers is about 23% lower than that of full-time working men (DeNavas-Walt et al. 2009). Even though a few women reach the highest levels, the labour market is not 'gender-equal'.

The second complication harks back to maternal ecology—the costs of being a female mammal. Highly successful women tend to pay a significant reproductive cost. In the US Senate in 2005, the 87 male senators averaged three children; the 13 female senators averaged 0.8 children. Among the women Supreme Court justices, only Ruth Bader Ginsberg has children. And children are largely not mentioned in the biographies of the Fortune 500 women leaders, suggesting a degree of childlessness.

Children require care; parents are likely to take more time off work than single employees, and more women do so than men. The price for taking time off, in terms of promotions and salary, is steep. In many cases, childless women do almost as well as men in salary terms—but not mothers (e.g. Joshi et al. 1998). In the US, this may be due in part to the American feminist movement having made a conscious choice to emphasize equal rights and equal opportunities—and not to address issues of family responsibilities (*New York Times* interview by David Leonhardt with Jane Waldfogel, 3 August 2010; see also Bertrand et al. 2010).

As a result, we have come much closer to having equal economic rights and opportunities; in a recent study, men and women had nearly identical labour incomes and weekly hours worked in the early years after graduation from the University of Chicago Business School (Bertrand et al. 2010)—but again, with a frequent reproductive cost. Taking time off from work, as parenthood often requires, has serious economic costs, and women pay those costs more often than do men. As a result, 15 years after graduation, men earned about 75% more than the women. Staying at home also reduces one's likelihood of moving up a career ladder. If we really seek sexual equity in the workplace, we may have to confront the joint need for, and career damage done by, flexible work schedules and part-time work.

Currently, the Scandinavian countries lead the world in most measures of gender equity: in job opportunities and salary, and in political leadership. And a recent British law that gives workers the right to request part-time or flexible schedules may be an idea for a start for other nations. Parental leave for both parents is common in a number of European countries.

In making individual choices and policy decisions, we need to understand the line between ecologically-sensible divisions of labour and ideologically-driven decisions. The paths to relative gender equity have in most countries been considered successful, torturous and long. The year 2010 is the ninetieth anniversary of women getting the vote in the United States (the 19th Amendment of the US Constitution prohibited women from voting). Equal rights were seen as the larger issue underlying women's voting rights by many women suffragettes, such as Alice Paul, who introduced the first Equal Rights Amendment in 1923. The passage of the current Equal Rights Amendment did not occur until 1972, and the fight was long and bitter. The formal approval of women's voting rights and of the Equal Rights Amendment were not events, but long and painful processes.

Clearly, gender roles can become politically charged, and difficult to shift—something to remember as women in other countries seek gender equity today. A final caveat: in some arenas, seeking equity occasionally can have unexpected consequences. In the US, equity in vehicle insurance means women younger than 25 years old pay the same rates as men under 25, even though they have far fewer accidents. Or consider a well-meaning attempt to reduce female infanticide in India by empowering women. The idea was that better-educated mothers would at least not discriminate against their daughters, and might invest more in them. Sadly, after the interventions, the more educated or wealthy mothers (and their marital families) were, the more daughters were discriminated against, including female infanticide (Mahalingham 2007; Das Gupta and Visaria 1996).

The UNFPA is clear on its goal[11] of empowering women: 'Gender equality is, first and foremost, a human right. Women are entitled to live in dignity and in freedom from want and from fear. Empowering women is also an indispensable tool for advancing development and reducing poverty'. Gender equality is included as one of the eight Millennium Development Goals. In some developing nations, there are difficult embedded conditions reflected by the analyses above, including violence against women, economic discrimination, and frequent sexual assaults during armed conflicts (e.g. WuDunn and Kristof 2010). These are serious problems. As Inglehart and Baker (2000) note, economic development and industrialization are associated with shifts from traditional thinking to more secular-rational thinking—which correlates with greater gender equity. These are long processes, but we must beware of attempting to bring our values into others' preferences through assertion. Consider the cover picture on *Time* magazine (9 August 2010): it features a woman whose nose and ears have been cut off for sexual transgression. It horrifies people in many Western countries, for example. In some countries stoning remains as punishment for unacceptable sexual activity. But who has the right to intervene directly to prevent such things? Cultural norms are historically influenced, and quirky, even harmful, things persist. Some cultural sensitivity may make interventions more successful.

Across developing nations, in an earlier and relatively small (seven-country) sample, most women desired the ability to control their own fertility. What made such control more possible or likely? Remunerative work of their own; and in some settings belonging to community groups helped. Women broadly bemoaned the inaccessibility and poor quality of health care

---

[11] See http://www.unfpa.org/gender/

(which frequently leads to needless infant death). Note that these desires are about maternal control, not simply about free choice of mate or sexual pleasure (see Petchesky 2001).

On the one hand, not every solution to gender inequality must look alike; on the other hand, ignoring inequalities that do real harm to women (and their children) can be literally fatal (e.g. the *Time* case; WuDunn and Kristof 2010). Approaching this problem requires more than the typical amount of real humility in beginning to solve a serious cross-cultural dilemma. While we may have 'come a long way, baby' we have far to go if we really seek culturally-attuned equal opportunities, equal protection, and equal access to resources for women.

References

Alexander R.D. (1979). *Darwinism and human affairs*. University of Washington Press, Seattle, WA.

Bertrand, M., Goldin, C., and Katz, L. (2010). Dynamics of the gender gap for young professionals in the financial and corporate sectors. *American Economic Journal: Applied Economics*, **2**, 228–55.

Borgerhoff Mulder, M. (1988). Kipsigis bridewealth payments. In: L. Betzig, *M. Borgerhoff Mulder, and P.W. Turke (eds), Human reproductive behavior: a Darwinian perspective*, pp. 65–82. Cambridge University Press, Cambridge.

Borgerhoff Mulder, M. (1989). Earlier-maturing Kipsigis women have higher lifetime reproductive success, and cost more. *Behavioral Ecology and Sociobiology*, **124**, 145–53.

Borgerhoff Mulder, M. (1998). The demographic transition: are we any closer to an evolutionary explanation? *Trends in Ecology and Evolution*, **13**, 266–70.

Borgerhoff Mulder, M., Bowles, S., Hertz, T., *et al.* (2009). The intergenerational transmission of wealth and the dynamics of wealth in pre-modern societies. *Science*, **326**, 682–8.

Braude, G.J. and Greene, S.J. (1976). Cross-cultural codes on twenty sexual attitudes and practices. *Ethnology*, **15**, 409–29.

Daly, M. and Wilson, M. (1988). *Homicide*. Aldine de Gruyter, Hawthorn, NY.

Das Gupta, M. and Visaria, L. (1996). Son preference and excess female mortality in India's demographic transition. In: *Korea Institute for Health and Social Affairs and United Nations Population Fund (eds), Sex Preference for Children and Gender Discrimination in Asia*, pp. 96–102. Oxford University Press, New Delhi, India.

DeNavas-Walt C., Proctor, B.D., and Smith, J.C. (2009). *Income poverty and health insurance coverage in the United States: 2008*. U.S. Government Printing Office, Washington, DC.

Dickemann, M. (1979). Female infanticide, reproductive strategies, and social stratification: a preliminary model. In: N.A. Chagnon and W. Irons (eds), *Evolutionary biology and human social behavior: an anthropological perspective*, pp. 321–67. Duxbury Press, North Scituate, MA.

Divale, W. and Harris, M. (1976). Population, warfare, and the male supremacist complex. *American Anthropologist*, **80**, 21–41.

Flinn, M.V. and Low, B.S. (1986). Resource distribution, social competition, and mating patterns in human societies. In: D. Rubenstein and R. Wrangham (eds), *Ecological aspects of social evolution*, pp. 217–243. Princeton University Press, Princeton, NJ.

Hartung, J. (1982). Polygyny and inheritance of wealth. *Current Anthropology*, **23**, 1–11.

Hrdy, S.B. (2000). *Mother Nature: maternal instincts and how they shape the human species*. The Ballantine Publishing Group, New York.

Hudson, V.M. and den Boer, A.M. (2004). *Bare branches: the security implications of Asia's surplus male population*. MIT Press, Cambridge, MA.

Hurtado, M., Hill, K., Kaplan, H., and Hurtado, J. (1992). Tradeoffs between food acquisition and child care among Hiwi and Ache women. *Human Nature*, **3**, 185–216.

Inglehart, R. (2004). *Islam gender culture and democracy: findings from the Values Surveys*. De Sitter Publications, Ontario.

Inglehart, R. and Baker, W.E. (2000). Modernization, cultural change, and the persistence of traditional values. *American Sociological Review*, **65**, 19–51.

Inglehart, R. and Welzel, C. (2005). *Modernization, cultural change, and democracy*. Cambridge University Press, New York.

Jha, P., Kumar, R., Vasa, P., Dhingra, N., Thiruchelvam, D., and Moineddin, R. (2006). Low male-to-female sex ratio of children born in India: national survey of 1·1 million households. *Lancet*, **367**, 211–18.

Joshi, H., Pierella, P., Makepeace, G., and Waldfogel, J. (1998). *Unequal pay for men and women: evidence from the British Birth Cohort Studies*. MIT Press, Cambridge.

Leacock, E. and Lee, R. (1982). *Politics and history in band societies*. Cambridge University Press, Cambridge.

Low, B. (1989). Human responses to environmental extremeness and uncertainty: a cross-cultural perspective. In: E. Cashdan (ed.), *Risk and uncertainty in tribal and peasant economies*, pp. 229–55. Westview Press, Boulder, CO.

Low, B. (1990a). Sex, power, and resources: ecological and social correlates of sex differences. *International Journal of Contemporary Sociology*, **27**, 49–74.

Low, B. (1990b). Land ownership, occupational status, and reproductive behavior in 19th-century Sweden: Tuna parish. *American Anthropologist*, **92**, 457–68.

Low, B. (1991). Reproductive life in nineteenth century Sweden: an evolutionary perspective on demographic phenomena. *Ethology and Sociobiology*, **12**, 411–48.

Low, B. (1992). Sex, coalitions, and politics in preindustrial societies. *Politics and the Life Sciences*, **9**, 1–18.

Low, B. (2000a). *Why sex matters: a Darwinian look at human behavior*. Princeton University Press, Princeton, NJ.

Low, B. (2000b). Sex, wealth, and fertility—old rules, new environments. In: L. Cronk, N.A. Chagnon, and W.G. Irons (eds), *Adaptation and human behavior: an anthropological perspective*, pp. 323–44. Aldine de Gruyter, New York.

Low, B. (2007). Ecological and sociocultural influences on mating and marriage systems. In: R.I.M. Dunbar and L. Barrett (eds), *The Oxford Handbook of Evolutionary Psychology*, pp. 449–62. Oxford University Press, Oxford.

Low, B. and Clarke, A.L. (1991a). Family patterns in 19th-Century Sweden – impact of occupational-status and land ownership. *Journal of Family History*, **16**, 117–38.

Low, B. and Clarke, A.L. (1991b). Occupational status, land ownership, migration, and family patterns in 19th-century Sweden. *Journal of Family History*, **16**, 117–38.

Low, B. and Clarke, A.L. (1992). Resources and the life course—patterns through the demographic-transition. *Ethology and Sociobiology*, **13**, 463–94.

Low, B., Hazel, A., Parker, N., and Welch, K. (2008). Influences on women's reproductive lives: unexpected ecological underpinnings. *Journal of Cross-Cultural Research*, **42**, 201–19.

Mahalingham, R. (2007). Culture, ecology, and belies about gender in son-preference caste groups. *Evolution and Human Behavior*, **28**, 319–29.

Manson, J. and Wrangham, R. (1991). Intergroup aggression in chimpanzees and humans. *Current Anthropology*, **32**, 369–90.

Murdock, G.P. (1957). World ethnographic sample. *American Anthropologist*, **59**, 195–220.

Murdock, G.P. (1967). *Ethnographic atlas*. University of Pittsburgh Press, Pittsburgh, PA.

Murdock, G.P. (1981). *Atlas of world cultures*. University of Pittsburgh Press, Pittsburgh, PA.

Murdock, G. and Provost, C. (1973). Factors in the division of labor by sex. *Ethnology*, **12**, 203–25.

Murdock, G. and White, D. (1969). Standard cross-cultural sample. *Ethnology*, **8**, 329–69.

Petchesky, R.P. (2001). Cross-country comparisons and political visions. In: R.P. Petchesky and K. Judd (eds), *Negotiating reproductive rights: women's perspectives across countries and cultures*, pp. 295–323. Zed Book, London and New York.

Pew Research Center (2010). Gender equality universally embraced, but inequalities acknowledged: 22-Nation Pew Global Attitudes Survey. Pew Research Center, Washington, DC.

Reichard U. and Boesch, C. (2003). *Monogamy mating strategies and partnerships in brds, humans and other mammals*. Cambridge University Press, Cambridge.

Shenk, M., Borgerhoff Mulder, M., Bowles, S., et al. (2009). Intergenerational wealth transmission among agriculturalists. *Current Anthropology*, 51, 65–82.

Smock, P. and Noonan, M.C. (2005). Intersections: gender, work, and family research in the United States. In: S.M. Bianchi, L.M. Casper and R.B. King (eds), *Work, family, health and well-being*, pp. 343–360. Erlbaum, Mahwah, NJ.

Smuts, B. (1985). *Sex and friendship in baboons*. Aldine de Gruyter, New York.

Whyte, M.K. (1978). Cross-cultural codes dealing with the relative status of women. *Ethology*, 17, 211–37.

Whyte, M.K. (1979). *The status of women in pre-industrial society*. Princeton University Press, Princeton, NJ.

WuDunn, S. and Kristof, N. (2010). *Half the sky: turning oppression into opportunity for women worldwide*. Knopf Doubleday, New York.

Yanca, C. and Low, B. (2004). Female allies and female power: a cross-cultural analysis. *Evolution and Human Behavior*, 25, 9–23.

Chapter 10

The evolution of charitable behaviour and the power of reputation

Pat Barclay

Introduction

Humans are arguably the most cooperative species on the planet when it comes to non-kin interactions. Humans regularly help non-kin in both formal and informal settings. For example, in my home country of Canada in 2007 alone, 84% of people made charitable donations for a value of almost $10 billion, and 46% of people volunteered an average of 166 hours (Hall et al. 2009). Of course these figures are underestimates of the real total, as they do not include other ubiquitous forms of informal help such as favours, advice, exchange of benefits, restaurant tips, or restraint from overharvesting resources and selfish competition. Although helping rates certainly vary, Canada is far from alone in such behaviour.

From an evolutionary perspective, such charitable behaviour is very puzzling. Why would an organism ever do something to benefit another if it were costly to do so? Shouldn't such behaviour be naturally selected against? If non-helpers save themselves the cost of helping, we might expect them to have an advantage in competition against those who possess more charitable sentiments. Yet many organisms do help others, and the existence of such behaviour is one of the central puzzles in the study of evolution and behaviour (e.g. Vogel 2004; Pennisi 2009).

It has been several decades since scientists realized that behaviour does not evolve 'for the good of the species' (Hamilton 1964; Williams 1966; Dawkins 1976), and much theoretical and empirical work has identified the selective pressures that can select for and maintain such apparently selfless behaviour. Given that this is a chapter on *applied* evolutionary psychology, I will not attempt a formal review of all models of non-kin cooperation, but will instead present the basics on some of them before proceeding to potential applications and limitations on those applications. Many of these applications are tentative, either because there has not yet been formal research on their use in applied charitable settings, or in the case of using reputational pressures, because the theories themselves are still incomplete and first require more basic research. Nevertheless, it is hoped that the following will give evolutionary researchers some ideas on how to apply their research and will give non-evolutionary applied researchers a reason to start thinking about evolutionary psychology.

The basics of an evolutionary approach to charitable behaviour

In evolutionary theory, a psychology for helping will evolve whenever possessing such a psychology returns net benefits in terms of inclusive fitness. Several processes could cause that to happen, all of which ultimately rely on either direct benefits to self or indirect benefits to close relatives (Reeve 2000; Foster et al. 2006; West et al. 2007). Whatever form these benefits take, they function to reinforce the behaviour. This reinforcement can cause a person to learn to be generous, for example by internalization of the norms which promote helping. Over evolutionary time, such reputational benefits can cause the evolution of cooperative sentiment. In the following sections I will outline some of the major selective pressures that could have selected for helping, and will speculate on some possible applications of this knowledge for increasing the amount of charitable behaviour observed.

It is very easy to misunderstand evolutionary approaches and predictions, given that evolutionary thinkers use different terminology and ask different types of questions than other fields. Even among evolutionists, many bitter arguments have resulted from nothing more than differences in terminology or levels of analysis. As such, it is well worthwhile to clarify some basic points before proceeding, as this can help avoid later misunderstandings.

Avoiding confusion: what phenomena am I referring to?

By 'charitable behaviour', I refer to all types of behaviour that benefit others at a cost to self, regardless of whether the helper later receives benefits for doing so.[1] Different disciplines use different terms for behaviours that benefit others at a cost to self, including 'altruism', 'cooperation', 'generosity', and 'pro-social behaviour'. Because of these differing definitions, these terms are somewhat loaded with unintended implications and controversies. Rather than debate the 'correct' use of such terms, here I will attempt to use the more neutral 'helping' as a general purpose term for all behaviours that benefit others at a cost to self, and I will simply invite readers to substitute whatever term in their field is most appropriate for such behaviours.

I am lumping together many types of helping because evolutionary psychology usually treats different types of help with the same theoretical models and experimental games, whether that help is the reciprocal exchange of aid, coalitional support, emergency helping, courtship giving, tipping, charitable donations, conservation of resources, production of public goods, or pro-environmental behaviour (to name just a few). Different types of help occur in different currencies with different costs and benefits and will be performed by different people towards different recipients, but from an evolutionary perspective there is no a priori reason to predict that different types of help will respond in *qualitatively* different ways to changes in such parameters. As such, evolutionary psychology generally assumes some generality of its models.

Avoiding confusion: levels of analysis

Most people would agree that many body parts exist because they have an evolutionary function. For example, hearts pump blood, eyes extract visual information about the world, and the visual cortex processes that information to make it usable. The presence of these organs increases survival, and hence, reproductive success. Behaviour can also affect survival and reproductive success, so it is also subject to natural selection, along with whatever psychological mechanisms and cognitive processes produce the behaviour. Behaviour is not inherited directly—it is produced by brains, which in turn are produced by developmental interactions between genes and environments. If a gene interacts with the developmental environment in such a way as to grow a brain which produces beneficial behaviour more often, then that gene will be selected for because of those positive consequences.

If a particular behaviour reliably results in a specific positive outcome, then the production of that outcome can be said to be the 'function' of that behaviour. For example, the function of

---

[1] I am defining 'costs' as relative to the global population, not just relative to one's local group (West et al. 2006). For a discussion of this important distinction, see West et al. (2007).

eating is the acquisition of energy and essential nutrients. Of course, most of us don't think about energy acquisition when we eat: we think about our hunger. However, the reason that hunger exists is because it leads us to eat, which results in the acquisition of energy and nutrients. Thus, there is a difference between the psychological mechanism (hunger) and the function of that mechanism (acquisition of energy): the evolutionary function is the reason that the mechanism exists. If we look at charitable behaviour, we must similarly distinguish between the psychological mechanisms (e.g. empathy, guilt, concern for reputation) and their potential evolutionary functions (e.g. acquisition of social benefits such as future reciprocation); we experience only the mechanism, but its function is the reason that the mechanisms exists.

In fact, there are four different types of question that we need to distinguish between in order to fully understand behaviour (Tinbergen 1963). We must ask about *psychological mechanisms*, and ask whether helping is caused by empathy, a warm glow, conformity to norms, or something else within the individual (i.e. 'What went on in my brain that made *me* help?'). Secondly, we must ask about the *development* of those mechanisms, and whether they are present at birth, learned via reinforcement or internalization of norms, or otherwise acquired within an individual's developmental history (i.e. 'How did my genes and environment interact to produce that psychological mechanism?'). Thirdly, we must ask about *evolutionary function*, and whether helping exists because it invites reciprocation, brings social prestige, wards off punishment, or raises inclusive fitness in other ways (i.e. 'What are the benefits of the behaviour, i.e. why does *anyone* have that psychological mechanism and developmental response?'). Finally, we must understand the *phylogeny* or *evolutionary history* of different helping behaviours, whether they are common to all our ancestors or are recently derived, and from what (i.e. 'How and when did that psychological mechanism evolve?'). Most behavioural scientists seek to understand the psychological mechanisms and their development, but we must also understand why those mechanisms exist at all and how they evolved. Thus, these four types of questions are complementary, not mutually exclusive. Maternal helping behaviour is a good illustrative example. Maternal concern (*mechanism*) gets triggered in certain circumstances and involves specific neurobiological processes. Mothers are capable of such concern because of the way their genes and past social environments interacted (*development*). Maternal concern is present in all primates and many mammals, and uses similar brain pathways, indicating that it is evolutionarily ancient (*evolutionary history*). This concern exists because it causes mothers to help their offspring and thus contributes to offspring survival and reproduction (*function*). All four questions must be addressed for a complete understanding.

It is very easy to confuse these different levels of analysis, especially when dealing with helping behaviour. No one would suggest that people are always consciously concerned with energy acquisition whenever they eat, but many have assumed that people are consciously concerned with the acquisition of benefits whenever they help (e.g. conscious concern for reputation) or they have interpreted other researchers as making that argument (e.g. Fehr and Henrich 2003; Gintis et al. 2003). This has led to much confusion in the literature. Instead, we must realize that mechanisms, development, function, and phylogeny are simply four different and complementary questions. By investigating function and phylogeny instead of just mechanism and development, it gives us a greater understanding of helping behaviour and how to promote it, for example by changing our social circumstances to ensure that the function is fulfilled.

Avoiding confusion: adaptive on average

Evolutionary researchers propose that helping behaviour exists because it increases inclusive fitness on average. However, not every instance of helping will do so, nor does it need to. For a trait to evolve, it needs to bring its bearer more net benefits on average than a different trait would (or would have done so in ancestral environments), even if that trait occasionally results in instances of non-beneficial behaviours. For example, empathy might have evolved because it causes people to behave better towards associates, which in turn results in the empathic person receiving better behaviour (Frank 1988). However, it might also cause people to sometimes help total strangers who will never be in a position to reciprocate. If the overall benefits from helping the 'right' people (e.g. friends and associates) outweigh the cost of helping both 'right' and 'wrong' people (e.g. the occasional stranger), then being empathic pays better than being non-empathic, so empathy will evolve. Of course, rather than using a fixed 'help' or 'no-help' strategy, evolution refines our psychological mechanisms to make adaptive distinctions between different people and situations, for example by making us more likely to help kin versus non-kin (Daly and Wilson 1988; Stewart-Williams 2007), friends versus strangers (e.g. Majolo et al. 2006), attractive people versus unattractive people (Farrelly et al. 2007), and non-competitors versus competitors (West et al. 2006).

Despite this fine-tuning of our psychological adaptations, there is no evolutionary theory that predicts that every single instance of helping should increase fitness, because errors are inevitable in any decision-making process (Haselton and Buss 2000; Nesse 2005). Rather than focus on specific instances of helping and asking whether they are adaptive, one must focus on the *mechanism*: is this particular 'decision rule' adaptive? Does it bring more benefits in the real world than an alternative decision rule? By increasing your likelihood of helping, you simultaneously increase the likelihood of helping in the 'right' circumstances (i.e. when you can benefit from it) and of helping in the 'wrong' circumstances (i.e. when you cannot benefit); the optimal level of helping strikes a balance between the former and the latter (see Figure 10.1). If the 'right' circumstances are much more frequent, or if failing to help in those circumstances is very costly (e.g. lost opportunities for reciprocity or other reputation), then our psychological mechanisms will cause us to err on the side of more helping. This will occasionally result in helping when it brings no benefits, but this cost is outweighed by the overall benefit. Conversely, as the 'wrong' circumstances increase in frequency, or as the cost of helping in those circumstances increases, then we should err on the side of low helping. This will result in more missed opportunities to receive benefits, but these are outweighed by the overall lower costs. Thus, although errors are inevitable, an evolutionary approach helps to predict whether people will err on the side of helping too much versus too little, and this knowledge can be used to increase people's likelihood of helping.

Inclusive fitness and vested interests

Inclusive fitness theory

Any gene can propagate copies of itself either by aiding the reproduction of its host or the reproduction of those who are statistically likely to carry copies of that same gene, especially close kin (Hamilton 1964). This is known as inclusive fitness theory, and it predicts that all else being equal, people will be more likely to help those who are close kin, as opposed to distant kin or non-relatives. This prediction has been abundantly confirmed in non-humans and in humans from many societies (for one textbook review, see Barrett et al. 2002). Inclusive fitness theory is the cornerstone of modern studies of evolution and behaviour. All else being equal, people can expect much better treatment from kin than non-kin, and interactions with kin are generally more cooperative than interactions with non-kin in the same situations (Daly and Wilson, 1988). Despite its importance, I will spend little time discussing kinship because: 1) it is much more obvious to most evolutionists, and 2) helping non-kin is more difficult to understand and some aspects remain unresolved.

	Decision	
	Help	Don't help
Helping would increase fitness (e.g. recipient is kin/friend, people notice, vested interest in outcome, etc.)	**Cost**: pay to help **Benefit**: reciprocity, gain reputation, direct benefit to self or to kin, etc. **Total**: net benefit	**Cost**: punishment, lose reputation, etc. **Benefit**: none (though also no cost paid to help) **Total**: net cost
Helping wouldn't increase fitness (e.g. recipient is stranger, no one sees, etc.)	**Cost**: pay to help **Benefit**: none **Total**: net cost	**Cost**: none **Benefit**: none (though also no cost paid to help) **Total**: no cost or benefit

Fig. 10.1 Fitness costs and benefits for helping and not-helping. Given that all organisms cannot know with 100% certainty what circumstances they are in, errors are inevitable (i.e. occasionally helping in the 'wrong' circumstances or not-helping in the 'right' circumstances). Rather than look at the payoffs for specific outcomes, we need to look at the overall payoff for helping versus not-helping averaged across all 'right' and 'wrong' circumstances (weighted by their frequency). Our psychology mechanisms and emotions evolved to be adaptive on average: they will make us err on the side of helping when the *average* payoff for helping (column 1) outweighs the *average* payoff for not-helping (column 2) and vice versa, which in turn depends on the magnitude of costs/benefits for each outcome and the frequency of the 'right' circumstances.

Before leaving the topic, it is worth mentioning that kinship need not be real to have an effect on helping: presenting people with unconscious cues of kinship can be enough to trigger cooperative behaviour. For example, facial resemblance is a cue of kinship (DeBruine 2005), and DeBruine (2002) showed that laboratory participants entrusted more money to people whose pictures had been 'morphed' to slightly resemble their own faces. Using a similar technique, Krupp and colleagues (2008) showed that people became increasingly more cooperative as the number of perceived kin in the group increased. Oates and Wilson (2002) found that people were more willing to help people who had the same family name as themselves, especially if it was a rare name (and thus a more reliable cue of kinship). Salmon (1998) found that the use of kinship terms in political speeches was effective at getting people to agree with the speaker, especially for firstborns who are typically closer to their families. This research all suggests that cues of kinship can be effective at eliciting charitable behaviour, whether the cues are verbal, facial, nominal, or merely kinship terms. Of course, overt use of such fictive kinship cues might trigger hostile reactions from those who are aware of their function, just as my evolutionary classmates in graduate school and I used to get angry at one particular campus political persona who would refer to us as 'brothers and sisters'.

Vested interests

Kinship is not the only way for one individual to have a fitness stake in the well-being of another individual (Roberts 2005). For example, people have a vested interest in the well-being of their friends and allies, and in the continued existence of their social groups (Tooby and Cosmides 1996; Kokko et al. 2001; Lahti and Weinstein 2005). If these friends or social groups were to die, disappear or disband, then those involved would be worse off. As such, it is sometimes worth incurring an unreciprocated cost in order to ensure their well-being. Even extreme generosity towards one's in-group can be adaptive if the alternative is extinction at the hands of hostile out-groups (Reeve and Hölldobler 2007; Barclay and Benard, submitted). Many cases of apparent 'altruism'—such as so-called 'altruistic punishment' of non-cooperators in a group—are in fact beneficial to the person performing the act (West et al. 2007).

Most perspectives would predict that people will display more charitable behaviour if they have a vested interest in the outcome, such as cancer victims donating towards cancer research. Evolutionary theory reminds us that people can also have a vested interest in the well-being of other people or even entire groups, and that unreciprocated helping should occur towards such recipients. Political leaders are well aware of this when helping allies or bailing out failing companies that perform necessary national services. Anything that increases the interdependency of people can also increase people's stake in one another's well-being, and thus the charitable behaviour observed towards others.

Direct reciprocity

Trivers (1971) noted that two individuals who exchange help can be better off than individuals who neither give nor receive help. His paper inspired much empirical and theoretical work on the evolution of pairwise reciprocal exchange of aid, which was most famously demonstrated by the success of the 'Tit-for-Tat' cooperative strategy in Axelrod's computer simulations on the evolution of cooperation (Axelrod 1984). Three general conclusions from this broad literature are: a) it pays to help others if there is a chance that they can reciprocate; b) many people are 'conditional cooperators' who are willing to cooperate if and only if others follow suit; and c) cooperation is more likely to arise when the costs are low, the benefits are high, and the chance of interacting again is high.

Structural solutions to enhance direct reciprocity

Since the costs, benefits, and probability of future interaction all affect the evolution of cooperation, changing these simple factors can have strong effects on the rates of helping. Changing actual costs and benefits may be difficult, but sometimes all that is necessary is to correct false beliefs about these factors. The stability of partnerships and probability of future interaction (or perceptions thereof) can also sometimes be changed: by making individuals more likely to interact again, it makes it more worthwhile for them to engage in reciprocal exchange of aid.

A less-known but very important structural solution is to change the scale of competition between people (West et al. 2006). West and colleagues used a cooperative game (a Prisoner's Dilemma) where people were divided into subgroups of three within a classroom, and they could cooperate within those subgroups to earn points. They gave a prize to either the highest-scorer in each subgroup or to the highest scorers within the whole classroom. When the prize went to the highest-scorers in the classroom, there was an incentive for people to cooperate within the subgroups in order to better compete against the rest of the classroom ('global competition'), so cooperation flourished. However, when the prize went to the highest-scorer within each subgroup, there was less reason to cooperate within that subgroup because doing so would directly aid one's competitors ('local competition'), and cooperation plummeted. This has many implications for the structure of human groups (Crespi 2006): many business and academic groups are set up such that people compete for promotions or grades against the same people they could be cooperating with. Such competitive structures will inhibit cooperation. If instead groups were designed to shift competition from a local scale (within-group) to a global scale (against the broader population), then it could greatly increase the level of within-group cooperation. One way to do this is to eliminate incentive programmes that rate people only relative to their local group (e.g. choosing or promoting only the best member of each group), because this local competition can encourage hostility and sabotaging within the groups instead of cooperation (see also Sober and Wilson 1998).

Conditional cooperation and expectations of others' behaviour

Some of the most obvious ways to promote charitable behaviour involve people's evolved tendencies towards reciprocation and conditional cooperation. This can involve inducing reciprocation, creating perceptions that others are cooperating, and reducing the negative effects of others who fail to help. For example, many charities induce reciprocity by giving small gifts, thus inducing a feeling of obligation to reciprocate. Other charities announce a large capital contribution (Andreoni 2006), which is like having someone else make the first cooperative move in a cooperative venture.

Evolutionary researchers have noted that people can occasionally make errors by failing to help when they intend to, or by misinterpreting whether others have helped. These errors reduce helping because they trigger retaliation from partners, who assume that the non-cooperation occurred and was deliberate (e.g. Axelrod 1984; Nowak and Sigmund 1992). As such, successful cooperative strategies must be able to recover from these errors by being forgiving of the occasional defection (e.g. Nowak and Sigmund 1992) or by being slightly more generous than one's partner so that one's errors don't make one appear stingy (e.g. Van Lange et al. 2002; Van der Bergh and Dewitte 2006). As such, error-correction mechanisms can help maintain cooperation, as will promoting forgiveness of the occasional failure to help.

Evolutionary theory predicts that organisms will try to avoid cooperating with those who do not cooperate (Trivers 1971; Axelrod 1984: Cosmides and Tooby 1992), so any changes in the perceived or expected cooperativeness of others should change people's likelihood of cooperation. There is much evidence—much of it not explicitly evolutionary—to support this prediction (for a review, see Pruitt and Kimmel 1977). For example, people are much more likely to cooperate if given the opportunity to communicate with each other beforehand (e.g. Caldwell 1976; Ostrom et al. 1992; Davis and Holt 1993); this can occur either because they can make promises to each other or simply because communication gives them an opportunity to assess others' cooperativeness. People are more likely to cooperate with members of an in-group or when in-group membership is made salient (e.g. Messick and Brewer 1983), and Toshio Yamagishi and colleagues have shown that this is because people have higher expectations of cooperation from in-group members (Yamagishi and Kiyonari 2000; Yamagishi 2003). Furthermore, the threat of punishment increases people's contributions towards a group fund because of both a direct effect on people who can receive the punishment and an indirect effect on the resulting expectations of cooperation from others (Shinada and Yamagishi 2007).

There are multiple ways to change people's expectations about others' cooperation. Changing the structural parameters (costs, benefits, length of interactions) should do so. People are more cooperative if given the opportunity to gradually build and escalate trust rather than immediately commit to high helping (e.g. Roberts and Sherratt 1998; Kurzban et al. 2001; Roberts and Renwick 2003). People are also expected to be sensitive to the perceived frequency of helpers and non-helpers in the population (e.g. McNamara and Houston 2002; Barclay 2008; McNamara et al. 2009), so one could work to change this perception or correct any misperceptions. Observing non-cooperation has a greater effect on people's helping behaviour than observing cooperation

(Monteresso et al. 2002; Offerman 2002), so it is better to focus attention on the cooperators rather than the non-cooperators.

One relatively simple way to change people's expectations about others' cooperation is to change how situations are presented to them. If different framings can change what is considered appropriate, then they will change not only what people feel they 'should' do, but will also change people's expectations about others' behaviour and thus make it 'safe' to cooperate. For example, people are more likely to donate money to their group if the group is framed as a 'community' and the money is framed as belonging primarily to the community (Rege and Telle 2004). Similarly, people are more likely to cooperate with partners when a cooperative scenario is framed as the 'Community Game' versus the 'Wall Street Game' (Liberman et al. 2004). The practical implications of this are straightforward: pay attention to presentation. Further research should investigate the relative effects of different types of pro-social framings, for example whether presenting helping as a 'social obligation' is more effective than presenting it as a 'nice thing to do', and whether very obvious framings will be perceived as being manipulative and cause people to react negatively.

Generalized helping and public reputation

Humans are especially known for their willingness to help people outside of their immediate circle of kin, friends, and allies. This psychology is often believed to be have evolved because of various forms of reputational costs and benefits, such as indirect reciprocity, punishment, and costly signalling; I have previously reviewed these in greater detail elsewhere (Barclay, 2010c). There is considerable overlap of predictions and applications arising from these, so I will discuss applications from them together in the following section after briefly introducing each in this section.

Indirect reciprocity

Direct reciprocity occurs when individuals exchange helping acts, but sometimes such acts are reciprocated by someone other than the recipient—this is termed 'indirect reciprocity' (Alexander 1987) and is reviewed in detail by Nowak and Sigmund (2005). In the best-known form of indirect reciprocity (termed 'downstream reciprocity'), those who help tend to acquire a good reputation, and this increases the probability that someone else will help them when they need it. If someone is observed refusing to help, this harms their reputation and the chances that others will help them. Such indirect reciprocity can evolve provided that reputations can be tracked, and people distinguish between a justified failure to help a 'bad' person and an unjustified failure to help a 'good' person (see Nowak and Sigmund 2005). Much experimental evidence shows that people who give help are more likely to receive help, even from those who have never received anything from them and never will (e.g. Gurven et al. 2000; Wedekind and Milinski 2000; Milinski et al. 2001; Seinen and Schram 2006).

In addition to this 'downstream reciprocity', it is worth briefly mentioning one other form of indirect reciprocity called 'upstream reciprocity', where those who are helped become more likely to help others whom they later encounter (Nowak and Sigmund 2005). This is popularly known as 'paying it forward', and can evolve because it triggers direct reciprocity from those helped (Nowak and Roch 2007). Experimental evidence shows that after people receive help, they are indeed more likely to help others that they later encounter (Bartlett and DeSteno 2006; Fowler and Christakis 2010); similar effects have also been found in rats (Rutte and Taborsky 2007). Although this 'upstream reciprocity' is not actually related to reputation—people are basing their helping decisions on their own receipt of help rather than the reputation of others—it still deserves mention because it too can be harnessed to promote charitable behaviour. If people are more likely to help others after receiving help, then one simple idea is to start a wave of charitable behaviour and simply let it propagate itself through social networks (Christakis and Fowler 2009). If the structure of a social network can be changed to better propagate these waves, so much the better.

Punishment

Indirect reciprocity shows that the rewards of a good reputation can sustain charitable behaviour. Having a bad reputation can have explicit costs too, as people are very willing to pay to impose costs upon those who do not pitch in their fair share in cooperative situations (e.g. Yamagishi 1986; Ostrom et al. 1992). People will seemingly punish even if they themselves receive no benefit from the punishee's future cooperative behaviour (Fehr and Gächter 2002). The presence of punishment makes it costly to refuse to help, thus providing a selective pressure for helping (e.g. Boyd and Richerson 1992). Whereas rewards and indirect reciprocity are efficient for eliciting charitable behaviour from just a few group members, punishment is more efficient if cooperation is required from all members of a population, because one only needs to use sanctions on rare non-cooperators (Oliver 1980). Fostering such peer-to-peer punishment may be one way to promote unanimous cooperation.

Despite its power to promote cooperation, peer-to-peer punishment is theoretically puzzling because all group members benefit when an uncooperative group member is coerced into cooperating, but only the punisher pays the cost to deliver such peer-to-peer punishment (Oliver 1980; Yamagishi 1986). As such, the evolutionary function of such apparently spiteful behaviour is currently being hotly debated. For example, it could be maintained by punishers receiving a good or a tough reputation (Brandt et al. 2003; Barclay 2006), conformist imitation of group members (e.g. Guzman et al. 2007), or by punishers or their kin directly benefiting from the subsequent increased group cooperation (e.g. West et al. 2007). An understanding of the relative importance of these forces in maintaining peer-to-peer punishment will increase people's ability to regulate the cooperation in their own groups.

Costly signalling

Informational value

Helping others is costly, especially for those who do not have the resources to spare. In addition, those with little concern for others (and no intent to engage in later cooperation) will be particularly unwilling to incur the costs of helping because they receive no benefits from doing so. Thus, those who observe someone else helping can infer certain traits about the helper, such as his/her abilities, resources, and/or cooperative intent (Hawkes 1991; Boone 1998; Smith and Bliege Bird 2000; Gintis et al. 2001; Lotem et al. 2002). Helping thus functions as an honest signal of those traits, and the costs of helping make it not worth the cost for those who do not honestly possess those traits. Observers benefit from pairing, mating with, or befriending those who have high abilities, resources, and cooperative intent. As such, it is beneficial to be seen to help because this makes one a more desirable social partner or mate (e.g. Smith et al. 2003; Barclay 2004, 2010a). This explanation of helping is but one example of 'costly signalling', an idea which has been of great use in evolutionary biology for explaining exaggerated traits such as peacocks' tails and extravagant displays (for reviews, see Zahavi and Zahavi 1997; Searcy and Nowicki 2005). Applying costly signalling theory to explain human helping does not require that people are aware of what their generosity signals (though obviously some are), only that the reputational benefits can sustain the behaviour.

Evidence for costly signalling

Field and laboratory studies both provide evidence that helping functions as a signal of personal traits and that others respond to it as such. Hunting big game and widely sharing the meat has been described as a costly signal of physical abilities, and data show that good hunters have higher reproductive success and more extra-marital affairs than poorer hunters—even in societies where the hunters receive no more meat than anyone else (e.g. Kaplan and Hill 1985; Hawkes 1991; Smith et al. 2003; Smith 2004). Laboratory studies show that heroic male risk-takers are more desirable mates than non-heroes (Kelly and Dunbar 2001; Farthing 2005); similar effects have been found with other types of helping (Barclay 2010a). Hosting large feasts is an effective way to advertise one's resources because viewers have a vested interest in attending to the signal, not only because of its informational value but also to receive some of the food and gifts being distributed (e.g. Boone 1998; Smith and Bliege Bird 2000). Large monetary donations have also been interpreted as a means to acquire prestige (Harbaugh 1998).

With the former types of help, these acts signal the ability or resources to provide help that not everyone can provide. Helping can also signal concern for others, and can thus be a cue of one's likelihood of cooperating or helping in the future. Less theoretical work has been done on helping as an honest signal of cooperative intent and how the honesty of such signalling is maintained, but few people will be surprised to learn that helping at time A predicts later helping at time B (e.g. Clark 2002; Kurzban and Houser 2005); in fact, this is the basis of personality. Research also suggests that people treat helping as though it carries informational value. For example, people are more trusting of others who give money to other group members (Wedekind and Braithwaite 2002; Barclay 2004, 2006), to charity (Albert et al. 2007), or to their previous social partners (Keser 2003). Costly signalling theory has only recently been used to explain helping behaviour, so it is not yet clear when helping should signal qualities that not everyone possesses (e.g. physical abilities or monetary resources) versus behavioural traits such as a willingness to cooperate; though see André (2010) and Barclay and Reeve (submitted) for recent thoughts on this.

Some researchers argue that religious rituals may serve as costly signals towards members of one's religious group (see Sosis 2004). The cost of such rituals deters those who would take the benefits of group membership without actually valuing the group's position. This idea is important because religious donations comprise the largest type of all charitable donations (Hall et al. 2009). Hazing (initiation rituals involving humiliation, harassment, or abuse) may serve a similar function in fraternities, another group known for high rates of solidarity and monetary support from members.

When do we expect costly signalling?

When should observers pay most attention to signals of quality or cooperative intent? This is an important question, because it affects the size of the reputational benefits, and thus the effectiveness of reputation at increasing helping behaviour. Audiences should only attend to signals if they are informative, i.e. they carry information that is useful and is honest on average. If everyone has the same abilities or resources, then audiences gain no useful information from attending to costly signals of these qualities (Gintis et al. 2001). Attending to such signals only pays if there are enough low quality individuals to make them worth avoiding. Similarly, if everyone is cooperative, then there is no reason to demand costly signals of cooperative intent from potential partners because it is easier to simply accept all potential partners; it only pays to discriminate if there is a risk of being cheated on (McNamara and Houston 2002; McNamara et al. 2009). Thus, as variation in a population increases in terms of abilities, resources, and/or cooperative intent, more people will attend to helping behaviour as a cue of those traits, in turn increasing the reputational benefits for helping. Thus, reputation will be more effective at increasing charitable behaviour when there is high variation in the population.

The power of reputation

Preconditions and evidence

For cooperative sentiment to evolve via indirect reciprocity, punishment, or costly signalling, people must be able to acquire a reputation for their helping (or conversely, not helping), and these reputational benefits must outweigh the costs of providing help. These benefits can come from the recipients of the help, those who observed it, or from others who hear about it. Indirect reciprocity theory and costly signalling theory both make many similar predictions, to the point that disentangling them is often difficult (and possibly unnecessary: see André 2010). However, both of these theories—as well as theories of punishment—make the prediction that helping behaviour should correlate with the magnitude and certainty of the reputational consequences.

These three theories predict that people will be more willing to help others when they can acquire a reputation for doing so, and this prediction has been abundantly confirmed. For example, people give more money to others in experimental games when their behaviour is publicly announced (Rege and Telle 2004; Bereczkei et al. 2007), made available to people they will later interact with (Milinski et al. 2002; Barclay 2004; Hardy and Van Vugt 2006), talked about or commented on by fellow participants (Masclet et al. 2003; Piazza and Bering 2008), or even just known to the experimenters (Hoffman et al. 1994). People are more likely to contribute to a fund to educate others about climate change if they can acquire a reputation for this giving, and they tend to be rewarded for doing so (Milinski et al. 2006). In an experimental tax game, participants were less likely to cheat on their taxes if the pictures of tax-evaders were to be made public, and this effect was in addition to any deterrent effects of monetary sanctions (Coricelli et al. 2010). Sometimes this behaviour is quite strategic, in that the same people help when they can acquire a reputation but then shortly afterwards refuse to help in anonymous situations (Milinski et al. 2002; Semmann et al. 2004; Barclay and Willer 2007; see also Figure 10.2). However, adaptive responses to cues of reputation need not be conscious, and in real life are probably often driven by heightened emotions like guilt and shame in public.

Competitive altruism

When people can choose with whom to interact, people will even actively compete to give more than others, in what has been termed 'competitive altruism' (Roberts 1998; Barclay 2004; Hardy and Van Vugt 2006; Barclay and Willer 2007; Sylwester and Roberts, 2010). Such behaviour increases one's access to desirable social partners. Helping behaviour is sensitive to social comparison, which makes sense in the light of this market-based competition over cooperative partners. For example, people give more money to radio fundraising if given information other people's high donors (Shang and Croson, 2006), and are influenced by the perceived number of other people who have made pro-environmental decisions like conserving water by re-using hotel towels (for review, see Kazdin 2009).

Competitive altruism explains such why such social comparisons are effective: no one wants to lose out on partnership opportunities for being slightly less cooperative than others. Competitive altruism also explains how cooperative norms can escalate over time, as people try to outdo each other in terms of relative generosity. Fostering competitive giving could increase charitable behaviour, such as by naming university buildings after the highest donors. Other ways to

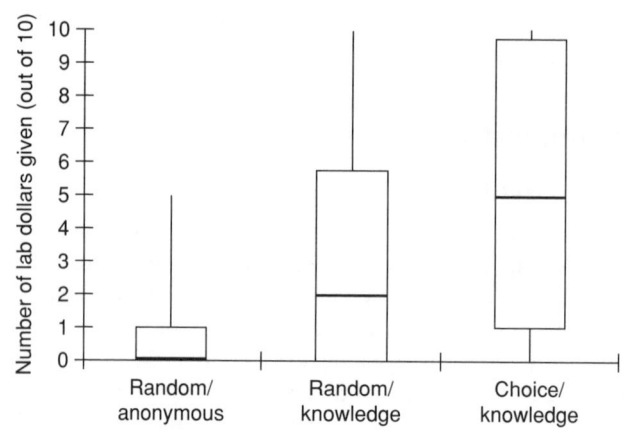

Fig. 10.2 Experimental evidence for competitive giving. Participants in an experimental game gave little money to their partners when no one finds out how much they gave, gave more money when observed by a third party who might then interact with them, and gave the most when that third-party observer could choose with whom to interact (thus giving an incentive to appear more cooperative than one's partner). Barclay and Willer, Partner choices creates competitive altruism in humans. Proceedings of the Royal Society B, 274, 749–753. Copyright (2007) The Royal Society.

foster competitive giving involve increasing people's ability to choose their social partners, thus allowing the best cooperators to pair with the best cooperators and giving everyone an incentive to be more helpful in order to pair up with better partners. Private clubs may use this by allowing opportunities for high donors to meet and interact with each other; political fundraising events (e.g. '$500 a plate' dinners) can serve a similar function.

Implicit cues of reputation

People are so sensitive about reputation that cues of reputation do not even need to be explicit. In one famous study, Haley and Fessler (2005) gave people the option of giving money to others in an experimental game, and then varied whether there were a set of stylized eyespots on the computer screen versus the laboratory's logo. Even though participants' decisions were completely anonymous in either case, they gave more money when eyes were present than when absent. These results have been replicated in other studies with slightly different experimental games and stimuli (Burnham and Hare 2007; Mifune et al. 2010; though see the discussion in the later section 'Situational potency of reputational cues'). Rigdon and colleagues (2009) were even able to show this effect with three dots presented in a 'watching-eyes configuration' (downward triangle) versus a neutral configuration (upward triangle). As evidence that this effect works outside the laboratory, Bateson and colleagues (2006) found that people gave more money to pay for coffee on an 'honour system' when there were eyes above the money jar than when there were flowers. These studies all show that people's reputational concerns can be triggered to increase charitable behaviour, and the triggers can be relatively minor.

People can also be influenced by being primed about other aspects of their reputation. For example, after being put into a mating mindset by viewing pictures of attractive people and imagining a perfect date with one of them, men and women reported being more willing to engage in helping acts that could signal their generosity, prestige, or heroism (Griskevicius et al. 2007). Being primed to think about their social status increases people's reported willingness to buy eco-friendly products (e.g. hybrid cars) relative to similarly priced luxury products (Griskevicius et al. 2010).

If such effects can occur outside the laboratory, then many types of helping behaviour could be affected by triggering the right reputational concerns.

Harnessing the power of reputation

If we can harness these powers of reputation, then we can potentially create powerful and long-lasting increases in charitable behaviour. Evolutionary models of cooperation show that cooperation is more likely when people are likely to interact again or when a person's reputation is more likely to come back to affect them. Currently, many interactions in developed countries are with strangers, such that there is less likelihood of one's reputation benefiting or harming oneself in the long run. On the one hand, this is a drawback, but on the other hand it is a great opportunity: if we can somehow increase the likelihood that people's reputation will follow them to situations where it matters, then we can greatly increase cooperation. This could be accomplished by creating stable groups that are likely to have repeated interactions or by increasing the transmission of reputational information. The Internet may be particularly useful for this information transfer, especially the increased use of Internet social networking sites. Although we value privacy, some things like energy use affect others, so perhaps we should consider making such information more public so that reputational pressures will entice people to act more sustainably.

We can also use people's reputational concerns to foster demand for 'green' or socially responsible products. Once an act (such as purchasing socially responsible products) becomes associated as being 'good', systems of indirect reciprocity will tend to support that act and reward those who do it (Milinski et al. 2002; Panchanathan and Boyd 2004). As long as companies follow their economic interests by responding to consumer demands, then this can influence what products get produced, and can potentially reduce unsustainable or exploitative production. To phrase it poetically: 'money talks, but reputation helps determine what that money says' (Barclay 2010b). Consumer boycotts have been successful before, but an outright boycott is not necessary because in the long run, the larger market share will go to whichever companies or products are being bought more. Rather than wait for long-term effects, companies should be made aware of why people are avoiding their products, because this will result in much faster changes in their practices. It is debatable whether it is more effective to focus on the reputations of individual consumers versus the reputations of Chief Executive Officers (CEOs) or other corporate decision-makers (Jennifer Jacquet, pers. comm. 19 June 19, 2010): focusing on the former can change the market forces such that it is in a company's interest to become socially responsible, whereas focusing on the latter has a greater per capita immediate effect because CEOs have more influence than the average consumer.

Creating reputational pressures

It is one thing to affect the anonymity of people's acts, but is another to make sure that reputational forces will actually act to promote helpful or socially responsible behaviour (i.e. instead of people simply ignoring a given behaviour). Beyond making acts more public, how do we ensure that the right reputational pressures will be there to promote socially responsible acts? This is much easier asked than answered. Once an act is perceived as 'good', then those who actually are 'good' will tend to do it, thus sustaining the correlation between that act and good character, such that reputation can promote the act. But what creates this perception in the first place? Unless that perception already exists, we need a way to link certain acts (e.g. buying only socially responsible products) with a 'good' reputation. Doing so will involve changing public perceptions to make some acts or purchases seem socially desirable and other acts seem socially undesirable, but this is not always easily accomplished and there has not been much evolutionary research on this question.

Although it may be sufficient to simply assert that 'good people do X and bad people do Y', it is probably more effective to point out exactly how X benefits others and Y hurts others. Common knowledge of these benefits and harms is probably important, because it allows people to infer something about the character of whoever is performing the act. Someone who is genuinely concerned for others' welfare will not readily do something that harms another. As such, anyone who knowingly performs such a harmful act can be inferred to not have such concern—but only if they should reasonably know that the act is socially harmful. One potential example of how the norms have changed is that of campaigns against drinking and driving: these have been very successful at changing social norms, such that in many areas the social pressures are more effective at reducing impaired driving than police enforcement. By focusing on the social dangers of impaired driving, it changed people's perceptions of those who drink and drive and made it socially unacceptable to do so. If it becomes more commonly known that certain acts (e.g. pollution, unsustainable consumption) are socially harmful, then people will be more willing to sanction those who do them. In Canada, anti-smoking advertisements are beginning to focus on smoking's harm to others rather than to oneself, which subtly changes the social pressures. In a more extreme case, one can imagine a sort of 'Asshole Campaign' to shame those whose behaviour negatively affects others, such as the use of 'Gas guzzling = Asshole' bumper stickers on highly inefficient vehicles.

One can attempt to link harmful behaviours with negative reputation by focusing on the social costs, or one can link positive behaviours with good reputation by focusing on the social benefits. Thus, education about positive effects can foster a reputation system to support that behaviour. For example, someone who cares about others' welfare and has the opportunity to help at low cost will not knowingly refrain from doing. As such, whether or not someone performs a helping act will become a cue about that person's concern for others—as long as the helper and the observer both know that the act actually does help others. One example of this is how Milinski and colleagues (2006) showed that educating people about the dangers of climate change helps increase the social pressure to fight against climate change. In that experiment, education without reputation had marginal effects on people's willingness to fight climate change, but the combination of education and reputation was particularly effective. Thus, educating people about the effects of their behaviour is not separate from reputational effects. Instead, they are linked: education is a crucial component in creating the reputational pressures to sustain charitable behaviour, and reputational forces give people a reason to care about what they have learned and to act on it.

Limitations

Despite the potential for reputation to increase charitable behaviour, it has several important limitations that need to be overcome in order to promote sustained helping. If we attempt to use reputations to increase charitable behaviour without first giving thought to these issues, the net effect could be marginal, transitory, or possibly even counterproductive. For each of the following limitations—which are presented in increasing order of importance—I also attempt to provide solutions.

Costs of helping may outweigh reputational benefits

It is easy to make the simplistic argument that charitable behaviour will increase when people can acquire a reputation. However, that prediction is based on people receiving social benefits for helping others, so we must actually assess whether the expected reputational benefits will outweigh the costs of helping. When the expected reputational benefits are small relative to the expected costs (e.g. when most people do not associate a particular helping act as being 'good'), then reputation should have a negligible effect on helping, and may even decrease helping (see later section 'Explicit incentives reduce intrinsic motivations for helping'). This point seems extraordinarily obvious, but is also surprisingly easy to forget once we try to incorporate multiple kinds of delayed costs and benefits, multiple observers, and competition over reputational benefits.

Potency of reputational cues versus situations

Sometimes reputation manifests itself with strong and tangible consequences, such as monetary sanctions (e.g. Yamagishi 1986) or the presence of mating opportunities (Griskevicius et al. 2007; Barclay 2010a). However, other cues—such as the use of eyespots to subconsciously trigger reputational concerns (e.g. Haley and Fessler 2005)—are much more subtle and are often not actually followed by real opportunities for reputation. The presence of reputation and non-anonymity may have effects across situations, but weak cues of reputation will be more easily overwhelmed by other features of the situation such as internalized norms regarding appropriate behaviour or other costs and benefits. Consistent with this idea, two recent studies have found that weak reputational cues did little to change giving behaviour in economic games which had strong situational cues and norms: offers and responses in Ultimatum Games are unaffected by the presence of others in a room (Lamba and Mace 2010), and amounts reciprocated in a Trust Game are unaffected by eyespots on a computer (Fehr and Schneider 2010). In these cases, the normative expectations of fairness—and internalization of these norms—may have outweighed the relatively weak reputational cues in those experiments. Thus, reputation may be more effective at eliciting cooperation in some situations than others, such that stronger reputational incentives are necessary in situations where there are other countervailing factors.

Habituation to non-reinforced reputational cues

People may be predisposed to be sensitive to their reputation, but they are also excellent learners whose behaviour is shaped by reinforcement and non-reinforcement. Many animals eventually habituate or acclimatize to stimuli that carry no useful information, and they cease performing non-reinforced behaviours (e.g. Domjan and Burkhard 1993). As such, we should expect humans to similarly habituate to stimuli that lose their usefulness in predicting the presence of important reputational opportunities, and responding to uninformative cues should be extinguished. For example, the presence of eyespots are a subtle cue of being watched and thus increase cooperation (e.g. Haley and Fessler 2005; Burnham and Hare 2007), but if eyespots are repeatedly presented to people without any tangible reputational opportunities, then people should eventually come to ignore them—just as predators eventually come to ignore the fake eyespots flashed by prey species to scare off predators (Stevens 2005). Consistent with this prediction, Soetevent (2005) found that reputational cues had only temporary effects on charitable donations in churches. To resolve this problem, would-be social engineers should ensure that when they cue reputational concerns, those cues should be occasionally followed by real opportunities for non-trivial reputational benefits. Otherwise, the cues will eventually lose their effectiveness with repeated use.

Not everyone cares about reputation

Some experiments show a non-zero proportion of people who continue to act selfishly even when they are being conspicuously observed. For example, Rege and Telle (2004) found that people donate more money towards a group project when they have to announce their contributions to the entire group than when such donations are anonymous, but the proportion of people giving zero is unchanged by the presence of non-anonymity. It is as if there is some low proportion of 'rugged individualists' or psychopaths who care little for the respect or disdain of others. If some people do not need or desire a reputation for helping, then it is not worth the cost for them to help others (Barclay and Reeve submitted). For example, in attractiveness terms, some people are already so desirable that they can get away with being selfish (Takahashi et al. 2006).

These persistent cheaters may drag down the cooperation of others, given that most people cooperate only if others also do so (Fischbacher et al. 2001). Furthermore, some types of cooperation require unanimity (Oliver 1980), and would thus fail to work if there were some people who cared little about social pressures and rewards. When unanimous cooperation is required, tangible punishment of non-cooperators may be required to ensure that everyone cooperates (Oliver 1980), but that punishment must be strong enough to be effective (e.g. Gneezy and Rustichini 2000a).

Devaluing the future benefits of reputation

Investing in future reputational benefits is an investment in the future, so it only pays off for individuals who will be around to reap the long-term benefits. Humans and other organisms value the future less than the present, such that there is a strong temptation to avoid the immediate costs of helping others because future benefits are perceived as less valuable (Frank 1988; Stephens 2000; Stephens et al. 2002). Supporting this idea, people who place more value on the future are more likely to cooperate with each other in experimental games (Harris and Madden 2002).

This devaluing of future benefits is potentially problematic. However, the degree to which people value the future is not static and can vary adaptively according to local circumstances and life history (Wilson and Daly 1997, 2004). Thus, increasing people's 'time horizons' is predicted to increase their charitable behaviour. This will take more than rhetorical exhortations for people to invest in the future; it will take the presence of cues indicating that investing in the future will actually pay off and that each given person is likely to be around for it. Such cues can include cues of long life expectancy, better futures, or realistic life options (Wilson and Daly 1997), though obviously these are major structural changes that are more easily proposed than implemented.

Explicit incentives reduce intrinsic motivations for helping

Whereas some people only help others when there are explicit incentives for doing so, others have internalized norms about helping and seem to have an intrinsic desire to help others (Simpson and Willer 2008). Some have argued that it is this intrinsic motivation which causes much human cooperation (e.g. Fehr and Fischbacher 2004). Unfortunately, much research shows that the presence of extrinsic rewards can undermine intrinsic motivation for tasks (for reviews, see Deci et al. 1999; Bowles 2008). Similarly, when punishment of non-cooperators is present, it undermines trust and reciprocity, possibly because it becomes unclear whether others are cooperating out of intrinsic motivation or out of fear of punishment (Fehr and Rockenbach 2003; Mulder et al. 2006). This reduction in intrinsic motivation can even outweigh any increased cooperation due to explicit incentives: two classic studies found that the presence of fines made parents *more* likely to be late to pick up their children from daycare than when no fines were present (Gneezy and Rustichini 2000a) and that giving small rewards for good performance on IQ tests makes people perform *worse* than if they receive no rewards at all (Gneezy and Rustichini 2000b). Bowles (2008) reviews a number of similar cases.

This undermining of intrinsic motivation is a very serious problem that must be considered whenever using incentives for helping. One potential solution is to rely on implicit incentives, such as verbal rewards, which do not undermine intrinsic motivation (Deci et al. 1999). Even better would be to simply create reputational opportunities for people to apply their own informal social sanctions: these can still be effective at changing behaviour (Barr 2001; Fessler, 2002;

Masclet et al. 2003), but they do not rely on the type of top-down explicit institutional incentives which would undermine intrinsic motivation to cooperate. Eventually, these informal social pressures can cause people to internalize the desire to help. Explicit or institutional incentives could be reserved for situations where reputational pressures are insufficient (see earlier section 'Not everyone cares about reputation').

Identifying the reputational benefits may reduce them

If helping is designed to signal one's good character, then it may be detrimental to call attention to any benefits that people receive. Doing so makes it unclear whether people are helping because they are concerned for others or because they are merely concerned with their own reputation—people tend to like the former type of person and be ambivalent or hostile towards the latter. Some people may even use this ambiguity to downplay the generosity of others: by calling attention to the reputational benefits someone else receives for helping, it limits the reputational advantage he/she receives. This is an example of 'do-gooder derogation' (Monin 2007) that is expected when competition over reputation is high, and is especially expected from non-cooperators who seek to justify their lack of helping (and thus prevent a reputational loss).

This problem is presently tricky to resolve because there is not much theoretical or empirical research on signals of cooperative intent, and even less research on how the effectiveness of such signals varies as a function of the apparent costs and benefits (e.g. audience size). One solution to this problem might be to create subtle opportunities for people to benefit from a reputation, but to strictly avoid any mention of people's reputational benefits. In this case, the social rewards and sanctions can occur on their own without it being commonly known that people do indeed benefit from helping others. A second solution would be to announce other people's generosity on their behalf so they don't have to personally flaunt it; this allows them to get a reputation without engaging in shameless self-promotion, and gives them an excuse for why their generosity becomes common knowledge. Thirdly, even if people are explicitly helping to gain a good reputation, it is worth comparing such people to others who refuse to help regardless of their reputation. It is also worth comparing people who seek status via helping versus those who seek status via costly consumption: someone who conspicuously gives extra meat away is nicer than someone who conspicuously burns it (Gintis et al. 2001), someone with an expensive pro-environmental product is nicer than someone with an equally expensive unsustainable product (Griskevicius et al. 2010), and someone who spends money to save the last whale is morally superior to someone who spends money to catch and eat that whale (Barclay 2010b). By using the appropriate comparisons, we can help ensure that the identification of reputational benefits does not completely eliminate those benefits.

Conclusion

An evolutionary approach focuses on the *function* of charitable behaviour. In addition to examining the psychological mechanisms underlying helping and the development of those mechanisms, evolutionists ask why humans possess psychological mechanisms in the first place, and why humans develop those mechanisms instead of just learning to be selfish. In this chapter, I have outlined some of the major selective pressures responsible for the evolution of helping, and I have suggested means by which this knowledge can be used to promote charitable behaviour. Focusing on the ultimate costs and benefits of helping can lead to research to create situations in which charitable behaviour pays off in the long run. This does not necessarily mean that people consciously seek to benefit from helping others, but instead means that these benefits will increase the frequency of helping either via learning (including internalization of norms) or via natural selection for cooperative sentiment.

In particular, I have focused on harnessing the power of reputation because it is a novel approach with great potential to increase helping behaviour towards non-kin. At the same time, I have attempted to point out some limitations that need to be overcome in order for reputational pressures to be effective. Without consideration of these limitations, harnessing the power of reputation can be ineffective, and at worst could even reduce helping—recall the daycare study by Gneezy and Rustichini, (2000a) discussed in the section 'Explicit incentives reduce intrinsic motivations for helping'. Theories of general reputation are far from complete, and there are many issues that still require resolving, such as how the magnitude of reputational benefits are affected by the apparent costs of helping, the frequency of helpers, the timescale, the type of reward, and especially from the public knowledge that there *are* benefits to having a good reputation. Hopefully the identification of these issues will inspire future research and lead to a more mature science of reputation. And hopefully such a science of reputation will have great power to increase charitable behaviour in the world.

References

Albert, M., Güth, W., Kirchler, E., and Maciejovsky, B. (2007). Are we nice(r) to nice(r) people? An experimental analysis. *Experimental Economics*, **10**, 53–69.

Alexander, R.D. (1987). *The biology of moral systems*. Aldine de Gruyter, New York.

André, J.B. (2010). The evolution of reciprocity: social types or social incentives? *The American Naturalist*, **175**, 197–210.

Andreoni, J. (2006). Leadership giving in charitable fund-raising. *Journal of Public Economic Theory*, **8**, 1–22.

Axelrod, R. (1984). *The evolution of cooperation*. Basic Books, New York.

Barclay, P. (2004). Trustworthiness and competitive altruism can also solve the 'tragedy of the commons'. *Evolution and Human Behavior*, **25**, 209–20.

Barclay, P. (2006). Reputational benefits for altruistic punishment. *Evolution and Human Behavior*, **27**, 325–44.

Barclay, P. (2008). Enhanced recognition of defectors depends on their rarity. *Cognition*, **107**, 817–28.

Barclay, P. (2010a). Altruism as a courtship display: some effects of third-party generosity on audience perceptions. *British Journal of Psychology*, **101**, 123–35.

Barclay, P. (2010b). 'Harnessing the power of reputation.' Invited talk at the TEDxGuelphU conference (independently licensed TED conference), April 2010, University of Guelph, Guelph, ON.

Barclay, P. (2010c). *Reputation and the evolution of generosity*. Nova Science Publishers, Hauppauge, NY.

Barclay, P. and Benard, S. (submitted). 'Power corrupts but competition for power corrupts more': manipulation of perceived threats to preserve rank in cooperative groups.

Barclay, P. and Reeve, H.K. (submitted). Extravagant and everyday helping: the varying relationship between helping and individual quality.

Barclay, P. and Willer, R. (2007). Partner choices creates competitive altruism in humans. *Proceedings of the Royal Society B*, **274**, 749–53.

Barr, A. (2001). *Social dilemmas and shame-based sanctions: experimental results from rural Zimbabwe*. University of Oxford, Institute of Economics and Statistics, Centre for the Study of African Economies, Oxford.

Barrett, L., Dunbar, R.I.M., and Lycett, J. (2002). *Human evolutionary psychology*. Princeton University Press, Princeton, NJ.

Bartlett, M.Y. and DeSteno, D. (2006). Gratitude and prosocial behavior: helping when it costs you. *Psychological Science*, **17**, 319–25.

Bateson, M., Nettle, D., and Roberts, G. (2006). Cues of being watched enhance cooperation in a real-world setting. *Biology Letters*, **2**, 412–14.

Bereczkei, T., Birkas, B., and Kerekes, Z. (2007). Public charity offer as a proximate factor of evolved reputation-building strategy: an experimental analysis of a real-life situation. *Evolution and Human Behavior*, **28**, 277–84.

Boone, J. L. (1998). The evolution of magnanimity: when is it better to give than to receive? *Evolution and Human Behavior*, **9**, 1–21.

Bowles, S. (2008). Policies designed for self-interested citizens may undermine 'the moral sentiments': evidence from economic experiments. *Science*, **320**, 1605–9.

Boyd, R. and Richerson, P.J. (1992). Punishment allows the evolution of cooperation (or anything else) in sizable groups. *Ethology and Sociobiology*, **13**, 171–95.

Brandt, H., Hauert, C., and Sigmund, K. (2003). Punishment and reputation in spatial public goods games. *Proceedings of the Royal Society B*, **270**, 1099–104.

Burnham, T. and Hare, B. (2007). Engineering human cooperation: does involuntary neural activation increase public goods contributions? *Human Nature,* **18**, 88–108.

Caldwell, M.D. (1976). Communication and sex effects in a five-person Prisoner's Dilemma game. *Journal of Personality and Social Psychology*, **33**, 273–80.

Christakis, N.A. and Fowler, J.H. (2009). *Connected: the surprising power of our social networks and how they shape our lives*. Little, Brown and Company, New York.

Clark, J. (2002). Recognizing large donations to public goods: an experimental test. *Managerial and Decision Economics*, **23**, 33–44.

Coricelli, G., Joffily, M., Montmarquette, C., and Villeval, M.C. (2010). Cheating, emotions, and rationality: an experiment on tax evasion. *Experimental Economics*, **13**, 226–47.

Cosmides, L. and Tooby, J. (1992). Cognitive adaptations for social exchange. In: J. Barkow, L. Cosmides, and J. Tooby (eds), *The adapted mind: evolutionary psychology and the generation of culture*, pp. 163–228. Oxford University Press, New York.

Crespi, B. (2006). Cooperation: close friends and common enemies. *Current Biology*, **16**, R414–15.

Daly, M. and Wilson, M. (1988). *Homicide*. Aldine de Gruyter, New York.

Davis, D.D. and Holt, C.A. (1993). *Experimental economics*. Princeton University Press, Princeton, NJ.

Dawkins, R. (1976). *The selfish gene*. Oxford University Press, Oxford.

DeBruine, L.M. (2002). Facial resemblance enhances trust. *Proceedings of the Royal Society B*, **269**, 1307–12.

DeBruine, L.M. (2005). Trustworthy but not lust-worthy: context-specific effects of facial resemblance. *Proceedings of the Royal Society B*, **272**, 919–22.

Deci, E.L., Koestner, R., and Ryan, R.M. (1999). A meta-analytic review of experiments examining the effects of extrinsic rewards on internal motivation. *Psychological Bulletin*, **125**, 627–8.

Domjan, M., and Burkhard, B. (1993). *The principles of learning and behavior (3rd Edition.)*. Brooks/Cole Publishing, Pacific Grove, CA.

Farrelly, D., Lazarus, J., and Roberts, G. (2007). Altruists attract. *Evolutionary Psychology*, **5**, 313–29.

Farthing, G.W. (2005). Attitudes toward heroic and nonheroic physical risk takers as mates and friends. *Evolution and Human Behavior*, **26**, 171–85.

Fehr, E. and Fischbacher, U. (2004). Social norms and human cooperation. *Trends in Cognitive Sciences*, **8**, 185–90.

Fehr, E. and Gächter, S. (2002). Altruistic punishment in humans. *Nature*, **415**, 137–40.

Fehr, E. and Henrich, J. (2003). Is strong reciprocity a maladaptation? On the evolutionary foundations of human altruism. In: P. Hammerstein (ed.), *Genetic and cultural evolution of cooperation*, pp. 55–82. MIT Press, Cambridge, MA.

Fehr, E. and Rockenbach, B. (2003). Detrimental effects of sanctions on human altruism. *Nature*, **422**, 137–40.

Fehr, E. and Schneider, F. (2010). Eyes are on us, but nobody cares: are eye cues relevant for strong reciprocity? *Proceedings of the Royal Society B*, **277**, 1315–23.

Fessler, D.M.T. (2002). Windfall and socially distributed willpower: the psychocultural dynamics of rotating savings and credit associations in a Bengkulu village. *Ethos*, **30**, 25–48.

Fischbacher, U., Gächter, S., and Fehr, E. (2001). Are people conditionally cooperative? Evidence from a public goods experiment. *Economics Letters*, **71**, 397–404.

Foster, K.R., Wenseleers, T., Ratnieks, F.L.W., and Queller, D.C. (2006). There is nothing wrong with inclusive fitness. *Trends in Ecology and Evolution*, **21**, 599–600.

Fowler, J.H. and Christakis, N.A. (2010). Cooperative behavior cascades in human social networks. *Proceedings of the National Academy of Sciences of the USA*, **107**, 5334–8.

Frank, R.H. (1988). *Passions within reason*. Norton, New York.

Gintis, H., Bowles, S., Boyd, R., and Fehr, E. (2003). Explaining altruistic behavior in humans. *Evolution and Human Behavior*, **24**, 153–72.

Gintis, H., Smith, E.A., and Bowles, S. (2001). Cooperation and costly signaling. *Journal of Theoretical Biology*, **213**, 103–19.

Gneezy, U. and Rustichini, A. (2000a). A fine is a price. *Journal of Legal Studies*, **29**, 1–17.

Gneezy, U. and Rustichini, A. (2000b). Pay enough or don't pay at all. *The Quarterly Journal of Economics*, **3**, 791–810.

Griskevicius, V., Tybur, J.M., Sundie, J.M., Cialdini, R.B., Miller, G.F., and Kenrick, D.T. (2007). Blatant benevolence and conspicuous consumption: when romantic motives elicit strategic costly signals. *Journal of Personality and Social Psychology*, **93**, 85–102.

Griskevicius, V., Tybur, J.M., and Van den Bergh, B. (2010). Going green to be seen: status, reputation, and conspicuous conservation. *Journal of Personality and Social Psychology*, **98**, 392–404.

Gurven, M., Allen-Arave, W., Hill, K., and Hurtado, A.M. (2000). 'It's a Wonderful Life': signaling generosity among the Ache of Paraguay. *Evolution and Human Behaviour*, **21**, 263–82.

Guzman, R.A., Rodríguez-Sickert, C., and Rowthorn, R. (2007). When in Rome, do as the Romans do: the coevolution of altruistic punishment, conformist learning, and cooperation. *Evolution and Human Behavior*, **28**, 112–17.

Haley, K.J. and Fessler, D.M.T. (2005). Nobody's watching? Subtle cues enhance generosity in an anonymous economic game. *Evolution and Human Behavior*, **26**, 245–56.

Hall, M., Lasby, D., Ayer, S., and Gibbons, W.D. (2009). *Caring Canadians, involved Canadians: highlights from the 2007 Canada survey of giving, volunteering and participating*. Statistics Canada, Ottawa, ON.

Hamilton, W.D. (1964). The genetical evolution of social behaviour (I and II). *Journal of Theoretical Biology*, **7**, 1–52.

Harbaugh, W.T. (1998). What do donations buy? A model of philanthropy based on prestige and warm glow. *Journal of Public Economics*, **67**, 269–84.

Hardy, C. and Van Vugt, M. (2006). Giving for glory in social dilemmas: the competitive altruism hypothesis. *Personality and Social Psychology Bulletin*, **32**, 1402–13.

Harris, A.C. and Madden, G.J. (2002). Delay discounting and performance on the Prisoner's Dilemma game. *The Psychological Record*, **52**, 429–40.

Haselton, M.G. and Buss, D.M. (2000). Error management theory: a new perspective on biases in cross-sex mind reading. *Journal of Personality and Social Psychology*, **78**, 81–91.

Hawkes, K. (1991). Showing off: tests of an hypothesis about men's foraging goals. *Ethology and Sociobiology*, **12**, 29–54.

Hoffman, E., McCabe, K., Schachat, K., and Smith, V. (1994). Preferences, property rights, and anonymity in bargaining games. *Games and Economic Behavior*, **7**, 346–80.

Kaplan, H. and Hill, K. (1985). Food sharing among Ache foragers: tests of explanatory hypotheses. *Current Anthropology*, **26**, 223–46.

Kazdin, A.E. (2009). Psychological science's contributions to a sustainable environment: extending our reach to a grand challenge of society. *The American Psychologist*, **64**, 339–56.

Kelly, S. and Dunbar, R.I.M. (2001). Who dares, wins: heroism versus altruism in women's mate choice. *Human Nature*, **12**, 89–105.

Keser, C. (2003). Experimental games for the design of reputation management systems. *IBM Systems Journal*, **42**, 498–506.

Kokko, H., Johnstone, R.A., and Clutton-Brock, T.H. (2001). The evolution of cooperative breeding through group augmentation. *Proceedings of the Royal Society B*, **268**, 187–96.

Krupp, D.B., DeBruine, L.M., and Barclay, P. (2008). A cue of kinship promotes cooperation for the public good. *Evolution and Human Behavior*, **29**, 49–55.

Kurzban, R. and Houser, D. (2005). Experiments investigating cooperative types in humans: a complement to evolutionary theory and simulations. *Proceedings of the National Academy of Sciences of the USA*, **102**, 1803–7.

Kurzban, R., McCabe, K., Smith, V.L., and Wilson, B.J. (2001). Incremental commitment and reciprocity in a real-time public goods game. *Personality and Social Psychology Bulletin*, **27**, 1662–73.

Lahti, D.C. and Weinstein, B.S. (2005). The better angels of our nature: group stability and the evolution of moral tension. *Evolution and Human Behavior*, **26**, 47–63.

Lamba, S. and Mace, R. (2010). People recognise when they are really anonymous in an economic game. *Evolution and Human Behavior*, **31**, 271–8.

Liberman, V., Samuels, S.M., and Ross, L. (2004). The name of the game: predictive power of reputations versus situational labels in determining Prisoner's Dilemma game moves. *Personality and Social Psychology Bulletin*, **30**, 1175–85.

Lotem, A., Fishman, M.A., and Stone, L. (2002). From reciprocity to unconditional altruism through signaling benefits. *Proceedings of the Royal Society B*, **270**, 199–205.

Majolo, B., Ames, K., Brumpton, K., Garratt, R., Hall, K., and Wilson, N. (2006). Human friendship favours cooperation in the Iterated Prisoner's Dilemma. *Behaviour*, **143**, 1383–95.

Masclet, D., Noussair, C., Tucker, S., and Villeval, M.C. (2003). Monetary and nonmonetary punishment in the voluntary contributions mechanism. *The American Economic Review*, **93**, 366–80.

McNamara, J.M. and Houston, A.I. (2002). Credible threats and promises. *Philosophical Transactions of the Royal Society B*, **357**, 1607–16.

McNamara, J.M., Stephens, P.A., Dall, S.R.X., and Houston, A.I. (2009). Evolution of trust and trustworthiness: social awareness favours personality differences. *Proceedings of the Royal Society B*, **276**, 605–13.

Messick, D.M. and Brewer, M.B. (1983). Solving social dilemmas. In: L. Wheeler and P. Shaver (eds), *Review of personality and social psychology*, Vol. **4**, pp. 11–44. Sage Publishing, Beverly Hills, CA.

Mifune, N., Hashimoto, H., and Yamagishi, T. (2010). Altruism towards in-group members as a reputation mechanism. *Evolution and Human Behavior*, **31**, 109–17.

Milinski, M., Semmann, D., Bakker, T.C.M., and Krambeck, H.J. (2001). Cooperation through indirect reciprocity: image scoring or standing strategy? *Proceedings of the Royal Society B*, **268**, 2495–501.

Milinski, M., Semmann, D., and Krambeck, H.J. (2002). Reputation helps solve the 'tragedy of the commons'. *Nature*, **415**, 424–6.

Milinski, M., Semmann, D., Krambeck, H.J., and Marotzke, J. (2006). Stabilizing the Earth's climate is not a losing game: supporting evidence from public goods experiments. *Proceedings of the National Academy of Sciences of the USA*, **103**, 394–8.

Monin, B. (2007). Holier than me? Threatening comparison in the moral domain. *Revue Internationale de Psychologie Sociale*, **20**, 53–68.

Monteresso, J., Ainslie, G., Toppi Mullen, P, and Gault, B. (2002). The fragility of cooperation: a false feedback study of a sequential iterated prisoner's dilemma. *Journal of Economic Psychology*, **23**, 437–48.

Mulder, L.B., van Dijk, E., DeCremer, D., and Wilke, H.A.M. (2006). Undermining trust and cooperation: the paradox of sanctioning systems in social dilemmas. *Journal of Experimental Social Psychology*, **42**, 147–62.

Nesse, R.M. (2005). Natural selection and the regulation of defenses: a signal detection analysis of the smoke detector principle. *Evolution and Human Behavior*, **26**, 88–105.

Nowak, M.A. and Roch, S. (2007). Upstream reciprocity and the evolution of gratitude. *Proceedings of the Royal Society B*, **274**, 605–9.

Nowak, M.A. and Sigmund, K. (1992). Tit for tat in heterogenous populations. *Nature*, **355**, 250–3.

Nowak, M.A. and Sigmund, K. (2005). Evolution of indirect reciprocity. *Nature*, **437**, 1291–8.

Oates, K. and Wilson, M. (2002). Nominal kinship cues facilitate human altruism. *Proceedings of the Royal Society B*, **269**, 105–9.

Offerman, T. (2002). Hurting hurts more than helping helps. *European Economic Review*, **46**, 1423–37.

Oliver, P. (1980). Rewards and punishments as selective incentives for collective action: theoretical investigations. *American Journal of Sociology*, **85**, 1356–75.

Ostrom, E., Walker, J., and Gardner, R. (1992). Covenants with and without a sword: self-governance is possible. *American Political Science Review*, **86**, 404–17.

Panchanathan, K. and Boyd, R. (2004). Indirect reciprocity can stabilize cooperation without the second-order free rider problem. *Nature*, **432**, 499–502.

Pennisi, E. (2009). On the origin of cooperation. *Science*, **325**, 1196–9.

Piazza, J. and Bering, J.M. (2008). Concerns about reputation via gossip promote generous allocations in an economic game. *Evolution and Human Behavior*, **29**, 172–8.

Pruitt, D.G. and Kimmel, M.J. (1977). Twenty years of experimental gaming: critique, synthesis, and suggestions for the future. *Annual Reviews of Psychology*, **28**, 363–92.

Reeve, H.K. (2000). Multi-level selection and human cooperation. *Evolution and Human Behavior*, **21**, 65–72.

Reeve, H.K. and Hölldobler, B. (2007). The emergence of a superorganism through intergroup competition. *Proceedings of the National Academy of Sciences of the USA*, **104**, 9736–40.

Rege, M. and Telle, K. (2004). The impact of social approval and framing on cooperation in public good situations. *Journal of Public Economics*, **88**, 1625–44.

Rigdon, M., Ishii, K., Watabe, M., and Kitayama, S. (2009). Minimal social cues in the dictator game. *Journal of Economic Psychology*, **30**, 358–67.

Roberts, G. (1998). Competitive altruism: from reciprocity to the handicap principle. *Proceedings of the Royal Society B*, **265**, 427–31.

Roberts, G. (2005). Cooperation through interdependence. *Animal Behaviour*, **70**, 901–8.

Roberts, G. and Renwick, J.S. (2003). The development of cooperative relationships: an experiment. *Proceedings of the Royal Society B*, **270**, 2279–83.

Roberts, G. and Sherratt, T.N. (1998). Development of cooperative relationships through increasing investment. *Nature*, **394**, 175–9.

Rutte, C. and Taborsky, M. (2007). Generalized reciprocity in rats. *PLoS Biology*, **5**, 1421–5.

Salmon, C. (1998). The evocative nature of kin terminology in political rhetoric. *Politics and the Life Sciences*, **17**, 51–7.

Searcy, W.A. and Nowicki, S. (2005). *The evolution of animal communication: reliability and deception in signaling systems*. Princeton University Press, Princeton, NJ.

Seinen, I. and Schram, A. (2006). Social status and group norms: indirect reciprocity in a helping experiment. *European Economic Review*, **50**, 581–602.

Semmann, D., Krambeck, H.-J., and Milinski, M. (2004). Strategic investment in reputation. *Behavioral Ecology and Sociobiology*, **56**, 248–52.

Shang, J. and Croson, R. (2006). The impact of social comparisons on nonprofit fundraising. *Research in Experimental Economics*, **11**, 143–56.

Shinada, M. and Yamagishi, T. (2007). Punishing free-riders: direct and indirect promotion of cooperation. *Evolution and Human Behavior*, **28**, 330–9.

Simpson, B. and Willer, R. (2008). Altruism and indirect reciprocity: the interaction of person and situation in prosocial behavior. *Social Psychology Quarterly*, **71**, 37–52.

Smith, E.A. (2004). Why do good hunters have higher reproductive success? *Human Nature*, **15**, 343–64.

Smith, E.A. and Bliege Bird, R. (2000). Turtle hunting and tombstone opening: public generosity as costly signaling. *Evolution and Human Behavior*, **21**, 245–62.

Smith, E.A., Bliege Bird, R., and Bird, D.W. (2003). The benefits of costly signaling: Meriam turtle hunters. *Behavioral Ecology*, **14**, 116–26.

Sober, E. and Wilson, D.S. (1998). *Unto others: the evolution and psychology of unselfish behavior*. Harvard University Press, Cambridge, MA.

Soetevent, A.R. (2005). Anonymity in giving in a natural context—a field experiment in 30 churches. *Journal of Public Economics*, **89**, 2301–23.

Sosis, R. (2004). The adaptive value of religious ritual. *American Scientist*, **92**, 166–72.

Stephens, D.W. (2000). Cumulative benefit games: achieving cooperation when players discount the future. *Journal of Theoretical Biology*, **205**, 1–16.

Stephens, D.W., McLinn, C.M., and Stevens, J.R. (2002). Discounting and reciprocity in an Iterated Prisoner's Dilemma. *Science*, **298**, 2216–18.

Stewart-Williams, S. (2007). Altruism among kin vs. nonkin: effects of cost of help and reciprocal exchange. *Evolution and Human Behavior*, **28**, 193–8.

Stevens, M. (2005). The role of eyespots as anti-predator mechanisms, principally demonstrated in the Lepidoptera. *Biological Reviews*, **80**, 573–88.

Sylwester, K. and Roberts, G. (2010). Cooperators benefit through reputation-based partner choice in economic games. *Biology Letters*, 6, 659–62.

Takahashi, C., Yamagishi, T., Tanida, S., Kiyonari, T., and Kanazawa, S. (2006). Attractiveness and cooperation in social exchange. *Evolutionary Psychology*, **4**, 315–29.

Tinbergen, N. (1963). On aims and methods of ethology. *Zeitschrift fur Tierpsychologie*, **20**, 410–33.

Tooby, J. and Cosmides, L. (1996). Friendship and the Banker's Paradox: other pathways to the evolution of adaptations for altruism. *Proceedings of the British Academy*, **88**, 119–43.

Trivers, R. (1971). The evolution of reciprocal altruism. *Quarterly Review of Biology*, **46**, 35–57.

Van den Bergh, B. and Dewitte, S. (2006). The robustness of the 'Raise-the-Stakes' strategy: coping with exploitation in noisy Prisoner's Dilemma games. *Evolution and Human Behavior*, **27**, 19–28.

Van Lange, P.A.M., Ouwerkerk, J.W., and Tazelaar, M.J.A. (2002). How to overcome the detrimental effects of noise in social interaction: the benefits of generosity. *Journal of Personality and Social Psychology*, **82**, 768–80.

Vogel, G. (2004). The evolution of the Golden Rule. *Science*, **303**, 1128–31.

Wedekind, C. and Braithwaite, V.A. (2002). The long-term benefits of human generosity in indirect reciprocity. *Current Biology*, **12**, 1012–15.

Wedekind, C. and Milinski, M. (2000). Cooperation through image scoring in humans. *Science*, **288**, 850–2.

West, S.A., Gardner, A., Shuker, D.M., *et al.* (2006). Cooperation and the scale of competition in humans. *Current Biology*, **16**, 1103–6.

West, S.A., Griffin, A.S., and Gardner, A. (2007). Social semantics: altruism, cooperation, mutualism, strong reciprocity and group selection. *Journal of Evolutionary Biology*, **20**, 415–32.

Williams, G.C. (1966). *Adaptation and natural selection*. Princeton University Press, Princeton, NJ.

Wilson, M. and Daly, M. (1997). Life expectancy, economic inequality, homicide, and reproductive timing in Chicago neighbourhoods. *British Medical Journal*, **314**, 1271–4.

Wilson, M. and Daly, M. (2004). Do pretty women inspire men to discount the future? *Biology Letters*, **271**, S177–9.

Yamagishi, T. (1986). The provision of a sanctioning system as a public good. *Journal of Personality and Social Psychology*, **51**, 110–16.

Yamagishi, T. (2003). The group heuristic: a psychological mechanism that creates a self-sustaining system of generalized exchanges. *Paper prepared for workshop on 'The Co-evolution of Institutes and Behavior'*, Sante Fe Institute, 10–12 January, 2003.

Yamagishi, T. and Kiyonari, T. (2000). The group as the container of generalized reciprocity. *Social Psychology Quarterly*, **63**, 116–32.

Zahavi, A. and Zahavi, A. (1997). *The handicap principle: a missing piece of Darwin's puzzle.* Oxford University Press, New York.

Chapter 11

Altruism as showing off: a signalling perspective on promoting green behaviour and acts of kindness

Wendy Iredale and Mark van Vugt

…cooperation evolves because it constitutes an honest signal of the member's quality as a mate, coalition partner or competitor, and therefore results in advantageous alliances for those signaling in this manner
Gintis et al. (2001, p.103)

Introduction

Compared with other animals, humans are unusual in their altruism and generosity towards unrelated strangers in sometimes very large groups. Other highly social animals (such as bees, termites, or prairie dogs) are altruistic only towards kin, whereas humans are also highly altruistic towards genetic strangers. We donate money to charities, build hospitals to help the sick, make efforts to save the environment, and even risk our lives to save people we have never met before (Alexander 1979; Becker and Eagly 2004; Van Vugt et al. 2000). Why do we engage in these seemingly irrational acts? Why do we go out of our way to help genetic strangers at sometimes considerable cost to ourselves?

From an evolutionary perspective, stranger helping is somewhat puzzling. Darwin's theory of natural selection posits that organisms should be concerned about fostering their personal welfare rather than that of others. So what advantage does helping strangers give, especially if it involves a significant cost? More broadly, if individuals are primarily concerned with their own welfare, how could any selfless behaviour have evolved? This question puzzled Darwin and other great scientists in biology, philosophy, and psychology for many centuries. Darwin (1871) pondered: 'He who was ready to sacrifice his life, as many a savage has been, rather than betray his comrades, would often leave no offspring to inherit his noble nature' (p. 163). Although the puzzle of human altruism has not yet been completely solved, researchers are getting closer to an answer (Barclay 2006; Fehr and Fischbacher 2003; Gintis 2000; Nowak and Sigmund 2005; Van Vugt and Van Lange 2006; Zahavi 1995). In this chapter we discuss some recent insights into the altruism puzzle, focusing on why humans are sometimes motivated to help strangers in very large groups, such as in the case of charity-giving, bystander helping, volunteering, foreign aid, and environmental conservation. Our main perspective is that altruism may have evolved because of its reputation benefits—altruism as showing off.

Survival of the kindest

Traditional evolutionary explanations for altruism suggest that humans help either because they receive pay-offs by helping those with whom they share genes (kin selection or inclusive fitness; Hamilton 1964; Kruger 2003; Madsen et al. 2007; Maynard Smith 1964; Okasha 2002) or by helping those who can return help in the future (reciprocal altruism; Allen-Arave et al. 2008; Axelrod and Hamilton 1981; Rutte and Pfeiffer 2009; Trivers 1971; Wilkinson 1984). The problem with these explanations is that they do not explain all, or even most accounts of human altruism (Lotem et al. 2003; Van Vugt et al. 2007). Altruistic behaviour directed towards non- kin or non-reciprocators has been termed 'disinterested altruism' (Dawkins 1976), and human history is teeming with examples, for example, making blood donations (Alexander 1987), aiding in natural disasters (Van Vugt 2001; Van Vugt and Samuelson 1999), risking life and limb at times of war (Stern 1995), bystander emergency interventions (Latané and Darley 1970), donations to panhandlers (Goldberg 1995; Latané 1970), and public goods provision and charity-giving (Van Vugt et al. 2000). Charitable organizations are based on the notion that people are willing to help complete strangers, and they do! Over the past 25–30 years, for example, people in Britain have raised well over a billion pounds for the charities Comic Relief and BBC Children in Need, showing just how extensive this stranger helping can be.

Survival of the showiest: prestige and reputation-based helping

How could 'disinterested altruism' have evolved? What possible pay-offs could there be for helping genetic strangers in large groups? Some recent evolutionary theories of cooperation assume that being *seen* as an altruist may be an important incentive for offering help. Alexander (1987) explains that there may be indirect benefits for the altruist from others who have observed or heard about the altruistic act. Building on Alexander's ideas, indirect reciprocity models (Nowak and Sigmund 2005) assume that individuals who have an altruistic reputation are more likely to receive benefits from people who they have not helped but who have seen or heard about the act. As an empirical demonstration of indirect reciprocity, Wedekind and Milinski (2000) found that people tend to help individuals who have been shown to help others in earlier interactions. The costs of helping may be offset by the benefits of getting an altruistic reputation. Displaying altruism in front of an audience could be a very cost-effective way of promoting reputation. Because altruists get a positive reputation, they may be perceived as good coalition partners who can be trusted not to defect in future reciprocal helping. An individual would therefore be more likely to help someone who has previously been seen to be an altruist even if they have not received a direct benefit from them. Indeed through the method of image scoring (measuring someone's reputation), indirect reciprocity has been shown to be evolutionary stable (Nowak and Sigmund 1998).

Extending the notion that altruism may have indirect benefits though reputation building, evolutionary signalling theories assume that altruism may have signalling powers. Signalling theories assume that attributes of organisms, that cannot be observed directly, can be conveyed to others through signalling (Alvard 2003; Bliege-Bird and Smith 2005; Otte 1974; Sosis and Alcorta 2003). Signals may convey an individual's quality as a potential mate, ally, or even as a competitor. Unlike indirect reciprocity however, signalling theory is less concerned with 'keeping score', but more concerned about what an individual's behaviour says about their underlying quality. For example, gazelles have been seen to irregularly jump around near potential predators (stotting). It is suggested, since stotting is correlated with physical condition, that this behaviour signals to predators that it would be a waste of their time and effort chasing an individual who can escape capture (Caro et al. 1995). Signals are therefore perceivable indicators of unobservable qualities. Signalling in humans can take a variety of different forms. Veblen's (1899 [1973]) theory of 'conspicuous consumption' suggests that consumption of luxury goods signals a person's underlying financial quality. Acts such as driving a fast car, ordering an expensive bottle of wine, perhaps even wearing a charity's badge, may signal some underlying positive 'quality' (Miller 2000).

Does altruism have a signalling function and, if so, what does it signal? Zahavi's costly signalling theory (1990, 1995) argues that animals appear to 'use' altruism to show off their qualities as coalition partners or sexual mates. Studying Arabian babblers (*Turdoides squamiceps*), he found they behave altruistically by sharing food with non-relatives, communally nest, and take risks to protect others from predators. He argued that birds receive benefits for these altruistic displays through gaining *social prestige* which enhances their coalition and mating opportunities (Zahavi and Zahavi 1997; see also Benabou and Tirole 2006; Bliege-Bird et al. 2001).

Evolutionary signalling theories assume that for a trait to evolve as a signal it must be beneficial to both the signaller and receiver—if they did not benefit, they would simply ignore it (Searcy and Nowicki 2005). Because there are often both convergent and conflicting interests within a relationship, the fundamental problem of signalling is dishonesty and cheating. If a signal is not a true indication of the signaller's underlying quality it would disrupt the communication between signaller and receiver, causing the system to collapse. Therefore, receivers should ignore signals if they are not honest indicators of underlying qualities, and signallers should not invest in signals if it does not alter the behaviour of receivers in ways that benefits them. How can receivers tell whether a signal is an honest indication of some hidden quality? Costly signalling theory (CST) argues that only costly behaviours can emerge as *signals* of quality—they are therefore referred to as handicaps (Zahavi 1975). Those who can carry the handicap have greater biological fitness than those who cannot. Applying this logic to altruism, only a healthy and fit babbler can carry the costs of being generous. Similarly, someone who earns £100,000 a year can theoretically afford to give £50,000 to charity, but someone who earns £50,000 cannot. Through mathematical modelling, costly signals as honest indicators of a signaller's quality have been shown to be evolutionary stable (Grafen 1990).

Competitive altruism

That altruism can be a signal is supported by both modelling and behavioural data on humans. Roberts' (1998) theory of competitive altruism suggests that, because altruism is a signal of quality, individuals competing for partners should compete to display themselves as the most altruistic. Researchers found, through economic games, that the most generous people were chosen more often as cooperative partners, and, because of this, people would actively compete to be more generous than others (Barclay and Willer 2007; Hardy and Van Vugt 2006).

A real-life example of competitive altruism can be seen in a competition run by two Malaysian newspapers; *The New Straits Times* and *The Star*. While raising money for the 2004 Boxing Day Tsunami Relief Fund, they advertised that the individual who raised most for the fund would receive front page coverage. As a consequence, the fund benefited through a large number of very high donations and the highest donor (who gave M$24 million) benefited through the advertisement of his generosity (Grace and Griffin 2009).

But how do we know that altruism is really a signal of some hidden quality? Theorists (e.g. Smith and Bliege-Bird 2000; Hardy and Van Vugt 2006) hold that for a trait such as altruism to be considered a true signal of quality it must fulfil the following criteria: 1) it must be costly to the signaller in terms of energy, time, economic resources, risk, or some other some other domain, and the costlier the behaviour the more honest the signal; 2) it must be conspicuous and therefore easily observable to others; 3) it should be an indicator of traits that are important to others, so they pay attention to it; and 4) it must increase the signaller's fitness in some way. Does human altruism fulfil these criteria?

With respect to costliness, there are various examples of humans engaging in conspicuous displays of generosity. For example, the Meriam people of Melanesia engage in costly displays of food sharing and gift giving, without expectation of reciprocity. Two to five years after a death, the family of the deceased puts on an elaborate feast for the rest of the village (this feast also coincides with the erection of an expensive and showy tombstone). Along with gifts, one of the main courses is usually turtle meat. Bliege-Bird et al. (2001) argue that because turtle meat is often dangerous and time consuming to obtain, the ability to supply many turtles for the funeral feast serves as an honest signal of the physical quality of the males in the family. Festival ceremonies known as potlatch gatherings offer a classic example of the ways in which humans may use altruism as a costly signal (Bliege-Bird and Smith 2005). The potlatch is a cultural practice among tribes of the North Pacific Coast of North America where chiefs organize parties where they give away huge numbers of items such as food, blankets, and ornaments. However, they are not interested in whether or not these goods are useful to recipients (sometimes these goods are even burnt) but in the prestige associated with these 'showing-off' displays of wealth, which helps to forge useful future alliances with other chiefs.

Second, people tend to behave generously when they know they are being observed by others (Barclay 2004; Hardy and Van Vugt 2006; Milinski et al. 2001; Van Vugt and Hardy 2010). For instance, Hardy and Van Vugt (2006) conducted a public goods game in which participants were asked to invest money to a group fund where all members would gain an equal division of the group's investment, regardless of whether or not they made a contribution. If all individuals acted for the good of the group, all would gain greater benefits. However, because individuals kept any money they did not invest in the group, it was in people's self-interest to free-ride on others' contributions, because they would still get a pay-off from everyone else's contributions. The study found that participants gave considerably more when they knew their contribution would be made public. This is also reflected in charity research, which suggests that people donate more when donations are made visible (Caporeal et al. 1989). Indeed, donors may have a taste for the 'prestige' of having their donations made public (Harbaugh 1998), for example, if given a pin or tag that advertises that they have donated (Low and Heinen 1993). Our desire to be seen as altruists often results in other public displays of compassion: 'conspicuous compassion encompasses a number of different behaviours, such as public weeping for deceased celebrities, demonstrations, apologies for historical misdemeanours, and, in terms of donation behaviour, the wearing of empathy ribbons or the like (e.g. plastic red noses)' (Grace and Griffin 2009, p.15). The importance of 'being seen to give' means that people may make donations even if their contribution makes no impact on the benefactor. In an experimental study, people in public conditions were found to contribute to the public good even when it was already provided or could not be provided at all (Van Vugt and Hardy 2010). People would wastefully donate on average 40–50% when publicly monitored, suggesting that people are sometimes more concerned about their reputation than about the efficacy of their helping act. This may explain why people still gave to charities which aided the relief of the 2005 Asian tsunami even once it was publicly announced that any extra money could not be used.

Third, do people pay attention to altruistic displays? Receivers might pay attention because they can gain a high quality partner but bystanders may also be able to benefit from these displays when they are looking for coalition partners or sexual mates. There is some evidence to suggest that charity donations serve as a signal of resources, social esteem (Milinski et al. 2002), cooperative intent (Gintis et al. 2001; McNamara and Houston 2002; Smith et al. 2003; Smith and Bliege-Bird 2005), or commitment.

Fourth and finally, there must be benefits conferred to people with altruistic reputations. Researchers have argued that individuals who signal altruism benefit in terms of better access to cooperative relationships and greater cooperation within those relationships (Bliege Bird et al. 2001; Gintis et al. 2001; Hawkes and Bliege-Bird 2002; Lotem et al. 2003; Zahavi and Zahavi 1997). For example, people are typically more trusting and cooperative towards altruistic individuals than selfish individuals (Barclay 2004, 2006; Milinski et al. 2002; Wedekind and Braithwaite 2002) because altruism signals an individual's social reputation (Benabou and Tirole 2006). Individuals who are more generous in cooperative tasks are often given a higher status by group members and are selectively preferred as interaction partners (Hardy and Van Vugt 2006). Research into charity giving among businesses shows that altruism can pay. For instance, the ice-cream company Ben and Jerry's annually donate 7.5% of pre-tax profit to social causes and this 'corporate social responsibility' enhances the company's reputation, allowing them to charge a premium price for their products, and facilitating the recruitment and retention of high quality workers (McWilliams and Siegel 2000). By advertising one's quality from altruism, an individual might therefore gain direct benefits through social prestige in terms of access to coalitions and sexual mates (Farrelly et al. 2007; Iredale et al. 2008; Phillips 2008; Roberts 1998).

The peacock's tail of altruism: men showing off

Signalling altruism may be particularly important in mate choice. The purpose of sexual signals is either to attract mates of the opposite sex or to deter same-sex rivals (Lindström and Kotiaho 2002). Sexual cues in many species have been shown to function as costly signals (e.g. Andersson 1994; Maynard Smith and Harper 2003). The classic example is the peacock's train, identified as a costly signal because it conveys to peahens that the male has good genes and can afford to carry such an elaborate ornament (Petrie 1994).

Could human altruism be a peacock's tail? Evolutionary psychologist Geoffrey Miller believes so (2000): 'We have the capacity for moral behaviour and moral judgments today because our ancestors favored sexual partners who were kind, generous, helpful, and fair (p. 292).' Is that still true in our modern society? The core theme of many fictional stories (novels, films) is how being generous gives the main character a sexual advantage (for a review see Tessman 1995). In Jane Austen's *Pride and Prejudice*, for example, Elizabeth's affection towards Mr Darcy increases when she hears about him helping her family.

One obvious question is whether altruism is considered to be a sexy quality for both men and women? Altruism signals many positive qualities, such as wealth, resourcefulness, generosity, and commitment which are all important qualities in a potential partner (Buss 1994). In both sexes, kindness is the most desired trait in a sexual partner and this is also what altruism signals (Buss 1989). However, it is also true that men and women value certain attributes differently in sexual partners. This difference stems from sex differences in lifetime reproductive success and parental investment, which is a feature of all mammalian species (Bateman 1948; Trivers 1972; see also Chapter 9, this volume) and it changes the pay-offs of the mating game, with implications for what men and women look for in a partner and thus what partners signal. Although both males and females are attracted to signs good genes and health (Gangestad and Scheyd 2005;

Roberts and Little 2008), men are particularly concerned with physical attraction, in particular cues of age and fertility (Borgerhoff Mulder 1988; Buss 1989; Kenrick and Keefe 1992; Waynforth and Dunbar 1995), whereas women seek men who are good providers and select males who display cues of high social status and wealth (Buss and Schmitt 1993). Furthermore, women are more strongly influenced by non-physical factors than males, in particular, a man's social status (Kniffin and Wilson 2004; Townsend and Levy 1990). Buss (1989) shows that these sex differences are universal; he found that across 37 cultures, women place greater emphasis on status and wealth and males place emphasis on sexual fidelity and physical attraction.

Men might signal their status and resource quality by driving a Porsche car or buying a Picasso painting. However, this does not give any information whether they are willing to share these resources with a potential mate; to signal their quality as a mate it may therefore be better for a man to engage in a conspicuous act of generosity.

What evidence is there that men use altruism as a signal of mate quality? Social psychological studies have examined sex differences among individuals giving money to street beggars and found that lone men disproportionately gave to female street beggars (Goldberg 1995). The presence of a female companion has been found to increase giving to both male and female street beggars (Latané 1970). In a similar study, Mulcahy (1999, cited in Barrett et al. 2002) found that in interviews afterwards, men in the early stages of their relationship were significantly more likely to give to beggars than those in long-term committed relationships, apparently to impress their new partner.

In a recent laboratory experiment we (Iredale et al. 2008) found essentially the same result. Men and women played various economic games for money, during which they were either not observed or there was an (attractive) same- or opposite-sex observer in the same room. After the games had finished, participants were asked what percentage of their earnings they would donate to charity. Men donated significantly more in the presence of an attractive woman, but women's donations did not increase in the presence of an attractive man.

What about helping acts that involve no money? One study examined whether public philanthropy could be elicited by romantic motives (Griskevicius et al. 2007). They found that priming men with mating motives increased conspicuous helping when the act signalled was heroic (e.g. diving into icy water after a stranger falls from a boat in a storm), but not mundane helping (e.g. teaching underprivileged youths to read). Thus, men are more likely to show off their qualities as potential sexual mates by engaging in conspicuous helping displays. Logically men should only show off their generosity if this is a trait that women find attractive in potential mates. Do they?

We know that women have a greater preference towards affectionate and compassionate mates (Howard et al. 1987), and prefer males who talk positively about giving help (Jensen-Campbell et al. 1995). A recent study found that altruistic acts such as 'donates blood regularly' and 'volunteered to help out in a local hospital' were considered by women to be attractive traits in a male sexual partner (Phillips et al. 2008). However, the extent to which altruistic men are preferred as mates depends upon whether women are looking for a long- or short-term partner. Although altruistic males may be preferred as long-term mates, heroes (voluntary risk takers) are preferred over altruists at least for short-term mating purposes (i.e. having consensual sex) (Kelly and Dunbar 2001).

One of our recent studies directly tested whether females are attracted towards altruistic men (Iredale 2010). Female students watched a video clip of an attractive male actor displaying either costly altruism (donating £30), not-so-costly altruism (donating £1), or no altruism (donating nothing), as they walked past a street beggar. Females rated the male actor as more attractive when he gave more money, presumably because giving £30 to a street beggar is an honest signal of some underlying trait. However, when this study was reversed, male ratings of a female actor did not change. Indeed, compared to females, males actively avoid altruistic mates as one-night stands (Barclay 2010). Barclay (2010) argues that this may be because men expect less success with 'good girls' in short-term relationships.

Why is altruism sexy?

Why might females be attracted to displays of altruism in men? What does altruism say about the underlying mate qualities of men? As in most attractive traits, candidates are that altruistic males provide direct benefits (through providing high-quality parental care) or indirect benefits (by providing high-quality genes). In a recent study (Iredale 2010), we tested whether there are there certain types of altruistic acts that are more attractive than others and whether females associate altruism with any important mate qualities such as good genes, resources, or relationship commitment. Women watched a video clip of the same male either donating blood (volunteering), donating money to beggar (financial altruism), rescuing a stolen bag (heroic altruism), or not displaying any of these behaviours. First, women rated him in terms of sexual attractiveness. Next, they were asked to place six lonely heart advertisements in order, according to the perceived likelihood that the man had written them. The adverts varied in terms of good genes quality (using the descriptors 'healthy male', 'good at sports', or 'weak, generally unwell'), resource potential ('high earning ambition', 'no interest in becoming rich'), and relationship quality ('looking to commit to a relationship', 'looking for a bit of fun'). As expected, women rated the male actor significantly more attractive when he displayed altruism. Furthermore, women ranked the male actor rescuing the bag as significantly physically stronger and healthier, supporting the good gene hypothesis that risky and physically costly altruistic acts signal gene quality. In addition, women rated the male as more committed to a relationship when he gave blood or donated money than when he rescued the bag. This research thus shows that altruism as a mate signal can work in two ways: brave, heroic displays can signal genetic quality whereas financial help and volunteering signal willingness to care and invest in the relationship (Iredale 2010).

Going green to be seen: solving the tragedy of the commons

If altruism is a signal of quality, then could other costly behaviours which involve helping the common good (e.g. 'green' behaviours) also work as a signal of quality? It would take us less time, energy, and effort if we didn't bother recycling, to cycle to work, to carry re-usable bags, or turn off the lights each time we leave the house. In essence, being 'green' incurs a cost to ourselves to help the common good and can therefore be regarded as altruistic (Van Vugt 2001). Could 'green' behaviours act as a signal of quality? By understanding environmental behaviour as a costly signal we might help promote 'green' behaviours.

Many local and global environmental problems are social dilemma problems in which private and collective interests collide. The social dynamics underlying many local and global environmental challenges was famously captured by Garrett Hardin's 'Tragedy of the Commons' (1968). The essay tells the story of how the management of a communal pasturage by a group of herdsmen turns into ecological disaster when each of them increases their herd size for personal benefit, thereby unintentionally causing the destruction of the Commons. Such an example highlights the difficulty of engaging people in environmental behaviours. For environmental conservation to be sustainable, everyone must engage in the behaviour. However, if there are personal costs for taking part in environmental care, then it may be difficult to encourage people to participate.

Research suggests that this tragedy is not inevitable, however. Individuals are really not indifferent to the welfare of others, their group, or the environment. In social dilemmas, people are not just motivated by economic self-interest but also by concerns about how they are being perceived by others (Van Vugt 2009). Could it be that, where people make public choices, they are more likely to act in a sustainable way, through recycling, conserving energy, and green purchasing, in order to show off their qualities?

Research also indicates that people do 'show off' conservation behaviour. For example, people are more likely to give money in a public goods game to preserve the environment when the giving was public rather than private (Milinski et al. 2006). This may be because being seen to help the common good can improve one's status (Hardy and Van Vugt 2006). One way in which people might 'show off' their green credentials is through purchasing green products. For example, people may be motivated to buy pro-environmental 'green' products because they signal something about a person's underlying quality and status (Griskevicius et al. 2010). Green products incur a cost to the self, as they often cost more and are of lower quality than their conventional counterparts. However, by buying these products, an individual helps others because green goods benefit the environment for everyone. In that respect, purchasing expensive green products may be a costly signal of quality. In a recent study, Griskevicius et al. (2010) highlight the Toyota Prius, a hybrid car, as an example of a product that acts as a costly 'green' signal. The Toyota Prius offers less luxury to the consumer and is thousands of dollars more expensive than other standard cars; however, it is more energy efficient and therefore kinder to the environment. Although the argument could be made that the owners of these cars are more environmentally conscious than other buyers, by driving such a car, these consumers also say something about their status and quality, indeed, 'Prius owners proudly reported that the number one reason for purchasing the car is because it "makes a statement about me"' (Griskevicius et al. 2010, p. 392).

To test the extent to which green products act as a signal of quality, Griskevicius et al. (2010) conducted three experiments. In their first experiment, participants were asked to imagine they were shopping for a car, a dishwasher, and a household cleaner. They were asked to choose between either a luxurious non-green option or a less high-performing, green option (both costing the same price). Before making their choice, participants were either primed to think about a status situation. This involved imagining they had just arrived for their first day at a high-powered job, where they meet their new boss, who informs them that there was a lot of competition for this job, and that in a year, one of them would move up into a fancy office. Or they were included as controls, in which they either had no story or imagined they had been searching their house for a lost concert ticket, which they found just before they had to leave. The results showed that status could influence the decision to purchase green products, as those who were primed with status were more likely to choose environmentally-friendly products than those who were not.

Similarly, if green purchasing is truly a costly signal then people should engage in these behaviours most when they have observers who can witness their behaviour (Smith and Bliege-Bird 2000). The second study therefore tested whether purchase decisions were influenced by whether the purchase was made in public or private (Griskevicius et al. 2010). Participants were placed in either a public or private condition. In the public condition, participants were told to imagine they were shopping in a store for three products (a backpack, batteries, table lamp). In the private condition participants were told to imagine they were online shopping, for the same products and by themselves, at home. Before they decided which type of product to buy (green or luxury), participants were further placed in a status group or control group as above. Status motivated preference for green products when observed by others, but not when alone, in contrast to preference for luxurious products when alone. This is consistent with costly signalling theory and shows that people forgo luxury products for environmental products when choices are observed and status can be signalled.

However, some research has found that conservation behaviours, such as recycling and taking public transport, are associated with lower, not higher status (Sadalla and Krull 1995), perhaps because these behaviours signal that the person does not have enough resources to behave otherwise. These behaviours may not work as costly signals because they are not costly to display (taking public transport is cheaper than buying a car). Griskevicius et al.'s (2010) third study supports this and shows that 'green' purchasing only works as a signal of quality if it is costly to display. Again, participants were asked to consider that they were shopping in a store for three products (a car, a backpack and a dishwasher), for each being able to choose from a green product or non-green, luxury product. However, unlike study one, the price of the green product varied between groups. Some participants had to make a choice between an expensive green item and a cheaper luxury item, whereas others chose between a cheap, green item and an expensive, luxury item. As above, participants were separated into status and control groups before they made their decision. Status motives had a different effect on the desirability of green products, depending on price. When status was not controlled for, green products were seen as more attractive when they were cheaper than their non-green counterparts. However, when status was controlled for, status motives increased the desire for green products when they were expensive, but not when they were cheap. This suggests that when people are thinking about status, they want to spend more to demonstrate not only that they are environmentally conscious, but also that they can afford to be.

Potential policy implications

The showing-off effect has important implications for charities, volunteer, and environmental organizations. If it is sexy to donate, help others, and go green, then perhaps organizations could make use of this to encourage public acts of altruism. They could emphasize to potential donors that acts of altruism is attractive to the opposite sex. To raise contributions among men in particular, they could use attractive female models to encourage altruism much in the same way that advertisements commonly use attractive female models to sell goods. Volunteering organizations could emphasize networking opportunities for meeting potential business partners, friends, and romantic partners.

Research suggests that applying costly signalling theory to the promotion of environmental behaviours could have real implications. For marketers selling 'green' goods, research suggests they could improve sales by clearly linking such products with status, especially when the product is relatively expensive. In addition, since costly signalling posits that green behaviours must be observable to others, marketers should ensure that green products should stand out from non-green products. Take, for example, recyclable bags in supermarkets: since they cost money to purchase, they are more costly than free, non-recyclable, plastic bags; however, they also display an element of environmental awareness and thus signal something about the consumer's quality. To encourage such behaviour, supermarkets need to make these bags distinct from regular plastic bags. However, the costly signalling framework suggests that it would be ineffective to associate green products to status when they are relatively cheaper than the alternatives, as non-costly products would not signal an underlying quality.

In addition, recent research suggests that people are more likely to act responsibly and help the common good when people are made to feel like they are being observed—even if there is no one present—'the eyes on the screen' effect (Bateson et al. 2006; Haley and Fessler 2005; see also Chapter 10, this volume). This begs the question of whether something as simple as a picture of a pair of eyes on signs reminding people to conserve energy (for example) could elicit this behaviour. If so, there could be a number of applications for businesses and government organizations, say for example posting photographs of eyes above slight switches reminding people to turn off the lights when they leave a room.

More research is needed in this area to fully understand the impact of costly signalling on a range of altruistic behaviours. Nevertheless, by applying costly signalling theory as an explanation of human altruism, we might further understand the origins and motivations for why we help strangers, and perhaps even offer up applicable solutions to encourage people to work for the common good.

References

Alexander, R.D. (1979). *Darwinism and human affairs.* University of Washington Press, Seattle, WA.

Alexander, R.D. (1987). *The biology of moral systems.* Aldine de Gruyte, Hawthorne, NY.

Allen-Arave, W., Gurven, M., and Hill, K. (2008). Reciprocal altruism, rather than kin selection, maintains nepotistic food transfers on an Ache reservation. *Evolution and Human Behavior,* **29,** 305–318.

Alvard, M. (2003). Kinship, lineage identity, and an evolutionary perspective on the structure of cooperative big game hunting groups in Indonesia. *Human Nature,* **14,** 129–63.

Andersson, M. (1994). *Sexual selection.* Princeton University Press, Princeton, NJ.

Axelrod, R. and Hamilton, W.D. (1981). The evolution of cooperation. *Science,* **211,** 1390–6.

Barclay, P. (2004). Trustworthiness and competitive altruism can also solve the 'tragedy of the commons'. *Evolution and Human Behavior,* **25,** 209–20.

Barclay, P. (2006). Dissertation abstract: Reputational benefits of altruism and altruistic punishment. *Experimental Economics,* **9,** 181–2.

Barclay, P. (2010). Altruism as a courtship display: Some effects of third party generosity on audience perceptions. *British Journal of Psychology,* **101,** 123–35.

Barclay, P. and Willer, R. (2007). Partner choice creates competitive altruism in humans. *Proceedings of the Royal Society B,* **274,** 749–53.

Bateman, A.J. (1948). Intra-sexual selection in Drosophila. *Heredity,* **2,** 349–68.

Bateson, M., Nettle, D., and Roberts, G. (2006). Cues of being watched enhance cooperation in a real-world setting. *Biology Letters,* **2,** 412–14.

Becker, S.W. and Eagly, A.H. (2004). The heroism of women and men. *American Psychologist,* **59,** 163–78.

Benabou, R. and Tirole, J. (2006). Incentives and prosocial behaviour. *American Economic Review,* **96,** 1652–78.

Bliege-Bird, R. and Smith, E.A. (2005). Signalling theory, strategic interaction, and symbolic capital. *Current Anthropology,* **46,** 221–48.

Bliege-Bird, R., Smith, E.A., and Bird, D.W. (2001). The hunting handicap: costly signaling in human foraging strategies. *Behavioral Ecology and Sociobiology,* **50,** 9–19.

Borgerhoff Mulder, M. (1988). Behavioural ecology in traditional societies. *Trends in Ecology and Evolution,* **3,** 260–4.

Buss, D.M. (1989). Sex differences in human mate preference. *Behavioral and Brain Sciences,* **12,** 1–49.

Buss, D. (1994). *The evolution of desire.* Basic Books, New York.

Buss, D. and Schmitt, D.P. (1993). Sexual strategies theory: an evolutionary perspective on human mating. *Psychological Review,* **100,** 204–32.

Caporeal, L.R., Dawes, R.M., Orbell, J-M., and van de Kragt, A. (1989). Selfishness examined: cooperation in the absence of egoistic incentives. *Behavioral and Brain Sciences,* **12,** 683–739.

Caro, T.M., Lombardo, L., Goldizen, A.W., and Kelly, M. (1995). Tail-flagging and other antipredator signals in white-tailed deer: new data and synthesis. *Behavioral Ecology,* **6,** 442–50.

Darwin, C. (1871). *The descent of Man and selection in relation to sex.* John Murray, London.

Dawkins, R. (1976). *The selfish gene.* Oxford University Press, Oxford.

Farrelly, D., Lazarus, J., and Roberts, G. (2007). Altruists attract. *Evolutionary Psychology,* **5**, 313–29.

Fehr, E. and Fischbacher, U. (2003). The nature of human altruism. *Nature,* **425**, 785–91.

Gangestad, S.W. and Scheyd, G.J. (2005). The evolution of human physical attractiveness. *Annual Review of Anthropology,* **34**, 523–48.

Gintis, H. (2000). Strong reciprocity and human sociality. *Journal of Theoretical Biology,* **206**, 169–79.

Gintis, H., Smith, E.A., and Bowles, S. (2001). Costly signalling and cooperation. *Journal of Theoretical Biology,* **213**, 103–19.

Goldberg, T. L. (1995). Altruism towards panhandlers: Who gives?. *Human Nature,* **6**, 79–89.

Grace, D. and Griffin, D. (2009). Conspicuous donation behaviour: scale development and validation. *Journal of Consumer Behaviour,* **8**, 14–25.

Grafen, A. (1990). Biological signals as handicaps. *Journal of Theoretical Biology,* **144**, 517–46.

Griskevicius, V., Tybur, J.M., Sundie, J.M., Cialdini, R.B., Miller, G.F., and Kenrick, D.T. (2007). Blatant benevolence and conspicuous consumption: when romantic motives elicit strategic costly signals. *Journal of Personality and Social Psychology,* **93**, 85–102.

Griskevicius, V., Tybur, J.M., and Van den Bergh, B. (2010). Going green to be seen: status, reputation, and conspicuous conservation. *Journal of Personality and Social Psychology,* **98**, 392–404.

Haley, J.J. and Fessler. C.M.T. (2005). Nobody's watching? Subtle cues affect generosity in an anonymous economic game. *Evolution and Human Behavior,* **26**, 245–56.

Hamilton, W.D. (1964). The genetic evolution of social behaviour I and II. *Journal of Theoretical Biology,* **7**, 1–52.

Harbaugh, W.T. (1998). The prestige motive for making charitable transfers. *American Economic Review,* **88**, 277–82.

Hardin, G. (1968). The tragedy of the commons. *Science,* **162**, 1243–8.

Hardy, C. and Van Vugt, M. (2006). Nice guys finish first: the competitive altruism hypothesis. *Personality and Social Psychology Bulletin,* **32**, 1402–13.

Hawkes, K. and Bliege-Bird, R. (2002). Showing off, handicap signalling, and the evolution of men's work. *Evolutionary Anthropology,* **11**, 58–67.

Howard, J.A., Blumstein, P., and Schwartz, P. (1987). Social or evolutionary theories? Some observations on preferences in human mate selection. *Journal of Personality and Social Psychology,* **53**, 194–200.

Iredale, W. (2010). 'Altruism as a mate signal in humans: The role of sexual selection in the evolution and social psychology of human altruism.' Thesis (PhD), University of Kent, UK.

Iredale, W., Van Vugt., M., and Dunbar, R. (2008). Showing off in humans: Male generosity as mate signal. *Evolutionary Psychology,* **6**, 386–92.

Jensen-Campbell, L., Graziano, W.G., and West, S.G. (1995). Dominance, prosocial orientation, and female preferences: do nice guys really finish last?. *Journal of Personality and Social Psychology,* **68**, 427–40.

Kelly, S. and Dunbar, R.I.M. (2001). Who dares wins: heroism versus altruism in women's mate choice. *Human Nature,* **12**, 89–105.

Kenrick, D.T. and Keefe, R.C. (1992). Age preferences in mates reflect sex differences in human reproductive strategies. *Behavioral and Brain Sciences,* **15**, 75–133.

Kniffin, K.M. and Wilson, D.S. (2004). The effect of nonphysical traits on the perception of physical attractiveness: three naturalistic studies. *Evolution and Human Behavior,* **25**, 88–101.

Kruger, D.J., Fisher, M., and Jobling, I. (2003). Proper and dark heroes as dads and cads: alternative mating strategies in British Romantic literature. *Human Nature,* **14**, 305–17.

Latané, B. (1970). Field studies of altruistic compliance. *Representative Research in Social Psychology,* **1**, 49–61.

Latané, B. and Darley, J.M. (1970). *The unresponsive bystander: why doesn't he help?.* Appleton-Century Crofts, New York.

Lindström, L. and Kotiaho, J.S. (2002). Signalling and reception. In: *Encyclopaedia of Life Sciences*. Nature Publishing Group, London.

Lotem, A., Fishman, M.A., and Stone, L. (2003). From reciprocity to unconditional altruism through signalling benefits. *Proceedings of the Royal Society B*, **270**, 199–205.

Low, B.S. and Heinen, J.T. (1993). Population, resources, and environment: implications of human behavioral ecology for conservation. *Population and Environment*, **15**, 7–41.

Madsen, E.A., Tunney, R.J., Fieldman, G., et al. (2007). Kinship and altruism: a cross-cultural experimental study. *British Journal of Psychology*, **98**, 339–59.

Maynard Smith, J. (1964). Group selection and kin selection. *Nature*, **201**, 1145–47.

Maynard Smith, J. and Harper, D. (2003). *Animal signals*. Oxford University Press, Oxford.

McNamara, J.M. and Houston, A.I. (2002). Credible threats and promises. *Philosophical Transactions of the Royal Society B*, **357**, 1607–16.

McWilliams, A. and Siegel, D. (2000). Corporate social responsibility and financial performance: correlation or misspecification?. *Strategic Management Journal*, **21**, 603–9.

Milinski, M., Semmann, D., Bakker, T.C.M., and Krambeck, H.J. (2001). Cooperation through indirect reciprocity: image scoring or standing strategy?. *Proceedings of the Royal Society B*, **268**, 2495–501.

Milinski, M., Semmann, D., and Krambeck, H.J. (2002). Reputation helps solve the 'tragedy of the commons'. *Nature*, **415**, 424–6.

Milinski, M., Semmann, D., Krambeck, H.J., Marotzke, J. (2006). Stabilizing the Earth's climate is not a losing game: supporting evidence from public goods experiments. *Proceedings of the National Academy of Sciences of the USA*, **103**, 3994–8.

Miller, G.F. (2000). Aesthetic fitness: how sexual selection shaped virtuosity as a fitness indicator and aesthetic preference as mate choice criteria. *Bulletin of Psychology and the Arts*, **2**, 20–5.

Nowak, M.A. and Sigmund, C. (1998). Evolution of indirect reciprocity by image scoring. *Nature*, **393**, 573–7.

Nowak, M.A. and Sigmund, C. (2005). Evolution of indirect reciprocity. *Nature*, **437**, 1291–8.

Okasha, S. (2002). Genetic relatedness and the evolution of altruism. *Philosophy of Science*, **69**, 138–49.

Otte, D. (1974). Effects and functions in the evolution of signalling systems. *Annual Review of Ecology and Systematics*, **5**, 385–417.

Petrie, M. (1994). Improved growth and survival of offspring of peacocks with more elaborate trains. *Nature*, **317**, 290–1.

Phillips, T., Barnard, C., Ferguson, E., and Reader, T. (2008). Do humans prefer altruistic mates? Testing a link between sexual selection and altruism towards non-relatives. *British Journal of Psychology*, **99**, 555–72.

Roberts, G. (1998). Competitive altruism: from reciprocity to the handicap principle. *Proceedings of the Royal Society B*, **265**, 427–31.

Roberts, S.C., and Little, A.C. (2008). Good genes, complementary genes and human mate choice. *Genetica*, **132**, 309–21.

Rutte, C. and Pfeiffer, T. (2009). Evolution of reciprocal altruism by copying observer behaviour. *Society and Science—Interdisciplinary Exchanges*, **97**, 1573–8.

Sadalla, E.K. and Krull, J.L. (1995). Self-presentational barriers to resource conservation. *Environment and Behavior*, **27**, 328–53.

Searcy, W.A. and Nowicki, S. (2005). *The evolution of animal communication: reliability and deception in signalling systems*. Princeton University Press, Princeton, NJ.

Smith, E.A., Bird, R., and Bird, D.W. (2003). The benefits of costly signalling: Meriam turtle hunters. *Behavioral Ecology*, **14**, 116–26.

Smith, E.A. and Bliege-Bird, R.L. (2000). Turtle hunting and tombstone opening: public generosity as costly signalling. *Evolution and Human Behavior*, **21**, 245–61.

Smith, E.A. and Bliege-Bird, R. (2005). Costly signalling and cooperative behaviour. In: H. Gintis, S. Bowles, R. Boyd, and E. Fehr (eds), *Moral sentiments and material interests: the foundations of cooperation in economic life*, pp. 115–48. MIT Press, Cambridge, MA.

Sosis, R. and Alcorta, C. (2003). Signalling, solidarity, and the sacred: the evolution of religious behavior. *Evolutionary Anthropology*, **12**, 264–74.

Stern, P.C. (1995). Why do people sacrifice for their nations?. *Political Psychology*, **16**, 217–35.

Tessman, I. (1995). Human altruism as a courtship display. *Oikos*, **74**, 157–8.

Townsend, J.M. and Levy, G.D. (1990). Effects of potential partners' physical attractiveness and socioeconomic status on sexuality and partner selection. *Archives of Sexual Behaviour*, **19**, 149–64.

Trivers, R.L. (1971). The evolution of reciprocal altruism. *Quarterly Review of Biology*, **46**, 35–57.

Trivers, R.L. (1972). Parental investment and sexual selection. In: B. Campbell (ed.), *Sexual selection and the descent of Man, 1871–1971*, pp. 136–79. Aldine, Chicago, IL.

Van Vugt, M. (2001). Community identification moderating the impact of financial incentives in a natural social dilemma: a water shortage. *Personality and Social Psychology Bulletin*, **27**, 1440–9.

Van Vugt, M. (2009). Averting the tragedy of the commons: using social psychological science to protect the environment. *Current Directions in Psychological Science*, **18**, 169–73.

Van Vugt, M. and Hardy, C. (2010). Cooperation for reputation: wasteful contributions as costly signals in public goods. *Group Processes and Intergroup Relations*, **13**, 1–11.

Van Vugt, M. and Samuelson, C.D. (1999). The impact of metering in a natural resource crisis: a social dilemma analysis. *Personality and Social Psychology Bulletin*, **25**, 731–45.

Van Vugt, M. and Van Lange, P. (2006). Psychological adaptations for prosocial behaviour: the altruism puzzle. In: M. Schaller, D. Kenrick, and J. Simpson (eds), *Evolution and social psychology*, pp. 237–61. Psychology Press, New York.

Van Vugt, M., Snyder, M., Tyler, T., and Biel, A. (2000). *Cooperation in modern society: promoting the welfare of communities, states, and organisations*. Routledge, London.

Van Vugt, M., Roberts, G., and Hardy, C. (2007). Competitive altruism: reputation based cooperation in groups. In R.I.M. Dunbar and L. Barrett (eds), *Handbook of evolutionary psychology*, pp. 531–541. Oxford University Press, Oxford.

Veblen, T. (1973). *The theory of the leisure class* [originally published in 1899]. Houghton Mifflin, Boston, MA.

Waynforth, D. and Dunbar, R.I.M. (1995). Conditional mate choice strategies in humans: evidence from 'Lonely Hearts' advertisements. *Behaviour*, **132**, 755–79.

Wedekind, C. and Braithwaite, V.A. (2002). The long-term benefits of human generosity in indirect reciprocity. *Current Biology*, **12**, 1012–15.

Wedekind, C. and Milinski, M. (2000). Cooperation through image scoring in humans. *Science*, **288**, 850–2.

Wilkinson, G.S. (1984). Reciprocal food sharing in vampire bats. *Nature*, **308**, 181–4.

Zahavi, A. (1975). Mate selection: a selection for a handicap. *Journal of Theoretical Biology*, **53**, 205–14.

Zahavi, A. (1990). Arabian babblers: the quest for social status in a cooperative breeder. In: P.B. Stacey and W.D. Koenig (eds), *Cooperative breeding in birds: longterm studies of ecology and behavior*, pp. 103–30, Cambridge University Press, Cambridge.

Zahavi, A. (1995). Altruism as a handicap—the limitations of kin selection and reciprocity. *Journal of Avian Biology*, **26**, 1–3.

Zahavi, A. and Zahavi, A. (1997). *The handicap principle: a missing piece of Darwin's puzzle*. Oxford University Press, Oxford.

Chapter 12

Evolutionary perspectives on intergroup prejudice: implications for promoting tolerance

Justin H. Park

Introduction

Can we achieve universal peace and tolerance? There are reasons to be optimistic. Around the world, people of diverse ethnicities, nationalities, and faiths coexist peacefully. Tolerance has become a social norm—if not a moral ideal—for the first time in many societies. Traditionally oppressed groups of individuals are attaining equal rights in many important spheres of life. Many university students today are so unprejudiced that sophisticated equipment is needed to measure their subconscious biases.

At the same time, we should not be complacent. Blatant and violent forms of intolerance remain ever-present, even in nominally tolerant societies. Recent surges of support for nationalist political parties in Western Europe betray the fragility of multiculturalism. Fundamentalism of all stripes is spreading in many parts of the world and fuelling intolerance. War and genocide—the ultimate forms of intolerance—show no auspicious signs of fading into history.

Explaining the gulf between such extremes of human conduct will require multidisciplinary efforts and judicious integration of ideas. This chapter is about how evolutionary perspectives have contributed to our understanding of intergroup prejudice—particularly of the underlying fundamental psychological processes—and about the implications of our enhanced understanding for efforts to reduce prejudice. Prejudice has been a staple of social psychology for decades; how did evolution enter the discussion?

Brief background

It has long been observed that humans congregate into groups and hold overly flattering views of their in-group relative to out-groups (Sumner 1906). Decades of field research yielded abundant data firmly establishing this phenomenon (Brewer and Campbell 1976; LeVine and Campbell 1972; Sherif et al. 1961). Why do people exhibit this tendency? *Social identity theory* offered one explanation: individuals identify with groups, and they derive esteem by positively distinguishing their group over other groups; negativity toward out-groups is, therefore, a by-product of motivations to exalt the in-group (Tajfel and Turner 1979, 1986). A key proposition of this theory was that intergroup bias is cognitively basic—in other words, even in completely de-contextualized settings, mere classification into groups should make people exhibit in-group favouritism (see Brewer 1979; Tajfel and Turner 1979).

The elegance of social identity theory made it one of the most influential ideas in social psychology. Nevertheless, a critical question remained: can it account for the varied forms and intensities of prejudices in the real world? Here is how Gil-White (2001, p. 535) put the problem:

> Why [is there] ethnic hatred and warfare but, for example, no architect/lawyer riots? Why don't students from different universities attack each other with rocks and assault weapons and protest interuniversity marriages? Clearly, social categories are not all the same, and a purported bias against any and all out-groups cannot, by its very generality, explain why conflict should be so different across different kinds of group boundaries.

Indeed, why are some groups of individuals (e.g. ethnic/national/religious out-groups) more likely to inspire hatred and aggression, compared with others (e.g. elderly people, lawyers) who elicit little more than mild aversion and perhaps behavioural avoidance? More fundamentally, what qualifies as a *group*? Are all possible social categories *groups* in a psychologically meaningful sense? Why do people form groups at all? And why identify with groups?

Such questions exposed the limits of social identity theory, which explained different social identities in terms of 'psychological salience' (for example, the reason that ethnicity matters more to people than, say, handedness, is that the former is psychologically more salient). Escaping this tautology and answering the more fundamental questions required a new footing, and evolutionary perspectives provided one. The basic starting point of evolutionary perspectives is that there may exist naturally selected psychological tendencies toward intergroup conflict (Darwin 1871; Richerson and Boyd 1998; Tooby and Cosmides 1988).

As often happens with any truly novel way of thinking, evolutionary perspectives encountered resistance (see Kenrick et al. 2006). As some of the resistance appears to stem from misunderstandings, it is useful to clarify what evolutionary perspectives do and do not imply.

Some misunderstandings

One elementary misunderstanding is the belief that identifying evolved psychological processes underlying prejudice may somehow excuse prejudice. Of course, it does not. Deciding that prejudice, war, and genocide are morally wrong is logically separate from illuminating their causes, including evolved psychological processes.

Another elementary misunderstanding is the belief that 'evolved' implies 'hardwired' which implies 'immutable'. Since human behaviour is clearly not immutable, the reasoning goes, much of human psychology must not be evolved. The truth is the exact opposite. Humans are more flexible than chimpanzees (and dogs, ants, and bacteria) precisely because humans have a richer set of evolved psychological processes, including the capacities to plan and to anticipate the consequences of their actions. These processes are what allow humans—but not other territorial animals—to overcome their antagonistic instincts and achieve, if only precariously, tolerant societies of diverse groups. To make social progress, we must understand the antagonistic instincts, not deny them.

A more sophisticated misunderstanding is the belief that evolutionary psychological approaches imply that people are driven—consciously or unconsciously—towards gene propagation. However, not only are gene-propagation motives theoretically unlikely (Tooby and Cosmides 1990), there is little empirical evidence for them. What actually seem to occupy people's heads are psychological mechanisms (constellations of cognitions, emotions, and motives) that are geared towards more specific problems such as finding food, mates, and allies, and avoiding disease, injury, and exclusion—mechanisms that exist because they happen to serve the ultimate function of gene propagation (Neuberg et al. 2010; Park and Buunk 2010; Pinker 1997; Tooby and Cosmides 1992).

Evolutionary approaches to intergroup prejudice

Few would deny that humans are a social species—no person is an island. A more important observation is that humans are a tribal species, readily forming coalitional alliances for the purposes of besting other alliances (Eibl-Eibesfeldt 1974; Richerson and Boyd 2005). This coalitional psychology of 'us' versus 'them', whose origins likely predate *Homo sapiens*, is deeply ingrained, as the social identity researchers discovered. But there is more to this psychology than deriving esteem by favouring one's in-group. Evolutionary analyses imply a number of specific hypotheses pertaining to how people draw in-group–out-group categorizations and what people think and feel about in-group and out-group members (Kurzban and Leary 2001; Schaller and Neuberg 2008; Van Vugt and Park 2010). Empirical investigations testing those hypotheses—most carried out in the past ten years—have reshaped our understanding of intergroup prejudice. In fact, it has become apparent that the concept of *intergroup* prejudice should not be loosely applied to all varieties of prejudice. In the remainder of this chapter, the term *intergroup prejudice* refers to phenomena that can reasonably be traced to coalitional psychology. As discussed below, many common prejudices do not fit the template.

There are three key ideas from evolutionary psychology that have been particularly useful in mapping out psychological processes underlying intergroup prejudice: specific threats posed by out-groups, error management theory, and functional flexibility. These are discussed in turn.

Specific threats posed by out-groups

Organisms constantly face challenges to survival and reproduction; many structural and behavioural traits are adaptations that evolved in response to those challenges. Tribal humans needed to deal with adaptive challenges posed by out-group members, including the threat of injury from intergroup hostility and the threat of disease from contact with immunologically-segregated out-groups (Diamond 1997; Schaller et al. 2003a; Van Vugt and Park 2009). Behavioural tendencies that helped humans neutralize those threats—via either avoidance or aggressive exclusion of threat-posing individuals—would have been adaptive. As adaptive behaviours require specific cognitions, emotions, and motives to drive them (Buss 1995; Nesse 1990), the psychological mechanisms driving those behaviours likely comprise specific cognitive processes (e.g. ascription of stereotypes), as well as specific emotional–motivational processes (e.g. arousal of fear or disgust). An important implication is that prejudices associated with different kinds of threat are likely to be qualitatively distinct, involving different psychological substrates (Cottrell and Neuberg 2005; Schaller et al. 2003a).

Error management theory

To respond adaptively to threat, it must first be detected. But the presence—and magnitude—of threat is often uncertain and must be inferred from imperfect cues, which means that any particular inference strategy entails inference errors (this is an example of a signal-detection problem). This raises an interesting question: to what extent should the subjective inference of threat correspond with the objective probability of threat suggested by available cues? A concrete example may help. Suppose that a soldier—say, a helicopter pilot—needs to quickly identify hostile enemies amongst civilians (in order to kill them), and he regularly encounters groups of strangers on the ground whose coalitional alliance is ambiguous. There are four possible scenarios, with four possible signal-detection consequences: 1) the strangers are enemy combatants and the soldier correctly infers this ('hit'); 2) the strangers are civilians and the soldier correctly infers this ('correct rejection'); 3) the strangers are civilians and the soldier incorrectly infers that they are enemy combatants ('false alarm'); and 4) the strangers are enemy combatants and the soldier incorrectly infers that they are civilians ('miss'). The soldier can try to be unbiased, balancing out false alarms and misses; or he can adopt biases toward false alarms or misses. In terms of his welfare, misses are more costly than false alarms. Consequently, the soldier will benefit by adopting a bias toward false alarms across these situations, which may be accompanied by a tendency to over-perceive threats given the available evidence.

To the extent that analogous situations recurred during human evolution (i.e. having to infer threats on the basis of imperfect cues), it is likely that humans evolved predispositions to over-perceive threats. This is an implication of *error management theory*, which proposes that across a variety of signal-detection contexts, perceivers should evolve to be biased toward committing the less reproductively costly inference error (Haselton and Buss 2000; Haselton and Nettle 2006). Thus, people may be biased not only toward 'seeing' threatening out-group members, but also toward overestimating the threat they pose.

Functional flexibility

The tendencies to over-perceive threats and to respond with specific aversive cognitions and emotions may have evolved for adaptive reasons, but they are not unconditionally adaptive. Excessive over-perception of threat and excessive aversive responses can impose net costs on reproductive fitness if the actual probability of threat is low. Over evolutionary history, it is likely that individuals encountered fluctuating levels of threat. An important implication is that psychological mechanisms may have evolved to be *functionally flexible*, contingently varying their responses as a function of threat-connoting factors (Schaller et al. 2007). In other words, defensive and aversive responses may be flexibly amplified when the (perceived) threat is greater, when their benefits are more likely to outweigh their costs.

It is perhaps not surprising that modern warfare conditions create many false alarms, with soldiers killing and imprisoning huge numbers of civilians (e.g. Burnham et al. 2006). Police officers also must make quick decisions, often on the basis of ambiguous cues: deciding to open fire if they think they saw a weapon, for instance. Does heightened threat increase the likelihood that they will wrongly perceive a weapon (and wrongly pull the trigger)? This appears to be the case. In fact, laboratory experiments have shown that threat may be suggested simply by the colour of a person's skin. An experimental task designed to simulate police officers' circumstances revealed that participants were more likely to perceive harmless objects as weapons when the objects were held by black men (who are, because of cultural stereotypes, associated with threat) than when they were held by white men, as indicated by a greater tendency to shoot weaponless black men (Correll et al. 2002).

Psychological processes underlying intergroup prejudice

This section presents a review of research on fundamental psychological processes underlying intergroup prejudice. There are two important components of prejudice: *in-group–out-group categorization* and *biased attitudes and cognitions*. The three key ideas described earlier in the 'Evolutionary approaches to intergroup prejudice' section have helped researchers gain greater understanding of why certain prejudicial tendencies are especially ubiquitous and violent; they have also led researchers to novel hypotheses and findings pertaining to specific cognitive and emotional processes, as well as factors that moderate those processes. Perhaps most importantly, the newfound knowledge gives us clearer clues regarding what we might do to reduce intergroup prejudice and what trade-offs may be inherent in this endeavour.

In-group–out-group categorization

Genocide Watch (www.genocidewatch.org) is an organization that 'exists to predict, prevent, stop, and punish genocide and other forms of mass murder'. Drawing on historical and sociological evidence, Gregory Stanton (its president) outlined eight successive stages of genocide, each describing human propensities and/or external factors that increase the likelihood of genocide. The first, according to Stanton, is *classification*—the tendency to divide the social world into 'us' and 'them'. The significance of classification is something that few social psychologists would dispute. As alluded to above, however, some classifications may matter more than others, and matter for different reasons.

In principle, people can be categorized along an endless number of variables: male–female, black–white, righty–lefty, taller–shorter than 1.67 metres, ad infinitum. In reality, a fairly small number of categories are robust and substantially influence social organization (for instance, genocides tend to be committed against ethnic, national, or religious out-groups; Jones 2006). One approach to the study of social categories is to examine which ones people tend to use when forming impressions of others—that is, which have psychological priority over others. Social psychologists have proposed that there are three 'primitive categories': race, gender, and age (Fiske and Neuberg 1990; Hewstone et al. 1991; Messick and Mackie 1989; Taylor et al. 1978).

Why gender and age should be prioritized makes sense from an evolutionary perspective. Knowing whether a person is male or female affords a number of useful, adaptively-relevant inferences, as does knowing whether a person is a child, youth, or adult. For instance, to identify potential romantic partners or rivals, gender and relative age are key pieces of information. Indeed, all known cultures divide their social world according to gender and age in similar ways (Brown 1991).

Does knowing a person's race provide adaptively relevant information? Intuitively it might seem to do so, and few social psychologists seem to have questioned this intuition. However, Kurzban et al. (2001) scrutinized race categorization from an evolutionary perspective and came away with a different view. They proposed that because humans rarely came into contact with individuals of other 'races' for much of evolutionary history, there are unlikely to be evolved psychological mechanisms dedicated to perceiving and responding to race per se. But because humans did regularly come into hostile contact with members of coalitional out-groups (bands, tribes), psychological mechanisms specialized for tracking coalitional alliances are likely to have evolved (see also Cosmides et al. 2003; Kurzban and Leary 2001). Using an experimental paradigm that social psychologists frequently use to identity basic categories, Kurzban et al. (2001) found that when others' coalitional alliances (membership in basketball teams) were uncorrelated with race, perceivers were more likely to encode coalitional alliance than race (put differently, one is more likely to remember a player's team membership than his race). They concluded that coalitional alliance—not race—is the real 'primitive'; race only appears as a primitive in many social psychological experiments because it serves as a heuristic cue for coalitional alliance in contemporary multiracial societies that are not completely integrated, perhaps because 'racial' features are readily perceptible. In fact, there is little biological basis for race categories (Cosmides et al. 2003), and a number of other important social categories—nationality, language, religion—may constitute more valid cues for coalitional alliance.

One intriguing set of studies has shown that young children may be particularly attentive to language as a cue for coalitional alliance: 6-month-old infants showed a preference for strangers who spoke their native language and in the native accent, 10-month-old infants were more likely to accept a toy offered by someone who spoke their native language, and 5-year-old children were more likely to choose to befriend children who spoke in their native accent (Kinzler et al. 2007). An even more intriguing finding was that although 5-year-old children were more likely to choose same-race children when the target children were silent, pitting linguistic accent against race eliminated this bias: children were more likely to choose other-race children speaking in their native accent than same-race children speaking in a foreign accent (Kinzler et al. 2009). Similar to Kurzban et al. (2001), these results indicate that race serves as a weak cue for coalitional alliance, one that is easily overridden by other, more valid cues. There is also evidence that 12-month-old infants appreciate behavioural indicators of alliance, such as helping and hindering (Kuhlmeier et al. 2003). More generally, other researchers have found that humans have a propensity to perceive different 'kinds' of people with distinct 'essences' (Gil-White 2001), a propensity which emerges at an early age and is distinct from the propensity to perceive gender and age categories (e.g. Hirschfeld 1996).

When people's coalitional alliances are not obvious, there may be classification errors. As discussed above, error management theory implies that people may be biased toward erroneously perceiving ambiguous others as hostile out-group members; the logic of functional flexibility suggests that this bias may be further exaggerated when perceivers feel vulnerable to harm. Miller et al. (2010) tested this hypothesis in a series of experiments. They found that white individuals were more likely to categorize racially ambiguous targets as black when perceptions of threat were heightened (when the target moved toward the perceiver, when the perceiver felt fearful). Even when the targets were racially unambiguous, the level of threat had a noticeable effect: when perceivers were under time pressure, white male faces holding angry (threatening) expressions were more likely to be categorized as black, than vice versa. Such threat-induced tendency to categorize others as out-group members can also occur in more systematic ways. In the months leading up to World War II, French Jews, who had all but achieved the status of full members of French society, began to be seen as outsiders by mainstream society. Cognizant of their predicament, some French Jews tried—unsuccessfully—to emphasize their coalitional alliance with France by setting themselves apart from the newly-arrived Jewish refugees (Friedländer 2007).

People can also vary in the extent to which they perceive in-group–out-group boundaries to be permeable, and the logic of functional flexibility suggests that heightened threat may lead to sharper in-group–out-group distinctions, increasing the probability that potentially harmful out-group members are kept classified as out-group members. Functional flexibility also applies at a cross-societal level. One type of threat that is pervasive yet variable across societies is that of communicable pathogens: some parts of the world, for various reasons, are characterized by higher prevalence of disease-causing pathogens. According to the logic of functional flexibility, people in those regions may benefit from behavioural tendencies that inhibit contact with out-group members who may harbour diseases to which the in-group members have no immunity. Collectivism (as opposed to individualism) is a cultural pattern associated with more strongly defined in-group–out-group distinctions and reduced intergroup contact (i.e. greater xenophobia). Fincher et al. (2008) hypothesized that regions with higher pathogen prevalence may also tend to have more collectivistic cultures; they found a strong correlation between pathogen prevalence and collectivism.

Understanding the conditions under which people categorize others as coalitional out-group members—and do so especially robustly—is important, as coalitional out-group members tend to be the targets of specific kinds of biased attitudes and cognitions.

Biased attitudes and cognitions

Attitudes—or degrees of liking/disliking—are among the most widely researched topics in social psychology. For prejudice researchers within social psychology, attitudes toward specific social categories constitute the core of the matter (Allport 1954; Brewer 1999; Messick and Mackie 1989). Within this tradition, prejudice has usually been conceptualized as general antipathy— non-specific feelings of dislike. Recall, for instance, the assumption of social identity theory that 'intergroup bias' (i.e. tendency to favour the in-group) should emerge even in completely de-contextualized settings. Evolutionary perspectives, which effectively re-contextualized prejudice, have added greater nuance to this picture. As noted above, humans may be predisposed to perceive distinct kinds of social categories for different adaptive reasons; hence, attitudes toward those different categories may involve qualitatively distinct thoughts and feelings. Evidence supports this conjecture (Cottrell and Neuberg 2005). Moreover, because aversive responses (e.g. emotional reactions, ascription of stereotypes) are expected to be contingent upon threat-connoting factors (e.g. hostile environments), attitudes toward specific kinds of out-groups may be moderated by specific kinds of threat. This too is supported by evidence (Schaller et al. 2007). To be fair, many social psychologists have noted the importance of different emotions in different social contexts (e.g. Smith et al. 2007); however, such research has been largely descriptive and lacks an explanatory framework.

With regard to *intergroup prejudice*, evolutionary perspectives suggest a number of specific hypotheses. Most fundamentally, social contexts that resemble tribal intergroup contexts should be characterized especially by competitiveness and hostility; when taken to the extreme, such contexts may engender motivations to exterminate the out-group (there is no shortage of historical—and contemporary—examples of this). Moreover, hostile and exterminatory behaviours with evolutionary origins are likely to be accompanied by specific psychological tendencies that facilitate those behaviours—including dehumanizing out-group members and associating them with disease, and ascribing danger-connoting stereotypes to out-group members; there is evidence for all of those tendencies (Friedländer 2007; Leyens et al. 2000; Markel and Stern 2002; Schaller et al. 2003a), some of which is described in greater detail below.

As already mentioned, part of intergroup prejudice may be explained by disease-avoidance processes. For people of a given culture, certain out-groups may be particularly threatening in terms of spreading disease. Because each culture has its own set of practices for preventing infection, cultures with different practices—especially in food preparation and hygiene—may be perceived as posing disease threats. One series of studies found that heightened threat of disease led to more strongly negative attitudes toward cultural out-group members, but only those out-groups perceived to be especially foreign in their practices (Faulkner et al. 2004). Another study found that higher disgust sensitivity (an emotional response that may facilitate disease-avoidance behaviour) is associated with more strongly ethnocentric attitudes (Navarrete and Fessler 2006). It was mentioned above that regional variation in pathogen prevalence has been found to be associated with xenophobic cultural tendencies. Along these lines, there is evidence suggesting that disease-avoidance behaviours—such as reduced air travel during a pandemic—are more pronounced in pathogen-prevalent regions (Hamamura and Park 2010). That is, when faced with the same threat, people from more 'xenophobic' cultures may react more aversively to out-group members.

Evolutionary reasoning implies another set of important hypotheses, pertaining to sex differences. Specifically, because tribal conflict has been primarily a male activity during much of human (and probably prehuman) evolutionary history (Chagnon 1988; Goodall 1986), coalitional psychological tendencies may be exaggerated in males. This idea, called the *male warrior hypothesis*, has received empirical support. For instance, one set of experiments found that intergroup competition increased the tendency among men, but not women, to identify and cooperate with their in-group (Van Vugt et al. 2007). This sort of effect may extend to group-relevant moral attitudes. Intergroup prejudice is associated with the tendency to moralize in-group-enhancing values such as loyalty, obedience, and purity (Haidt 2007). Perceptions of threat have been found to be associated with upholding of the in-group-enhancing morals (Van Leeuwen and Park 2009); and there is preliminary evidence suggesting that this effect is stronger in men (Van Leeuwen and Park, unpublished data). Not only are males more likely to exhibit intergroup biases, they are also more likely to be targets of hostile and exterminatory actions during intergroup conflict (see below for cognitive biases with respect to out-group males); females are more likely to be victims of sexual exploitation (Kurzban and Leary 2001).

Prejudice is often accompanied by stereotypes about members of the social category in question (e.g. 'professors are boring'). The important thing for prejudice researchers is that the contents of those stereotypes may have behavioural consequences (e.g. you might prefer to avoid sitting next to a professor on a transatlantic flight).

In contemporary social contexts, individuals categorized as members of coalitional out-groups (e.g. those of other ethnicities) may be associated especially with danger-connoting traits, and perceivers' tendencies to activate such stereotypes may be more pronounced when they perceive higher degrees of threat. A series of studies (among Canadian students) found that ambient darkness—an environmental cue suggesting heightened threat—increases tendencies to associate out-groups (Iraqis, Africans) with danger-connoting (but not danger-irrelevant) stereotypic traits (Schaller et al. 2003a,b). These findings further reveal the uniqueness of coalitional psychology: danger-connoting stereotypic traits usually are not ascribed to women, the elderly, gay people, obese people, and many other targets of prejudice (see also Fiske et al. 2007).

In addition to ascribing danger-relevant stereotypes to out-group members, another sort of functional response is to (mis)perceive aggressive intent in such people (inferring non-existent aggressive intent is less costly than failing to perceive existent aggressive intent). Maner et al. (2005) termed this tendency *functional projection*, and they proposed that people may tend to perceive anger in the faces of out-group members (especially male out-group members), even if those people are holding neutral expressions. They found that experimentally heightened self-protective motives (achieved via showing a scene from a scary movie) increased the tendency among white American participants to perceive anger in the faces of black and Arab men (but not in the faces of white men or women).

In sum, several lines of theory and research indicate that humans possess psychological mechanisms specialized for managing antagonistic in-group–out-group relations (e.g. Richerson and Boyd 1998; Schaller et al. 2003a; Sidanius and Pratto 1999; Tooby and Cosmides 1988; Van Vugt and Park 2009). Humans seem predisposed to carve the social world in specific ways, and coalitional in-group–out-group categorization appears to be psychologically special, lying at the root of the competitive and hostile kinds of intergroup bias. Due in large part to evolutionarily guided research, the notion of a generic, 'purely cognitive group' (Tajfel and Turner 1979) has been replaced by a more textured picture of intergroup psychology. An important implication is that it may be scientifically counterproductive to apply a common concept 'group' to all possible social categories, even the other 'primitive' ones. Sexism and ageism—no matter how vicious—simply do not involve intergroup conflict. Men and women, the young and the elderly, do not wage war against each other.

These findings are reshaping our understanding of prejudice and opening up additional research avenues. What implications do they have for efforts to reduce prejudice and promote tolerance?

Implications for promoting tolerance

Allport (1954) is credited with the most famous idea for reducing prejudice, commonly known as the *contact hypothesis*. He proposed that contact may reduce prejudice between antagonistic groups if the parties involved have 'equal status' and are 'in the pursuit of common goals' (among other things). Many researchers have examined this hypothesis and have come away with mixed views on the effectiveness of contact as well as the utility of the hypothesis (Brewer 1996; Hewstone and Brown 1986; Pettigrew 1998; Pettigrew and Tropp 2000). For instance, although contact is often found to be associated with lower prejudice, it is not always clear whether contact causes prejudice reduction or whether existing variation in prejudice causes degree of contact. Also, many moderating variables have been implicated but there is lack of agreement regarding which are essential and what their effects are.

The perspective presented in this chapter offers a different take on the contact hypothesis. Cooperation is a defining feature of a coalitional alliance. If two 'groups' fully overlapped in their status and goals—if they were fully cooperative—they would be virtually indistinguishable in terms of coalitional alliance. The key to Allport's (1954) vision of reducing prejudice, it seems, is not contact per se but meeting the preconditions that supposedly make contact an effective strategy. In other words, the challenge is to induce people to treat each other as 'in-group' members.

Given our evolved coalitional psychology, then, there would seem to be two general strategies for reducing intergroup prejudice. One is to induce people to psychologically assimilate the parties involved into a single coalitional group so that, as far as our minds are concerned, there is no 'out-group' to which prejudice can be directed—what some might call the 'melting pot' model. Social psychologists have proposed a similar strategy called *decategorization* (Dovidio and Gaertner 1999). The perspective presented in this chapter suggests that while certain categories may be amenable to decategorization, others may be resistant. For social categories built upon coalitional psychology, assimilation or decategorization might be facilitated by eliminating heuristic cues that connote alternative alliances. In fact, this strategy is implemented in many modern nations. Policies requiring immigrants to speak the local language and to shed conspicuous markers of foreignness, such as religious attire, are often justified as a means to aid integration. Findings on the importance of linguistic (Kinzler et al. 2007, 2009) and physical/behavioural signs of alliance (Kurzban et al. 2001) suggest that such methods may effectively decrease prejudice, although the evidence also suggests that immigrants would have to speak in the native accent in order to be fully accepted (see also Gluszek and Dovidio 2010). Of course, an important limitation of the melting-pot strategy is that it works only within its own confines—the strategy does nothing to reduce prejudice against those outside the pot. For this assimilative strategy to be applied on a global scale, we would have to get every human to perceive every other human as a coalitional in-group member, which is not only practically difficult, but perhaps undesirable if it required mass cultural extinction.

A second general strategy is to leave existing groups intact and find ways to achieve true peaceful coexistence, what some might call the 'salad bowl' model. Past research has identified useful ways of achieving intergroup cooperation, such as highlighting superordinate goals (Gaertner et al. 1999; Sherif et al. 1961). Other ways are to maintain existing categories while creating new crosscutting categories—*recategorization* (Dovidio and Gaertner 1999)—or emphasizing broader, more inclusive categories (Dovidio and Gaertner 1999). To the extent that the newly created categories tap into coalitional psychology, these strategies may be effective. For example, whenever the Football World Cup comes around every 4 years, strong coalitional alliances to local teams are temporarily suspended, while alliance to the national team, which clearly constitutes a coalitional group competing against other coalitional groups, overrides the lower-level intergroup antipathies, among players and fans alike. It is, of course, having a common enemy (i.e. having a superordinate goal) that stimulates cooperation among members in the new category. An implication is that creating a more inclusive category is less likely to be effective if it fails to tap into coalitional psychology. Appeals such as 'we are all humans' seem impotent.

The research presented in this chapter suggests additional strategies for maintaining tolerance among diverse groups. For instance, we must combat the error-management tendency to over-infer threats posed by out-groups: when we think we perceive a certain degree of threat, we must recognize that our perception does not correspond to reality. We must also combat tendencies to dehumanize out-group members and to associate out-group members with diseases and dangers, especially when ambient cues seem to suggest heightened threat.

Of course, given the evolutionary *raison d'être* of human coalitional groups, preserving diverse groups while maintaining peaceful coexistence will stretch human ingenuity. Adding to the challenge is the fact that our prejudices are often deliberately inflamed by groups with esoteric interests. Atrocities such as wars and genocides are not simply excesses of human intolerance; they are orchestrated by small groups of people and sold to the masses via fear-arousing and victim-dehumanizing propaganda (e.g. Anderson and Bushman 2002; Chomsky 2003; Friedländer 2007).

So there seem to be two different kinds of lessons to reflect upon. The first is that we should do what we can to disarm environmental circumstances that fuel people's prejudices. The second is that we should learn to protect ourselves against those who would deliberately attempt to inflame our prejudices. And because the mass media are frequent purveyors of propaganda (Edwards and Cromwell 2009; Herman and Chomsky 1988), we might cultivate healthy scepticism regarding information disseminated by the media, to guard against over-inferring and over-reacting to threats. It is well-known that people tend to overestimate the frequency of more salient events (Tversky and Kahneman 1974), with the consequence that people's fears correspond poorly to objective risks (Glassner 1999). The media contribute to the problem by their over-coverage of what are already psychologically salient events. (As an example, the September 11 event, which triggered endless media coverage and large-scale 'defensive' measures, took 2995 lives. In that same year, motor vehicle accidents in the US took 42,196 lives. While there obviously are good reasons for people to be more attentive to deliberate attacks than to accidents, this attentional bias is exaggerated by arguably disproportionate media coverage.)

It should be clear by now that intergroup prejudice (prejudice against 'tribal' out-groups) differs in important ways from other kinds of prejudice, such as sexism and ageism. Indeed, these rarely involve 'intergroup' conflict and are often psychologically ambivalent, comprising both subjectively positive and negative thoughts and feelings (Cuddy and Fiske 2002; Glick and Fiske 1996). This is not to imply that sexism and ageism are less damaging; in fact, they may be especially persistent because their subjectively positive aspects blind people to their evil. A broader implication of evolutionary perspectives is that there is unlikely to be a panacea for reducing every kind of prejudice. The best methods will differ for prejudices against ethnic out-groups, women, elderly people, gay people, obese people, disabled people, and nonhuman animals, because the roots of these prejudices differ (Cottrell and Neuberg 2005; Fiske et al. 2007; Kurzban and Neuberg 2005; Schaller and Duncan 2007; Singer 1975). This is not to say that there are no domain-general psychological processes underlying prejudice (there surely are). Domain-general prejudice-reduction methods, such as promoting individuated rather than categorized social perception (Miller and Brewer 1986), might usefully complement more focused interventions.

Another avenue that is likely to bear intellectual fruit involves integration of evolutionary, social, and developmental perspectives. Social psychological approaches have tended to assume that individuals acquire prejudices passively during socialization, which implies the simplistic notion that prejudice can be prevented simply by cutting off damaging input at early ages. However, developmental psychologists have noted that the passive socialization view is unsound, as what children learn is substantially influenced by their pre-existing assumptions (see Dunham and Degner 2010). Research with infants and children has produced intriguing data that help us better understand pre-existing psychological priorities. Understanding what children are innately equipped with, what they learn, and how that learning translates into prejudice is a task that will require the cooperation of social, developmental, and evolutionary psychologists (Kinzler et al. 2010).

Conclusion

With ever-increasing levels of migration, traditional group boundaries are shifting, dissolving, and yielding to new boundaries. Even though humans may possess tribal tendencies, contemporary cosmopolitan life is evidence that tolerant, pluralistic societies are not only possible but can be deeply satisfying. However, many old prejudices remain, and many new ones are being created. Therefore, any scholarly work that helps us get a handle on the problem of prejudice is valuable. Social psychology has enjoyed a long tradition of developing hypotheses and testing them rigorously. Evolutionary perspectives have extended this tradition by generating more specific, more nuanced hypotheses, which have led to a more interesting and complex picture of the psychology of intergroup prejudice. The more we know about the psychological hurdles, the better we will be able to manage them, as we continue our difficult climb towards universal peace and tolerance.

Acknowledgements

I would like to thank Raphael Baroni, Luke Conway, David Herring, and Florian van Leeuwen for their helpful comments.

References

Allport, G.W. (1954). *The nature of prejudice*. Addison-Wesley, Cambridge, MA.

Anderson, C.A. and Bushman, B.J. (2002). Human aggression. *Annual Review of Psychology*, **53**, 27–51.

Brewer, M.B. (1979). In-group bias in the minimal intergroup situation: a cognitive–motivational analysis. *Psychological Bulletin*, **86**, 307–24.

Brewer, M.B. (1996). When contact is not enough: social identity and intergroup cooperation. *International Journal of Intercultural Relations*, **20**, 291–303.

Brewer, M.B. (1999). The psychology of prejudice: ingroup love or outgroup hate? *Journal of Social Issues*, **55**, 429–44.

Brewer, M.B. and Campbell, D.T. (1976). *Ethnocentrism and intergroup attitudes: East African Evidence*. Wiley, New York.

Brown, D.E. (1991). *Human universals*. Temple University Press, Philadelphia, PA.

Burnham, G., Lafta, R., Doocy, S., and Roberts, L. (2006). Mortality after the 2003 invasion of Iraq: a cross-sectional cluster sample survey. *Lancet*, **368**, 1421–8.

Buss, D.M. (1995). Evolutionary psychology: a new paradigm for psychological science. *Psychological Inquiry*, **6**, 1–30.

Chagnon, N.A. (1988). Life histories, blood revenge, and warfare in a tribal population. *Science*, **239**, 985–92.

Chomsky, N. (2003). *Hegemony or survival: America's quest for global dominance*. Metropolitan Books, New York.

Correll, J., Park, B., Judd, C.M., and Wittenbrink, B. (2002). The police officer's dilemma: using ethnicity to disambiguate potentially threatening individuals. *Journal of Personality and Social Psychology*, **83**, 1314–29.

Cosmides, L., Tooby, J., and Kurzban, R. (2003). Perceptions of race. *Trends in Cognitive Sciences*, **7**, 173–9.

Cottrell, C.A. and Neuberg, S.L. (2005). Different emotional reactions to different groups: a sociofunctional threat-based approach to 'prejudice'. *Journal of Personality and Social Psychology*, **88**, 770–89.

Cuddy, A.J.C. and Fiske, S.T. (2002). Doddering but dear: process, content, and function in stereotyping of older persons. In: T.D. Nelson (ed.), *Ageism: stereotyping and prejudice against older persons*, pp. 3–26. MIT Press, Cambridge, MA.

Darwin, C. (1871). *The descent of man, and selection in relation to sex.* Murray, London.

Diamond, J. (1997). *Guns, germs, and steel: the fates of human societies.* Norton, New York.

Dovidio, J.F. and Gaertner, S.L. (1999). Reducing prejudice: combating intergroup biases. *Current Directions in Psychological Science*, **8**, 101–5.

Dunham, Y. and Degner, J. (2010). Origins of intergroup bias: developmental and social cognitive research on intergroup attitudes. *European Journal of Social Psychology*, **40**, 563–8.

Edwards, D. and Cromwell, D. (2009). *Newspeak in the 21st Century.* Pluto Press, London.

Eibl-Eibesfeldt, I. (1974). The myth of the aggression-free hunter and gatherer society. In: R. L. Holloway (ed.), *Primate aggression, territoriality, and xenophobia*, pp. 435–57. Academic Press, New York.

Faulkner, J., Schaller, M., Park, J.H., and Duncan, L.A. (2004). Evolved disease-avoidance mechanisms and contemporary xenophobic attitudes. *Group Processes and Intergroup Relations*, **7**, 333–53.

Fincher, C.L., Thornhill, R., Murray, D.R., and Schaller, M. (2008). Pathogen prevalence predicts human cross-cultural variability in individualism/collectivism. *Proceedings of the Royal Society B*, **275**, 1279–85.

Fiske, S.T. and Neuberg, S.L. (1990). A continuum of impression formation, from category-based to individuating processes: influences of information and motivation on attention and interpretation. In: M.P. Zanna (ed.), *Advances in experimental social psychology*, vol. **23**, pp. 1–74. Academic Press, New York.

Fiske, S.T., Cuddy, A.J.C., and Glick, P. (2007). Universal dimensions of social cognition: warmth and competence. *Trends in Cognitive Sciences*, **11**, 77–83.

Friedländer, S. (2007). *The years of extermination: Nazi Germany and the Jews, 1939–1945.* Weidenfeld and Nicolson, London.

Gaertner, S.L., Dovidio, J.F., Rust, M.C., et al. (1999). Reducing intergroup bias: elements of intergroup cooperation. *Journal of Personality and Social Psychology*, **76**, 388–402.

Gil-White, F.J. (2001). Are ethnic groups biological 'species' to the human brain? *Current Anthropology*, **42**, 515–54.

Glassner, B. (1999). *The culture of fear: why Americans are afraid of the wrong things.* Basic Books, New York.

Glick, P. and Fiske, S.T. (1996). The ambivalent sexism inventory: differentiating hostile and benevolent sexism. *Journal of Personality and Social Psychology*, **70**, 491–512.

Gluszek, A. and Dovidio, J.F. (2010). The way *they* speak: a social psychological perspective on the stigma of non-native accents in communication. *Personality and Social Psychology Review*, **14**, 214–37.

Goodall, J. (1986). *The chimpanzees of Gombe: patterns of behavior.* Harvard University Press, Cambridge, MA.

Haidt, J. (2007). The new synthesis in moral psychology. *Science*, **316**, 998–1002.

Hamamura, T. and Park, J.H. (2010). Regional differences in pathogen prevalence and defensive reactions to the 'Swine Flu' outbreak among East Asians and Westerners. *Evolutionary Psychology*, 8, 506–15.

Haselton, M.G. and Buss, D.M. (2000). Error management theory: a new perspective on biases in cross-sex mind reading. *Journal of Personality and Social Psychology*, **78**, 81–91.

Haselton, M.G. and Nettle, D. (2006). The paranoid optimist: an integrative evolutionary model of cognitive biases. *Personality and Social Psychology Review*, **10**, 47–66.

Herman, E.S. and Chomsky, N. (1988). *Manufacturing consent: the political economy of the mass media*. Pantheon, New York.

Hewstone, M. and Brown, B. (eds). (1986). *Contact and conflict in intergroup encounters*. Basil Blackwell, Oxford.

Hewstone, M., Hantzi, A., and Johnston, L. (1991). Social categorization and person memory: the pervasiveness of race as an organizing principle. *European Journal of Social Psychology*, **21**, 517–28.

Hirschfeld, L.A. (1996). *Race in the making: cognition, culture, and the child's construction of human kinds*. MIT Press, Cambridge, MA.

Jones, A. (2006). *Genocide: a comprehensive introduction*. Routledge, London.

Kenrick, D.T., Schaller, M., and Simpson, J.A. (2006). Evolution is the new cognition. In M. Schaller, J.A. Simpson, and D.T. Kenrick (eds), *Evolution and social psychology*, pp. 1–13. Psychology Press, New York.

Kinzler, K.D., Dupoux, E., and Spelke, E.S. (2007). The native language of social cognition. *Proceedings of the National Academy of Sciences of the USA*, **104**, 12577–80.

Kinzler, K.D., Shutts, K., DeJesus, J., and Spelke, E.S. (2009). Accent trumps race in guiding children's social preferences. *Social Cognition*, **27**, 623–34.

Kinzler, K.D., Shutts, K., and Correll, J. (2010). Priorities in social categories. *European Journal of Social Psychology*, **40**, 581–92.

Kuhlmeier, V., Wynn, K., and Bloom, P. (2003). Attribution of dispositional states by 12-month-olds. *Psychological Science*, **14**, 402–8.

Kurzban, R. and Leary, M.R. (2001). Evolutionary origins of stigmatization: the functions of social exclusion. *Psychological Bulletin*, **127**, 187–208.

Kurzban, R. and Neuberg, S. (2005). Managing ingroup and outgroup relationships. In: D.M. Buss (ed.), *The handbook of evolutionary psychology*, pp. 653–75. Wiley, Hoboken, NJ.

Kurzban, R., Tooby, J., and Cosmides, L. (2001). Can race be erased? Coalitional computation and social categorization. *Proceedings of the National Academy of Sciences of the USA*, **98**, 15387–92.

LeVine, R.A. and Campbell, D.T. (1972). *Ethnocentrism: theories of conflict, ethnic attitudes, and group behavior*. Wiley, New York.

Leyens, J.P., Paladino, P.M., Rodriguez-Torres, R., et al. (2000). The emotional side of prejudice: the attribution of secondary emotions to ingroups and outgroups. *Personality and Social Psychology Review*, **4**, 186–97.

Maner, J.K., Kenrick, D.T., Becker, D.V., et al. (2005). Functional projection: how fundamental social motives can bias interpersonal perception. *Journal of Personality and Social Psychology*, **88**, 63–78.

Markel, H. and Stern, A.M. (2002). The foreignness of germs: the persistent association of immigrants and disease in American society. *The Milbank Quarterly*, **80**, 757–88.

Messick, D.M. and Mackie, D.M. (1989). Intergroup relations. *Annual Review of Psychology*, **40**, 45–81.

Miller, N. and Brewer, M.B. (1986). Categorization effects on ingroup and outgroup perception. In: J.F. Dovidio and S.L. Gaertner (eds), *Prejudice, discrimination, and racism*, pp. 209–30. Academic Press, San Diego, CA.

Miller, S.L., Maner, J.K., and Becker, D.V. (2010). Self-protective biases in group categorization: threat cues shape the psychological boundary between 'us' and 'them'. *Journal of Personality and Social Psychology*, **99**, 62–77.

Navarrete, C.D. and Fessler, D.M.T. (2006). Disease avoidance and ethnocentrism: the effects of disease vulnerability and disgust sensitivity on intergroup attitudes. *Evolution and Human Behavior*, **27**, 270–82.

Nesse, R.M. (1990). Evolutionary explanations of emotions. *Human Nature*, **1**, 261–89.

Neuberg, S.L., Kenrick, D.T., and Schaller, M. (2010). Evolutionary social psychology. In: S.T. Fiske, D.T. Gilbert, and G. Lindzey (eds), *Handbook of social psychology* (fifth edition), pp. 761–96. Wiley, Hoboken, NJ.

Park, J.H. and Buunk, A.P. (2010). Interpersonal threats and automatic motives. In: D. Dunning (ed.), *Social motivation*, pp. 11–35. Psychology Press, New York.

Pettigrew, T.F. (1998). Intergroup contact theory. *Annual Review of Psychology*, **49**, 65–85.

Pettigrew, T.F. and Tropp, L.R. (2000). Does intergroup contact reduce prejudice? Recent meta-analytic findings. In: S. Oskamp (ed.), *Reducing prejudice and discrimination*, pp. 93–114. Erlbaum, Mahwah, NJ.

Pinker, S. (1997). *How the mind works*. Norton, New York.

Richerson, P.J. and Boyd, R. (1998). The evolution of human ultrasociality. In: I. Eibl-Eibesfeldt and F.K. Salter (eds), *Indoctrinability, ideology, and warfare: evolutionary perspectives*, pp. 71–95. Berghahn Books, New York.

Richerson, P.J. and Boyd, R. (2005). *Not by genes alone: how culture transformed human evolution*. University of Chicago Press, Chicago, IL.

Schaller, M. and Duncan, L.A. (2007). The behavioral immune system: its evolution and social psychological implications. In: J.P. Forgas, M.G. Haselton, and W. von Hippel (eds), *Evolution and the social mind: evolutionary psychology and social cognition*, pp. 293–307. Psychology Press, New York.

Schaller, M. and Neuberg, S.L. (2008). Intergroup prejudices and intergroup conflicts. In: C. Crawford and D.L. Krebs (eds), *Foundations of evolutionary psychology*, pp. 399–412. Erlbaum, Mahwah, NJ.

Schaller, M., Park, J.H., and Faulkner, J. (2003a). Prehistoric dangers and contemporary prejudices. *European Review of Social Psychology*, **14**, 105–37.

Schaller, M., Park, J.H., and Mueller, A. (2003b). Fear of the dark: interactive effects of beliefs about danger and ambient darkness on ethnic stereotypes. *Personality and Social Psychology Bulletin*, **29**, 637–49.

Schaller, M., Park, J.H., and Kenrick, D.T. (2007). Human evolution and social cognition. In: R.I.M. Dunbar and L. Barrett (eds), *The Oxford handbook of evolutionary psychology*, pp. 491–504. Oxford University Press, Oxford.

Sherif, M., Harvey, O.J., White, B.J., Hood, W.R., and Sherif, C.W. (1961). *Intergroup conflict and cooperation: the Robbers Cave Experiment*. Wesleyan University Press, University Press of New England, Hanover, NH.

Sidanius, J. and Pratto, F. (1999). *Social dominance: an intergroup theory of social hierarchy and oppression*. Cambridge University Press, New York.

Singer, P. (1975). *Animal liberation: a new ethics for our treatment of animals*. Avon Books, New York.

Smith, E.R., Seger, C.R., and Mackie, D.M. (2007). Can emotions be truly group level? Evidence regarding four conceptual criteria. *Journal of Personality and Social Psychology*, **93**, 431–46.

Sumner, W.G. (1906). *Folkways: a study of the sociological importance of usages, manners, customs, mores, and morals*. Ginn, Boston, MA.

Tajfel, H. and Turner, J.C. (1979). An integrative theory of intergroup conflict. In: W.G. Austin, and S. Worchel (eds), *The social psychology of intergroup relations*, pp. 33–47. Brooks/Cole, Monterey, CA.

Tajfel, H. and Turner, J.C. (1986). The social identity theory of intergroup behavior. In: S. Worchel, and W.G. Austin (eds), *Psychology of intergroup relations*, pp. 7–24. Nelson-Hall, Chicago, IL.

Taylor, S.E., Fiske, S.T., Etcoff, N.L., and Ruderman, A.J. (1978). Categorical and contextual bases of person memory and stereotyping. *Journal of Personality and Social Psychology*, **36**, 778–93.

Tooby, J. and Cosmides, L. (1988). *The evolution of war and its cognitive foundations* (Institute for Evolutionary Studies Tech. Rep. No. 88-1). Institute for Evolutionary Studies, Palo Alto, CA.

Tooby, J. and Cosmides, L. (1990). The past explains the present: emotional adaptations and the structure of ancestral environments. *Ethology and Sociobiology*, **11**, 375–424.

Tooby, J. and Cosmides, L. (1992). The psychological foundations of culture. In: J.H. Barkow, L. Cosmides, and J. Tooby (eds), *The adapted mind: evolutionary psychology and the generation of culture*, pp. 19–136. Oxford University Press, New York.

Tversky, A. and Kahneman, D. (1974). Judgment under uncertainty: heuristics and biases. *Science*, **185**, 1124–31.

Van Leeuwen, F. and Park, J.H. (2009). Perceptions of social dangers, moral foundations, and political orientation. *Personality and Individual Differences*, **47**, 169–73.

Van Vugt, M. and Park, J.H. (2009). Guns, germs, and sex: how evolution shaped our intergroup psychology. *Social and Personality Psychology Compass*, **3**, 927–38.

Van Vugt, M. and Park, J.H. (2010). The tribal instinct hypothesis: evolution and the social psychology of intergroup relations. In: S. Stürmer and M. Snyder (eds), *The psychology of prosocial behavior: group processes, intergroup relations and helping*, pp. 13–32. Wiley-Blackwell, Chichester.

Van Vugt, M., De Cremer, D., and Janssen, D.P. (2007). Gender differences in cooperation and competition: The male-warrior hypothesis. *Psychological Science*, **18**, 19–23.

Chapter 13

The evolutionary psychology of criminal behaviour

Aurelio José Figueredo, Paul Robert Gladden, and Zachary Hohman

Introduction

Aetiological theories of the origin and nature of criminal behaviour are reviewed, compared, and contrasted with each other and with the empirical evidence. These are first classified into: 1) standard social science theories, and 2) evolutionary social science theories. The standard social science theories are then further classified into: 1a) classical theories, 1b) positivist theories, 1c) functionalist theories, 1d) cultural, sub-cultural, and social learning theories, 1e) control theories, 1f) cognitive theories, and 1g) traditional personality theories. The evolutionary social science theories are further classified into: 2a) behavioural genetic theories, 2b) reactive heritability and epigenetic theories, 2c) sexual selection theories, 2d) differential parental investment theories, 2e) competitive disadvantage theories, 2f) frequency-dependent selection theory, 2g) pathogen stress theory, and, finally, 2h) life history theories.

We propose that most of these theories are mutually contradictory to a minimal degree, mostly differing on matters of detail as well as in the conflation of proximate and ultimate levels of causation. As an alternative to this chaotic state of affairs, we propose a cross-disciplinary integration based on the inclusive framework provided by life history theory. An array of empirical evidence is provided in support of this view as the most inclusive and integrative framework currently available, and the most useful framework for explaining previous findings within an evolutionary context. A specific model is presented from our own ongoing research that illustrates the kind of integration we envision, and provides empirical evidence in favor of the heuristic utility of the proposed approach.

Standard social science theories of criminal behaviour

Classical theories

Modern criminological theory originated in 17th-century Europe as a prescriptive application of the egalitarian philosophies of scholars such as Voltaire and Montesquieu. The founder of the 'classical' school, Cesare Beccaria (1995 [1767]), used the concept of social contract to explain that individuals might be convinced to relinquish certain personal or social liberties if such constraints led to a lawful and peaceful society in which they would be able to flourish. Beccaria advocated minimal legal restrictions, and emphasized that laws should be clearly defined so as to be equally applied. Though his major work in 1767 consisted mostly of defining a just legal system, he also discussed the aetiology of criminal behaviour. In his view, rational self-interest leads individuals to act selfishly, therefore the formation of social contracts restricts individual behaviour via punishment, to deter selfish behaviour that harms other individuals or damages the structure of society as a whole.

Jeremy Bentham (1988 [1789]) developed this idea further by providing a predictive theory specifying a hedonistic utilitarian logic. In his view, individuals base their actions on the relative amounts of pain and pleasure derived from a given act, engaging in a behaviour only if pleasurable outcomes outweigh painful ones. Therefore, to deter behaviour that society deems undesirable, the aversive consequences should be made greater than the benefits of such behaviour. Codifying these consequences would impose a deterrent by imposing a threat of punishment that would presumably constrain the would-be criminal's behaviour. Bentham's theory had a classical behaviourist flavour in emphasizing the extrinsic consequences of behaviour, with little importance given to individual differences in utility function parameters.

Positivist theories

The theories of the 19th-century 'positivist' school contrasted sharply with the social egalitarianism of the classicists, as the emphasis was shifted towards an individual-level explanation of the causes of criminal behaviour. Instead of assuming that even criminal behaviour is the result of rational choice, they looked for differences among individuals to explain and predict crime. The Italian positivists were strongly influenced by the then-recent work of Charles Darwin, and argued for the existence of biological differences between criminals and law-abiding citizens. Cesare Lombroso (2006 [1876]) published *Criminal Man* less than a decade after Darwin's seminal work, and in it he unabashedly claimed that criminals are physiologically primitive, or 'atavistic', and therefore evolutionarily inferior to modern, enlightened humans. Lombroso cited dubious phenotypic evidence (that criminals have a more ape-like appearance) to support his claims, and his work has been much maligned as a result. However, current research (discussed below) may indicate that there are in fact heritable differences in cognition, if not appearance, between criminals and non-criminals. Although this kind of criminological theory has been publicly ridiculed for many years (e.g. Lewontin et al. 1984), researchers have recently tested the hypothesis that untrained observers are able to detect a propensity for violence in other people, based only on a brief (2 second) look at photographs of their faces. Estimated likelihoods of violence were significantly related to the actual violent histories of a sample of registered sex offenders, suggesting that violent tendencies can indeed be accurately inferred from phenotypic facial cues (Stillman et al. 2010).

Enrico Ferri, a student of Lombroso, while retaining the individual-level theoretical perspective on criminal behaviour, emphasized biological (race, age, sex) as well as ecological/environmental factors, including geographical location, the nature of social and cultural institutions, and economic climate. Ferri's (1895) treatise explicitly denied the role of rational choice in criminal behaviour, and declared instead that 'a man commits this or that crime only when he lives in definitely determined conditions of personality and environment which induce him to act in a certain way'. In modern terminology, the proposed determinants that cause criminal (and indeed all) behaviour are the main effects of and the interactions between the person (personality, cognition, etc.) and the situation (natural and social environment).

Functionalist theories

The 'functionalist' theories posit that criminal behaviour is a part of the fabric of society, and therefore serves a function within that framework. Emile Durkheim (1964) noted that crime was present in every human society. Within each, there is a distribution of individuals who are more or less collectivist, and those individuals who are less collectivist are more likely to engage in criminal behaviour. The increase in individualism and the corresponding decline in collectivism in certain societies leads to a condition of 'anomie' (lawlessness) marked by the inability of social institutions to constrain the personal ambitions of individuals, such as the poor and disenfranchised. This leads to a dynamic, evolving system in which the social structure must adapt to those novel conditions, and so crime serves a function as a catalyst for cultural and social evolution (Durkheim, 1964). There is recent evidence, however, that individualism is associated with a decrease in criminal behaviour (see our discussion below of the work of Thornhill 2010).

Merton (1938) further developed the idea of anomie (referred to as 'strain' theory), defining it as a pervasive condition of society resulting from the ubiquitous goal of material wealth without the necessary social structures in place to allow for the lawful acquisition thereof. This theory is reminiscent of the rational self-interest defined by the classical criminologists—crime results from the ultimate, selfish goal of self-enhancement, accomplished in modern societies via material gains. An important distinction, however, is that Merton was explicit in attributing the causes of self-interested (and therefore criminal) behaviour to sociological causes (consumerism, individualism), as opposed to an intrinsic aspect of 'human nature'.

A more recent development of these early theories of anomie and strain is present in Agnew's 'general strain theory'. As is suggested by the name, this theory more broadly defines strain not just in terms of striving for monetary or material resources, but as a disparity between aspirations and expectations, or between individual desires and their fulfilment. Three types of strain are identified in a language that parallels the description of operant contingencies: 1) the failure to attain positively-valued stimuli, 2) the removal (or anticipated removal) of positively-valued stimuli, and 3) the introduction (or anticipated introduction) of negative stimuli (Agnew 2006). Empirical evidence supporting the theory shows that a variety of measures of strain are positively correlated with levels of delinquency (Agnew and White 1992).

Cultural, sub-cultural, and social learning theories

Several cultural theories of criminal behaviour arose from the 'Chicago school' of sociology in the early 20th century, which sought to explain the social development of rapidly expanding urban areas. Shaw and McKay (1931, 1942) proposed that the characteristics of inner city environments lead to higher levels of both juvenile and adult delinquency. Factors such as a decreasing population, low rates of home ownership, and high unemployment were all significantly related to various measures of criminal activity, with the level of 'social disorganization' as the theorized causal link. This is directly analogous to Durkheim's theory: both anomie and social disorganization are phenomena caused by the failure of social institutions to adequately respond to or prevent social unrest, eventually leading to increases in crime.

As intellectual descendants of the Chicago school, Sutherland and Cressey proposed (1966) a 'differential association' theory, which essentially represents a social learning model accounting for the regional (urban) concentrations of crime. The first assumption of the theory is that criminal behaviour is not inherited and must be learned. Criminal ideology and behaviour is learned mainly in intimate interpersonal settings (not from the media, etc.); therefore associating with criminals will lead individuals to value law-breaking over law-abiding behaviour. While this theory is cohesive as a learning model, the dismissal of any heritable component of delinquency or criminal behaviour is not supported by current research (see below). Furthermore, the theory is based upon a false dichotomy between biological influences and social learning.

Control theory

Originally proposed by Travis Hirschi (1969), 'control' theory continues to be one of the most substantial and impactful theories of the causes of delinquency. According to Hirschi, a weak or broken bond between an individual and society may result in delinquent behaviour. There are several elements to this bond: *attachment* is the strength of a person's connection to other members of society. This includes family, friends, co-workers, and members of the community. *Commitment* is the common sense component of law-abiding behaviour. Societies tend to be organized in such a way that a person's self-interests would be in jeopardy if they were to violate the laws set forth by society. *Involvement* in rewarding pro-social activities (e.g. a legitimate job, community involvement) also facilitates the development of a strong social bond, thereby precluding criminal behaviour. In terms of modern evolutionary theory, this recognizes that mutualistic and antagonistic social strategies tend to be incompatible. *Belief* refers to a common system of norms and values that are shared by the members of a society, and may be thought of as a moral basis for behaviour.

A major fault with control theory is that it fails to give an adequately detailed mechanistic explanation of the processes described above. Also, certain types of crime do not fit well into the theory; for example, white collar crime is not accounted for by the involvement component. Addressing these issues, Gottfriedson and Hirschi (1990) later developed a more detailed and comprehensive general theory, referred to as 'self-control' theory. In this view, it is the lack of behavioural self-regulation that causes criminal behaviour, and there are many social and biological factors that influence an individual's level of self-control. The most notable biological factor identified by this theory is age: delinquency tends to decrease with age due to developmental factors such as hormonal regulation and brain development, as well as increased opportunity costs for older individuals. Social factors centred on learning self-control skills (initially from parents) are also of central importance. Self-control theory dismisses innate differences between individuals, such as racial characteristics; however, there is some consideration of ethnic or racial differences in quality of parenting resulting from cultural or sociological differences. In sum, self-control theory describes certain proximate biological and psychological factors that account for differences in delinquency; however, it does not provide an integrated adaptive explanation of individual differences.

Cognitive theories

'Cognitive' theories of crime postulate that individual or group-level differences in cognitive ability may explain some of the variance in delinquent behaviour. Several studies in different cultures have demonstrated a link between lower scores on tests of general cognitive ability (IQ) and criminal behaviour. For example, Denowski and Denowski (1985) found that the number of severely cognitively impaired individuals within the US prison population was significantly greater than that within the general population. One historically plausible explanation for the relationship between IQ and crime is that there might exist a spurious relationship with socioeconomic status (SES) as the common causal link. However, a Danish longitudinal study (Moffitt et al. 1981) concluded that the relationship between IQ scores and delinquency exists independently of the effect of SES. Furthermore, Kandel et al. (1988) proposed that IQ may act as a protective factor that decreases the risk of criminality. They found that the sons of fathers who were at high risk for criminal outcomes, but who nevertheless avoided delinquent behaviour, had on average a higher IQ than other risk groups. Moffitt et al. (1981) went on to postulate that deficiencies in verbal abilities might restrict the options available to impaired individuals, leading to higher rates of criminal behaviour.

In their influential book *The Bell Curve*, Herrnstein and Murray (1994) reviewed the mounting corpus of evidence in support of the cognitive theory, in which heritable differences in IQ account for individual differences in criminal behaviour. However, as they concede, there is some debate over the precise developmental processes involved in the heritability of IQ, including the likelihood of complex epigenetic effects such as gene-environment and gene-gene interactions (see further discussion of heritability in the 'Competitive disadvantage theory' section).

Personality theories

Some criminologists propose that criminal behaviour is linked to personality traits. A 'criminal personality' may have its roots early in development. For example, Bowlby (1951, 1970) showed that insecure attachment bonds early in life may lead to subsequent difficulty in forming healthy relationships and thereby a pattern of antisocial behaviour. In this view, individuals who experience a lack of affection (especially maternal affection) during development, or come from broken homes, may be at higher risk of delinquency.

Hans Eysenck (1964) devoted his book *Crime and Personality* to personality theories of crime, employing a three-factor model of personality, each a continuous dimension: neuroticism-stability (N), extroversion-introversion (E), and psychoticism-superego (P). Relating these traits to learning via operant conditioning, there are certain combinations that make aversive conditioning easier or more difficult. In general, individuals high in all three traits (neurotic extroverts high in psychoticism) are the most resistant to aversive conditioning, and are therefore the most likely to be criminals (see also Eysenck and Eysenck 1971; McGurk and McDougall 1981).

Evolutionary social science theories of criminal behaviour

Behavioural genetics and evolutionary psychology

Behavioural genetics and evolutionary psychology have been perceived by some researchers as distinct fields that overlap little. The former emphasizes the influences of genetic differences between individuals on behavioural traits, whereas evolutionary psychology typically focuses on our purportedly universal evolved psychological mechanisms (Tooby and Cosmides 1990). Contrary to the claims of its critics, evolutionary psychology emphasizes that human neurocognitive mechanisms are exquisitely adapted to respond effectively to recurrent environmental challenges of survival and reproduction. Thus, for many evolutionary psychologists, variable environmental conditions (developmental cues and situations) encountered by individuals, rather than genetic variability across individuals, is the preferred explanation for individual differences in behaviour. In this view, heritable individual differences in personality traits, which are of primary interest in behaviour genetics, are considered relatively minor variants on a universally shared evolved psychology. And heritable personality variation is sometimes considered selectively neutral 'genetic noise', which is presumed to be inconsequential for reproductive fitness (Tooby and Cosmides 1990).

However, 'all human behavioural traits are heritable' has been labelled the first law of behavioural genetics (Turkheimer 2000). (Heritability is the amount of phenotypic variance in a trait that is statistically accounted for by genotypic variance.) All individual difference traits are indeed heritable to some degree, including each of the following, which are relevant to criminality: juvenile delinquency, antisocial personality disorder, psychopathy, general intelligence, executive functions, each of the Big Five personality traits, religiosity, and tendencies toward drug abuse. In short, essentially all continuous phenotypic individual differences in personality traits and behavioural dispositions appear to be partly influenced by genetic variance. This fact suggests that if personality traits correlate with survival and reproductive outcomes, which they do, personality variation is not merely selectively neutral genetic noise (Figueredo et al. 2005).

Reactive heritability theory and epigenetic influences

The simple fact that 'everything is heritable' suggests almost nothing about the actual mechanisms underlying the development of behavioural outcomes (Turkheimer 2000): statistically partitioning sources of between-individual variance associated with a behavioural outcome like criminality cannot be confused with providing an explanation about concrete within-individual developmental processes (Turkheimer 1998, 2000). Instead, developmental mechanisms may often entail complex epigenetic interactions (gene–environment and gene–gene interactions). For example, certain microevolutionary processes, such as genetic assimilation (Waddington 1957) and genetic accommodation (West-Eberhard 2003), may produce developmental biases favouring the development of certain phenotypes over others in response to specific environmental contingencies. Thus, even within the context of conditional adaptive strategies, genetic polymorphism in the relative strength of these developmental biases could be generated in a population of otherwise facultative strategists. Furthermore, learning need not be totally *de novo*, but is instead based on evolved behavioural programmes (e.g. Waddington 1957; Seligman 1970; Garcia et al. 1974; Mayr 1974; Pinker 1994).

Studies of heritability cannot distinguish individual characteristics 'that are relatively direct products of the genes' from developmentally 'downstream' traits that are more indirect products of genotype–environment interactions (Pinker 2002). For example, the chance individual possession of independently heritable (but strategically relevant) characteristics may bias the selection of adaptive strategies (Figueredo 1995). Because interaction with the environment determines which behavioural strategy works best for each individual, other individual differences also matter. An individual assesses not only its external environment, but also itself within that environment. Gibson (1979) refers to similar transactional contingencies as *affordances*. The recognition of these complex gene–environment interactions has led to the development of theories about the so-called 'reactive heritabilities' of traits (e.g. Tooby and Cosmides 1990). The heritability of complex behavioural traits like criminality does not imply that there are 'genes for criminality' per se. Rather, theories of reactive heritability suggest that some complex traits are heritable (at least partly) because of reaction to other heritable characteristics (themselves the product of epigenetic processes, such as one's temperament or physical attractiveness). In other words, some heritable characteristics may lead to particular cues that shape individual differences in personality.

For example, since physically attractive women are desirable to men, such women may feel less concerned about the possibility of their partner defecting from their relationship, and thereby less jealous; this is reinforced by experience of easily attracting and maintaining male interest. Conversely, decreased physical attractiveness may lead to experiences that adaptively shape increased levels of romantic jealousy. If physical attractiveness is heritable, then levels of jealousy will also be indirectly heritable, at least in part, by virtue of the fact that they are reactively calibrated by the differential experiences associated with attractiveness. Thus, psychosexual development involves a self-assessment of sociosexual capabilities and opportunities, calibrating optimal utilization of physical assets such as size, strength, health, and attractiveness, as well as psychosocial assets such as intelligence, self-efficacy, social skills, personality, and SES, and financial prospects (cf. Hunter and Figueredo 2000; Figueredo and Jacobs 2000).

In the same way, a criminal disposition could be heritable because other, more directly heritable characteristics bring about differences in the environmental stimuli or conditions that an individual experiences. For example, if social intelligence is heritable and if low social intelligence in men leads to experience of rejection by highly desirable women, thus creating feelings of anger or resentment which in turn motivate violence against women, then violence will be heritable because of its functional relationship with social intelligence. Likewise, if social intelligence leads to low SES and low SES leads to inability to attract desirable women as long-term mates and increases the likelihood of criminality, then criminality will be heritable because it is associated with other heritable characteristics. An evolutionary perspective on criminality suggests that our psychological mechanisms will be adaptively structured (prepared) to respond appropriately to these varying environmental cues. Figueredo et al. (2000) applied this framework to address the ultimate causes of adolescent sex-offending behaviour by proposing a Brunswikian Evolutionary Developmental (BED) theory, wherein an inability to use more socially acceptable sexual strategies lead to more coercive and aggressive ones. Because some adolescents suffer psychosocial problems and consequent competitive disadvantages in the sexual marketplace, sex-offending behaviour may represent the culmination of a cascade of failing sexual and social strategies, leading from psychosocial deficiencies to sexual deviance, thence to antisocial deviance, and finally to sexual criminality.

Sexual selection theory

Sex, risk-taking, and differential parental investment

Parental investment theory predicts that sexual selection pressures are stronger on the sex that invests less in raising offspring (Trivers 1972). In humans, males have generally undergone the stronger selection pressures of intersexual mate choice and intrasexual competition. Because the potential number of offspring (i.e. *reproductive potential*) that women can produce is inherently limited by physiological constraints (e.g. lengthy gestation and lactation), whereas men's reproduction is limited instead by the number of women they are able to impregnate, sexual selection theory predicts that men will compete more with other men for sexual access to fertile women (for further discussion, see Chapter 9, this volume). Similarly, parental investment theory predicts that men will have greater *variance* in reproductive success. Since humans are not completely monogamous: if a man impregnates more than one woman, then for each additional woman he impregnates, another man will be prevented from reproducing for a period of time (some men are prevented from reproducing entirely). Thus, parental investment theory predicts that increased sexual selection pressure among men will result in an increase in any tactics which enhance reproductive success relative to competitors. Men who took risks to enhance social status, or otherwise out-competed men by obtaining more high-quality mates, tended to have more offspring. Under some circumstances (e.g. when low in status), risk-taking and violent aggression may have been effective tactics at securing relevant resources including fertile mates. These actions, which we now restrict with criminal laws, may have an adaptive logic. In sum, sexual selection theory predicts that men will be more violently aggressive to obtain and defend reproductively relevant resources from other men.

The age-crime curve revisited

Young men, in particular, are most likely to both commit and become a victim of crime and lethal violence (Hirschi and Gottfedson 1983; Wilson and Daly 1985; Daly and Wilson 1988). In what has been called the 'young male syndrome', Margo Wilson and Martin Daly argued that, historically at least, the sexual selection pressures of male competition and female mate choice are relatively higher among men entering adulthood (late adolescence through their 20s) and that this accounts for these elevated levels of violent aggression.

Reputation and culture of honour

Consistent with the 'young male syndrome', concern with reputation, trivial insults, dominance, and social status appear to motivate many homicides (Daly and Wilson 1988; Hiraiwa-Hasegawa 2005).

Evolutionary psychologists had previously explored the possible origins of such related factors as 'indirect reciprocity', 'blood revenge', and 'family honour' (Alexander 1987; Daly and Wilson 1988) as evolved adaptive strategies, but it was two social psychologists that produced an entire volume on the so-called 'culture of honour' (Nisbett and Cohen 1996; for an evolutionary psychological interpretation, see Shackelford 2005). They argued that displaying a willingness to retaliate violently (perhaps risking one's own death) against those that threatened one's reputation for toughness (through minor insults, slights) is adaptive under some conditions. Specifically, they focused on the ecological differences between historical societies of farmers living at relatively high population densities (e.g. the northern USA) and of herdsmen living at low densities (e.g. the southern USA). Herders have to consistently display toughness to deter those that might take their livestock. Nisbett and Cohen (1996) argued that this difference in culture of honour accounts for differences in levels of violence between these two regions. In societies with a culture of honour, trivial slights and altercations may be treated as worth risking one's own life and may lead to increased violent aggression. Figueredo et al. (2004) confirmed the predictions of culture of honour theory in a cross-cultural comparison of traditionally farming and herding Spanish-American societies (although Nisbett and Cohen's (1996) prediction of the lack of a culture of honour for foraging societies was not found in sedentary fishing communities).

Competitive disadvantage theory

Individuals at risk of being excluded from reproduction should make stronger efforts to prevent this by taking more risks to out-compete their rivals. Wilson and Daly (1985; Daly and Wilson 1988) found that both perpetrators and victims of homicide are more likely than expected to be unemployed (low SES) and unmarried. Furthermore, income inequality, rather than absolute income, is a particularly robust predictor of male-male homicide (Daly et al. 2001). This makes evolutionary sense because selection operates on relative rather than absolute reproductive fitness. In short, violence tends to occur among men lacking the resources and status necessary to attract high-quality mates.

The Competitive Disadvantage Theory (CDT) of aggression predicts that criminal aggression generally, and specifically sexual coercion, are increased when individuals perceive themselves to be at a competitive disadvantage in obtaining and retaining desirable mates (Thornhill and Thornhill 1983; Figueredo and McCloskey 1993; Figueredo et al. 2000). The CDT of sexual coercion predicts that the use of sexually aggressive mating tactics is developmentally conditional on cues that signal a disadvantage in some domain of mate value (e.g. failure to acquire status, low social intelligence, psychosexual difficulties). Figueredo et al. (2000) suggested a developmental model of criminal sexual offending consistent with the CDT, which pointed toward a history of social and sexual failure as important cues that lead to the adoption of progressively more extreme criminal behavioural tactics (both nonsexual and sexual forms). It is important to note that, according to CDT, low social status and psychosocial deficits are likely positively related to one another, and both may be signals of competitive disadvantage that lead to criminal behaviour.

Frequency-dependent selection theories of psychopathy

Psychopathy refers to a set of interrelated behavioural traits including relatively early onset of aggression, lack of emotional empathy for others, lack of remorse or guilt for one's actions, impulsivity, dishonesty, and unrestricted sociosexuality (Lalumiere et al. 2008). Mealey (1995) theorized that the cluster of traits characterizing psychopathy is a specialized set of tactics that were selected together as a 'cheater' strategy. That is, psychopaths selfishly capitalize on the fact that there are many cooperators around to be exploited. The more cheaters present in a given population, the lower the relative payoff of adopting such a strategy because cheaters themselves are not easily cheated by others and may be especially prone to retaliate aggressively against being cheated. In other words, Mealey (1995) argued that low levels of psychopathy are maintained through frequency-dependent selection. This theory depends on the assumption that psychopathic individuals experience high levels of short-term mating success, perhaps at least partially due to their use of coercive mating tactics. Consistent with this, self-reported psychopathy is positively associated with self-reported mating effort, number of sexual partners, and use of sexual coercion to obtain mates (Lalumiere and Quinsey 1996; Rowe et al. 1997; Gladden et al. 2008). Thus, psychopathy could be conceptualized as a short-term mating strategy specialized for 'cheating' cooperators by using deceptive and aggressive tactics.

Disease and criminal behaviour

Pathogen stress, cultural collectivism, and violence

A recent evolutionary analysis on the causes of violent crime suggested that both lethal and non-lethal violence, including male-male 'honour' or 'argument-based' homicide, spousal homicide, and non-lethal romantic partner abuse, result from two related factors: infectious disease stress and cultural collectivism (Thornhill 2010). Thornhill reported positive associations between: 1) rates of infectious disease and 2) cultural collectivism with each of these types of violence across states in the USA. Further, each of these relationships remains strong after statistically controlling for income inequality, indicating that the relationship is not due to relative social status or competitive disadvantage. He suggested that features of cultural collectivism (e.g. willingness to defend the honour of one's in-group) may promote the eruption of violent conflict.

Infectious disease stress and cultural collectivism are strongly positively correlated cross-nationally (Fincher et al. 2008). Fincher et al. argue that cultural collectivism and features associated with it (e.g. strong kinship-based bonds, out-group avoidance, assortative sociality) are adaptive responses to effectively reduce infectious disease stress by inhibiting pathogen transmission between groups. Further, risk of infectious disease may promote the fractioning of a single parent group into multiple daughter groups. This group splitting could be associated with intrastate conflict and wars. Consistent with this idea, the number of religions present in a particular region of the world (Fincher and Thornhill 2008) and levels of armed within-state group conflict are both associated with infectious disease stress (Letendre et al. 2010). All this evidence suggests that disease stress and cultural collectivism are key variables of interest to evolutionary perspectives on criminology.

Infectious disease, cognitive abilities, and violent crime

Consistent with the idea that infectious disease risk contributes to levels of violent criminality, Eppig et al. (2010) reported a negative correlation between cross-national parasite prevalence and the average levels of general cognitive ability, that remains strong after controlling for a number of potential confounds (e.g. education, temperature, gross domestic product (GDP) per capita). They suggested that in populations that regularly face the bioenergetically costly task of fighting infectious disease, trade-offs must be made between allocations toward the immune system and developing an expensive brain. As previously noted, since low general cognitive ability is implicated in contributing to criminality (e.g. Herrnstein and Murray 1994), the negative relationship between disease risk and cognitive ability suggests that cognitive ability may be one possible mediator of the relationship between infectious disease stress and criminal interpersonal aggression. A related possibility is that the heritability of general intelligence may be due, in part, to genetic variation in immunocompetence (Eppig et al. 2010). Individuals that inherit low-quality immune systems may require higher bioenergetic resource allocations toward fighting off pathogens, in effect reducing allocations toward brain development.

Similarly, infectious disease stress could inhibit the development of other complex cognitive abilities, such as executive functioning. Executive functions are composed of several interrelated yet distinct cognitive abilities. Some (e.g. inhibition of prepotent responses implicated in 'self-control', shifting between multiple tasks) appear largely independent of general cognitive ability (Friedman et al. 2006). But like general intelligence, executive functions are also implicated in contributing to criminal behaviour by enabling 'self-control' (Gottfredson and Hirschi 1990). If correct, the strong relationships between infectious disease and violence may be mediated by either: 1) decreased general cognitive ability, 2) impaired executive functioning, or 3) a combination of both (Wenner et al. 2007).

Life history theory and criminal behaviour

Life history (LH) describes the strategic allocation of resources among the competing demands of the two major components of fitness, survival and reproduction (represented by somatic and reproductive effort, respectively). Species lie on a fast-slow (r–K) continuum, representing a covarying range of reproductive behavioural strategies, inversely relating life-history traits such as fecundity and parenting. LH theory predicts that species living in harsh (high risk of extrinsic morbidity and mortality), unpredictable, and uncontrollable environments evolve clusters of 'fast' life history traits, which are associated with high reproductive rates (including minimal parental investment and relatively brief inter-generation times). In contrast, species living in relatively safe, predictable, stable, and controllable environmental conditions evolve clusters of 'slow' life history traits (extensive parental investment, long inter-generational times). Hence, the fast-strategist is a short-term planner, taking benefits opportunistically with little regard for long-term consequences. In contrast, the slow-strategist is a long-term planner, delaying immediate gratification in the service of future eventualities. It is important to recognize, however, that a 'slow' strategy is neither superior nor inferior to a 'fast' one. In purely Darwinian terms, 'superiority' (meaning better adaptation) depends on the ecology in which the organism is situated.

Personality and attitudes

LH theory suggests that a wide variety of personality traits will tend to form adaptively coordinated behavioural clusters. Among traits relevant to criminal behaviour that have been implicated as possible facets of fast LH strategy are: 1) a desire for casual sex, 2) high mating effort, 3) low emotional attachment to romantic partners, 4) tendencies toward risk-taking or impulsive behaviour, 5) decreased social and moral rule-following, and 6) decreased law-abidingness (Rushton 1985; Thornhill and Palmer 2000; Figueredo et al. 2006; Gladden et al. 2008, 2009). These traits tend to co-occur rather than to appear independently in individuals. For example, because several of these LH traits have been shown to predict sexual coercion (Malamuth et al. 2005), it was hypothesized that a fast LH strategy might facilitate the disproportionate use of coercive sexual strategies. Indeed, a variety of LH traits that clustered into a single 'Protective' (meaning slow) LH factor, predicted decreased frequencies of sexually coercive behaviours among college students and fully mediated the relationship between the biological sex of participants and frequencies of sexually coercive behaviour (Figueredo et al. in press; Gladden et al. 2008). This result was interpreted as suggesting that the traits predicted by CDT (and by other evolutionary theories of sexual coercion) were not independent of each other, and that a fast LH strategy may underlie a wide variety of socially deviant behaviour and explain why men are more delinquent or criminal than women (see also Rowe et al. 1997; Rowe 2001).

In addition, slow LH strategy was also found to be highly associated ($r=0.49$) with a general trait of positive 'Evaluative Self-Assessment' (Gladden et al. 2010), composed of several theoretically heterogeneous measures of collective and individual self-esteem, and perceived mate value and mating success (Kirsner et al. 2003, 2009). This result was interpreted as suggesting that faster LH individuals correctly perceive themselves as being lower in phenotypic quality. This is reasonable because they characteristically invest less of their bioenergetic and material resources in somatic effort and receive less parental and nepotistic effort from parents and kin, as well as less reciprocal altruism and mutualistic support from other associates. Fast LH strategists seem to possess the traits implicated by CDT, whereas slow LH strategists may instead possess traits that inhibit coercive tactics, conferring greater competitive advantage.

Fast LH strategists also exhibit increased racism and sexism, and perhaps, increased socially hostile attitudes towards others generally. Cross-culturally, slower LH individuals exhibit decreased negative ethnocentrism towards multiple perceived social outgroups, including Arab and Mexican immigrants (in the US), and Afro-Costa Ricans and Nicaraguan immigrants (in Costa Rica) (Figueredo, Andrzejczak et al. 2010). Another study found that slower LH individuals exhibit decreased negative androcentrism (Gladden et al. 2009). Relationships between negative androcentrism (hostility towards women, sexist attitudes, acceptance of rape-myths) and both sexual coercion and intimate partner violence have been demonstrated by much previous research (Malamuth 1996, 1998). It is possible that negative androcentrism is indeed a mediator of sexually coercive behaviour and slow LH strategists are more able to inhibit socially undesirable or deviant behaviour such as sexual coercion.

Such hostile and aggressive attitudes make sense from a strategic LH perspective because fast LH strategists are characterized by high mating effort, a preference for short-term sexual relationships, and low attachment to sexual partners; a suite of traits selected to produce a high number of genetically diverse offspring (Figueredo and Wolf 2009). The strategic interests of fast LH individuals are therefore expected to produce more direct conflict with both members of one's same sex (intrasexual competition) and with members of the opposite sex (intersexual conflict).

Malamuth (1998) suggested that a generalized disposition towards either concordant or adversarial relationships with members of the opposite sex ultimately drives both of the major convergent pathways (hostility and promiscuity) that have been found to predict sexually-coercive behaviour. This disposition can be described in terms of two strategies anchoring opposite ends of a continuum: a convergent interest sexual strategy and a divergent interest sexual strategy. Specifically, males following the former see their reproductive interests and those of the female as mutualistic, and base their intersexual relationships on this perceived common ground. In contrast, males following the divergent interest strategy perceive their reproductive interests and those of the female as mutually inconsistent, or *antagonistic*, and base their intersexual relationships on this perceived conflict of interests. Malamuth (1998) further suggested that the development of these strategies might be biased by different LH strategies: slow LH strategists are more prone to adopt convergent interest (mutualistic) sexual strategies, which are consistent with long-term sexual relationships and cooperative biparental care, whereas fast LH strategists are more prone to adopt divergent (antagonistic) sexual interest strategies, which are clearly inconsistent with these long-term reproductive tactics.

Figueredo and Jacobs (2010) proposed an extension of this model, beyond the sexual and into the general social domain, in which slow LH strategists are more prone to adopt otherwise equivalent convergent interest social strategies (being more likely to engage in reciprocally altruistic relationships), while fast LH strategists tend to adopt otherwise equivalent divergent interest social strategies. This is because slow LH strategists clearly prefer long-term and cooperative social as well as sexual relationships, which are evidently easier and more profitable to maintain in the more stable, predictable, and controllable environments in which slow LH strategists typically evolve and develop (e.g. Brumbach et al. 2007; Ellis et al. 2009).

Fast LH strategists are expected to exhibit increased hostility both towards their own sex (due to increased intrasexual competition) and the opposite sex (due to increased strategic interference by potential mates). The increased mating effort associated with faster LH strategies has been shown cross-culturally to predict increased amounts of certain intrasexually competitive displays (Weiss et al. 2004; Egan et al. 2005). Faster LH strategies have also been shown to be associated with problematic levels of intrasexual competitiveness in women (Salmon et al. 2009). In contrast, a shared slow LH strategy has been identified as a very powerful predictor of relationship satisfaction, both cross-sectionally and longitudinally, in romantically-involved heterosexual couples (Olderbak and Figueredo 2009, 2010).

Both slow LH and general mental ability (IQ) is positively associated with enhanced levels of executive functioning (e.g. Wenner et al. 2007). The ability to set goals, plan, sequence, prioritize, organize, initiate, inhibit, pace, shift, monitor, control, and complete actions all involve executive functions (cf. Lezak et al. 2004). As in 'self-control' theory (Gottfredson and Hirschi 1990), executive functions inhibit psychopathic attitudes, and these psychopathic attitudes, which consist of lack of feelings of empathy or guilt for antisocial actions, and the willingness to lie, cheat, manipulate, or use aggression against others instrumentally for selfish gain or reactively, then contribute positively to socially deviant behaviours across a variety of intrasexual, intersexual, and general social situations (Wenner et al. 2007).

Figueredo et al. (2010) tested a factor-analytic structural equations model incorporating a set of theoretically-specified predictors that examine the hypothesis that intimate partner violence is practiced disproportionately by faster LH strategists. In this model, slow LH strategy was associated with decreased intimate partner violence through multiple causal pathways, and all of these modelled pathways were indirect effects. The various effects were mediated through two indirect causal pathways: 1) lower psychopathic and aggressive attitudes, themselves indirectly influenced by slow LH through enhanced executive functioning, lower short-term mating, and lower culture of honour revenge ideology; and 2) higher mate value among slow LH individuals. The total of all these effects in the model explained 32% of the variance in intimate partner violence.

Developing a predictive model

More recently, Figueredo et al. (2010) tested a similar model to test whether it would predict interpersonal aggression in general.[1] We hypothesized that the construct representing an antagonistic social schema should be able to take the place formerly held by the culture of honour revenge ideology in our previous intimate partner violence model, the latter being relegated to a subsidiary role as a by-product of the antagonistic social schema and no longer modelled as a direct causal influence. The authors constructed the *Interpersonal Relations Rating Scale* (IRRS) as a measure of psychological and physical aggression towards individuals they have encountered. The IRRS asks participants how often each action occurred during the past 12 months, containing parallel items for same-sex or opposite-sex victims of interpersonal aggression (respectively, these were aggregated into the IRRS-S Perpetration and IRRS-O Perpetration scales). These items were constructed to be otherwise equivalent in form and content to the items of the *Relationship Behaviour Rating Scale—Revised* (RBRS-R; O'Hara and Beck 2009), the measure previously used

---

[1] Some of the results and related graphics that were presented in this paper (including Figure 13.1), representing our previous and ongoing work, are featured in some of the cited articles, including several currently in submission and in preparation.

to assess psychological and physical victimization by relationship partners (Figueredo et al. in press); the major difference was that the IRRS asked participants to report their perpetration rather than their victimization, and that the questions were not limited to interactions with romantic partners.

To test the evolutionary psychological interpretation that most interpersonal aggression is ultimately about sex, Figueredo et al. (2010) included four additional scales to be able show psychometric convergence with the IRRS-O and IRRS-S Perpetration Scales, and provide a link to their previous work on intimate partner violence: 1) the *Mating Aggression Scales: Competitor Derogation Tactics Scale* (Buss and Dedden 1990), 2) the *Intrasexual Competition Scale* (Buunk and Fisher 2009), 3) the *Mate Guarding Scale* (Buunk 1997), and 4) the *Mate Retention Inventory* (Buss 1988). Although some specific items were modified to specify same-sex or opposite-sex immediate targets of these behaviours, it was not possible to disaggregate these 'Mating Aggression Scales' by the sex of the victim, as was done with the IRRS-O and IRRS-S Perpetration Scales. Most of the aggressive mating behaviours sampled in these scales clearly constitute *strategic interference* with both one's intended sexual object and one's perceived sexual rival, who should *both* be considered victims of these aggressive behaviours. These strategies are therefore not directed exclusively to either same-sex or opposite-sex targets. As predicted, the four mating aggression scales comprised a single common factor. The results of this factor-analytic structural equations model are displayed graphically in Figure 13.1.

Two of the higher-order latent variables were constructed explicitly within the model: 1) the Psychopathic and Aggressive Attitudes factor, and 2) the Interpersonal Aggression factor. This was done because the construction of these higher-order factors was of special theoretical interest because they showed convergence across multiple heterogeneous measures, whereas the aggregation of subscales within each of battery of homogeneous measures described above was relatively uncontroversial.

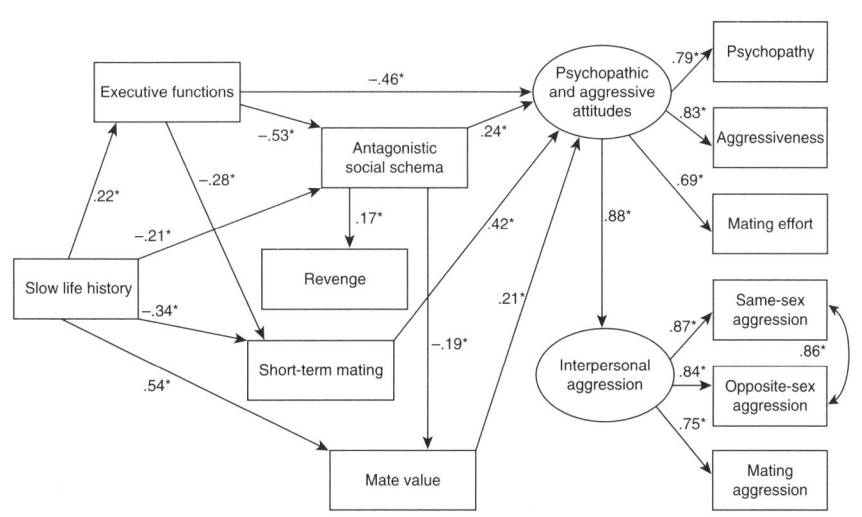

Fig. 13.1 Factor-analytic structural equations model for interpersonal aggression and life history strategy. All path coefficients (effect sizes) that are significantly different ($p < 0.05$) from zero are indicated by the asterisks (*). Standardized regression coefficients (λ-weights or β-weights) for the measurement and structural pathways are reported. All model parameters were estimated by maximum likelihood (ML).

Our explicit measurement model for the Interpersonal Aggression factor includes the IRRS-S Perpetration scale, the IRRS-O Perpetration scale, and the combined Mating Aggression scales. A residual correlation was also specified a priori between the IRRS-S and the IRRS-O Perpetration scales to account for the shared method (test-specific) variance associated with both components of the IRRS but not shared with the Mating Aggression scales when estimating the common general Interpersonal Aggression factor. Our explicit measurement model for the Psychopathic and Aggressive Attitudes factor includes the Levenson Self-Report Psychopathy, the Reactive-Proactive Aggression Questionnaire, and the Mating Effort scale.

According to this factor-analytic structural equations model, slow LH contributes directly to increasing both mate value and executive functions. These findings are consistent with the findings of Gladden et al. (2010) for mate value, as well as those of Salmon et al. (2009) for executive functioning. Slow LH and executive functions both serve to inhibit a short-term mating strategy. This is consistent with the view that long-term mating requires both a behavioural disposition towards that lifestyle and sufficient mental ability to inhibit and control competing tendencies, such as short-term mating urges (see Figueredo and Jacobs 2010). As predicted, slow LH and executive functions both contribute negatively to an antagonistic social schema. An antagonistic social schema, in turn, contributes positively to the culture of honour revenge ideology, indicating that the latter might represent a special case of a generally antagonistic social schema, at least in a society having a low culture of honour. In contrast, an antagonistic social schema contributes negatively to mate value, indicating that individuals with an antagonistic social schema might actually perceive themselves as competitively disadvantaged in the mating market by their adversarial relationships with others.

High executive functions tend to greatly decrease psychopathic and aggressive attitudes. In contrast, a highly antagonistic social schema, a short-term mating orientation, and a high mate value appear to contribute positively to psychopathic and aggressive attitudes. This latter positive effect might seem to be somewhat anomalous, but might be at least partially attributable to a narcissistic response bias in self-reported mate value that was found previously in high-mating effort individuals (Rowe et al. 1997). High psychopathic and aggressive attitudes contribute positively to interpersonal aggression. These findings are also consistent with those of Wenner et al. (2007) for generalized social deviance.

The major results of this study can be summarized concisely as follows: 1) slower LH strategy contributes to increasing both executive functioning and mate value; 2) slower LH strategy and higher executive functioning both contribute to decreasing short-term mating; 3) slower LH strategy and higher executive functioning both contribute to decreasing an antagonistic social schema; 4) a more antagonistic social schema contributes to increasing culture of honour revenge ideology; 5) higher executive functioning contributes to decreasing psychopathic and aggressive attitudes, whereas higher short-term mating, a more antagonistic social schema, and higher mate value tend to increase them; and 6) higher psychopathic and aggressive attitudes contribute to increasing interpersonal aggression.

The various negative indirect effects of slow LH strategy on interpersonal aggression were therefore mediated through lower psychopathic and aggressive attitudes, which are themselves indirectly influenced by slow LH though enhanced executive functioning, higher mate value, lower short-term mating, and a less antagonistic social schema. The total of all these effects in the model explained 67% of the variance in psychopathic and aggressive attitudes and 78% of the variance in interpersonal aggression. Slow LH strategy was thus associated with decreased interpersonal aggression through multiple causal pathways. All of these modelled pathways were indirect effects and were ultimately mediated through psychopathic and aggressive attitudes.

In conclusion, Figueredo et al. (2010) found that an extremely similar structural model indeed fit the interpersonal aggression data as well as the original model fit the intimate partner violence data. One minor difference was that the protective effect previously found for higher mate value on intimate partner violence was not obtained for interpersonal aggression in the present model. In addition, a more antagonistic social schema directly depressed mate value in a way that the culture of honour revenge ideology previously did not in the intimate partner violence model.

A secondary aim of this study was to determine what role, if any, was played by an antagonistic social schema in this structural model as a partial mediator between LH strategy and interpersonal aggression. One of the main differences between the present interpersonal aggression model and the prior intimate partner violence model was that, as hypothesized, the construct representing an antagonistic social schema took the place formerly held by the culture of honour revenge ideology, the latter being relegated to a subsidiary role as a by-product of the antagonistic social schema and no longer modelled as a direct causal influence. At least, this was the result in the sample from Arizona, US, and awaits cross-validation in societies known to be higher in culture of honour, such as Sonora, Mexico, where an ideology of revenge is more socially normative (Figueredo et al. 2004). In addition, higher short-term mating did not contribute to a more antagonistic social schema in the way that it previously did to the culture of honour revenge ideology in the intimate partner violence model. In any event, an antagonistic social schema did not have a direct effect on interpersonal aggression, and influenced interpersonal aggression only indirectly through psychopathic and aggressive attitudes. Furthermore, the magnitude of the effect of an antagonistic social schema on psychopathic and aggressive attitudes was only trivially greater than that previously found for the culture of honour revenge ideology, whose place it took in the interpersonal aggression model.

In spite of these differences, the results of the two models were surprisingly similar. If anything, the explanatory power of the present model was even greater, increasing from explaining 32% of the variance in intimate partner violence to explaining 78% of the variance in interpersonal aggression, with only these minor modifications.

Furthermore, our results indicate that the frequencies of interpersonally aggressive behaviours are highly proportional when comparing those targeting victims of the same-sex with those targeting victims of the opposite-sex as the perpetrator. This suggests that this generalized disposition towards aggressive behaviour against social and sexual partners is not particularly 'gendered' or sexually-specific. In addition, the high degree of convergence of the Mating Aggression composite with the two Interpersonal aggression (IRRS Perpetration) scales also suggests that there is little unique about either the aetiology or the nature of intimate partner violence, and that it appears to be little more than a special case of interpersonal aggression. This supports the evolutionary psychological interpretation that most interpersonal aggression is ultimately about sex, whether it is directed at the same- or opposite-sex (in contrast to intimate partner violence which is usually opposite-sex). We understand that this conclusion stands in sharp contrast to the views of many who have historically proposed theories regarding the 'specialness' of intimate partner violence, as distinct from interpersonal aggression in general. All we can say is that we are among those people, in that the first author has published extensively in the field and proposed domain-specific theories of his own, but that when the empirical evidence is not consistent with these theories, we are willing to admit that we might have been wrong.

At this point in our continuing programme of research, individuals with antagonistic social schemata appear to recognize only two types of other people in the world: sex objects (opposite-sex conspecifics) and sex rivals (same-sex conspecifics). Individuals with antagonistic social schemata appear to have adversarial relationships with both.

Conclusions and limitations: old wine in new bottles?

One of the most striking results in this model is the large negative effect of 'Executive Functions' on 'Psychopathic and Aggressive Attitudes' (Figure 13.1). Is this model representing nothing more than a repackaged version of the General Theory of Crime (Gottfredson and Hirschi 1990)? We believe that it does not, for several reasons.

First, our theoretical formulation differs fundamentally from the General Theory of Crime in that 'self-control' theory strongly implies that the failure to control criminal impulses occurs at the point immediately prior to actually performing the criminal behaviour, when the opportunity randomly arises. Our theoretical formulation instead models the inhibitory effects of enhanced executive functioning as operating indirectly through another latent construct, Psychopathic and Aggressive Attitudes, with no direct effect of enhanced executive functioning on interpersonal aggression. This implies that control of aggressive impulses occur further back in the hypothesized causal sequence: it suppresses any covert cognitions as well as overt behaviours that might be socially deviant.

Second, our theoretical formulation also features several individual difference variables that are excitatory, rather than just inhibitory, with respect to psychopathic and aggressive attitudes. These are higher mate value, lower short-term mating, and a less antagonistic social schema, which together explained 67% of the variance in psychopathic and aggressive attitudes. 'Self-control' theory relies almost exclusively on the lack of impulse control as the primary cause of criminal behaviour, and does not sufficiently address individual differences in the nature and the strength of the criminal impulses themselves.

Third, our theoretical formulation identifies slow LH as the ultimate common causal influence behind the enhanced executive functioning, the higher mate value, the lower short-term mating, and the less antagonistic social schema that collectively inhibit psychopathic and aggressive attitudes, and indirectly inhibit interpersonal aggression. As with the effects of LH strategy in general (Figueredo et al. 2007), these effects are small to moderate, but they are cumulative and summate. LH strategy exerts a pervasive influence on many of our cognitions and behaviours, many of which either facilitate or inhibit criminal behaviour. To our knowledge, no other theory has been able to adequately and comprehensively account for the suite of behavioural adaptations required to pursue either a mutualistic or antagonistic social strategy.

Fourth, our theoretical formulation differs fundamentally from the General Theory of Crime in that executive functions are not identical to impulse control (although they do also encompass this purely inhibitory function), but also include ability to set goals, plan, sequence, prioritize, organize, initiate, pace, shift, monitor, control, and complete actions. Executive functions are also not identical to general mental ability (IQ), although they are somewhat correlated and contribute jointly to suppressing socially deviant behaviours (Wenner et al. 2007). Therefore, our theoretical formulation is not just a repackaged version of the cognitive (IQ) theory of criminal behaviour.

The one way in which the evidentiary base supporting the General Theory of Crime is currently superior to that supporting our own theoretical formulation is that the former encompasses 'crimes against property' as well as the 'crimes against persons' included in our measures of interpersonal aggression. However, that situation might be a temporary one, due to the history of the Figueredo et al. (2010) study of interpersonal aggression having been patterned on the prior Figueredo et al. (in press) study of intimate partner violence. We believe that we can confidently predict that this same model should also account for 'crimes against property' as well as 'crimes against persons'. In particular, an integrative theory of 'white-collar' crime, which encompasses both, is consistent with this evolutionary model.

Finally, our theoretical formulation traces back the aetiology of criminal behaviour to evolution within harsh, unpredictable, and uncontrollable environments, which are the ultimate selective pressure underlying fast LH strategies. This leaves us some hope for intervention at the longer-term ecological level, rather than attributing criminal behaviour exclusively to relatively permanent and stable individual traits.

We must recall, however, that fast LH is not logically equivalent to socially deviant or criminal behaviour. Because LH requires that an overall strategy consist of a coordinated set of behavioural tactics that are mutually consistent and reinforcing, selection acts to eliminate any individual tactics that interferes with the others in the suite. Slow LH is therefore generally inconsistent with socially deviant or criminal behaviour. For example, if a middle-aged slow LH strategist who is currently married, an investing parent, and employed in a high-status profession, decided to rob an off licence or rape a woman, the consequences of his life and career would be devastating. If the same was done by a juvenile delinquent in the inner city, with dismal job prospects, and no significant romantic, family, or community attachments, the consequences would be substantially less catastrophic. Even if an individual does in fact rape women or rob off licences, it is by no means mandatory: it represents no more than a tactic that is potentially consistent with a fast LH strategy. This means that a slow LH strategy should be expected to *inhibit* socially deviant or criminal behaviours, whereas a fast LH strategy should merely be expected to *permit* them, perhaps contingently upon environmental triggers. In behavioural endocrinology, this is what is known as a 'permissive' effect, rather than one of compulsion. Exactly what turns an otherwise peaceful and law-abiding fast LH strategist into a social deviant remains a mystery, but an interesting question for future research.

References

Agnew, R. (2006). *Pressured into crime: an overview of general strain theory.* Roxbury, Los Angeles, CA.

Agnew, R. and White, H. (1992). An empirical test of general strain theory. *Criminology*, **32**, 475–99.

Alexander, R.D. (1987). *The biology of moral systems.* Aldine de Gruyter, New York.

Beccaria, C. (1995 [1767]). *On crimes and punishments and other writings.* Cambridge University Press, Cambridge.

Bentham, J. (1988 [1789]). *The principles of morals and legislation.* Prometheus, Amherst, NY.

Bowlby, J. (1951). *Maternal care and mental health.* World Health Organization, Geneva.

Bowlby, J. (1970). Disruption of affectional bonds and its effects on behavior. *Journal of Contemporary Psychotherapy*, **2**, 75–86.

Brumbach, B.H., Walsh, M., and Figueredo, A.J. (2007). Sexual restrictedness in adolescence: a life history perspective. *Acta Psychologica Sinica*, **39**, 481–8.

Buss, D.M. (1988). The evolution of human intrasexual competition: tactics of mate attraction. *Journal of Personality and Social Psychology*, **54**, 616–28.

Buss, D.M., and Dedden, L.A. (1990). Derogation of competitors. *Journal of Social and Personal Relationships*, **7**, 395–422.

Buunk, B.P. (1997). Personality, birth order and attachment styles as related to various types of jealousy. *Personality and Individual Differences*, **23**, 997–1006.

Buunk, A.P. and Fisher, M. (2009). Individual differences in intrasexual competition. *Journal of Evolutionary Psychology*, **7**, 37–48.

Daly, M. and Wilson, M. (1988). *Homicide.* Aldine de Gruyter, Hawthorne, NY.

Daly, M., Wilson, M., and Vasdev, S. (2001). Income inequality and homicide rates in Canada and the United States. *Canadian Journal of Criminology*, **43**, 219–46.

Denowski, G. and Denowski, K. (1985). The mentally retarded offender in the state prison system: identification, prevalence, adjustment, and rehabilitation. *Criminal Justice and Behavior*, **12**, 55–70.

Durkheim, E. (1964). *The rules of sociological method.* The Free Press of Glencoe, New York.

Egan, V., Figueredo, A.J., Wolf, P., et al. (2005). Sensational interests, mating effort, and personality: evidence for cross-cultural validity. *Journal of Individual Differences*, **26**, 11–19.

Ellis, B.J., Figueredo, A.J., Brumbach, B.H., and Schlomer, G.L. (2009). Mechanisms of environmental risk: the impact of harsh versus unpredictable environments on the evolution and development of life history strategies. *Human Nature*, **20**, 204–68.

Eppig, C., Fincher, C.L., and Thornhill, R. (2010). Parasite prevalence and the worldwide distribution of cognitive ability. *Proceedings of the Royal Society B*, **277**, 3801–8.

Eysenck, H. (1964). *Crime and personality.* Routledge, London.

Eysenck, S. and Eysenck, H. (1971). Crime and personality: item analysis of questionnaire responses. *British Journal of Criminology*, **11**, 49–62.

Ferri, E. (1895). *Criminal sociology.* Unwin, London.

Figueredo, A.J. (1995). The epigenesis of sociopathy. *Behavioral and Brain Sciences*, **18**, 556–7.

Figueredo, A.J. and Jacobs, W.J. (2000). Strategic sexual pluralism through a Brunswikian lens. *Behavioral and Brain Sciences*, **23**, 603–4.

Figueredo, A.J. and Jacobs, W.J. (2010). Aggression, risk-taking, and alternative life history strategies: the behavioral ecology of social deviance. In: M. Frias-Armenta and V. Corral-Verdugo (eds), *Bio-psycho-social perspectives on interpersonal violence*, pp. 3–28. Nova Science Publishers, Hauppauge, NY.

Figueredo, A.J. and McCloskey, L.A. (1993). Sex, money, and paternity: the evolutionary psychology of domestic violence. *Ethology and Sociobiology*, **14**, 353–79.

Figueredo, A.J. and Wolf, P.S.A. (2009). Assortative pairing and life history strategy: A cross-cultural study. *Human Nature*, **20**, 317–30.

Figueredo, A.J., Sales, B.D., Becker, J.V., Russell, K., and Kaplan, M. (2000). A Brunswikian evolutionary-developmental model of adolescent sex offending. *Behavioral Sciences and the Law*, **18**, 309–29.

Figueredo, A.J., Tal, I.R., McNeill, P., and Guillén, A. (2004). Farmers, herders, and fishers: the ecology of revenge. *Evolution and Human Behavior*, **25**, 336–53.

Figueredo, A.J., Vásquez, G., Brumbach, B.H., Sefcek, J.A., Kirsner, B.R., and Jacobs, W J. (2005). The K-factor: individual differences in life history strategy. *Personality and Individual Differences*, **39**, 1349–60.

Figueredo, A.J., Vásquez, G., Brumbach, B.H., et al. (2006). Consilience and life history theory: from genes to brain to reproductive strategy. *Developmental Review*, **26**, 243–75.

Figueredo, A.J., Vásquez, G., Brumbach, B.H., and Schneider, S.M.R. (2007). The K-factor, covitality, and personality. *Human Nature*, **18**, 47–73.

Figueredo, A.J., Gladden, P.R., and Beck, C.J.A. (2010). 'Interpersonal violence, cognitive schemata, and life history strategy.' Human Behavior and Evolution Society meeting, Eugene, Oregon.

Figueredo, A.J., Andrzejczak, D.J., Jones, D.J., Smith-Castro, V., and Montero-Rojas, E. (2011). Reproductive strategy and ethnic conflict: Slow life history as a protective factor against negative ethnocentrism in two contemporary societies. *Journal of Social, Evolutionary, and Cultural Psychology*, 5, 14–31.

Figueredo, A.J., Gladden, P.R., and Beck, C.J.A. (in press). Intimate partner violence and life history strategy. In: A. Goetz, and T. Shackelford (eds), *The Oxford handbook of sexual conflict in humans.* Oxford University Press, New York.

Fincher, C.L. and Thornhill, R. (2008). Assortative sociality, limited dispersal, infectious disease and the genesis of the global pattern of religion diversity. *Proceedings of the Royal Society B*, **275**, 2587–94.

Fincher, C.L., Thornhill, R., Murray, D. and Schaller, M. (2008). Pathogen prevalence predicts human cross-cultural variability in individualism/collectivism. *Proceedings of the Royal Society B*, **275**, 1279–85.

Friedman, N.P., Miyake, A., Corley, R., Young, S.E., DeFries, J.C., Hewitt, J.K. (2006). Not all executive functions are related to intelligence. *Psychological Science*, **17**, 172–9.

Garcia, J., Hankins, W.G., and Rusiniak, K.W., (1974). Behavioral regulation of the *milieu interne* in man and rat. *Science*, **185**, 824–31.

Gibson, J.J. (1979). *The ecological approach to visual perception.* Houghton Mifflin, Boston, MA.

Gladden, P.R., Sisco, M., and Figueredo, A.J. (2008). Sexual coercion and life history strategy. *Evolution and Human Behavior*, **29**, 319–26.

Gladden, P.R., Figueredo, A.J., Andrzejczak, D.J., Jones, D.N., and Smith-Castro, V. (2009). 'Life history strategy, executive functioning, and negative androcentrism.' Human Behavior and Evolution Society meeting, Fullerton, CA.

Gladden, P.R., Figueredo, A.J., and Snyder, B. (2010). Life history strategy and evaluative self-assessment. *Personality and Individual Differences*, **48**, 731–5.

Gottfredson, M.R. and Hirschi, T. (1990). *A general theory of crime.* University of Stanford Press, Stanford, CA.

Herrnstein, R.J. and Murray, C. (1994). *The Bell curve.* The Free Press, New York.

Hiraiwa-Hasegawa, M. (2005). Homicide by men in Japan, and its relationship to age, resources, and risk taking. *Evolution and Human Behavior*, **26**, 332–43.

Hirschi, T. (1969). *Causes of delinquency.* University of California Press, Berkeley, CA.

Hirschi, T. and Gottfedson, M. (1983). Age and the explanation of crime. *American Journal of sociology*, **89**, 552–84.

Hunter, J.A. and Figueredo, A J. (2000). The influence of personality and history of sexual victimization in the prediction of juvenile perpetrated child molestation. *Behavior Modification*, **29**, 259–81.

Kandel, E., Mednick, S.A., Kirkegaard-Sorensen, L., *et al.* (1988). IQ as a protective factor for subjects at high risk for antisocial behavior. *Journal of Consulting and Clinical Psychology*, **56**, 224–6.

Kirsner, B.R., Figueredo, A.J., and Jacobs, W.J. (2003). Self, friends, and lovers: structural relations among Beck Depression Inventory scores and perceived mate values. *Journal of Affective Disorders*, **75**, 131–48.

Kirsner, B.R., Figueredo, A.J., and Jacobs, W.J. (2009). Structural relations among negative affect, mate value, and mating effort. *Evolutionary Psychology*, **7**, 374–97.

Lalumiere, M.L. and Quinsey, V.L. (1996). Sexual deviance, antisociality, mating effort, and the use of sexually coercive behaviors. *Personality and Individual Differences*, **21**, 33–48.

Lalumiere, M., Mishra, S., and Harris, G.T. (2008). In cold blood: the evolution of psychopathy. In: J. Duntley and T. Shackelford (eds), *Evolutionary forensic psychology*, pp. 176–197. Oxford University Press, New York.

Letendre, K., Fincher, C.L. and Thornhill, R. (2010). Does infectious disease cause global variation in the frequency of intrastate armed conflict and civil war? *Biological Reviews*, **85**, 669–83.

Lewontin, R.C., Rose, S., and Kamin, L.J. (1984). *Not in our genes: biology, ideology and human nature.* Pantheon, New York.

Lezak, M.D., Howieson, D.B., Loring, D.W. Hannay H.J., and Fischer J.S. (2004). *Neuropsychological assessment* (4th Edition.). Oxford University Press, New York.

Lombroso, C. (2006 [1876]). *Criminal man.* Duke University Press, London.

Malamuth, N.M. (1996). Sexually explicit media, gender differences and evolutionary theory. *Journal of Communication*, **46**, 8–31.

Malamuth, N.M. (1998). The confluence model as an organizing framework for research on sexually aggressive men: risk moderators, imagined aggression, and pornography consumption. In: R.G. Geen, and E. Donnerstein (eds), *Human aggression: theories, research, and implications for social policy*, pp. 229–45. Academic Press, San Diego, CA.

Malamuth, N., Huppin, M., and Paul, B. (2005). Sexual coercion. In: D.M. Buss (ed.), *Handbook of evolutionary psychology*, pp. 394–418. Wiley, Hoboken, NJ.

Mayr, E. (1974). Behavioral programs and evolutionary strategies. *American Scientist*, **62**, 650–9.

McGurk, B. and McDougall, C. (1981). A new approach to Eysenck's theory of criminality. *Personality and Individual Differences*, **2**, 338–40.

Mealey, L. (1995). The sociobiology of sociopathy: an integrated evolutionary model. *Behavioral and Brain Sciences*, **18**, 523–99.

Merton, R. (1938). Social structure and anomie. *American Sociological Review*, **3**, 672–82.

Moffitt, T.E., Gabrielli, W.F., Mednick, S.A. and Schulsinger, F. (1981). Socioeconomic status, IQ, and delinquency. *Journal of Abnormal Psychology*, **90**, 152–6.

Nisbett, R.E. and Cohen, D. (1996). *Culture of honor: the psychology of violence in the South*. Westview Press, Boulder, CO.

O'Hara, K.L. and Beck, C.J. (2009). 'The Relationship Behavior Rating Scale: validation and expansion.' Paper presented at the annual meeting of the American Psychology—Law Society, San Antonio, TX.

Olderbak, S.G. and Figueredo, A.J. (2009). Predicting romantic relationship satisfaction from life history strategy. *Personality and Individual Differences*, **46**, 604–10.

Olderbak, S.G. and Figueredo, A.J. (2010). Life history strategy as a longitudinal predictor of relationship satisfaction and dissolution. *Personality and Individual Differences*, **49**, 234–9.

Pinker, S. (1994). *The language instinct: how the mind creates language*. William Morrow, New York.

Pinker, S. (2002). *The blank slate: the modern denial of human nature*. Viking New, York.

Rowe, D.C. (2001). *Biology and crime*. Roxbury Press, Los Angeles, CA.

Rowe, D.C., Vazsonyi, A.T., and Figueredo, A.J. (1997). Mating effort in adolescence: conditional or alternative strategy? *Journal of Personality and Individual Differences*, **23**, 105–15.

Rushton, J.P. (1985). Differential K theory: the sociobiology of individual and group differences. *Personality and Individual Differences*, **6**, 441–52.

Salmon, C., Figueredo, A.J., and Woodburn, L. (2009). Life history strategy and disordered eating behavior. *Evolutionary Psychology*, **7**, 585–600.

Seligman, M.E.P., (1970). On the generality of the laws of learning. *Psychological Review*, **77**, 406–18.

Shackelford, T.K. (2005). An evolutionary psychological analysis of cultures of honor. *Evolutionary Psychology*, **3**, 381–91.

Shaw, C. and McKay, H. (1931). *Social factors in juvenile delinquency*. Government Printing Office, Washington, DC.

Shaw, C. and McKay, H. (1942). *Juvenile delinquency and urban areas*. University of Chicago Press, Chicago, IL.

Stillman, T.F., Maner, J.K., and Baumeister, R.F. (2010). A thin slice of violence: distinguishing violent from nonviolent sex offenders at a glance. *Evolution and Human Behavior*, **31**, 298–303.

Sutherland, E. and Cressey, D. (1966). *Principles of criminology*. Lippencott, Philadelphia, PA.

Thornhill, R. (2010). 'Margo Wilson's research continues to inspire new investigations of homicide.' Paper presented at Human Behavior and Evolution Society meeting 2010, Eugene, OR.

Thornhill, R. and Palmer, C. (2000). *A natural history of rape: biological bases of sexual coercion*. MIT Press, Cambridge, MA.

Thornhill, R. and Thornhill, N. (1983). Human rape: an evolutionary analysis. *Ethology and Sociobiology*, **4**, 137–173.

Tooby, J. and Cosmides, L. (1990). On the universality of human nature and the uniqueness of the individual: the role of genetics and adaptation. *Journal of Personality*, **58**, 17–68.

Trivers, R. (1972). Parental investment and sexual selection. In: B. Campbell (ed.), *Sexual selection and the descent of man: 1871–1971*, pp. 136–179. Aldine, Chicago, IL.

Turkheimer, E. (1998). Heritability and biological explanation. *Psychological Review*, **105**, 782–91.

Turkheimer, E. (2000). The three laws of behavioral genetics and what they mean. *Current Directions in Psychological Science*, **9**, 160–4.

Waddington, C.H., (1957). *The strategy of genes*. Allen and Unwin, London.

Weiss, A., Egan, V., and Figueredo, A.J. (2004). Sensational interests as a form of intrasexual competition. *Personality and Individual Differences*, **36**, 563–573.

Wenner, C., Figueredo, A.J., Rushton, J.P, and Jacobs, W.J. (2007). 'Executive functions, general intelligence, life history, psychopathic attitudes, and deviant behavior.' International Society for Intelligence Research meeting, Amsterdam, The Netherlands.

West-Eberhard, M.J. (2003). *Developmental plasticity and evolution*. Oxford University Press, Oxford.

Wilson, M. and Daly, M. (1985). Competitiveness, risk-taking, and violence: the young male syndrome. *Ethology and Sociobiology*, **6**, 59–73.

Chapter 14

War, martyrdom, and terror: evolutionary underpinnings of the moral imperative to extreme group violence

Scott Atran

> The noble man's soul has two goals
> To die or to achieve its dreams
> What is life if I don't live
> Feared and what I have is forbidden to others
> When I speak, all the world listens
> And my voice echoes among people
> I see my death, but I rush to it
> This is the death of men....
> I will throw my heart at my enemies' faces
> And my heart is iron and fire!
> I will protect my land with the edge of the sword
> So my people will know that I am the man.
> *Abdelrahim Mahmud, The Martyr, 1937*

Glory is priceless

In *The Descent of Man*, Charles Darwin (1871, pp. 164–166) wrote that:

> The rudest savages feel the sentiment of glory... A man who was not impelled by any deep, instinctive feeling, to sacrifice his life for the good of others, yet was roused to such action by a sense of glory, would by his example excite the same wish for glory in other men, and would strengthen by his exercise the noble feeling of admiration.

Glory is the promise to take and give up life in the hope of giving greater life to some group of genetic strangers who believe they share an imagined community under God (or under His modern secular manifestations, like the Nation and Humanity). It is the willingness of at least some to give their last full measure of devotion to the imaginary that makes the imaginary real.

The official website for John McCain's candidacy for President of the United States, whose motto was 'Country First!' had as its banner a quote from his book *Faith of My Fathers* (1999):

> Glory is not a conceit. It is not a prize for being the most clever, the strongest, or the boldest. Glory belongs to the act of being constant to something greater than yourself, to a cause, to your principles, to the people on whom you rely, and who rely on you in return. No misfortune, no injury, no humiliation can destroy it.

My point here is not to relativize morality, or to argue that the patriot, 'the rudest savage', and the jihadi are just the same. What I want to suggest is that sacrificing life for God or group (which, as Durkheim (1912) noted, are basically the same) is not an exception in human history and cultural life, but a general means by which groups form, vie for survival, and thrive. People sacrifice self-interest to gods to make groups of common interest that can compete against others. That is what got us out of the caves, created competitive cultures and civilizations, and enabled us to conduct ever-widening spheres of commerce and war across the world. In nearly all cultures and throughout human history, social sacrifice is noblest in war. In war, sacrifice becomes glory, bringing forth the greatest esteem from friends with whom glory is shared, and eternal respect from the community and God for whom glory is sought. At least, that is what tales of heroes tell.

The moral imperative for war

> The Moral Law causes the people to be … undismayed by any danger.
> (Sun Tzu, The Art of War, c.540 B.C.)

Altruism is the sacrifice of one's own interests for the sake of others, as in giving to charity, lending a helping hand, or just taking time to offer directions to a stranger. Parochial altruism, especially bravery and heroism in war, involves sacrifice for one's own group to the detriment of rival groups (Choi and Bowles 2007). Parochial altruism is a basic aspect of the evolutionary imperative of human populations to 'cooperate to compete'. In all cultures, parochially altruistic acts are considered noble and good (though what is good and noble in one culture and time can be evil and ignoble for another). Individuals within a society may also differ widely in their appreciation of the value of an altruistic act, such as suicide bombing or the struggle for national liberation (which also can be parochial in the sense of being against those who would deny such a right).

Charles Darwin (1859), gathering an astounding amount of data from his voyage around the world as a naturalist aboard Her Majesty's ship *Beagle*, and from other people's observations, tried to show that all living kinds are basically competitive and selfish. The different forms of life, including humans and their cultural shells, develop through a process of natural selection that favours survival of the best competitors for resources. This, he argued in *The Origin of Species*, promotes adaptations only for the individual's own use in its struggle to gain resources to produce offspring: 'good for itself,' but 'never. . . for the exclusive good of others.'

Under Darwin's original theory, if we give to charity, or help children, strangers, and the infirm, it is because we seek enhanced social status, or a heightened sense of self-worth, or whatever else may serve our interests in the short or long run. But heroism, martyrdom, and other forms of self-sacrifice for the group appear to go beyond the principles of reciprocity, such as *quid pro quo* or even the Golden Rule. Darwin puzzled mightily over what would motivate people 'who freely risked their lives for others?' (Darwin 1871, pp. 163–165). Of course, he acknowledged that the brave warrior who survives the fight will often gain more power, wealth, social worth, or mates, and so improve his chances for producing healthy and successful offspring in greater numbers. But if the risk of death is very high, the material prospects for victory low, or if the odds of success are too difficult to calculate, then it is very doubtful that gain would outweigh loss. Indeed, risk assessments about war are difficult even in simple contexts and the effects of miscalculation, a common occurrence, are extremely severe: frequent inter-group conflict leads to chronic underutilization of resources such as land (Kelly 2005), and historically war leads to high numbers of casualties, with the losing group often being decimated (Bowles 2006). Even if accurate calculation about the relative strength of two sides in a conflict is possible, the underdogs often prevail (Arreguín-Toft 2001). Moreover, evidence for selective benefits accruing to individual participants in warfare is inconsistent at best. While some societies can deal with the free-rider problem by enforcing participation in war, this does not hold true for less complex societies or for self-radicalizing small groups of people who volunteer to kill themselves in acts of inter-group violence (Sageman 2008; Atran 2010a).

How, then, could self-interest alone account for man's aptitude for self-sacrifice to the point of giving his life—the totality of his self-interests—for his extended family, tribe, nation, religion, or for humanity? The puzzle led Darwin to modify his view that natural selection only produces selfish individuals. In *The Descent of Man*, he suggests (p. 166) that humans have a naturally selected propensity to moral virtue, that is, a willingness to sacrifice self-interest in the cause of group interest. Humans are above all moral animals because they are creatures who love their group as they love themselves.

> It must not be forgotten that although a high standard of morality gives but a slight or no advantage to each individual man and his children over the other men of the same tribe, yet that an advancement in the standard of morality and in increase in the number of well-endowed men will certainly give an immense advantage to one tribe over another.

Human warfare is vastly more lethal than inter-group conflict in other primate species. Genocide, the extermination of one group by another, is a frequent method of 'conflict resolution' that humans have practiced since prehistoric times (Keeley 1996). We do not know how frequent genocide was in human prehistory, but even its occasional occurrence could have favoured the emergence of moral virtue including self-sacrifice for the group: survival would depend on unquestioning commitment from each member of the group.[1] Moral obligations are often far more powerful and durable glue than a mere social contract. For, if some better social contract is likely to be available somewhere down the line, then, reasoning by backward induction, there is no more justified reason to accept the current contract than convenience.

Anthropologist Roy Rappaport has argued that group-level moral obligations, such as religious beliefs and prescriptions, reinforce cooperative norms by conferring on them 'sacredness'. Sacred assumptions are 'ineffable' in the sense that, unlike secular social contracts, they cannot be fully expressed and analyzed because they include a logic of moral appropriateness that are—at least in part—immune to instrumental calculations. Indeed, to be effective such 'sacred propositions' must be immune to instrumental calculation lest they be undermined by free-rider effects (Rappaport 1999). This becomes particularly important in times of vulnerability and stress, when social deception and defection in the pursuit of self-preservation is more likely to occur, as Ibn Khaldûn (1958 [1318]) noted long ago. Examining different waves of invasion in the Maghreb in what is arguably the first comparative study of history, he found that enduring dynastic power stems from moral commitment and 'group feeling', with its ability to unite desires, inspire hearts,

---

[1] I make no claims regarding whether moral virtue, including a willingness to sacrifice for the group, was naturally selected as a biological adaptation of certain individuals to groups or whether it was culturally selected in the course of inter-group competition and warfare (see Atran and Axelrod 2008).

and support mutual cooperation. Recent studies in social psychology suggest that such group attachments can even blind committed members to the availability of an exit strategy (Van Vugt and Hart 2004).

Science of the sacred

Sacred values often have their basis in religion, but such transcendent core secular values as a belief in the importance of individual morality, fairness, reciprocity, and collective identity ('justice for my people') can also be sacred values. These values will often trump the economic thinking of the marketplace or considerations of *realpolitik*. Rational choice involves selecting and ordering the best means for achieving given goals in the future. The more distant a goal is, the less its real value here and now, and the less committed a person is to implement the means to realize it. But sacred values upset these calculations. In many cases, sacred values are concerned with sustaining tradition for posterity. In other cases, the future takes on a transcendent value, the dream of what ought to be rather than what is, as in the fight for liberty or justice. Sometimes, sacred values take on aspects of both tendencies: say, to regain the freedom that should have been, or the dream of a righteous Caliphate. In all of these cases, there is no discounting of the future. In fact the opposite: on the basis of sacred values, people may purposely choose to live and act now for a remote end, and to value the traditions of a distant past more than the trappings of the present or a probable future.

Devotion to some core values may represent universal responses to long-term evolutionary strategies that go beyond short-term individual calculations of self-interest but that advance individual interests in the aggregate and long run (Lim and Baron 1997). Matters of principle, or 'sacred honour', are enforced to a degree far out of proportion to any individual or immediate material payoff when they are seen as defining 'who we are'. Revenge, 'even if it kills me', between whole communities that mobilize to redress insult or shame to a single member go far beyond individual 'tit-for-tat' (Axelrod and Hamilton 1981) and may become the most important duties in life. This is because such behaviour defines and defends what it means to be, say, a Southern gentleman (Nisbett and Cohen 1996), a Solomon Islander (Havemeyer 1929), or an Arab tribesman (Peters 1967). The Israeli army has risked the lives of many soldiers to save one as a matter of 'sacred duty', as have certain elite US military units (Bowden 2000).

Of course, sincere displays of willingness to avenge at all costs can have the long-term payoff of thwarting aggressive actions by stronger but less committed foes. Likewise, a willingness to sacrifice for buddies can help create greater *esprit du corps* that may lead to a more formidable fighting force. But these acts far exceed the effort required for any short-term payoff and offer no immediate guarantee for long-term success.

Seemingly intractable political conflicts—in the Middle East, Central and South Asia, Kashmir, and beyond—and the extreme behaviours often associated with these conflicts, such as suicide bombings, are often motivated by sacred values. Consider, for example, the view of martyrdom as a sacred duty expressed to me by Hamas spiritual guide Sheikh Hamed al-Betawi: 'A martyr fights and dies for dignity, nation, religion, and *Al "Aqsa"*. In the Koran, the book of *Al-Tauba*, verse 111, tells us that Allah brings souls to Paradise killing the enemy and getting killed – that is the sacred principle of Jihad'.[2]

Across the world, people believe that devotion to sacred or cultural values that incorporate moral beliefs—such as the welfare of their family and country, or commitment to religion,

---

[2] In an interview with S. Atran, Nablus, West Bank, 8 September 2004.

honour, and justice—are, or ought to be, absolute and inviolable. Research with colleagues (Atran et al. 2007; Ginges et al. 2007) suggests that people will reject any type of material compensation for dropping their commitment to their sacred values and will defend their sacred values regardless of the costs. We surveyed nearly 5000 Palestinians and Israelis between 2004 and 2009, questioning citizens from across the political spectrum including refugees, supporters of Hamas, Israeli settlers, and national leaders from the major Israeli and Palestinian political factions. We asked them to react to hypothetical but realistic compromises in which their side would be required to give away something it valued in return for lasting peace.

In one cycle of studies, we used a 'between-subjects' experimental design where we randomly chose some subjects to respond to a deal with an added material incentive such as financial compensation, while a third group responded to a deal where the other side made a symbolic sacrifice over one of their own sacred values. In another cycle we used a 'within-subjects' design, where all subjects would be exposed to the same set of deals. First they would be given a straight-up offer in which each side would make difficult concessions in exchange for peace; next they were given a scenario in which their side was granted an additional material incentive; and last came a proposal in which the other side agreed to a symbolic sacrifice of one of its sacred values. Results were much the same for both the between-subjects and within subjects designs, indicating that the order in which deals were presented didn't matter and that people responded the same way to deals give singly or as part of a set.

Each set of trade-offs included an original offer that we pre-tested as likely to be rejected (taboo), the same trade-off with an added material incentive (taboo +), and the original trade-off with an added symbolic concession (a separate test showed that 'tragic' held no material value for participants). For example, a typical set of trade-offs offered to Palestinians might begin with this (taboo) premise: 'Suppose the United Nations organized a peace treaty between Israel and the Palestinians; Palestinians would be required to give up their right to return to their homes in Israel; and there would be two states, a Jewish state of Israel and a Palestinian state in the West Bank and Gaza'. Second, we would sweeten the pot (with the taboo +): 'In return, the US and the European Union would give Palestine one billion dollars a year for 100 years'. Then the symbolic (tragic) concession: 'For its part, Israel would apologize for suffering caused by the displacement and dispossession of civilians in the 1948 war'.

Many of the respondents insisted that the values involved were sacred to them. For example, nearly half the Israeli settlers we surveyed said they would not consider trading any land in the West Bank—territory they believe was granted them by God—in exchange for peace. More than half the Palestinians considered full sovereignty over Jerusalem in the same light, and more than four-fifths felt that the 'right of return' was a sacred value, too. Among Palestinians, the greater the material incentive offered the greater the disgust registered, and the more joyful the reaction to the idea. In one scenario, Israeli settlers were offered a deal to give up the West Bank to Palestinians in return for an American subsidy to Israel of $1bn a year for 100 years. For those who had chosen to live in the Occupied Territories for reasons of economy or quality of life, the offer led to increased willingness to accept land for peace, a decrease in disgust and anger at the deal, and a corresponding reduction in willingness to use violence to oppose it. But for those settlers who believe the Occupied Territories to be God's ancient trust to them, expressions of anger and disgust and willingness to use violence rose markedly.

This sort of 'moral absolutist' sentiment runs directly counter to prevailing economic theories of rational choice and to political science theories of rational play in negotiation. Our results imply that using the standard approaches of business-like negotiations in such seemingly intractable conflicts will only backfire, with material offers and sweeteners interpreted as morally taboo and insulting (like accepting money to sell your child or sell out your country). Given the closeness of elections in Palestine and Israel, even a small group of absolutists can thwart any peace.[3]

In recent studies in Indonesia (Ginges and Atran 2009), we found that support for violence among moderate as well as radical madrassah students was significantly greater in response to a deal that involved a significant material incentive to give up the struggle to have the country 'ruled strictly according to *Sharia*' (Muslim law). In India, Sachdeva and Medin (2009) show that the sacred and secular do not mix in the conflict over Kashmir, although the issue appears to be more symbolically charged for Muslims than for Hindus. Hindus and Muslims are equally likely to disapprove of a material compromise over Kashmir ('taboo' trade-off: 'Instead of the current two to one split of Kashmir, it would be evenly divided between Pakistan and India'), and to envisage rioting over the issue. Muslims, however, are much more likely than Hindus to approve a deal and to downplay rioting were the other side to make a symbolic concession ('tragic' trade-off: 'India would recognize the sacred and historic right that Muslims have to Kashmir and apologize for all the wrongs done over the years').

In a 2009 Internet experiment designed by Morteza Dehghani and Rumen Illiev, our research team asked Iranians living both inside and outside Iran to imagine these hypothetical situations:

Iran will give up its nuclear program; Israel in return will give up their nuclear program and destroy any existing nuclear weapons.

Iran will give up its nuclear program; Israel in return will give up their nuclear program and destroy any existing nuclear weapons. In addition, the EU will pay $40 billion to Iran.

There was a clear difference between the first hypothesis (taboo) and the second (taboo +): Iranian subjects were generally approving of added material incentives, whereas a minority (11%) was strongly disapproving. For at least some Iranians, acquiring a nuclear capability had perhaps become something of a 'sacred value' that cannot simply be bought off with material incentives (Dehghani et al. 2009). One political analyst cautioned the US administration that 'You don't bring down a quasi-holy symbol – nuclear power – by cutting off gasoline sales'.[4] Indeed, Iranian officials claim that 'we cannot have any compromise with respect to the Iranian nation's inalienable right' to acquire a nuclear capability. In fact, our results suggest that a 'carrots (or sticks) approach', which is favored by the US, European Union, and United Nations, may actually backfire for those who identify most closely with the Iranian regime. Is 11% too few Iranians to matter? Perhaps not: even a minority, if it is committed enough, can carry the day if it is associated

---

[3] So far, the findings I've described make prospects for peace seem very dark. Many on the outside looking in on these clearly expressed 'irrational' preferences simply ignore them because in a sensible world they ought not to exist. Seemingly, the only realistic alternative is to fall back on a business-like approach and leave 'value issues' for last. The hope, in the meantime, is that concrete moves on material matters (electricity, water, agriculture, and so on), however small at first, will eventually accumulate enough force to dissolve the harder and more heartfelt value issues. But in reality, this is only a recipe for another Hundred Year's War, as progress on everyday material matters only heightens attention to heavy value-laden issues of who we are and want to be. Fortunately, our work also suggests another, more optimistic course. Absolutists who violently rejected profane offers of money or peace for sacred land were considerably more inclined to accept deals that involved their enemies making the symbolic but difficult gesture of conceding respect for the other side's sacred values. For example, Palestinian hardliners were more willing to consider recognizing the right of Israel to exist, if the Israelis apologized for suffering caused to Palestinian civilians in the 1948 war (which Palestinans call *Nakba*, 'The Catastrophe'). See Atran et al. (2007); also S. Atran, J. Ginges. *How words could end a war*. New York Times, 25 January 2009.

[4] H. Jaseb, F. Dahl. *Iran signals no compromise on nuclear issue*. Reuters, 12 September 2009.

with a power structure that is willing to do most anything to stay in power (like the Alawites in Syria, who account for 10–20% of the population, yet have ruled for decades).

One obvious issue is that while people often recognize their own side's sacred values, they often ignore or downplay the importance of the other side's values. In Afghanistan, Pashtun tribesmen will defend to the death the ancient code of honour known as *pashtunwali*, which requires protecting valued guests at the risk of one's own life.[5] After 9/11, the Taliban leader, Mullah Omar, assembled a council of clerics to judge his claim that Bin Laden was the country's guest and could not be surrendered. The clerics countered that because a guest should not cause his host problems, Bin Laden should leave. But instead of keeping pressure on the Taliban to resolve the issue in ways they could live with, the US ridiculed their deliberation and bombed them into a closer alliance with Al Qaeda. Pakistani Pashtuns then offered to help out their Afghan brethren, as a matter of honour, despite the fact that few of the notoriously unruly frontier tribes were initially sympathetic to the Afghan Taliban programme of homogenizing tribal custom and politics under one rule, much less to Qaeda's global ambitions (Atran 2010b).

Moreover, in many Middle Eastern, Central Asian, and North African societies, political, economic, and social structure is organized according to patrilineal descent (exclusively through the father's line), where the patriline's 'honour' depends on the enforced modesty and protection of women (Atran 1985). Thus, even a small gesture that impugns the esteem of a senior male, or the modesty of a post-pubescent woman, can rouse a whole patriline to implacable hostility, along with people from the entire community who feel their culture's most sacred values threatened.[6]

War makes men men

> 'Tat, tat, tat, tat, tat, bad position, bad position,' the old Afridi tribesman sputtered as he pointed an invisible rifle at the rugged and barren hills of the Khyber Pass.
>
> 'Why bad position?' I asked a Pakistani army man, a dentist who happened to be watching the exchange with delighted curiosity and who spoke fairly good English. He explained that, here, the luckless English soldiers had passed through in 1919 on their way to losing (their third and last war in) Afghanistan.
>
> 'Ah but good position with Nadir Khan in Kabul, tat, tat, tat, tat, tat, good position,' gleefully squealed the man's equally ancient tea partner, a Wazir tribesman who had joined the father of the last king of Afghanistan in sacking the capital ten years later.
>
> The Afridi lowered his head: 'No good position now. No good the fight now. Bad position.'
>
> 'Why bad position now?' I asked.
>
> The army dentist queried the two white-bearded gentlemen and came back with a laugh: 'They say it's been so calm since [Kabul was sacked in 1929] that a man has no opportunity to become a man!'
>
> 'Bad position,' the Wazir nodded in sad agreement. 'Bad position.'
>
> I continued shooting the breeze with the Afridi tribesman and his Wazir companion (I suspected they had become friends simply because they were old rather than because the traditional

---

[5] S. Atran. *To beat Al Qaeda look to the East*. New York Times, 13 December 2009.

[6] But even small gestures of respect towards elders and obvious attempts to maintain social distance towards women can broadcast intent to cooperate with a community in a surprisingly effective way. Here, an almost no-cost act becomes amplified through the local value system to great benefit. See Atran and Axelrod (2008).

enmity between their tribes had lessened), as four young boys somehow managed to unhinge the engine block from the back of our van and were struggling to lift the thing away. The army dentist stopped them with a stern word, shooed them away, and with a smile that showed pleasure both at his thought and his command of English, threw up his hands and said, 'Boys will be boys.'

Author's field notes from Landi Kotal, Khyber Pass, July 1976

Now, after years of almost constant war, I imagine the old Afridi and Wazir would see that such boys had indeed become 'men's men'.

Until very recently, hunting and warfare have been occupations of men. Indeed, in most cultures, on every inhabited continent, initiation into manhood has required proof of skill in hunting savage beasts or men. Among the Maasai of Kenya, a boy would become a man by killing a lion and then going to war. Frontiersmen of British and French North America got manly reputations from hunting bears and scalping Native Americans. For many Native American tribes, a man had to fight in war before he could be called a man (Lowie 1954; Turney-High 1949, p. 145). Among the Dorzé of Ethiopia, to become a man meant taking as a trophy an enemy's testicles—the larger the testicles, the greater a man he would become. Among the Naga of Assam (India), the Iatmul of New Guinea and the Shuar of Peru, to become a man was to take off the head of another (Atran 2002). And, of course, warfare is often linked to male sexuality, from the strategic planning and vocabulary of invasions, to the ecstasy of conquest, to the thrusting and penetrating motions of much weaponry. Not to mention modern 'missile envy'.

In *The Descent of Man*, Darwin noted that most human violence is committed by young men. Across the world, about 80–90% of all human killing is committed by males aged 14–35 (Wilson and Daly 1988; Buss 2005). In the US, for example, men were responsible for 88% of all homicides between 1976 and 2004, and nearly three-quarters of these involved men killing other men. The peak period for murders in recent US history was between 1990 and 1994, when the homicide rate exceeded nine people killed per 100,000 (it has been between five and six for the last few years[7]). In those years, the number of killers ranged from 23–30 per 100,000 for male teens aged 14 to 17, from 34–41 per 100,000 for young men aged 18–24, and 15–18 per 100,000 for men aged 25–35 (Hellmuth 2000). These trends closely follow trends for killing in war, except that in war nearly 100% of killing is by men. Only in the last few decades of human history have women played any appreciable killing role in war, a dubious advance for women's rights.

Controlled experiments by Van Vugt et al. (2007) show that men sacrifice more for their group when competing with other groups (as in war) than in the absence of inter-group rivalry. Women generally scored higher than men on measures of cooperation within the group, and their levels of cooperation were not significantly affected by inter-group competition. Men scored higher than women on cooperation only when competition with other groups was on the horizon. Men kill much more than women in competitive situations, and with and for small bands of buddies.

The data on suicide bombing bear out these general trends. Less than 15% of suicide bombers are women, among groups that allow female suicide bombers. (Because women are generally less suspicious than men, it is surprising that women are not vetted for greater use.) Overall, the rate of female suicide bombing corresponds to the general rate of female homicide and participation in war, which is much less than male involvement.

Revenge for close family members seems to be highly significant for female suicide bombers (at least this appears to be so where data is available, for Chechnya, Iraq, and Palestine). There are cases of male suicide bombers out for revenge, but studies by Ariel Merari in Palestine suggest this is not a significant factor overall, nor can it be so for the several hundred foreign volunteers in the

---

[7] Crime in the US, 1986–2005; http://www.fbi.gov/ucr/05cius/data/table_01.html

'Sinjar' group for Jihad and martyrdom in Iraq.[8] A recent study of captured Al Qaeda volunteers by the Saudi Ministry of Interior also fails to indicate revenge or death of a close relative as a significant factor.[9]

According to Lindsay O'Rourke, who has studied female suicide attacks across the world since 1981, 'surprisingly similar motives' drive men and women to blow themselves up: 'The primary motivation of male and female suicide bombers is a deep loyalty to their communities combined with a variety of personal grievances ... it is simply impossible to say one sex cares more about the others'.[10] She argues that '95 percent of female suicide attacks occurred within the context of a military campaign against occupying forces,' that is, in a situation of intergroup conflict.

However the issue is not who cares more about the group, but how they care (Van Vugt et al. 2007). Cross-cultural research on killing by women indicate that women kill most often in self-defence and in defence of family, especially children (Wilson and Daly 1988; Buss 2005). Between June 2000 and April 2010, more than one-third of the 60 or so Chechen suicide bombers were women. Known as 'Black Widows', they mostly volunteered to revenge the deaths of family members. For example, on 29 November 2001, Elza Gazueva volunteered for a bombing mission after her husband and brother were tortured and killed by Russian forces. She went to the Russian military headquarters and managed to get close enough to the commandant who was responsible for taking her husband and brother from her home and who had ordered their torture and death. Gazueva approached the commandant asking, 'Do you remember me?' before exploding herself and killing him. Most of the other cases are quite similar, from the first Chechen suicide bombers, Khava Baraeva and Luiza Magomadova in June 2000, to Roza Nogaeva and Mariam Tuburova in September 2004 (Speckhard 2005).

Chechen suicide bombings fell off in the wake of popular revulsion to the September 2004 Beslan massacre, when Islamists seized a school in North Ossetia (Russian Federation), demanding an end to the Russian occupation of Chechnya. (In the ensuing gun battle between the hostage-takers and Russian security forces, over 300 hostages were killed, including nearly 200 children.) But bombing resumed in 2007 when the pro-Russian Chechen government moved to crush remaining militants in coordination with Russia's newly picked president of neighbouring Dagestan. Dzhanet Abdullayeva, the 17-year-old widow of a Dagestani rebel she met through the Internet and who was killed in 2009, was one of two 'Black Widows' who bombed the Moscow subway in March 2010, when 40 people died.[11]

As O'Rourke notes, women, although occasionally involved in mass bombings, are five times more likely than men to target specific individuals for assassination. Male suicide bombers rarely act against singled-out individuals, and even more rarely act alone. In almost all cases, males form part of a small group of friends that becomes a 'band of brothers' whose members die for one another as much or more than for any cause. The Saudi Ministry of the Interior study (see footnote 8) finds that 64% join through friends and 24% through family, a result consistent with Marc Sageman's (2004) research on how volunteers across the world join the Al Qaeda-inspired movement. There are a few examples of two, three, or four Chechen, Uzbek, and Iraqi women striking in a coordinated suicide bombing, but scant evidence that these operations involved close friends.

---

[8] Data available from the Combating Terrorism Center, West Point.
[9] Presented to the Riyadh Meeting on Terrorism, Security Forces Officers Club, 25–28 January 2008.
[10] L. O'Rourke: *Behind the woman behind the bomb*. New York Times, 2 August 2008.
[11] P. Pan. *One Moscow suicide bomber was teenage widow of Islamist rebel*. Washington Post, 3 April 2010.

Although the Jihad discriminates against frontline participation by women (as most military organizations and groups still do), their occasional participation does little to lessen the general lesson that Jihad is a team and blood sport for underachieving and glory-seeking young men.

The noble beast

> But of the cities of these people, which the Lord thy God doth give thee for an inheritance, thou shalt save alive nothing that breatheth. But thou shalt utterly destroy them; namely, the Hittites, and the Amorites, the Canaanites, and the Perizzites, the Hivites, and the Jebusites.... I will make mine arrows drunk with blood, and my sword shall devour flesh; and that with the blood of the slain and the captives, from the beginning of revenges upon the enemy.
>
> Deuteronomy 20:16–17; 32:42

There were likely two broad and overlapping epochs in human prehistory: one in which men primarily hunted animals, and another in which men primarily hunted men. The passage from one to the other may be the most important advance in human social evolution. Humans cooperate to compete first against the elements of Nature, and then against each other. Our ancestors lived in a world inhabited by rivals far more numerous, strong, and savage than themselves. It was a competition that humans very nearly lost. Human salvation lay in persistent reliance on a social band of kin for collective strength, and a special form of primate wit that made it work as a winning team.

The flipside of human teamwork is group competition. Our first major competitors were the prey we hunted and the animal predators that preyed on us. Then other groups of people became humans' most feared predators and choicest prey. No other species primarily preys on itself. The senses of all hunting animals are most stimulated in the hunt and humans are no exception.

Warfare, argues Barbara Ehrenreich (1997), may be a continuation of the evolutionary drive to hunt and to avoid being hunted, but where human groups alone are predators and prey. The rituals of war, of blood sacrifice and savaging of flesh, credit this view. Ancient Greeks and Carthaginians would ritually sacrifice their own and offer burnt offerings of animals (*hecatomb*) to ensure good outcomes in war. The ancient Semites substituted animal for human sacrifice when they found the one true God (Robertson Smith 1894). But like the pagan gods of the Celts (Davidson 1988), the Hebrew God demanded that his Chosen People slaughter every man, woman and child of His people's enemies, as well as all their pigs, chickens, goats and cows. 'Thou Shalt not kill' commands God, but that applies only for His Chosen People. Against rivals, mere killing is not enough.

Hundreds of years ago, in the American southwest, the Anasazi of Chaco Canyon killed, butchered, and ate other human beings they warred against. Then they ritually defecated the remains into the hearths of their victims' dwellings, probably to signal dominance and disdain (Holden 2000). The 19th-century Native American tribes of the lower Colorado River did much the same:

> The Mohave cut off the heads of every one [of the Maricopa] and carried them to the village where they camped. They scalped the heads. They built a fire in a hole, and when it had burned to coals, placed the heads around the fire, as one bakes pumpkins, as an insult. They warmed them on each side, raked away the coals, and baked the heads on the ground like pumpkins.
>
> (Cited in Spier 1933, p. 175.)

The 16th-century Aztecs of Mexico (Clendinnen 1991) and the 19th-century Dahomey of West Africa (Law 1985) conducted wars for the primary purpose of capturing sacrificial victims, who they ritually ripped apart and ate. In this 21st century, in the Ituri forest district of the Democratic Republic of the Congo, rebel forces of the Ugandan-backed Movement for the Liberation of the Congo (MLC) have been repeatedly denounced by the UN Security Council[12] for grilling people on spits, boiling young girls alive, cutting open the chests of Pygmies and other non-combatants, and ritually 'ripping out their hearts, livers and lungs, which they ate while still warm'.[13] At the trial of eight men accused of planning to bomb trans-Atlantic airliners with home-made liquid explosives in 2006, the 'martyrdom video' of one of the men, Abdulla Ahmed Ali, was played. There was relish in the prospect of shredding flesh:

> I'm doing this... to punish and to humiliate the Kuffar [non-believer], to teach them a lesson that they will never forget.... Leave us alone. Stop meddling in our affairs and we will leave you alone. Otherwise expect floods of martyr operations against you and we will take our revenge and anger, ripping amongst your people and scattering the people and your body parts and your people's body parts responsible for these wars and oppression, decorating the streets.[14]

In all societies, moral norms strongly constrain and punish murder, rape, torture, desecrating the body, pillage, and plunder. But that applies mainly within the group. In war, all may be allowed, even encouraged. A man who kills many in our own society is a mass murderer. In war, mass killing may be rewarded by military honours. America's best-known World War I war hero, Sergeant Alvin York, religiously rejected all forms of violence at home but received a ticker-tape parade for attacking a machine gun nest and killing 28 German soldiers.

In today's world, adherence to human rights and the Geneva Convention place limits on the carnage that nations agree to allow, though for powerful states these are easily suspended or ignored in the name of 'national security'. In secret meetings between Central Intelligence Agency (CIA) and military lawyers in the fall of 2002 at Guantánamo Bay, a position on torture was laid out that the Bush administration later approved (in another secret meeting):[15] 'It [torture] is basically subject to perception', said CIA lawyer Jonathan Friedman, according to meeting minutes released at a Senate hearing in June 2008: 'If the detainee dies, you're doing it wrong'.[16] Lawyers on both sides believed they had found a legal loophole permitting 'cruel, inhuman or degrading' methods overseas as long as they didn't directly lead to death. Arguably, this is still some sort of an advance over cannibalizing an enemy.

War is a supremely moral act. In all cultures, war and warriors are considered noble and good (though what is good and noble in one culture and time may be evil and ignoble for another, and individuals within a society may also differ widely in their appreciation of the value of war). Hatred and dehumanization of the enemy as animals, or at least 'barbarians', is almost a constant of war that is fought in defence of the group. 'Barbarian' is one way the US government designates Al-Qaeda[17], as Al-Qaeda does the US.

'The art of war', wrote Adam Smith (1993) in *The Wealth of Nations*, 'is certainly the noblest of all arts'. The reason, he argues, is that it has allowed the progressive advance of commerce and

---

[12] BBC News (bbc.co.uk). *UN condemns DR Congo cannibalism*. 15 January 2003; E. Isango. *Cannibalism shock as atrocities revealed*. theage.com, 18 March 2005.
[13] R. Ngowi. Associated Press wire, 19 May 2003.
[14] BBC News (bbc.co.uk) '*Suicide videos': what they said*. 4 April 2008.
[15] J. Warrick. CIA tactics endorsed in secret memos. Washington Post, 15 October 2008.
[16] US Senate Armed Services Committee Hearing, 17 June 2008. Supporting documents: Counter Resistance Strategy Minutes (pages 2–3 of 5) 13:40, 2 Oct 2002, p. 15.
[17] National Security Strategy of the United States, September 2003, p. 11: http://www.whitehouse.gov/nsc/nss.html

civilization, bringing the greatest benefits of peace and prosperity to the most people for the longest time. People are most cooperative and creative when they fight others in war. 'War alone brings up to their highest tension all human energies and imposes the stamp of nobility upon the peoples who have the courage to make it', exulted Benito Mussolini (1935), the Italian Dictator and founder of modern fascism. This sentiment has been felt and proclaimed throughout history, regardless of whether war was to conquer or set people free. 'Gentlemen may cry, Peace, Peace', exhorted the American revolutionary Patrick Henry, 'but there is no peace. The war is actually begun. . . . Is life so dear, or peace so sweet, as to be purchased at the price of chains and slavery? Forbid it, Almighty God! I know not what course others may take; but as for me, give me liberty or give me death!'

For cause and comrades: a potent and explosive mix

Among American military psychologists and historians, the conventional wisdom on why soldiers fight is that ideology is not important. Most of the studies focus on measures of 'fighter spirit' among American soldiers in World War II (WWII), Korea, and Vietnam (Smith 1983a). Only leadership and group loyalty appear to be consistently important. In WWII, for example, solidarity and loyalty to the group helped mightily to sustain combat soldiers, while personal commitment to the war and ideology were much less meaningful. American soldiers 'ain't fighting for patriotism' and the British soldier 'never gave democracy a thought' (Stouffer et al. 1949, especially the chapter entitled 'The general characteristics of ground combat'). Soldiers' belief in the legitimacy of a cause worth fighting for steadily increased during WWII and steadily decreased during the Korean War, yet fighting spirit remained fairly constant (Smith 1983b).

In *The Deadly Brotherhood*, John McManus (2003) argues that the American combat soldier in WWII did not fight and die for abstract concepts, such as democracy or love of country, but for his devoted 'fraternity' or 'band of brothers' with whom he shared dangers and hardship on the front line. A rifleman in the 32nd Infantry Division wrote: '. . .survival for one's self was the first priority by far. The second priority was survival for the man next to you and the man next to him. So, right or wrong, love of country and pride. . . was a good bit behind'.

American analysts dismiss the idea of a semi-mystical bond of camaraderie in favour of rational self-interest:

> [T]he intense primary-group ties so often reported in combat groups are best viewed as mandatory necessities arising from immediate life-and-death exigencies. Much like the Hobbesian description of primitive life, the combat situation can be nasty, brutish and short. To carry the Hobbesian analogy a step further, one can view primary-group processes in the combat situation as a kind of rudimentary social contract which is entered into because of advantages to individual self-interest. Rather than viewing soldiers' primary groups as some kind of semi-mystical bond of comradeship, they can better be understood as pragmatic and situational responses.
>
> (Moskos 1975)

In Vietnam, falling morale, desertion, and fragging (killing officers) increased long after popular support for the war collapsed, and only after soldiers began feeling that 'Vietnamization' (handing over security to South Vietnamese forces) was a lost cause that no soldier wanted to be the last to die for (Moskos 1975).

American soldiers said that the cause of democracy was 'crap' and a joke' in Vietnam, yet they described the selfless bravery of the North Vietnamese 'because they believed in something' and 'knew what they were fighting for' (Moskos 1970, p. 148; Spector 1994, p. 71). Maybe others do die for a cause and not only for comrades.

In *Frontsoldaten: The German Soldier in WWII*, Stephen Fritz argues that strong unit cohesion *and* ideological commitment made the German rank and file the best soldiers of the war:

> The Germans consistently outfought the far more numerous Allied armies that eventually defeated them.... On a man for man basis the German ground soldiers consistently inflicted casualties at about a 50 percent higher rate than they incurred from the opposing British and American troops under all circumstances. This was true when they were attacking and when they were defending, when they had a local numerical superiority and when, as was usually the case, they were outnumbered.
>
> (Fritz 1995, p. 24)

By D-Day (6 June 1944), 35% of German soldiers had been wounded at least once and 21% twice or more. The impressive unit cohesion and fighting spirit of the *Landser*, the ordinary German infantryman, was most remarkable on the Russian front, where 80% of Germans fought. Most important were the bonds of military friendship inherited from Prussian tradition, which raised comradeship to the level of strategic doctrine. German soldiers were apparently far more committed to the war aims of National Socialism than previously thought, which allowed for any action, no matter how beastly, against enemies of the German people.

Hitler promised a 'social revolution' that would merge the all-for-one values of the combat soldier with the whole civilian population, thus creating the *Volksgemeinschaft*. The Hitler Youth was not a brainwashing factory but a place where children were encouraged to ignore social status and think about the group before the individual. The ideological movement was emotionally rewarding for children as well as the soldiers the children would become. Rank and file German soldiers could readily quote philosophers and details of European history in their letters back home. The young men who marched to war under the Swastika were well-trained, deeply bonded to their peers, and highly motivated by the cause of National Socialism to take the fight to Germany's enemies.

In *Kameradschaft*, Thomas Kühne (2006) focuses on the concept of camaraderie among German soldiers in WWII. He traces it to medieval notions of chivalry through the French Revolution and Napoleonic era and the developing idea of general war in the Romantic era, reinforced by training and military service in defence of the nation and national honour (allied to masculinity). In WWI, soldiers were enlisted and organized on a strongly regional basis, as in the American Civil War, bonding on the front lines into 'trench families: front-line groups united by daily routine, regional affinity, proximity and shared experience. This ethos of comradeship carried into WWII despite the high casualty rates on the Russian Front that eviscerated affinity groups. Hitler's social revolution turned longing for community, or *Gemeinschaft*, into a national passion. 'Good' was anything that strengthened the community. The highest prestige accrued to those who rejected norms—doubts, scruples, inhibitions—that impeded the fight against so-called enemies and threats to the community. The emotions generated in mutual caring, commitment, and sacrifice for others made it possible for soldiers to come to terms with war's inhuman face: destruction and killing. Even in the later stages of the war there was a sense of group empowerment 'strong enough in the group's later stages to approximate group immortality and the will to fight on against desperate odds'. It explains how 'never have men fought better for a worse cause' than the men who marched to war under the Swastika.

The regional affinity of the German army and US Civil War armies contrasts with American soldiers in WWII and afterwards. The peer cohesion may well have been a social contract, which broke up after the combat experience:

> The paramountcy of individual self-interest in combat units is also indicated by looking at the pattern of letter writing. Squad members who have returned to the United States seldom write to those remaining behind... those still in the combat area seldom attempt to initiate mail contact with a former squad

member. The rupture in communication is mutual despite protestations of lifelong friendship during the shared combat experience.

(Moskos 1975, p. 29)

In *For Cause and Comrades*, James McPherson (1997) notes that, unlike later American armies of mostly draftees and professional soldiers, Civil War armies on both sides were both composed mainly of volunteers who joined up and fought with family, friends, and neighbours from the same communities. Unlike the WWII or Vietnam veterans, they returned to the same affinity groups that they had fought with. And:

> Ideological motifs almost leap from many pages of these documents. A large number of the men in blue and gray were intensely aware of the issues at stake and passionately concerned about them. How could it be otherwise? This was, after all, a civil war. Its outcome would determine the fate of the nation... the future of American society and of every person in that society [unlike, say the Iraq war] Civil War soldiers lived in the world's most politicized and democratic country in the mid-nineteenth century. A majority of them had voted in the election of 1860, the most heated and momentous election in American history. When they enlisted, many of them did so for patriotic and ideological reasons – to shoot as they had voted, so to speak. These convictions did not disappear after they signed up... They needed no indoctrination lectures to explain what they were fighting for, no films like Frank Capra's 'Why We Fight' series in World War II.

(McPherson 1997, pp. 91–92.)

In the Civil War, ideology was given a particularly religious cast:

> Civil War armies were, arguably, the most religious in American history. Wars usually intensify religious convictions.... Many men who were at best nominal Christians before they enlisted experienced conversion to the genuine article by their baptism of fire.

(McPherson 1997, p. 62.)

'We look to God and trust him to sustain us in our just cause', wrote a Florida cavalry captain in 1863. A New York lieutenant, twice wounded and awarded the Congressional Medal of Honor, wrote: 'He who is the embodiment of humanity will bestow in great abundance his blessings upon his and our cause' (McPherson 1997, pp. 72–73).

In *Fear in Battle*, John Dollard (1944) interviewed veterans from the Abraham Lincoln Brigade, Americans who fought in the Spanish Civil War against Franco and Fascism. In response to the question 'What would you say are the most important things that help a man overcome fear in battle?' 77% cited belief in their ideology and 'the aims of war' versus: leadership (49%), ésprit de corps (28%), hatred of enemy (21%), distraction and keeping busy (17%). It appears, then, that American warfare from WWII to today may be the exception in warfare, rather than the rule. To die and kill for Jihad more nearly follows the general rule.

In sum, while overwhelming military production and technological superiority has assured the victory of American arms in frontal wars fought in Clausewitzian terms of advancing policy through power, others fight against greater odds inspired by their Cause, including revolutionary and guerrilla movements and jihadis. One reason resource-deficient revolutionary movements can compete with much larger armies and police is willingness to delay gratification and accept material sacrifice for a greater cause. Consider the founding of the United States. Without calculating the probability of success, a few poorly equipped rebels knowingly took on the mightiest empire in the world. The Declaration of Independence concluded with the words: 'And for the support of this Declaration, with a firm reliance on the protection of Divine Providence, we mutually pledge to each other our Lives, our Fortunes and our sacred Honor'. As Osama Hamdan, the ranking Hamas politburo member for external affairs, put it to me in Damascus: 'George

Washington was fighting the strongest military in the world, beyond all reason. That's what we're doing. Exactly'.[18]

If so, the implications for defence policy may be novel and significant. For instance, according to the US Quadrennial Defense Review, the chief aim of counterterrorism efforts is to 'minimize US costs in lives and treasure, while imposing unsustainable costs on the enemy'. To a significant degree, however, terrorists may not respond to utilitarian cost-benefit analysis. The conspirators in the summer 2006 plot to blow up airliners with liquid chemicals that they smuggled aboard, knowingly chose the targets most watched; in autumn 2007, plotters in Ulm, Germany knew they were under surveillance and flaunted this knowledge in a display of costly commitment to their cause. Committed terrorists may respond more strongly to moral values, and be more than willing to die for the cause. Rather than 'minimizing' the appeal and effect of radical ideology and action by raising their costs in lives, each death may inspire many more young people to join the cause. Indeed, a utilitarian policy may actually play into the hands of terrorists who turn it around to show that the NATO allies try to reduce people to material matter rather than moral beings.

Acknowledgements

Thanks to the U.S. Office of Naval Research for support (contents are entirely the author's responsibility).

References

Arreguín-Toft, I. (2001). How the weak win wars: a theory of asymmetric conflicts. *International Security*, **26**, 93–128.

Atran, S. (1985). Managing Arab kinship and marriage. *Social Science Information*, **24**, 659–96.

Atran, S. (2002). *In Gods we trust: the evolutionary landscape of religion*. Oxford University Press, New York.

Atran, S. (2010a). *Talking to the enemy: faith, brotherhood, and the (un)making of terrorists*. HarperCollins, New York.

Atran, S. (2010b). A question of honour: why the Taliban fight and what to do about it. *Asian Journal of Social Science*, **38**, 341–61.

Atran, S. and Axelrod R. (2008). Reframing sacred values. *Negotiation Journal*, **24**, 221–46.

Atran, S., Axelrod, R., and Davis, R. (2007). Sacred barriers to conflict resolution. *Science*, **317**, 1039–40.

Axelrod, R. and Hamilton, W. (1981). The evolution of cooperation. *Science*, **211**, 1390–6.

Bowden, M. (2000). *Black Hawk down: a study of modern war*. Penguin, London.

Bowles, S. (2006). Group competition, reproductive leveling, and the evolution of human altruism. *Science*, **314**, 1569–72.

Buss, D. (2005). *The murderer next door: why the mind is designed to kill*. Penguin, New York.

Choi, J.K. and Bowles, S. (2007). The coevolution of parochial altruism and war. *Science*, **318**, 636–40.

Clendinnen, I. (1991). *Aztecs*. Cambridge University Press, Cambridge, UK.

Darwin, C. (1859). *On the origin of species by means of natural selection*. John Murray, London.

Darwin, C. (1871). *The descent of man, and selection in relation to sex*. John Murray, London.

Davidson, H.E. (1988). *Myths and symbols in pagan Europe: early Scandanavian and Celtic religions*. Syracuse University Press, Syracuse, New York.

Dehghani, M., Iliev, R., Sachdeva, S., Atran, S., Ginges, J., and Medin, D. (2009). Emerging sacred values: the Iranian nuclear program. *Judgment and Decision Making*, **4**, 550–3.

Dollard, J. (1944). *Fear in battle*. The Infantry Journal, Washington, D.C.

---

[18] Interview with S. Atran, Damascus, Syria, 26 February 2010.

Durkheim, É. (1912). *Les formes élémentaires de la vie religieuse: le système totémique en Australie*. F. Alcan, Paris.

Ehrenreich, B. (1997). *Blood rites: origins and history of the passions of war*. Henry Holt, New York.

Fritz, S. (1995). *Frontsoldaten: the German soldier in World War II*. The University Press of Kentucky, Lexington, KY.

Ginges, J. and Atran, S. (2009). Noninstrumental reasoning over sacred values: an Indonesian case study. In: D. Ross, D. Bartels, C. Bauman, L. Skitka and D. Medin (eds), *Psychology of learning and motivation, Vol. 50: Moral judgment and decision making*, pp. 193–206. Academic Press, San Diego, CA.

Ginges, J., Atran, S., Medin, D., and Shikaki, K. (2007). Sacred bounds on rational resolution of violent political conflict. *Proceedings of the National Academy of Sciences of the USA*, **104**, 7357–60.

Havemeyer, L. (1929). *Ethnography*. Ginn and Co., Boston, MA.

Hellmuth, L. (2000). Has America's tide of violence receded for good? *Science*, **289**, 582–5.

Holden, C. (2000). Cannibalism: molecule shows Anasazi ate their enemies. *Science*, **289**, 1663.

Keeley, L. (1996). *War before civilization: the myth of the peaceful savage*. Oxford University Press, New York.

Kelly, R. (2005). The evolution of lethal intergroup violence. *Proceedings of the National Academy of Sciences of the USA*, **102**, 15294–8.

Khaldûn, I. (1958 [1318]). *The Muqaddimah*. Routledge and Kegan Paul, London.

Kühne, T. (2006). *Kameradschaft: die soldaten des nationalsozialistischen und das 20*. Vandenhoeck and Ruprecht, Göttinggen.

Law, R. (1985). Human sacrifice in pre-colonial West Africa. *African Affairs*, **84**, 53–87.

Lim, C.S. and Baron, J. (1997). Protected values in Malaysia, Singapore and the United States. [Internet]. http://www.sas.upenn.edu/~baron/papers.htm/lim.htm

Lowie, R. (1954). *Indians of the plains*. McGraw-Hill, New York.

McCain, J. (1999). *Faith of my fathers*. Random House, New York.

McManus, J. (2003). *The deadly brotherhood: the American combat soldier in World War II*. Presidio Press, Novato, CA.

McPherson, J. (1997). *For cause and comrades: why men fought in the Civil War*. Oxford University Press, New York.

Moskos, C. (1970). *The American enlisted man: the rank and file in today's military*. Russell Sage Foundation, New York.

Moskos, C. (1975). The American combat soldier in Vietnam. *Journal of Social Issues*, **3**, 25–37.

Mussolini, B. (1935). *The doctrine of Fascism*. Vallecchi Editore, Firenze.

Nisbett, R. and Cohen, D. (1996). *The culture of honor*. Westview Press, Boulder, CO.

Peters, E. (1967). Some structural aspects of the feud among the camel herding Bedouin of Cyrenaica. *Africa*, **37**, 261–2.

Rappaport, R. (1999). *Ritual and religion in the making of humanity*. Cambridge University Press, New York.

Robertson Smith, W. (1894). *Lectures on the religion of the Semites*. A. and C. Black, London.

Sachdeva, S. and Medin, D. (2009). 'Group identity salience in sacred value based cultural conflicts: an examination of Hindu-Muslim identitites in the Kashmir and Babri mosque issues.' Paper presented at the 31st annual conference of the Cognitive Science Society (CogSci), Amsterdam.

Sageman, M. (2004). *Understanding terror networks*. University of Pennsylvania Press, Philadelphia, PA.

Sageman, M. (2008). *Leaderless Jihad*. University of Pennsylvania Press, Philadelphia, PA.

Smith, A. (1993). *An inquiry into the nature and causes of the wealth of nations*. Oxford University Press, New York.

Smith, R. (1983a). Why soldiers fight: Part I. Leadership, cohesion, and fighter spirit. *Quality and Quantity*, **18**, 1–32.

Smith, R. (1983b). Why soldiers fight: Part II. Alternative theories. *Quality and Quantity*, **18**, 33–58.

Speckhard, A. (2005). *Chechen Russian Uzbek suicide terrorism study*. Interim Report. NATO, Brussels.

Spector, R. (1994). *After Tet: the bloodiest year in Vietnam*. Vintage, New York.

Spier, L. (1933). *Yuman tribes of the Gilga River*. University of Chicago Press, Chicago, IL.

Stouffer, S.A., Lumsdaine, A.A., Lumsdaine, M.H., *et al.* (1949). *The American soldier. combat and its aftermath*. Princeton University Press, Princeton, NJ.

Turney-High, H. (1949). *Primitive war: its practice and its concepts*. University of South Carolina Press, Columbia, SC.

Van Vugt, M. and Hart, C. (2004). Social identity as social glue: the origins of group loyalty. *Journal of Personality and Social Psychology*, **86**, 585–98.

Van Vugt, M., De Cremer, D., and Janssen, D. (2007). Gender differences in cooperation and competition: the male-warrior hypothesis. *Psychological Science*, **18**, 19–23.

Wilson, M. and Daly, M. (1988). *Homicide*. Aldine de Gruyter, New York.

Chapter 15

Evolutionary theory and behavioural biology research: implications for law

David J. Herring

Introduction

A growing number of legal scholars have begun to engage concepts and research from the field of evolutionary psychology. This chapter describes specific examples from this emerging area of applied evolutionary psychology in order to demonstrate its potential contribution to the development and application of law and public policy.

The work in this area exhibits two primary attributes. First, it demonstrates deep explanatory power, yielding useful insights for both legal scholarship and law in action. As to law in action specifically, this line of work provides useful guidance to legislators and public policy-makers who must address particular aspects of human behaviour. Second, this work provides foundations for the formulation of new hypotheses that researchers can test through empirical studies. These empirical research projects present exciting, new, and expanding opportunities for collaborative work involving evolutionary psychologists and legal scholars.

The legislative process is well suited to being informed by evolutionary psychology because it generally seeks to address human behaviour in a comprehensive manner. Through the legislative actions of its representatives, the people seek to channel, and possibly shape, behaviour in desired ways. Therefore, legislative action calls for sophisticated knowledge of behavioural tendencies. Legislative bodies can use such knowledge to assess the feasibility, and appreciate the costs, of possible initiatives (Jones and Goldsmith 2005). In its most positive role, this knowledge can help legislative bodies fashion measures that effectively harness behavioural tendencies in a way that achieves public goals.

Behavioural biology research, informed by evolutionary theory, produces knowledge of behavioural tendencies in many areas of public concern. This research has the capacity to assist legislators in identifying behavioural tendencies that could make it very difficult and costly to achieve particular goals. It also has the capacity to identify behavioural tendencies that may, if left unaddressed, undermine particular legislative schemes and approaches, or frustrate the achievement of other public goals. In addition, this research can identify behavioural tendencies that, if effectively evoked, could contribute to the achievement of particular legislative goals (Jones 2001; Jones and Goldsmith 2005).

Foster care legislation: two detailed illustrations

Federal legislative action that addresses family and child welfare in the United States provides a specific area in which to explore the potential contribution of evolutionary theory and research. As a condition for receiving federal child welfare funds, the US Congress requires states to design their child welfare systems in accord with a comprehensive federal scheme (Adoption and Safe

Families Act of 1997).[1] For example, Congress has strongly encouraged, if not mandated, foster care placements with kin (Allen and Davis-Pratt 2009). In addition, Congress has mandated that state child welfare agencies generally disregard race when placing children with foster and adoptive families (Bartholet 2006).

The actions of Congress in this area have been supported by a combination of child development theory, social work practice knowledge, and political ideology animated by personal anecdote (Herring 2007a; Harris and Skyles 2008). Many of the components of the legislative scheme have not been informed by rigorous empirical research, and certainly not by empirical research based on an evolutionary perspective. Such research could help inform the legislative approach in this area.

Kin as foster parents

Background to current practice

Congress has encouraged the placement of foster children with kin for more than a decade. Federal legislation now requires public child welfare agencies to inform individuals that a child in their extended family is in need of placement and to provide an opportunity for kin members to serve as foster parents. These agencies must also inform kin of the services available to them, including financial support, if they decide to serve as foster parents (Fostering Connections to Success and Increasing Adoptions Act of 2008).[2]

Congress has failed to explain clearly and fully the impetus for the preference for kin foster placements (Bartholet 2007) and in fact, this preference constitutes a break from past practice. Previously, child welfare agencies appeared reluctant to place children with kin based on a common sense fear that their extended kin group likely shared the dysfunction of their parents. The new approach of favouring kin placements appears to reflect a change in perception concerning the positive value of extended kin relationships. Child development theory and social work practice experiences have combined to enhance the perception that children removed from parental custody benefit from familiarity and continuity. Federal officials have acted to secure these perceived benefits based on an assumption that kin are likely to provide familiar, stable, committed, same-race relationships for foster children (Bartholet 2007; Allen and Davis-Pratt 2009) and resource-poor child welfare agencies have embraced the preference for kin as an easily applied placement decision rule (Bartholet 2007).

The articulation of this reasoning highlights federal lawmakers' reliance on general theories and experiences, as opposed to theory-based empirical research that supports a preference for kin placements. As the preference for kinship placements has taken hold in the field, several child welfare researchers have conducted studies that compare kinship placements with non-kin placements on numerous measures. These studies have been inconclusive, indicating only that kin foster placements achieve child well-being outcomes similar to those for non-kin foster placements despite the fact that, on average, kin foster parents are older, have less education, have lower incomes, and receive fewer support services (Ehrle and Geen 2002; Geen 2004).

Unfortunately, the empirical research to date has been observational in nature, unguided by a coherent theoretical framework. One result is that the operative definition of kin has included two distinct groups—non-genetic kin who have a prior relationship with the child and genetic kin who may or may not have a prior relationship with the foster child (Winokur et al. 2008; Zinn 2009).

---

[1] Adoption and Safe Families Act of 1997, Pub. L. No. 105–89, 111 Stat. 2115
[2] Fostering Connections to Success and Increasing Adoptions Act of 2008, Pub. L. No. 110–351, section 103, 122 Stat. 3956

(This situation is similar to that confronted by researchers examining cases of infanticide in that investigators classified genetic parents and step-parents simply as 'parents' (Jones 1997)). This type of definitional incoherence has the potential to obscure important research questions required for a rigorous comparison of genetic kin and non-kin placements. In addition, the research in this area is unsophisticated in that it pursues only a binary comparison of all kin to all non-kin, ignoring the possibility of differences among kin that may have important implications in the field (Herring 2008).

Insights from evolutionary theory

Evolutionary theory that addresses kinship can guide research in this area. For example, kinship theory delineates a clear definition of kin that is relevant to sophisticated empirical research on kinship placements. In addition, research conducted pursuant to kinship theory can provide useful insights for lawmakers and other child welfare decision-makers.

Evolutionary theory postulates that an individual will tend to provide more beneficial treatment to close kin than to more distant kin or non-kin (Park et al. 2008). Two related evolutionary concepts help explain this behavioural tendency for kinship altruism. The concept of inclusive fitness arises from the observation that in terms of passing genetic material to future generations, individuals benefit not only from their own reproductive success, but also from the reproductive success of kin (Hamilton 1964). This indirect reproductive benefit is explained by the concept of degree of relatedness, in that there is a higher probability that an individual will share a particular genetic segment, or allele, with a closely related kin member than with a more distant kin member or non-kin. For example, there is approximately a 50% probability that an individual shares a specific genetic segment (not already widely shared across the population) with his or her child or sibling (first-degree kin), compared to a 25% probability with regard to a grandchild or niece/nephew (second-degree kin). The probability declines as the kin relationship becomes more distant, with the probability for non-kin approaching zero (Hamilton 1964).

In light of these concepts, a mutation for kinship altruism will spread throughout a population if the reproductive costs (C) incurred by the altruist are less than the indirect reproductive benefits (B) realized through the reproductive success of the recipient, discounted by the degree of relatedness (r) between the altruist and the recipient ($C < Br$) (Hamilton 1964). Evolutionary theory postulates that this condition often existed within the ancestral environment, resulting (when given the genetic opportunity provided, for example, by a mutation occurring within the population) in the spread of a behavioural tendency to assist close kin (Park 2007). The conceptual result, relevant to an examination of kinship foster care placements, is that one reasonably expects higher levels of investment from close kin than from more distant kin and non-kin.

Animal and human research has found a behavioural tendency to favour close kin (Holmes and Sherman 1982; Fletcher 1987; Burnstein et al. 1994; Kruger 2003). In the area of kinship foster care specifically, research indicates a comparatively high level of parental investment, with kinship placements being significantly more stable than non-kin placements (Chamberlain et al. 2006). One child welfare researcher has examined aspects of stability and permanency for children in foster care, finding that the only significant differences were determined by degree of relatedness between foster child and foster parent (Testa 2005). Second-degree kin (i.e. grandparents and aunts/uncles) exhibited significantly higher levels of investment in their foster children than more distant relatives and non-kin.

Two additional evolutionary concepts support more sophisticated distinctions among kin in terms of expected level of parental investment. The first is the concept of paternity certainty. Whereas women can know with virtual certainty whether or not a child they care for is their child, men cannot be so sure (Buss 2008). This has implications for extended kin relations.

Reduced paternity certainty compounds as kinship links through males increase. For example, if within a population a man has a paternity certainty of 0.90, he has a grandpaternity certainty through his mate's sons of 0.81 (Gaulin et al. 1997). The overall implication is that one can reasonably expect a grandparent to favour daughters' children over sons' children. More generally, one can reasonably expect matrilineal kin to receive more favourable treatment than patrilineal kin.

The second additional concept is sexual selection. Women have a higher biological stake in each of their children than do men. Women are committed to investing in one child during 9 months of pregnancy, and in the human evolutionary environment, women were committed to nursing a child, and thus experienced a reduction in reproductive capacity for a significant period after giving birth (Buss 2008). In contrast, men have the option to move on readily in the pursuit of additional mating opportunities. As a result, natural selection appears to have favoured—in humans as in other mammals (see Chapter 9, this volume)—a tendency in males to invest less in childcare and more in mating effort than women (Buss 2008). This difference has implications for extended kin relations. Because women's capacity for direct reproductive success is limited in comparison to men's (both because of higher minimum investment per offspring and in lower maximum number of offspring), they have a higher stake in indirect reproductive success through kin (Gaulin et al. 1997; Huber and Breedlove 2007). Thus, one can reasonably expect women to tend to invest more in kin, on average, than men.

Researchers have addressed these two concepts in empirical studies of grandparental investment (for a fuller discussion of these effects, see Chapter 5, this volume). A series of historical population studies has found that maternal grandmothers have a significant positive effect on child survival and nutrition, while the presence of other grandparents has no positive effect (Sear et al. 2000; Jamison et al. 2002; Voland and Beise 2002; Ragsdale 2004). A second line of studies has used contemporary subjects to assess levels of grandparental investment. The results confirm the laterality effect, with maternal grandmothers tending to invest the most, followed in order by maternal grandfathers, paternal grandmothers, and paternal grandfathers, with a statistically significant difference between each category (Euler and Weitzel 1996; Laham et al. 2005).

Researchers have also examined levels of investment by aunts and uncles. These studies confirm both the sex effect, with aunts tending to invest at a higher level than uncles, and the laterality effect, with matrilateral aunts and uncles tending to invest at higher levels than patrilateral aunts and uncles (Gaulin et al. 1997; McBurney et al. 2002). A study of cousins has also confirmed the laterality effect (Jeon and Buss 2007).

In summary, the research based on evolutionary kinship theory indicates that individuals tend to favour kin over non-kin when conferring important benefits such as childcare. In addition, women tend to invest more in childcare than men, and matrilineal kin tend to invest more than patrilineal kin.

Applications

This research on differences among types of kin supports a ranking of second-degree kin by expected level of investment in kin, with maternal grandmothers occupying the highest investment position and paternal grandfathers the lowest (Gaulin et al. 1997; Herring 2008). Other categories of second degree kin would fall between these two extremes, but could also be distinguished from one another. Such a ranking would allow one to formulate hypotheses concerning types of kin and expected levels of investment in childcare. Researchers could test these hypotheses in the field. For example, researchers could design a comparative study that examines the level of care provided by maternal and paternal grandmothers, along with a comparison of child development outcomes, in order to produce new knowledge about kinship foster care and kin altruism.

This research on second-degree kin (those most often involved in kinship foster care placements) would provide insights for possible legislative action. For example, it would provide detailed information relevant to the possible construction of a hierarchy of legal preferences among kin. The research results would constitute new knowledge that could guide legislators in considering measures that, for example, favour placements with matrilineal kin in order to achieve, on average, higher levels of foster parent investment and better child development outcomes.

This research would also provide insights that inform child welfare legislation and policy that addresses the appropriate level of support for particular foster care placements. Findings could put legislators and policy-makers on notice of placements that may present a comparatively high risk of low investment and negative child outcomes (i.e. those with non-kin rather than kin, and with paternal rather than maternal relatives). These research findings, along with other considerations, may guide decisions concerning the provision of services so that legislators and child welfare administrators could use this information to allocate limited family support services more efficiently.

Race as a consideration in foster care placements

Current practice

Turning to federal legislation that prohibits the systematic consideration of race in making foster care placements, one sees the influence of both general theories of child development and political ideology. Federal child welfare legislation has long been propelled by permanency planning concepts developed within the field of social work. These concepts rest on child development theory that postulates the need for children to grow up in permanent, committed families. This theoretical perspective views foster care as a temporary living situation that should be avoided or limited strictly in duration (Herring 2000; Golden and Macomber 2009). However, many children languish for years in temporary foster care placements. This phenomenon of foster care drift is especially acute for minority children, with average stays in foster care being much longer for black compared to white youth (Roberts 2002; Harris and Skyles 2008).

In response to this situation, Congress enacted the Multiethnic Placement Act (MEPA)[3] in 1994. MEPA allowed child welfare agencies to consider race in deciding where to place a particular child, but prohibited delay in making placements based on attempts to match the child's race with that of adoptive or foster parents. Congress hoped to eliminate delays caused by attempts to match race (Herring 2007a). A key advocate in this legislative effort was Harvard Law Professor Elizabeth Bartholet. Through her personal story of different-race adoption, academic writings, and presentations to the public and to Congress, Bartholet supported a colour blind adoption placement policy and child welfare system (Bartholet 1991, 1999) which was embraced by members of Congress. The original version of the law was quickly perceived as too weak in that it allowed agencies to consider race as one factor among many, so the law was amended in 1996 to prohibit the consideration of race unless an agency could make an individualized showing of special need for a particular child.

Following the enactment of MEPA, several child welfare scholars speculated that children in different-race adoptive placements would suffer a loss of racial identity that would impair their healthy development and social functioning. In response, researchers have conducted several studies that compare functioning and outcomes for children in same-race and different-race

---

[3] Multiethnic Placement Act of 1994 (MEPA), Pub. L. 103–382, section 551, 108 Stat. 4056 (amended 1996)

adoptive homes (Brooks and Barth 1999). The results have been inconclusive. The findings indicate that children can function adequately, and even thrive, in different-race placements, but may suffer harm in terms of racial identity and belonging. These findings are generally viewed as not undermining MEPA's colour blind approach to adoptive placement decisions. However, evolutionary theory and research may challenge MEPA's approach to placement decisions in terms of child health and safety (Herring 2007a), on the basis of two key issues, which I now discuss.

Altruism and kinship heuristics

Research on cues of kinship may help to guide law and policy in addressing the complex situation of non-genetic parents of different race. The research has revealed two primary types of kinship cues: familiarity cues and similarity cues (Park et al. 2008). For example, if an individual has lived with another throughout early childhood, the individual is likely to perceive this familiar other as a kin member. Or if an individual perceives facial resemblance with another, the individual is likely to perceive the other person as kin (Lieberman et al. 2007; Park et al. 2008). Kinship cues operate as unconscious heuristic tools and the perception of kinship gives rise to a behavioural tendency to act altruistically toward the other person. These heuristic devices may operate in the absence of actual kinship, presenting a significant risk of error. As a result, one may mistakenly perceive another as kin and confer benefits on one who is not a kin member (Park et al. 2008).

A developing line of research examines facial resemblance as a kinship cue. This research has implications for parent/child relationships and expected levels of investment. The findings indicate that adults favour children who share their facial features (Platek et al. 2002, 2004; DeBruine 2004; Alvergne et al. 2010). They appear more willing to provide these children with more resources, to express an interest in adoption, and to discipline them less severely (Platek et al. 2002). Thus, the studies indicate that facial resemblance operates as a kinship cue that evokes altruistic behaviour generally and parental investment specifically, even in the absence of an actual genetic relationship. In addition, a separate study has examined attitude similarity as a kinship cue. The findings indicate that the perception of shared attitudes evokes altruistic behaviour (Park and Schaller 2005).

The research findings on the operation of heuristic kinship cues could be helpful in the context of making foster care placement decisions, possibly presenting an opportunity to evoke kin-like behaviour among non-kin. If child welfare personnel could successfully match foster parents and children in terms of facial resemblance and/or attitude similarity, it may result in foster families that, on average, exhibit comparatively high levels of altruism and parental investment. This approach would likely require an expensive and sustained effort, and therefore, is unlikely to be implemented within resource-poor child welfare systems.

However, the consideration of kinship cues may be of immediate use in identifying placements that present a comparatively high risk of low investment in foster children. Foster family situations characterized by a lack of familiarity and a high degree of dissimilarity between parents and child are likely to present this higher risk. To illustrate, consider a foster family in which the foster parents have no prior relationship with the foster child. The foster child has facial features that differ markedly from those of both foster parents, vividly signalling the absence of a genetic relationship. In addition, the foster child and the foster parents do not share attitudes about, among other things, reading books, playing games, participating in sports, or enjoyment of amusement parks. In the absence of kinship cues, the likelihood that this situation will evoke a high degree of parental investment is less than for a foster family whose members share a prior relationship, physical characteristics, and attitudes.

In a similar way, one could reasonably expect different-race foster care placements to present, on average, a comparatively high risk of low parental investment. The difference in race between foster parents and a foster child may, in light of the realities of a significantly racially segregated American society, stand as a robust proxy for a lack of prior relationship. More powerfully, the difference in race is likely to indicate a significant lack of physical similarity and kinship. Furthermore, because a difference in race may indicate divergent attitudes, especially between foster parents and foster children who have previously lived only in same-race families, a difference in race may minimize the likelihood that attitude similarity will serve as a cue that evokes high levels of parental investment.

The Cinderella effect, and adoptive versus foster placements

As described above (see section 'Insights from evolutionary theory'), the concept of inclusive fitness suggests that individuals in a parental role for a non-genetic child will tend to invest less in childcare than a genetic parent. However, this situation is more complex in terms of evolutionary theory than indicated in a simple consideration of inclusive fitness. For example, even though step-parental relationships (which were probably common in the human evolutionary environment) involve caring for non-kin, they are often likely to evoke significant investment in a partner's children because an exhibition of significant parental investment in a partner's child can be an important component of mating effort (Anderson et al. 1999). Another example of this complexity is provided by individuals who adopt an unrelated child. Such individuals may have an especially strong drive to parent, a possible by-product of an adaptation to provide parental care to genetic children. This strong drive is evidenced by their willingness to engage voluntarily and deliberately in a typically lengthy, difficult process in order to make a long-term, legally recognized commitment to parent a child—a clear exhibition of parenting effort. This drive is also reinforced by social mores that endorse those who serve as parents and question those who remain childless. Thus, one could reasonably expect adoptive parents to exhibit significant levels of parental investment (Gibson 2009).

As predicted by these considerations, studies indicate that adoptive parents' level of parental investment is similar to that of genetic parents (if not higher, because they are a selected and highly-motivated subset of parents), despite the lack of a genetic relationship with the children (Hamilton et al. 2007; Gibson 2009). In contrast, step-parents exhibit a behavioural tendency to invest in stepchildren at a comparatively low level. Step-parent mating effort appears to give rise to a weaker form of parental investment than that evoked by a genetic relationship, as evidenced by a significantly higher rate of child maltreatment for non-genetic children in the care of a step-parent (the Cinderella effect: Daly and Wilson 1988, 1994).

The relevant question here is whether foster parents tend to behave more like adoptive parents or step-parents in terms of parental investment. Although foster parents, like adoptive parents, voluntarily commit to parent a non-genetic child, their motivations may not be as clear or as strong. The commitment of foster parents to care for a child is only temporary—they are not establishing a permanent relationship akin to adoption. In addition, foster parents may be motivated by financial compensation that is often absent in the typical private adoption context. Thus, there are reasons to speculate that foster parents may exhibit behavioural tendencies that approach those of step-parents.

Implications for placements

This line of reasoning opens a possible new line of critique of MEPA that focuses on foster care rather than adoption placements. The possible differences between adoptive and foster parents in terms of expected levels of investment may call into question the use of different-race foster parents to secure foster care placements more quickly. In addition, because of the emergency situation surrounding many foster care placements, the risk of delay is less in the foster care placement context than in the adoption placement context. This diminishes one of the primary rationales for MEPA. Thus, the child welfare benefits that arguably result from MEPA are likely less in the foster care placement context.

Expanding on these points, the critique of MEPA in the context of foster care is more powerful because the nature of child welfare costs is different from the adoption context. Because the risk in the foster care context is comparatively low levels of parental investment, the costs are likely to manifest in concrete measures of child safety and health, rather than in the more ambiguous measure of loss of racial identity. Evolutionary insights allow a fuller recognition of the real costs of ignoring race in the foster care placement context. Put another way, they demonstrate that the pursuit of a race blind social policy may entail a sacrifice in terms of child safety and health. Thus, they assist policy makers in consciously considering the public policy tradeoffs presented by the MEPA approach.

This new critique of MEPA has implications for legislative action, possibly justifying the consideration of an amendment to MEPA as it applies to foster care placement decisions. The amendment may simply entail a return to MEPA's original approach that allowed child welfare agency personnel to consider race as one factor among many when making foster care placement decisions. Because children who experience foster care placements fare poorly in many areas, their need for a relatively high level of parental investment is apparent. A law that presents a heightened risk of low parental investment to these children is questionable and may demand change.

This critique of MEPA also calls for legislation to fund research. Studies that compare same-race and different-race placements would produce useful knowledge in this area. This research would test the hypothesis that, on average, foster children achieve better outcomes in same-race placements. If the research confirms this hypothesis, additional studies could examine which support services, if any, effectively address comparatively low levels of parental investment in different-race placements.

The insights provided by an evolutionary perspective may also have immediate implications for child welfare policy. While MEPA bars the consideration of race in making placement decisions, it does not prohibit caseworkers from recognizing the higher risk for low parental investment in different-race placements. They can decide to provide enhanced services to support these higher risk placements. In this way, child welfare agencies can target scarce public resources more effectively.

Child maltreatment

Step-parent relationships

The applications of evolutionary theory and research in the context of foster care indicate a potential to influence many areas of legal inquiry. In this vein, consider Owen Jones' groundbreaking work in which he develops a general approach to evolutionary analysis in law and applies it to child abuse, with a focus on infanticide (Jones 1997). Jones begins by recognizing an established public goal to reduce infanticide. He then explains the evolutionary concepts of reproductive success and inclusive fitness, and articulates four independent theories of how adaptive behaviour could contribute to infanticide. Jones draws on these theories to formulate a set of predictions concerning infanticide. He then presents the empirical evidence related to each prediction and concludes with several recommendations.

To illustrate Jones' approach, consider the theory of discriminative parental solicitude, which posits that parents discriminate among infants with different levels of parental investment. A parent will tend to invest in an infant in proportion to the infant's contribution to the parent's inclusive fitness. This leads to a hypothesis that an infant who must rely on parental investment from an unrelated adult is somewhat more likely to die of neglect or abuse by the adult care taker (Daly and Wilson 1988). Therefore, Jones (following work by Daly and Wilson) recounts the prediction that 'children will be at greater risk of infanticide in step-parent households than in [genetic] parent households' (Jones 1997, p. 1207). As to empirical evidence related to the prediction, Jones cites a series of studies that indicate that young children are many times more likely to experience lethal abuse at the hands of step-parents than of genetic parents (Daly and Wilson 1988, 1994).

Jones uses this line of analysis to suggest several legal strategies to reduce infanticide if (and indeed only if) society weighs the cost of infanticide to be higher than the cost of treating step-parents different than parents. These include the modification of risk assessment tools so as to weigh step-parent presence more heavily as a risk factor, the development of a more stringent legal standard for the disciplinary privilege for step-parents, and increasing deterrence through harsher punishment for step-parents who engage in abusive behaviour. Jones also suggests that the law encourage additional relevant research by mandating that state agencies collect child abuse data that differentiate between genetic and non-genetic parents.

Disabled and needy children

Two other legal scholars have also used evolutionary theory and research to formulate a prediction about child maltreatment (Brinig and Buckley 1999). Brinig and Buckley employ the theory of discriminative parental solicitude, along with reproductive access theory, which posits that a parent may tend to mistreat a child who poses a threat to the parent's romantic relationship. They draw on these theories to predict that disabled or especially needy children will be at greater risk of maltreatment than healthy children.

Brinig and Buckley conducted an empirical study that tested the hypothesis. Using a national sample of abused children and a matched sample of disabled children, they found that a child's disability status is predictive of both abuse and neglect (Brinig and Buckley 1999). The strongest association was for neglect, with a child's disability nearly doubling the likelihood of mistreatment. The findings also indicate that the presence of a step-parent is predictive of abuse, increasing the chance of mistreatment by more than 25% (although it simultaneously reduced the likelihood of child neglect by 13–14%, because a step-parent often increases family resources, thus reducing both economic stress and the likelihood of neglect).

Based on these findings, Brinig and Buckley propose significant changes to legal doctrine. Their primary recommendation is to relax the presumption of parental fitness for genetic parents of disabled children once maltreatment has occurred, with greater attention given to rescuing these children from their abusive parents. They note that this would constitute a significant exception to a line of court decisions that make it difficult to remove children from parental custody and very difficult to terminate parental rights. But they also note that their recommendation would fortify the federal law directive that a child's safety and health are paramount concerns in determining family preservation and reunification efforts.

Reducing the incidence of rape

Legal scholars have also used evolutionary theory and research to gain insights relevant to aspects of criminal law. The offence of rape provides an example. Owen Jones has noted our failure to reduce the incidence of rape despite both a public consensus to curb rape and the development of several theories of rape. We appear to lack a sufficient understanding of rape behaviour. Jones suggests that insights from the life sciences may contribute to our understanding in this area and to a reduction in rape (Jones 1999).

As Jones explains, the basic concepts of natural selection, reproductive success, and sexual selection have given rise to two evolutionary theories of rape behaviour. The first theory postulates that rape is a direct adaptation, in that a condition-dependent predisposition to rape behaviour in certain conditions had a net positive effect on a male's reproductive success in the human evolutionary environment (Thornhill 1999; Thornhill and Palmer 2000). Thus, this behavioural tendency spread through the human population and became one aspect in a broad array of male sexual strategies. The second theory of rape views the behavioural tendency to rape as a by-product of adaptive behaviours related to male sexual strategies (Thornhill and Palmer 2000). For example, rape could be a by-product of an adaptive male tendency to aggressively pursue willing sexual partners. Such aggressive pursuit may sometimes result in non-consensual sex.

Jones notes that these theories of rape implicate male sexual strategies (however unconsciously). This is a departure from prevailing social science theories of rape that focus on male aggression and violence toward women (Thornhill and Palmer 2000). This departure allows for the formulation of testable predictions based on the sexual nature of human rape behaviour. For example, one prediction is that 'the age of victims of attempted and completed rape will be overwhelmingly concentrated into the part of the female lifespan that is reproductive' (Jones 1999, p. 865). Jones notes that empirical studies provide strong support for this prediction, with victims of rape overwhelmingly likely to be in their peak reproductive years (Thornhill and Palmer 2000). Another example is provided by a prediction that 'a disproportionately high number of rapists will be young, sexually mature males' (Jones 1999, p. 870). This group of males is on average less likely to have the ability to attract willing sex partners, and thus, more likely to engage in rape behaviour. The empirical research confirms this prediction, with rapists having a median age of 25. Only a small fraction of rapists are over 30 (Amir 1971; Hall 1995).

Although further research in the area of rape is necessary, the evolutionary perspective allows one to see the nature of rape more clearly. This clarity has relevance for the consideration of aspects of rape law. Jones describes several possibilities (without advocating any). For one, the insights might support the inclusion of chemical castration among the sentencing options for rapists. If rape is, at least in part, a behavioural tendency related to male sexual strategies, a sentencing mechanism that reduces the sex drive may effectively reduce recidivism. Evolutionary insights may also help defend chemical castration from a constitutional attack as cruel and unusual punishment. The evolutionary perspective indicates that such a sentencing mechanism is not arbitrary, but rather is supported by both coherent theory and empirical research.

Another example is provided by Jones' discussion of the potential mismatch between the federal Violence against Women Act of 1994 (VAWA)[4] and the multiple causes of rape behaviour. The Act allows for the recovery of civil damages only if the victim can establish that the rapist acted based on animus toward the victim's gender. The law's focus on animus fails to reflect the complexity of causes underlying rape behaviour, and the failure to acknowledge the sexual aspect of rape behaviour may present the victim plaintiff with challenging, if not insurmountable, problems of proof. The result may be an overly narrow interpretation and application of federal law. Insights provided by an evolutionary perspective may prompt Congress to amend VAWA so that it is more effective in reducing the incidence of rape.

---

[4] Violence Against Women Act of 1994 (VAWA), Pub. L. No. 103–322. Sections 40111(a)-40611, 108 Stat. 1796, 1903–53

Fortification of sexual harassment legislation

Scholarship in the area of sexual harassment law also demonstrates the benefits of evolutionary theory and research. Julie Seaman's work provides an illustration (Seaman 2005). She begins by articulating the central goal of federal law—to prohibit 'discrimination against an individual on the basis of that individual's sex' (Seaman 2005, p. 409). She then examines a particular form and arena of sex discrimination—sexual harassment in the workplace. Seaman notes that the courts have had difficulty in developing a coherent explanation of when an employer is liable 'for abusive working conditions created by co-workers and/or supervisors of a complaining employee' (Seaman 2005, p. 327). The courts require a plaintiff to prove that the harassing behaviour was based on the plaintiff's sex. But explaining precisely why evidence of sexual harassment establishes discrimination 'because of sex' has proven difficult in many cases.

Seaman uses evolutionary theory and research to provide insights into how harassing behaviour constitutes discrimination because of sex. Initially, Seaman draws on the concepts of natural selection and sexual selection to acknowledge the possible existence of differences in behavioural tendencies between men and women. She avoids drawing any normative conclusions based on these differences, but notes one difference that may be relevant to sexual harassment: women are likely to perceive sexual behaviour differently than men, and as a result, such behaviour is likely to have a more harmful impact on women in the workplace.

Seaman examines specific categories of harassment cases. One pattern of behaviour she addresses has 'the purpose or effect of intimidating, humiliating, or otherwise "hazing" the victim so as to make that person either leave the workplace or submit to the demands (whether explicit or implicit) of the dominant group' (Seaman 2005, p. 397). This behaviour is not aimed at procuring sex, but it is often sexual or suggestive because this type of behaviour is often effective at achieving the exclusionary purpose. Seaman notes that nonsexual conduct is also often used to achieve this purpose.

Seaman draws on evolutionary theory to explain how such hazing or exclusionary behaviour is driven by, or 'because of', the sex of the victim. More specifically, she discusses the adaptive behaviours of social coalition building and bonding. These behaviours have given rise to gender norms as a universal aspect of human society. The enforcement of these gender norms is an important component of human behaviour, as illuminated by child developmental psychology research (Fagot et al. 2000). As evolutionary theory predicts and the relevant research indicates, males appear to have strong tendencies to group together and to exclude females and males who do not exhibit typical gender traits from their groups in order to increase group cohesion and bonding (Tiger 1969; Browne 2001).

This male behavioural tendency has implications for sexual harassment law. It helps to explain how 'non-sexual' conduct that excludes women and targeted men from a particular workplace constitutes harassment 'because of' the sex of the victim. In this way, evolutionary insights help to define the appropriate scope of federal sexual harassment law and to clarify legal doctrine that holds employers liable for such workplace conduct.

Seaman uses these evolutionary insights to critique several aspects of the law of sexual harassment. By identifying and describing the contexts in which this behaviour arises, she makes a strong case for employer liability for such workplace environments, whether or not individual employees are somehow deemed blameworthy. Her work reinforces and fortifies both feminist legal theory in this area and the judicial trend to expand the law's applicability to all situations where harassment occurs because of the victim's sex. Although she does not discuss this point in her article, Seaman's work also makes a case for additional research in this area. Her use of evolutionary theory to identify workplace environments that are likely to constitute or support harassment allows one to formulate predictions that researchers could test. This type of research would further guide the development of legal doctrine in this area.

The value of possession and the endowment effect

Several legal scholars have used evolutionary theory and research to illuminate the conception of property (Stake 2004; Krier 2009). They draw on the idea of evolutionarily stable strategies to recognize a behavioural heuristic—aggressively claim and defend an item when one is in possession and defer when another is in possession. This heuristic would likely enhance an individual's reproductive success because it asserts one's claims to selected items, but avoids physical fights over many items. As a result, this behavioural heuristic would likely spread within a population (Maynard Smith 1982; Stake 2004).

Such a behavioural heuristic makes possession meaningful, providing a foundational concept for the construct of property. The idea that possession has meaning in terms of behavioural tendencies is supported by research indicating that individuals often refuse to sell or trade an item they have just acquired at the maximum price they would have paid for the item just before they acquired it. It is as if the instant an individual comes to possess or own an item, the item acquires extra value to them for no apparent rational reason. This phenomenon of extra value appearing from nowhere is known as the 'endowment effect' (Jones and Brosnan 2008; see also Chapter 2, this volume).

Sarah Brosnan and Owen Jones have hypothesized that seemingly irrational behaviour in the current environment constituted rational adaptive behaviour in ancestral evolutionary environments for humans and other species. Jones terms this 'time-shifted rationality' (Jones 2001) and he speculates that the endowment effect may be the result of 'conditions in which the probable results of continued possession were less risky than the probable results of attempted exchanges' (Jones and Brosnan 2008, p. 1960).

Suspecting that the endowment effect has deep roots in primate evolution, Brosnan and Jones designed a chimpanzee experiment to test a prediction that 'the prevalence of the endowment effect will increase or decrease, respectively, with the increasing or decreasing evolutionary salience of the item in question' (Jones and Brosnan 2008, p. 1961). The results of the experiment support the prediction, with the chimpanzee subjects exhibiting a significantly stronger endowment effect for food items that have high evolutionary salience in terms of survival and successful reproduction than for toy items that have low evolutionary salience. Thus, the experiment suggests that evolutionary salience affects the prevalence and magnitude of the endowment effect.

This finding has implications for aspects of property and contract law. To illustrate, consider the contract law concept of efficient breach (Posner 2007). The law enables one party to breach a contract if she can compensate the other party and still be better off. The idea is to not tie up resources and effort in fulfilling past contractual obligations when the resources and effort can be redeployed to projects or activities that produce more overall value. The law seeks to promote this gain in the efficient allocation of resources by lowering transaction costs and requiring damages for the injured party that compensate only for what he expected to realize from the performance of the contract. This expectation is normally measured by what the injured party would have been willing to pay for the contract obligation before agreement had been reached. But if the endowment effect arises in the context of contractual obligations, the contract may have increased value for a party once an agreement is in place and the promise is possessed. For some contract obligations (i.e. ones that have a high level of evolutionary salience), damages in the amount the injured party was willing to pay before the contract was in place may be insufficient to avoid an inefficient allocation of resources.

The consideration of evolutionary concepts calls for more research in this area. As to efficient breach in particular, researchers need to determine the existence, nature, and strength of the endowment effect related to contract obligations. Findings from such research may guide legal decision makers in calibrating the appropriate measure of damages in terms of efficient breach. As to seemingly irrational behavioural tendencies in general, the formulation and testing of additional evolutionary-based hypotheses could provide useful insights for particular aspects of law and policy.

Conclusion

The examples included in this discussion provide only an indication of the numerous areas of law that are likely to benefit from evolutionary insights. From areas as diverse as child welfare law to contract law, one can use evolutionary concepts and research to inform legislative efforts and the formulation of public policy.[5] The best work in this vein will also provide a foundation for further applied research. Each of the examples discussed here calls for the formulation of new hypotheses that can be tested through empirical research. Thus, these works support a sustained dialogue between evolutionary psychologists and legal scholars that will provide additional useful insights.

But for all their illustrative power, the works discussed here provide examples only of efforts in the trenches of legal scholarship. It is important to acknowledge a developing literature that expresses a more comprehensive view of the use of evolutionary insights to inform law. For example, several leading legal scholars have entertained the idea that law and evolutionary psychology could play a central role in legal scholarship, lawmaking, and the application of the law—a role similar to that played by law and economics (Monahan 2000; Jones et al. 2011).

John Monahan discusses three fundamental attributes of economic theory that provide it with analytical power relevant to law and the potential for evolutionary theory to mirror these attributes. First, evolutionary concepts have the potential for great heuristic power in that they raise and address interesting questions that are relevant to the law and that no one has asked before. For example, he refers to step-parent investment theory and research that poses the original question of whether step-parents present a higher risk of maltreatment to children than genetic parents (Daly and Wilson 1996, 1997). Second, evolutionary theory has great breadth in that it possesses a set of core concepts that can be employed to answer the broad array of interesting questions that it poses. For example, Monahan points to overarching concepts such as parental investment and sexual strategies that provide insights into gender differences and aspects of criminal behaviour. Third, evolutionary analyses have depth in that they answer legal questions with the kind of specific detail necessary to guide both legal scholarship and law in action. For example, Monahan notes that evolutionary research indicates the age at which children are most vulnerable to homicide at the hands of a step-parent (0–2 years: Daly and Wilson 1988). He also notes the specific predictions concerning rape behaviour that are generated by evolutionary theory (e.g. rape victims are likely to be of reproductive age: Jones 1999).

That evolutionary theory displays these three fundamental attributes indicates its analytic power and potential influence. But its most significant contribution may be the encouragement of collaborative research that produces new knowledge relevant to law making and public policy (Herring 2007b).

---

[5] The Society for Evolutionary Analysis in Law (SEAL) supports this broad endeavour, with a comprehensive website: www.sealsite.org

The research addressing facial resemblance provides an illustration. A team of evolutionary psychology researchers have completed an interesting study of facial resemblance outside the context of foster care (Alvergne et al. 2010). The researchers used third-party judges to assess resemblance between children and parents within a family unit. They also had each parent separately assess his or her emotional closeness to each child. The study noted that 'emotional closeness as reported by fathers, but not by mothers, was found to be predicted by third-party assessments of resemblance between parent and child' (Alvergne et al. 2010, p. 7)

A team of psychology researchers and child welfare law scholars could attempt to replicate this study in the foster family context. Such a study could determine whether a third-party (i.e. a caseworker) could effectively assess the degree of facial resemblance between a foster child and a potential foster parent and whether facial resemblance predicts the level of foster parent investment. This project would produce new knowledge relevant to the development of foster care placement criteria that caseworkers and judges could apply in the field.

This final illustration indicates the potential contribution of evolutionary theory and research to legal scholarship and to law in action. Such an interdisciplinary research team approach could transform the nature of legal scholarship and contribute to the formulation and application of law based on human behavioural research conducted in the context of an established and coherent theoretical framework. This scholarly venture would constitute a significant contribution to the field of law.

References

Allen, M. and Davis-Pratt, B. (2009). The impact of ASFA on family connections for children. In: O. Golden and J.E. Macomber (eds), *Intentions and results: a look back at the Adoption and Safe Families Act*, pp. 70–82. The Urban Institute, Washington, DC.

Alvergne, A., Faurie, C., and Raymond, M. (2010). Are parents' perceptions of offspring facial resemblance consistent with actual resemblance? Effects on parental investment. *Evolution and Human Behavior*, **31**, 7–15.

Amir, M. (1971). *Patterns in forcible rape*. University of Chicago Press, Chicago, IL.

Anderson, K., Kaplan, H., and Lancaster, J. (1999). Paternal care by genetic fathers and stepfathers I: reports from Albuquerque men. *Evolution and Human Behavior*, **20**, 405–31.

Bartholet, E. (1991). Where do black children belong? The politics of race matching in adoption. *University of Pennsylvania Law Review*, **139**, 1163–256.

Bartholet, E. (1999). *Family bonds*. Beacon Press, Boston, MA.

Bartholet, E. (2006). Commentary: cultural stereotypes can and do die: it's time to move on with transracial adoption. *Journal of the American Academy of Psychiatry and Law*, **34**, 315–20.

Bartholet, E. (2007). The racial disproportionality movement in child welfare: false facts and dangerous directions. *Arizona Law Review*, **51**, 871–932.

Brinig, M.F. and Buckley, F.H. (1999). Parental rights and the ugly duckling. *Journal of Law and Family Studies*, **1**, 41–65.

Brooks, D. and Barth, R.P. (1999). Adult transracial and inracial adoptees: effects of race, gender, adoptive family structure, and placement history on adjustment outcomes. *American Journal of Orthopsychiatry*, **69**, 87–99.

Browne, K.R. (2001). Women at war: an evolutionary perspective. *Buffalo Law Review*, **49**, 51–247.

Burnstein, E., Crandall, C., and Kitayama, S. (1994). Some neo-Darwinian decision rules for altruism: weighing cues for inclusive fitness as a function of the biological importance of the decision. *Journal of Personality and Social Psychology*, **67**, 773–89.

Buss, D. (2008). *Evolutionary psychology: the new science of the mind*. Pearson/Allyn and Bacon, Boston, MA.

Chamberlain, P., Price, J.M., Reid, J.B., Landsverk, J., Fisher, P.A., Stoolmiller, M. (2006). Who disrupts from placement in foster and kinship care? *Child Abuse and Neglect*, **30**, 409–24.

Daly, M. and Wilson, M. (1988). *Homicide*. Aldine de Gruyter, Hawthorne, NY.

Daly, M. and Wilson, M. (1994). Some different attributes of lethal assaults on small children by stepfathers versus genetic fathers. *Ethology and Sociobiology*, **15**, 207–17.

Daly, M. and Wilson, M. (1996). Evolutionary psychology and marital conflict: the role of stepchildren. In: D. Buss and N. Malamuth (eds), *Sex, power, conflict: evolutionary and feminist perspectives*, pp. 9–28. Oxford University Press, New York.

Daly, M. and Wilson, M. (1997). Crime and conflict: homicide in evolutionary psychological perspective. In: M. Tonry (ed.), *Crime and justice: a review of research*, pp. 51–100. University of Chicago Press, Chicago, IL.

DeBruine, L.M. (2004). Resemblance to self increases the appeal of child faces to both men and women. *Evolution and Human Behavior*, **25**, 142–54.

Ehrle, J. and Geen, R. (2002). Kin and non-kin foster care—findings from a National Survey. *Children and Youth Services Review*, **24**, 15–35.

Euler, H.A. and Weitzel, B. (1996). Discriminative grandparental solicitude as reproductive strategy. *Human Nature*, **7**, 39–59.

Fagot, B.I., Rodgers, C.S., and Leinbach, M.D. (2000). Theories of gender socialization. In: T. Eckes and H.M. Trautner (eds), *The developmental social psychology of gender*, pp. 65–90. Lawrence Erlbaum Associates, Mahwah, NJ.

Fletcher, D.J.C. (1987). The behavioral analysis of kin recognition: perspectives on methodology and interpretation. In: D.J.C. Fletcher and C.D. Michener (eds), *Kin recognition in animals*, pp. 19–26. John Wiley and Sons, New York.

Gaulin, S.J.C., McBurney, D.H., and Brakeman-Wartell, S.L. (1997). Matrilateral bias in the investment of aunts and uncles: a consequence and measure of paternity uncertainty. *Human Nature*, **8**, 139–51.

Geen, R. (2004). The evolution of kinship care policy and practice. *The Future of Children*, **14**, 130–49.

Gibson, K. (2009). Differential parental investment in families with both adopted and genetic children. *Evolution and Human Behavior*, **30**, 184–9.

Golden, O. and Macomber, J.E. (2009). Framework paper: the Adoption and Safe Families Act (ASFA). In: O. Golden and J.E. Macomber (eds), *Intentions and results: a look back at the Adoption and Safe Families Act*, pp. 70–82. The Urban Institute, Washington, DC.

Hall, R. (1995). *Rape in America: a reference handbook*. ABC-CLIO, Santa Barbara, CA.

Hamilton, L., Cheng, S., and Powell, B. (2007). Adoptive parents, adaptive parents: evaluating the importance of biological ties for parental investment. *American Sociological Review*, **72**, 95–116.

Hamilton, W.D. (1964). The genetical evolution of social behavior. *Journal of Theoretical Biology*, **7**, 1–52.

Harris, M.S. and Skyles, A. (2008). Kinship care for African American children. *Journal of Family Issues*, **29**, 1013–30.

Herring, D.J. (2000). The Adoption and Safe Families Act: hope and its subversion. *Family Law Quarterly*, **34**, 329–58.

Herring, D.J. (2007a). The Multiethnic Placement Act: threat to foster child safety and well-being. *University of Michigan Journal of Law Reform*, **41**, 89–120.

Herring, D.J. (2007b). Legal scholarship, humility, and the scientific method. *Quinnipiac Law Review*, **25**, 867–85.

Herring, D.J. (2008). Kinship foster care: implications of behavioral biology research. *Buffalo Law Review*, **56**, 495–556.

Holmes, W.G. and Sherman, P.W. (1982). The ontogeny of kin recognition in two species of ground squirrels. *American Zoologist*, **22**, 491–517.

Huber, B.R. and Breedlove, W.L. (2007). Evolutionary theory, kinship, and childbirth in cross-cultural perspective. *Cross-Cultural Research*, **41**, 196–219.

Jamison, C.S., Cornell, L.L., Jamison, P.L., and Nakazato, H. (2002). Are all grandmothers equal? A review and a preliminary test of the 'grandmother hypothesis' in Tokugawa, Japan. *American Journal of Physical Anthropology*, **119**, 67–76.

Jeon, J. and Buss, D.M. (2007). Altruism toward cousins. *Proceedings of the Royal Society B*, **274**, 1181–7.

Jones, O.D. (1997). Evolutionary analysis in law: an introduction and application to child abuse. *North Carolina Law Review*, **75**, 1117–242.

Jones, O.D. (1999). Sex, culture, and the biology of rape: toward explanation and prevention. *California Law Review*, **87**, 827–941.

Jones, O.D. (2001). Time-shifted rationality and the law of law's leverage: behavioral economics meets behavioral biology. *Northwestern University Law Review*, **95**, 1141–205.

Jones, O.D. and Brosnan, S.F. (2008). Law, biology, and property: a new theory of the endowment effect. *William and Mary Law Review*, **49**, 1935–90.

Jones, O.D. and Goldsmith, T.H. (2005). Law and behavioral biology. *Columbia Law Review*, **105**, 405–502.

Jones, O.D., O'Hara, E.A., Stake, J.E. (2011). Economics, behavioral biology, and law. *Supreme Court Economic Review*, **19**, 8–38.

Krier, J.E. (2009). Evolutionary theory and the origin of property rights. *Cornell Law Review*, **95**, 139–59.

Kruger, D.J. (2003). Evolution and altruism: combining psychological mediators with naturally selected tendencies. *Evolution and Human Behavior*, **24**, 118–25.

Laham, S.M., Gonsalkorale, K., and von Hippel, W. (2005). Darwinian grandparenting: preferential investment in more certain kin. *Personality and Social Psychology Bulletin*, **31**, 63–72.

Lieberman, D., Tooby, J., and Cosmides, L. (2007). The architecture of human kin detection. *Nature*, **445**, 325–40.

Maynard Smith, J. (1982). *Evolution and the theory of games*. Cambridge University Press, New York.

McBurney, D.H., Simon, J., and Gaulin, S.J.C. (2002). Matrilateral biases in the investment of aunts and uncles: replication in a population presumed to have high paternity certainty. *Human Nature*, **13**, 391–402.

Monahan, J. (2000). Could 'law and evolution' be the next 'law and economics'? *Virginia Journal of Social Policy and the Law*, **8**, 123–8.

Park, J.H. (2007). Persistent misunderstandings of inclusive fitness and kin selection: their ubiquitous appearance in social psychology textbooks. *Evolutionary Psychology*, **5**, 860–73.

Park, J.H. and Schaller, M. (2005). Does attitude similarity serve as a heuristic cue for kinship? *Evolution and Human Behavior*, **26**, 158–70.

Park, J.H., Schaller, M., and Van Vugt, M. (2008). Psychology of human kin recognition: heuristic cues, erroneous inferences, and their implications. *Review of General Psychology*, **12**, 215–35.

Platek, S.M., Burch, R.L., Panyavin, I.S., Wasserman, B.H., and Gallup, G.G., Jr (2002). Reactions to children's faces: resemblance affects males more than females. *Evolution and Human Behavior*, **23**, 159–66.

Platek, S.M., Raines, D.M., Gallup, G.G., Jr, *et al.* (2004). Reactions to children's faces: males are more affected by resemblance than females are, and so are their brains. *Evolution and Human Behavior*, **25**, 394–405.

Posner, R.A. (2007). *Economic analysis of law*. Aspen Publishers, New York.

Ragsdale, G. (2004). Grandmothering in Cambridgeshire, 1770–1861. *Human Nature*, **15**, 301–17.

Roberts, D. (2002). *Shattered bonds: the color of child welfare*. Basic Civitas Books, New York.

Seaman, J.A. (2005). Form and (dys)function in sexual harassment law: biology, culture, and the spandrels of Title VII. *Arizona State Law Journal*, **37**, 321–433.

Sear, R., Mace, R., and McGregor, I.A. (2000). Maternal grandmothers improve nutritional status and survival of children in rural Gambia. *Proceedings of the Royal Society B*, **267**, 1641–7.

Stake, J.E. (2004). The property instinct. *Philosophical Transactions of the Royal Society B*, **359**, 1763–74.

Testa, M.F. (2005). The quality of permanence–lasting or binding? Subsidized guardianship and kinship foster care as alternatives to adoption. *Virginia Journal of Social Policy and Law*, **12**, 499–534.

Thornhill, R. (1999). The biology of human rape. *Jurimetrics Journal*, **39**, 137–47.

Thornhill, R. and Palmer, C.T. (2000). *A natural history of rape: biological bases of sexual coercion*. MIT Press, Cambridge, MA.

Tiger, L. (1969). *Men in groups*. Vintage Books, New York.

Voland, E. and Beise, J. (2002). Opposite effects of maternal and paternal grandmothers on infant survival in historical Krummhorn. *Behavioral Ecology and Sociobiology*, **52**, 435–43.

Winokur, M.A., Crawford, G.A., Longobardi, R.C., and Valentine, D.P. (2008). Matched comparison of children in kinship care and foster care on child welfare outcomes. *Families in society*, **89**, 338–46.

Zinn, A. (2009). Foster family characteristics, kinship, and permanence. *Social Service Review*, **28**, 185–213.

Section 4

Health

Chapter 16

Motivational mismatch: evolved motives as the source of—and solution to—global public health problems

Valerie Curtis and Robert Aunger

Introduction: evolutionary public health

While public health in most countries of the world is better now than it has ever been, a huge burden of preventable disease still remains. Solutions exist: we know that people at risk of HIV infection should use condoms, that children in malaria-rife countries should sleep under bednets, that we should wash our hands with soap, and that we should exercise more and eat a better diet. However, our behaviour does not seem to match our knowledge. Whilst we mostly *know* what is good for us, the problem is that we do not seem to want to *do* it. This disjuncture between what is desirable and what is health-promoting is the source of many of today's most pressing public health problems.

From an evolutionary perspective, this provides something of a puzzle. Humans have exquisitely complex brains that evolved to drive adaptive behaviour—behaviour that aided the survival and reproduction of our ancestors. These brains should be making us behave in ways that are healthy, not unhealthy. Of course, for the most part, they do. Our fear centres keep us away from predators and cliff edges, our disgust system keeps us away from parasite-ridden food and bodily wastes, our hunger motive keeps us seeking the nutrients we need, and our nurture system drives us to protect and care for our children. However, there are still many ways in which we harm our own health.

Evolutionary theory has much to contribute to public health and medicine. It helps us to understand the arms race between pathogens and hosts (e.g. HIV, malaria), the trade-offs between health and other fitness benefits (e.g. birth trauma versus infant cranial size), the functioning of our defence systems (e.g. fever and nausea), and the reasons for the diseases of aging, for example (Nesse and Williams 1995). Research in such areas is increasingly bearing fruit (Nesse et al. 2010). The proponents of Darwinian medicine have also pointed to the fact that there are major differences between our modern environments and those in which we evolved, which lead to what have been termed the 'diseases of civilization'—for example, modern diets causing obesity and excessive cleanliness causing allergies (Eaton et al. 1988; Nesse and Berridge 1997). This is called the 'mismatch' hypothesis, because it emphasizes that physiological and psychological characteristics which were adaptive in ancestral environments may be less favourable to survival and reproduction in current settings (Nesse 2004; Gluckman and Hanson 2006).

In this chapter we focus on the psychological mismatch between the environments in which we evolved and those in which we now live. We show that most current public health problems can be explained by maladaptive behaviour in the context of massive environmental changes, most having occurred since the Industrial Revolution, 150 years ago. We show how almost all of our major public health problems are associated with motivated behaviour, usually because we over- or under- use evolutionarily novel technologies. Hence, while we can trace suboptimal health to a lack of fit between our evolved motives and our current environment, understanding of these motivational drivers can help us to modify behaviour, environments, and technologies such that they generate healthier outcomes. We give an example of how ancient motives can be harnessed for the benefit of public health in the case of handwashing with soap—a novel health protective technology for which take-up is suboptimal. In short, we argue that, even if our ancestral motivations help to create public health problems, they can also help to solve them, once we understand how they work.

Public health: what's the problem?

Public health is better today than it ever has been. Infant mortality has been falling in nearly every country in the world[1] and there is no end in sight to improvements in longevity (Oeppen and Vaupel 2002). Yet we still die of avoidable causes. Today's big public health problems come in two varieties. First, a large proportion of the world still lives below the poverty line of one dollar per day and in circumstances of low public investment in healthy environments. Lacking in resources with which to grow or purchase food, with poor access to water and sanitation, and with no option but to burn solid fuels, people are prey to the deficiencies and diseases of poverty. In developing and resource-poor settings, infectious diseases are the major cause of premature mortality (65% of Africans die from infection compared to 35% of South Asians and 5% of those in Europe and the US).[2] Second, in contrast, people living in countries that have undergone the demographic transition (and increasingly, some population segments in developing countries) suffer primarily from diseases of affluence. People live longer and die of chronic, rather than infectious, causes, including cardiovascular disease, cancer, and diabetes (World Health Organization 2009).

Ezzati et al. (2002) comprehensively assessed the factors that could be modified to improve public health in high-, medium-, and low-mortality countries. Figure 16.1 shows their top 20 causes of loss of disability-adjusted life years (DALYs) globally. Top of the list came childhood and maternal underweight (associated with 9.5% of total DALYs lost), high blood pressure (4.4%), and alcohol (4.0%). However, as Figure 16.1 shows, there were major differences by region. For example, in developed regions the most important contributors to the burden of disease were tobacco (12.2%), high blood pressure (10.9%), alcohol (9.2%), high cholesterol (7.6%), and overweight (7.4%). In the high-mortality countries, which include sub-Saharan Africa and South East Asia, the leading causes of burden of disease included childhood and maternal undernutrition (14.9%), micronutrient deficiencies (3.1% for iron deficiency, 3.0% for vitamin A deficiency, and 3.2% for zinc deficiency), unsafe sex (10.2%), poor water, sanitation, and hygiene (5.5%), and indoor smoke from solid fuels (3.6%). However, high blood pressure, tobacco, and cholesterol were also in the 'top ten' in these countries.

In Table 16.1 we have collated Ezzati et al.'s (2002) top 20 global risk factors and the diseases they cause. To this, we have added columns on the factors which lead to such diseases, and, in particular, the novel technologies and behaviours underlying these factors.

Undernutrition is the leading cause of healthy life years lost across the world. This problem is primarily associated with a nexus of economic deprivation and repeated infection among mothers and children, and is improving, largely due to improvements in food availability and reductions in poverty worldwide. However, some major problems remain. Recent decades have

---

[1] See www.childinfo.org/mortality_imrcountrydata.php
[2] See www.globalhealth.org/infectious_diseases/

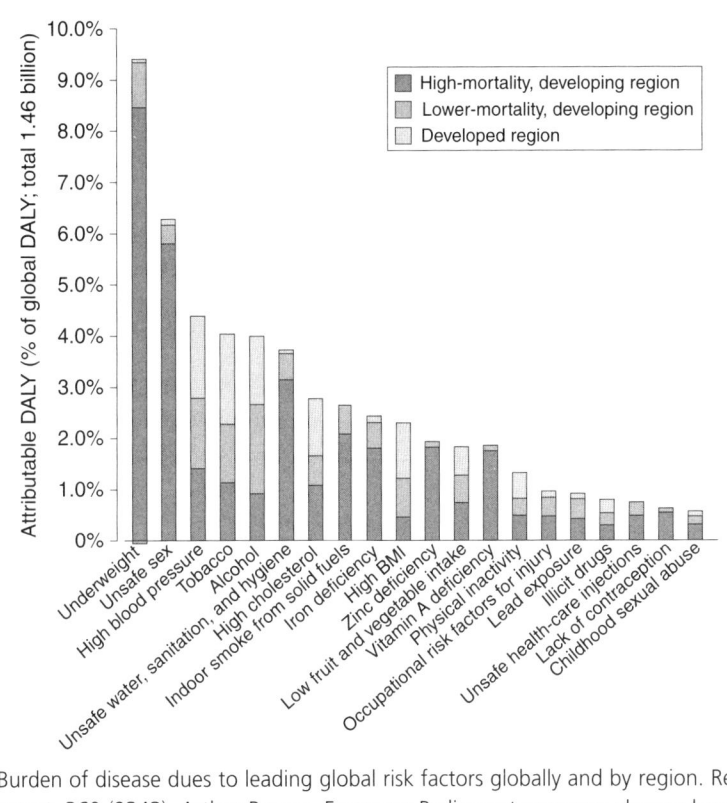

Fig. 16.1 Burden of disease dues to leading global risk factors globally and by region. Reprinted from The Lancet, 360 (9343), Arthur Rogers, European Parliament approves pharma law overhaul, pp. 1397–1398, copyright (2002), with permission from Elsevier.

seen major shifts away from exclusive breastfeeding and towards mass-produced milks and weaning foods. These more convenient novel food technologies deprive children of maternal immunoglobulins, vitamins, and protein, and expose them to environmental pathogens at an early age (Cousens et al. 1993; Black et al. 2008). Maladaptive behavioural responses to attractive new technologies (convenience foods) have been encouraged by marketing (Baumslag and Michels 1995), a recent cultural phenomenon which figures in many current public health problems.

Unsafe sex is the second most important cause of loss of DALYs, largely due to the HIV epidemic, which came about because of changes in sexual behaviour in recent decades (Quinn et al. 1986). The virus took advantage of changes in cultural values arising from rural–urban migration, driven by lifestyle changes and the use of illicit drugs (Udoh et al. 2009). Though a novel technology, the condom, is available to prevent the transmission of sexually-transmitted diseases, its use remains suboptimal, largely because condoms are unrewarding to use, interfering with the pleasure gained from our most basic reproductive motive (Valdiserri et al. 1989).

High blood pressure is the third biggest cause of loss of healthy life years. Contributory factors include diets high in refined salt, as well as inactivity and being overweight (Danaei et al. 2009). Tobacco and alcohol are the fourth and fifth causes of avoidable disease. While humans (and some animals) have always enjoyed consuming intoxicating substances, human ingenuity and the modern mass market have enabled unprecedented numbers to stimulate their reward centres cheaply and easily using refined products (Di Chiara et al. 1992; Kalivas and Volkow 2005).

Table 16.1 Risk factors for global burden of disease and contributory causes (from Ezzati et al. 2002)

Risk factor for burden of disease	Health outcomes	Contributory causes	Novel technologies	Novel behaviours	References
1. Underweight	Malnutrition, infection, low birthweight	Economic factors, recurrent infection, industrialization and mass production of food	Convenience foods*	Loss of traditional feeding practices (e.g. bottle feeding, weaning)	[1, 2]
2. Unsafe sex	STDs (HIV), cervical cancer	Rural-urban migration, social breakdown, sex industry, cultural factors	Condoms	Increased same and opposite sex promiscuity	[3, 4]
3. High blood pressure	Cardiovascular disease, stroke	Industrialization and mass production of food, sedentarization of work and leisure	Refined salt, sugar, oils, etc Labour saving and leisure technologies	Over-consumption, sedentary lifestyle	[5]
4. Tobacco	Cancer, heart disease, respiratory disease	Industrialization and mass production of cheap psychoactive drug	Tobacco high in available nicotine (cigarettes)	Smoking	
5. Alcohol	Cancer, heart disease, diabetes, depression, injuries	Industrialization and mass production of cheap psychoactive drug	Refined alcoholic drinks	Regular and binge drinking	
6. Water, Sanitation and Hygiene	Diarrhoeal disease, respiratory infection	Insufficient public/private investment in water supply and sanitation	Soap, toilet, water treatment devices	Handwashing, toilet and water filter use	[6]
7. High cholesterol	Cardiovascular disease, stroke	Industrialization and mass production of processed foods, sedentarization of work and leisure	Low density lipoproteins and trans fats	Use of processed foods, sedentary lifestyle	[5]
8. Indoor smoke	Respiratory disease	Cooking with solid fuels, house design	Improved (gas/electric) stoves	Use of solid fuels for cooking	

9. Iron; 11. Zinc; 13. Vitamin A deficiency	Anaemia, malnutrition, infection	Cereal-based diets, recurrent infection, helminth infection, early weaning	*Micronutrient supplements*	Consumption of cereals/ weaning foods
10. Overweight	Cardiovascular disease, stroke, diabetes, cancer	Industrialization and mass production of processed foods, sedentarization of work and leisure	*Refined salt/sugar/oils, labour-saving and leisure technologies*	Over-consumption, sedentary lifestyle	[7–9]
12. Low fruit and vegetable intake	Cardiovascular disease, stroke, cancer	Industrialization, mass production of processed foods	*Refined salt/sugar/oils*	Preferential consumption of processed foods
14. Physical inactivity	Cardiovascular disease, stroke, cancer	Sedentarization of work and leisure	*Labour-saving and leisure technologies*	Sedentary lifestyle
15. Occupational	Injury	Industrialization	*Industrial machinery*	Interaction with machinery
16. Lead exposure	Cardiovascular disease, mental retardation	Industrialization, mass production of automated transportation	*Cars, lorries*	Driving
17. Illicit drug use	HIV, overdose, injury, infection	Production and marketing of cheap psychoactive drugs	*Refined psychoactive compounds, syringes*	Drug consumption/ injection
18. Unsafe injections	Acute infection	Contaminated injections	*Syringes*	Syringe reuse
19. Lack of contraception	Maternal mortality	Cultural factors, lack of access	*Contraceptive technologies*	*Uptake of contraception*
20. Childhood sexual abuse	Depression, alcohol abuse	Cultural factors		

* Items in italics constitute technologies and behaviours that are, beneficial rather than detrimental to health (although they may be both). References: 1, Black et al. (2008); 2, Humphrey (2009); 3, Hamers and Downs 2004; 4, Quinn et al. (1986); 5, Danaei et al. (2009); 6, Bartram and Cairncross (2010); 7, Prentice (2006); Nesse and Williams (2005); Popkin (2003); Ezzati et al. (2003).

The communications technologies of the mass market have also allowed marketers to engineer cultural change towards making the use of such technologies socially normative. In the informal economy, the use of illicit psychoactive drugs is now widespread enough to be the seventeenth biggest burden of disease globally, again because they provide synthetic stimulus to the reward system. (Such psychoactive compounds may, however, also have had adaptive advantages, as recreational drugs can lead to promiscuity (Kurzban et al. 2010) and moderate drinking has been shown to be associated with improved longevity (Danaei et al. 2009)).

Environmental contamination is the sixth biggest cause of avoidable DALYs lost, due to a lack of clean water, sanitation, and hygiene, which affects mainly the poor in developing countries. These lead to morbidity and mortality from diarrhoea and acute respiratory infections, primarily in childhood. While public failure to invest in infrastructure is part of the problem, the failure of individuals to acquire and to use the novel technologies of soap, water filters, and toilets is also an important reason that both children and adults suffer widely from these conditions.

Regular exposure to indoor smoke, caused by cooking in poorly ventilated shelters with solid fuels like wood and coal, is eighth on the list. It is associated with avoidable respiratory ailments (Ezzati and Kammen 2001). While electrification is advancing rapidly and propane gas is becoming more widely available, the problem of failure to acquire these new technologies remains widespread in poor rural areas of developing countries.

Micronutrient deficiencies in iron, zinc, and vitamins are widespread, primarily in developing countries. They cause anaemia and malnutrition, and are attributable to multiple causes including recent changes in diet and repeated infection (Stoltzfus et al. 1997; Miller et al. 2002; Black 2003).

Being overweight is the tenth most significant avoidable risk in the world. This is caused by our novel 'obesogenic' environment which mass markets highly stimulating refined, energy-dense foodstuffs, coupled with technologies which facilitate inactive work and leisure patterns (Popkin 2003). Obesity is particularly prominent in developed countries, although it is spreading rapidly in developing countries. While these countries are still struggling with malnutrition and the diseases of poverty, modern problems of obesity, heart disease, and diabetes are also on the rise, threatening to overwhelm already stretched health services (Prentice 2006). A recent OECD (Organization for Economic Cooperation and Development) study found over 70% of Mexicans, and 50% of South Africans and Brazillians overweight, with rates in China and India increasing by as much as 1% annually (Cecchini et al. 2010).

Risk factor number 12, low vegetable and fruit intake, is at least partly due to alternative foodstuffs which are cheaper, more easily available, and more motivating to consume. If apples and cake are equally available, cake tends to be first choice (though preferences can be trained otherwise, with some effort). The 14th factor, lack of exercise, can, at least partially, be ascribed to novel technologies that make productive work less energy-consuming and sedentary leisure pursuits more attractive than active ones. Factor 15, occupational hazards, mainly concern injuries at the workplace, many of which concern the use of novel technologies (Leigh et al. 1999) (however, ancestral means of making a living may have been at least as hazardous). Factor number 16, lead exposure, can be ascribed to the use of novel transport technologies, especially private cars, which are now cheap and widely available (Fewtrell et al. 2004). Finally, syringes (number 18) are a novel form of, mainly beneficial, technology, whose re-use in resource-poor settings leads to a major burden of infection (Kermode 2004), and novel contraceptive technologies (number 19), if more widely used, could substantially reduce the burden of maternal mortality (Campbell and Graham 2006).

While one could take issue with how Ezzati et al. (2002) carve up what is a highly complex web of interlinked disease causation, and some of the factors that they leave out (e.g. risk factors for malaria, tuberculosis (TB), and depression) their widely-cited league table provides a snapshot of today's main public health problems. From Table 16.1 a striking pattern of technological and behavioural determinants of ill health emerges. The data shows that we endanger our health by consuming too much of some foods, too little of others, exercise too little, abuse psychoactive substances, and take risks with sex and cars.

Mismatched motivation

From an evolutionary perspective, the incidence and persistence of so much unhealthy behaviour is puzzling. Over evolutionary time, behavioural tendencies that lead to high morbidity and mortality should have been selected out of the gene pool (assuming they had no outweighing fitness benefits). Time, however, is exactly the problem. We have effected huge changes in the environments in which we live in just a few generations—particularly over the last 150 years or so—hardly enough time for there to have been genetic changes to affect our psychological make-up. Technological advances have made possible products such as refined sugars, edible oils and salt, psychoactive substances, labour-saving devices, and fast cars, and made them easily accessible to the populations of mass-market economies within the space of a century. Our motives have led us to create these technologies and a mass market of exchange of innovation and value creation, with ever-increasing efficiency of production and distribution (Ridley 2010). So while accelerating innovation and mass-production in formal and informal economies has brought huge gains in public health (water supply, disposable nappies, better nutrition), it has also had significant negative consequences (obesity, addiction, violent death).

The pattern is striking: almost all of the top twenty global causes of loss of DALYs can be ascribed to technological mismatches; either because of the widespread adoption of technologies with harmful effects (refined salt, sugars, oils, psychoactive compounds, sedentary leisure, guns, syringes) or the failure to adopt health-giving technologies (sanitation, soap, pills, bednets, bicycles). Why is this? Humans are equipped with a set of motives that cause us to behave in such a way as to meet our needs in the environments in which we evolved. Our ancestors—mammal, primate, pre-human, or *Homo sapiens*—needed to find food, mates, social partners, and other resources, and evolved brain systems to meet those needs (Aunger and Curtis 2008). However, those motives have also driven accelerating human innovation, leading to the invention and mass production of new technologies which have transformed living environments in all countries. Our motivated behaviour responds to these technologies and not to the ancestral objects and environments which 'designed' our brains.

The pattern is even more striking if we look at the top ten causes of burden of disease in developed economies, where nine of ten risk factors can be ascribed to this mismatch (tobacco, high blood pressure, alcohol, cholesterol, overweight, low fruit and vegetable intake, physical inactivity, illicit drugs, unsafe sex). A major feature of modern market economies is mass production. It reduces the cost of making products which are able to stimulate the senses in super-salient fashion (e.g. cigarettes, calorie-dense foods), and hence causes widespread abuse of highly rewarding products. In this way, new technologies such as refined foodstuffs and stimulants, vehicles, and communication devices, which have been designed to be attractive and motivating to use, become ubiquitous. However, acquiring such technologies to gratify these medium-term desires can have long-term health consequences, such as obesity, cardiovascular disease, addiction, and injury. Since hunger, lust, and comfort are fundamental motives, products that save energy and effort, and meet our appetites, thus providing reward, readily spread. And products that can provide reward directly, via the intake of ethanol, nicotine, or other psychoactive compounds, are especially attractive.

For example, people have smoked tobacco for thousands of years. However, there is little suggestion that lung cancer was a common problem in populations which smoked low-grade tobacco (e.g. among American Indians, Europeans). What turned the habit from a low-grade irritation into a primary carcinogen and hence public health problem was the widespread use of cigarettes after World War II (Boaz 2002). Cigarettes contain finely shredded tobacco leaves wrapped in paper. The increased surface area of tobacco being burned at higher temperature delivers a much larger nicotine hit to the lungs, which is physiologically distinct from slow-burning twists of tobacco. Nicotine releases dopamine in the brain, which makes tobacco smoke a psychoactive drug working on the brain's reward system.

What distinguishes psychoactive compounds from other classes of substances is that they provide psychological rewards independent of having achieved an evolutionary goal. Other behaviours, to produce rewards, must rely on positive feedback from the environment—either in terms of consumption of resources, or feedback in the form of recognized signs of success (e.g. a smile on the face of a fellow group member, suggesting a status improvement). With psychoactive substances, reward comes from consumption of the substance itself, which directly stimulates the reward system (Pomerleau 1997; Nesse and Williams 1998). The technology associated with cigarettes mimics the natural reward system, and subverts standard choice mechanisms in the brain.

Accelerating innovation has also made a number of health protective technologies widely available. Toilets, water filters, soap, and condoms are novel technologies and probably a good investment, even for the poorest; however, their take-up and use is suboptimal. It can be argued that this is because we have no intrinsic motivation to use them: had they been available ancestrally, the adaptive advantages they conferred might indeed have led to them becoming attractive in the same way that clean water is. The same argument could be made for the failure to comply with treatment for diseases such as leprosy or TB, to submit to influenza vaccination, or to sleep under a bednet. Getting an injection, remembering to take pills, or using a condom involves effort for which there is little immediate reward, and sometimes a disincentive—they can hurt, take time, and interfere with the joys of sex, for example. Such novel technologies were not a part of our ancestral environment, hence we have not undergone selection to find their use rewarding.

The market mirrors our motivations—where the demand is mainly for curative products which alleviate the discomfort of sickness, rather than products which prevent it. Hence, effort is invested in innovation for technologies of treatment rather than prevention, and people underutilize opportunities to vaccinate themselves, or to screen themselves for early signs of disease. The market has also failed to design and deliver technologies that can definitively rid us of many important infectious diseases, such as malaria, TB, and leishmaniasis.

The market also modifies our motivated responses. Modern marketing methods often exploit two other motives—status and affiliation. Technologies can be invested with status enhancing abilities, through celebrity endorsement for fast cars, for example (Miller 2009), and can become the norm to copy—when the cool guys or the majority of people in class seem to binge drink we may employ our 'copy the successful' and 'copy the frequent' heuristics, which also evolved for good evolutionary reasons (Richerson and Boyd 2005).

In a nutshell then, the reason that we do not behave optimally, as far as our health is concerned, is that there are alternatives to healthy behaviour that are more rewarding. Oily, salty, and fatty food is more rewarding than the alternatives and our desire to consume them preferentially has led the market to make such foods cheap and easy to access. If a healthy option exists, often it is not as intrinsically rewarding and is not widely taken up. Our once-adaptive preference for minimizing exertion has led to dramatic shifts away from energy-intensive occupational, leisure, transport, and domestic production activities (Popkin 2003). Our desires for cheap transport and for technologies of self-defence have unlooked for side effects, driving a rise in violent death due to accidents, anger, and depression. The use of alcohol, tobacco, and illicit drugs are on the rise because they are cheap and rewarding to use in the short term, but they damage health in the medium- to long-term. Condoms, bednets, soap, vaccinations, prescribed medicines, and toilets can be unrewarding to acquire and use in the short term, despite their long-term health benefits.

Evolutionary health promotion in practice

What then can be done about motivational mismatch? Are we doomed to face an ever-rising tide of obesity, addiction, and violence? Can we improve the uptake of health-giving technologies? First, we analyse one case study from this perspective—the problem of how to promote safe hygiene—and then reflect on the general applications of such evolutionary thinking to the public health problems that we have been discussing.

Much of our own work focuses on the prevention of diarrhoeal disease, the second biggest killer of children in developing countries, accounting for over 1.5 million deaths a year (Boschi-Pinto et al. 2008). Systematic reviews suggest that handwashing with soap (HWWS) is probably the most effective, and cost-effective, means of preventing this problem, cutting rates of diarrhoeal disease by 42–47% (Curtis and Cairncross 2003; Cairncross et al. 2010) and rates of respiratory infection by 23% (Ensink 2004; Rabie and Curtis 2006). One review of interventions to reduce the burden of disease in developing countries put hygiene promotion, including HWWS, as possibly the most cost-effective intervention of all (Jamieson et al. 2006).

Yet soap is a relatively novel technology, one that has only been mass-produced for about 150 years (Wilson 1954). Purchased regularly by almost all households of the world for the purposes of body and clothes washing, soap is still rarely employed on hands (Curtis et al. 2000) to prevent the faecal–oral transmission of diarrhoea-causing microbes (including *Escherichia coli*, *Salmonella*, *Shigella*, rotavirus, *Campylobacter*, *Vibrio cholera*, etc.). When asked, most people say that they wash hands with soap, however, we found that directly observed HWWS after toilet use stood at only 3% of mothers in Ghana and in rural India (Biran et al. 2009), 13% in rural China, 14% in Peru, 18% in Kyrgyzstan, and 29% in Kenya (Curtis et al. 2009). Handwashing is not so much better in the UK. In one study, we found that only 43% of mothers washed hands with soap after changing a dirty nappy. In a motorway service station, electronic counters revealed that only 32% of male and 64% of female toilet users used soap (Judah et al. 2009a). Of commuters in a sample of UK cities, 28% had bacteria of faecal origin on their hands (Judah et al. 2009b).

Over a period of 10 years we have been carrying out formative research studies to try to understand handwashing behaviour, so as to improve it. We have data from more than 12 countries in most geographical regions. A focus of the studies was to identify the motives that could be used to drive the use of soap for handwashing. We hypothesized that these would include disgust, fear, nurture, comfort, attraction, and affiliation. (Note that each motive has a technical definition according to its adaptive origins: for example, disgust as the driver of infection avoidance behaviour (Curtis et al. 2004); fear for harm avoidance from accident and violence; comfort for physiological equilibrium-seeking behaviour; nurture for child care behaviour; attraction for mate-seeking and adornment behaviour; status for social influence-seeking behaviour; and affiliation as the driver of group-adherence seeking.)

Most of the research has been qualitative (Curtis et al. 2009), but quantitative studies provide similar findings (Aunger et al. 2009). Key conclusions are surprisingly similar from country to country. Respondents almost always know of the health benefits of HWWS, but this fails to translate into practice. Key motives for HWWS were disgust, comfort, nurture, and affiliation. Physical settings, such as lack of easily available water, reduced, but did not prevent soap use.

A key role for disgust

Of all of the potential motives for HWWS, one in particular jumped out from the series of formative studies. Women everywhere said they washed their hands when they felt or smelled disgusting. They could only falteringly explain this: 'Because they are *yuk*, I can't explain, they are just *yuk*', went a typical interview. The most commonly mentioned contaminants were fish, excreta, and rotten or dead material, often of animal origin. Hands that had been in contact with faeces had to be washed. The fear of being perceived as dirty or disgusting by others was also a powerful motive for hygienic behaviour.

A series of studies on disgust helped to confirm our hypothesis that disgust evolved to drive the behaviours that prevent contact with infectious agents. The disgust system in the brain responds to cues indicating sources of infection risk in the environment, and orchestrates appropriate avoidance behaviour (Curtis and Biran 2001; Curtis et al. 2004). The system is tuned by exposure and cultural information; learning what it is best to avoid in local circumstances (Curtis et al. 2011).

Disgust should therefore be the motive that is most appropriate for the promotion of infection reduction behaviour, such as HWWS. This idea was fed into a commercial creative process to design national marketing campaigns (Curtis et al. 2007). In Ghana, the agency Lintas produced a powerful television commercial[3] depicting a mother emerging from a toilet with a purple stain on her hands—this was then transferred to the food that she prepared, and then to the child that ate it. In screenings, mothers found the advert powerful and shocking. After 6 months of a high-intensity nationwide media campaign, reported handwashing rates increased by 13% after using the toilet and by 41% before eating (Scott et al. 2007).

We further tested a variety of messages displayed at the entrance to public toilets in the UK, electronically monitoring the impact on soap use. Disgust-based messages such as 'soap it off or eat it later' and 'don't take the loo with you—wash with soap' worked significantly better than control messages (Judah et al. 2009a). Elsewhere, Porzig-Drummond et al. (2009) tested disgust-based handwashing interventions in the lab and in a public toilet, and in both situations found that the disgust motive worked better than hygiene education.

Other motives for handwashing with soap

The above studies suggested that motives other than disgust were also important drivers of hygiene behaviour. Mothers tended to do whatever everyone else in their village was doing; a typical comment was 'handwashing with soap is just not something we do around here'. Affiliation to local social norms of non-use of soap can therefore help to keep use rates low, but if the social norms support soap use, this can increase its uptake. Our public toilet experiment confirmed this effect. HWWS rates were higher at times when there were more people using the toilets and also when the message 'Is the person next to you washing hands with soap?' was displayed at the entrance. A key lesson was that public campaigns should never comment on how low soap use rates are, for fear of driving rates even lower, but should rather try to make HWWS appear common and the norm, because the affiliation motive will then drive it up (Perkins 2004).

Other potential motives that could drive increasing use of soap were status and attractiveness; however, as HWWS is not often visible in a social context, it is hard to use this motive to drive soap use. Mothers also wanted soap for its comfort value: even extremely poor families would often choose to purchase luxury bath soaps because they have a pleasing odour and do not dry the skin.

---

[3] See it at www.globalhandwashing.org/multimedia

Though traditional health education campaigns attempt to enlist fear of disease as a motivating factor, explaining to mothers the dire consequences of failure to improve their behaviour, our studies made it clear that this strategy was unlikely to work. Mothers already 'knew' of the health risks of poor hygiene, but regarded possible diarrhoea as a distant threat, one that was unlikely to be life-threatening, and one that was more often due to causes outside their control. Fear of disease only became relevant during local disease scares. HWWS rates in Kenya were unexpectedly high and were plausibly explained as a temporary response to a current cholera epidemic. Data from our public toilet research in the UK, collected during the recent H1N1 swine flu epidemic, suggested that HWWS peaked and then fell back to pre-epidemic levels, suggesting that fear responses in such epidemics may be short-lived.

While we have used evolutionary reasoning to seek for motives that might be key in driving the use of soap, necessary because it is a novel technology for which people have no intrinsic affinity, we are not the first to discover these drivers of soap use behaviour. The company Procter and Gamble (P&G) employed disgust in their early advertisements for Zest soap, where they claimed other soaps left a scummy residue, while Zest left skin 'truly clean'.[4] The comfort motive has been employed repeatedly by advertisers. For example, a 1957 advert for Unilever's Dove soap claimed that it 'doesn't dry your skin'.[5] Soap has also long been sold using the affiliation motive, suggesting you need soap to be an accepted member of society ('From your head down to your toe, a daily bath with Lifebuoy will stop B.O.'[6] or to be attractive ('Don't wait to be told, you need Palmolive Gold').[7] P&G's advertisements for Camay soap make the attraction motive even more explicit.[8] Finally, soap companies also recognized that soap could be sold using the nurture motive, mother's desire to care for and groom their children (see, for example, Johnson and Johnson's Indian baby milk soap advert).[9]

In many countries the conclusions of our formative research on HWWS were fed into a process based on commercial marketing, where creative professionals were briefed to develop interventions based on the motives that we identified, that could be applied on a mass, a community, or a family/individual basis, depending on the available channels of communication and budget (Curtis et al. 2007). Results were encouraging, with substantial measurable improvements in handwashing rates (Curtis et al. 2001; Scott et al. 2007).

The problem of novel technologies

Soap is an example of a novel health-enhancing technology. While hygienic behaviour evolved before humans did (being manifest throughout the animal kingdom), soap was invented only recently (by the Babylonians, Egyptians, or in the Middle Ages, depending on which authority you consult: Curtis 2007; Smith 2007), and has only become commercially available to the majority in the last hundred or so years. There is no intrinsic, evolved motivation to use soap—the advantage it confers is too recent to be reflected in brains. The problem is, therefore, how to make rubbing onto the skin a bar of sodium stearate (plus additives), then rinsing it off, a rewarding and hence motivating activity. Our efforts, and those of commercial marketers, have shown how soap use can become motivating to help avoid disgust and shame, as an aid to nurturing children,

---

[4] See www.youtube.com/watch?v=_96T_DRNNW8andNR=1
[5] See www.youtube.com/watch?v=SMtqXC20D8g
[6] See www.youtube.com/watch?v=astrjgUhc2Iandfeature=related
[7] See www.youtube.com/watch?v=cfP-wASMikQ
[8] E.g. http://www.youtube.com/watch?v=CLrNXz55k4wandNR=1
[9] See http://www.youtube.com/watch?v=PZTZIkC46Gk

and as an aid to affiliation via social norms. These efforts create new mental associations between ancient motives and new technologies. We are now engaged in an industrial design process in pursuit of new hand-cleansing technologies that are intrinsically more motivating to use (by making a more convenient or attractive product). Through a combination of an available, appealing product and effective promotion targeted at key motivations, as well as support from soap companies, the hope is that HWWS may become a normative behaviour in society, no longer dependent on the persistent efforts of health promoters.

Are there general lessons that can be drawn from this work that can be applied to other new health technologies, such as pills, injections, condoms, bednets, and toilets? We believe so. For example, consumer research showed that toilets might best be marketed, not for their health benefit (as governments and non-governmental organizations do at present), but for new values such as *status*, *comfort*, reduced *fear* of snakes or attacks at night, and through avoidance of *disgusting* faecal matter in open defecation fields (Jenkins and Curtis 2005). Research into the low uptake of insecticide-treated bednets showed that they might be better marketed as an aid to the *comfort* of a good night's sleep rather than for the health benefit they might confer (Guiguemde et al. 1994). Condom marketers have long realized that health messages are not the best way to sell condoms. The Durex company, for example, now aim to sell them as an aid, rather than as a deterrent, to sexual attractiveness.[10]

Beating mismatch

If it is possible to attach new motivations to products for which we have no intrinsic affinity, might it also be possible to use the same approach to discourage the unhealthy behaviours which are the source of most ill health in developed countries? Potentially. Novel technologies which have unhealthy consequences when consumed excessively (e.g. cigarettes) can also become associated with new motivational values. The British Heart Foundation advertisement series which associated disgust with cigarettes was thought to have been highly effective.[11] It has been suggested, however, that realization of the effects of smoking on the health of others was the main reason for its steep decline in recent years. Campaigners and legislators pointed out the injustice of harming others and imposed smoking bans in public areas, relying on the human need to affiliate to drive cessation (Christakis and Fowler 2008; although harmful technologies can, of course, also be made less attractive through public policy, such as by increasing sales tax or restricting access).

Take another public health problem: obesity. Most of the food types that dominate present diets were introduced quite recently: dairy products, cereal grains (especially refined grains that lack germ and bran), refined sugars (especially sucrose and fructose), refined vegetable oils (with low ω-3 and high ω-6 fatty acids), alcoholic beverages, refined salt, and ω-6 saturated, fatty, acid-rich, mammalian meats. These foods have displaced the wild plant and animal foods of our predecessors. Research shows that rats' brains react to these sweet, fatty foods in the same way that addicts' brains respond to cocaine. Thus 'conditioned hypereating' (Kessler 2009) works the same way as other 'stimulus response' disorders in which reward is involved, such as substance abuse. Furthermore, it has been suggested that some food companies are developing products that trigger compulsive overeating (Power and Schulkin 2009).

One solution that has been advocated is a return to ancestral diets (high in fibre, low in salt, carbohydrate, and fat: Milton 2000, 2002). However, such foods are outcompeted in supermarket

---

[10] See www.youtube.com/watch?v=yyahoTR1Otkandfeature=fvst
[11] See www.youtube.com/watch?v=ef3gofQcOKk

baskets by highly-motivating, highly-processed, super-stimulating, calorific foodstuffs, supported by sophisticated marketing employing motives such as status. Motivated by the threat of legislation and pressure from consumers, some global food companies are now investing effort in designing products that are both healthy and motivating. Drinks containing artificial sweeteners rather than sugar, and prepared fruit snacks that make fruit easier to consume, are early examples of what looks set to become a major trend.[12] Marketers can appeal to nurture and affiliation motives to make feeding healthier food (at least to children) both rewarding and normative.

Modern marketing is thus a social invention that can be used not only by commercial businesses to promote unhealthy products, but also by companies—and public health programmes—to make their health-promoting messages more effective at changing behaviour on a large scale. Industries can also turn their attention to developing products that meet the unmet needs of the poorest who are currently excluded from the benefits of modern technologies. Cheap (but still attractive) technologies such as water filters, soap substitutes, and insecticide-impregnated bednets can be designed and successfully marketed to the large consumer base at 'the bottom of the pyramid' (Prahalad 2005).

Conclusion: evolutionary public health

The idea that a 'mismatch' between ancestral conditions and modern lifestyles can lead to health problems is not new (Eaton et al. 1988; Williams and Nesse 1991). However, to date, there has been no systematic analysis of motivational mismatch as it applies to modern health problems, nor of the implications of this analysis for action to improve health.

Here we have seen that of the top ten risk factors for loss of DALYs, six (unsafe sex, high blood pressure, tobacco, alcohol, high cholesterol, overweight) are mainly due to mismatch and for the other four (underweight, iron deficiency indoor smoke, lack of water sanitation and hygiene), mismatch plays a part. Of the top 20 risk factors, 13 are directly due to mismatch and mismatch plays a part in the most of the rest. In developed market economies, fully nine out of ten of the main risk factors for loss of health can be attributed to motivational mismatch (tobacco, high blood pressure, alcohol, cholesterol, overweight, low fruit and vegetable intake, physical inactivity, illicit drugs, unsafe sex).

From an evolutionary public health perspective, these health problems come in three categories. The first is lack of uptake of health-giving technologies such as sufficient foods, micronutrient supplements, sanitation, soap, contraception, condoms, and cooking stoves. Poverty and underdevelopment is part of the reason why health-improving technologies are not more widely used, but another is that many of these technologies are evolutionarily novel and not intrinsically motivating to acquire or use, even if they are available. The second category is the overconsumption of highly-motivating novel technologies with direct ill-effects (tobacco, alcohol, psychoactive drugs, and foods high in salt, fat, and carbohydrate). These are intrinsically rewarding (or mimic the brain's reward system). The third category contains motivating novel technologies with harmful side effects such as labour-saving means of production, leisure, and transport, which reduce physical activity and sometimes cause injury (e.g. by producing environmental toxins, or by introducing infection on re-use in the case of syringes).

Societies are increasingly moving away from conditions that resemble the ancestral environments in which our motivational systems evolved, towards those with modern industrial

---

[12] See, for example www.sustainable-living.unilever.com/the-plan/nutrition/

economies and plentiful novel technologies, favouring the diseases of mismatch. Finding solutions to these problems is thus becoming more urgent. Poorer countries, particularly, increasingly face a double burden: not having yet cured the diseases of poverty, they simultaneously face inexorable rises in the panoply of modern health problems: cardiovascular disease, diabetes, cancer, and substance abuse.

So does this diagnosis, based in evolutionary public health, offer us new solutions to these intractable problems? We have suggested that they do. The link between novel technologies and psychological reward underpins our argument that public health interventions must either curb an evolved motivation exploited by a problematic novel technology, or associate the use of a health-beneficial technology with some new reward which increases its level of use. For example, cigarette smoking, which provides artificial reward directly to the brain, can be curbed by linking that practice to disgust or to disreputable people. Or use of a condom can be linked to the rewarding notion of being what a 'real man' does. In this way, a mismatch between some evolved motive and a novel technology which currently leads to a public health problem can be 'matched' with a different reward to help solve the problem.

Our analysis also suggests that health can be improved, not just by focusing on behaviour, but by improving the technologies on offer. For the poorest, more can be done to find cheap and attractive technologies that meet basic needs (protection from insect vectors of disease, improved simple toilets, new hand-cleansing technologies). For the better-off, more can be done to make healthy options more attractive (active sports, healthier food products, alternatives to smoking and drinking). Modern marketing techniques have much to offer public health practitioners (Curtis et al. 2007). Consumers and regulatory authorities will increasingly provide the carrots and sticks that will give a competitive advantage to those manufacturers of consumer products who strive to enhance health.

Taking an evolutionarily informed approach to public health thus has a number of benefits. It allows us to see public health in a long-term perspective, showing how patterns of disease and behaviour have changed as we have modified our settings. It highlights how our evolved motives have led us to create a world that is much better at meeting our needs than it ever has been (Ridley 2010). This has had major health benefits, but also given us the diseases of mismatch that now are amongst our biggest global health problems. We have argued that understanding the evolved motivational drivers of behaviour gives us a useful perspective, not just into the reasons why we fail to behave healthily, but also into means of promoting safer behaviour. Our lesson for the public health practitioner is this: motives got us into this mess but they can also get us out, if only we systematically understand the ways in which motivational mismatch works.

Acknowledgements

Thanks to Mícheál de Barra, Gaby Judah and Samantha Highsmith for helpful comments on a previous draft of this chapter.

References

Aunger, R. and Curtis, V. (2008). Kinds of behaviour. *Biology and Philosophy*, **23**, 317–45.

Aunger, R., Schmidt, W., Ranpura, A., *et al.* (2009). Three kinds of psychological determinants for hand-washing behaviour in Kenya. *Social Science and Medicine*, **70**, 383–91.

Bartram, J. and Cairncross, S. (2010). Hygiene, sanitation, and water: forgotten foundations of health. *PLoS Medicine*, **7**, e1000367.

Baumslag, N. and Michels, D. (1995). *Milk, money, and madness: the culture and politics of breastfeeding*. Bergin and Garvey, Westport, CT.

Biran, A., Schmidt, W.-P., Wright, R., et al. (2009). The effect of a soap promotion and hygiene education campaign on handwashing behaviour in rural India: a cluster randomised trial. *Tropical Medicine and International Health*, **14**, 1303–14.

Black, R.E. (2003). Zinc deficiency, infectious disease and mortality in the developing world. *Journal of Nutrition*, **133**, 1485S–9S.

Black, R., Allen, L., Bhutta, Z., Caulfield, L., de Onis, M., and Ezzati, M. (2008). Maternal and child undernutrition: global and regional exposures and health consequences. *Lancet*, **371**, 243–60.

Boaz, N. (2002). *Evolving health: the origins of illness and how the modern world is making us sick*. Wiley, New York.

Boschi-Pinto, C., Velebit, L., and Shibuya, K. (2008). Estimating child mortality due to diarrhoea in developing countries: a meta-analysis review. *Bulletin of the World Health Organization*, **86**, 710–17.

Cairncross, S., Hunt, C., Boisson, S., et al. (2010). Water, sanitation and hygiene for the prevention of diarrhoea. *International Journal of Epidemiology*, **39**, 193–205.

Campbell, O.M. and Graham, W.J. (2006). Strategies for reducing maternal mortality: getting on with what works. *Lancet*, **368**, 1284–99.

Cecchini, M., Sassi, F., Jeremy, A.L., Yong, Y.L., Veronica, G.-B., and Daniel, C. (2010). Tackling of unhealthy diets, physical inactivity, and obesity: health effects and cost-effectiveness. *Lancet*, **376**, 1775–84.

Christakis, N.A. and Fowler, J.H. (2008). The collective dynamics of smoking in a large social network. *New England Journal of Medicine*, **358**, 2249–58.

Cousens, S., Nacro, B., Curtis, V., et al. (1993). Prolonged breast-feeding: no association with increased risk of clinical malnutrition in young children in Burkina Faso. *Bulletin of the World Health Organization*, **71**, 713–22.

Curtis, V. (2007). Dirt, disgust and disease: a natural history of hygiene. *Journal of Epidemiology and Community Health*, **61**, 660–4.

Curtis, V. and Biran, A. (2001). Dirt, disgust, and disease: is hygiene in our genes? *Perspectives in Biology and Medicine*, **44**, 17–31.

Curtis, V. and Cairncross, S. (2003). Effect of washing hands with soap on diarrhoea risk in the community: a systematic review. *Lancet Infectious Diseases*, **3**, 275–81.

Curtis, V. A., Cairncross, S., and Yonli, R. (2000). Domestic hygiene and diarrhoea, pinpointing the problem. *Tropical Medicine and International Health*, **5**, 22–32.

Curtis, V., Kanki, B., Cousens, S., et al. (2001). Evidence for behaviour change following a hygiene promotion programme in West Africa. *Bulletin of the World Health Organization*, **79**, 518–26.

Curtis, V., Aunger, R., and Rabie, T. (2004). Evidence that disgust evolved to protect from risk of disease. *Proceedings of the Royal Society B*, **271**, S131–133.

Curtis, V., Garbrah-Aidoo, N., and Scott, B. (2007). Masters of marketing: bringing private sector skills to public health partnerships. *American Journal of Public Health*, **97**, 634–41.

Curtis, V., Danquah, L., and Aunger, R. (2009). Planned, motivated and habitual hygiene behaviour: an eleven country review. *Health Education Research*, **24**, 655–73.

Curtis, V., deBarra, M., and Aunger, R. (2011). Disgust as an adaptive system for disease avoidance behaviour. *Philosophical Transactions of the Royal Society B*, **366**, 389–401.

Danaei, G., Ding, E. L., Mozaffarian, D., et al. (2009). The preventable causes of death in the United States: comparative risk assessment of dietary, lifestyle, and metabolic risk factors. *PLoS Medicine*, **6**, e1000058.

Di Chiara, G., Acquas, E., and Carboni, E. (1992). Drug motivation and abuse: a neurobiological perspective. *Annals of the New York Academy of Sciences*, **654**, 207–19.

Eaton, S. B., Shostak, M., and Konner, M. (1988). *The Paleolithic prescription: a program of diet and exercise and a design for living*. Harper Collins, New York.

Ensink, J. (2004). *Health impact of handwashing with soap*. In: WELL Factsheets. London.

Ezzati, M. and Kammen, D.M. (2001). Indoor air pollution from biomass combustion and acute respiratory infections in Kenya: an exposure-response study. *Lancet*, **358**, 619–24.

Ezzati, M., Lopez, A.D., Rodgers, A., Vander Hoorn, S., and Murray, C.J.L. (2002). Selected major risk factors and global and regional burden of disease. *Lancet*, **360**, 1347–60.

Ezzati, M., Vander Hoorn, S., Rodgers, A., et al. (2003). Estimates of global and regional potential health gains from reducing multiple major risk factors. *Lancet*, **362**, 271–80.

Fewtrell, L., Prüss-Üstün, A., Landriganc, P., and Ayuso-Mateos, J.L. (2004). Estimating the global burden of disease of mild mental retardation and cardiovascular diseases from environmental lead exposure. *Environmental Research*, **94**, 120–33.

Gluckman, P. and Hanson, M. (2006). *Mismatch: the lifestyle diseases timebomb*. Oxford University Press, Oxford.

Guiguemde, T.R., Dao, F., Curtis, V., et al. (1994). Household expenditure on malaria prevention and treatment for families in the town of Bobo-Dioulasso, Burkina Faso. *Transactions of the Royal Society of Tropical Medicine and Hygiene*, **88**, 285–7.

Hamers, F. and Downs, A. (2004). The changing face of the HIV epidemic in western Europe: what are the implications for public health policies? *Lancet*, **364**, 83–94.

Humphrey, J.H. (2009). Child undernutrition, tropical enteropathy, toilets, and handwashing. *Lancet*, **374**, 1032–5.

Jamieson, D., Bremen, J., Measham, A., Alleyne, G., and Claeson, M. (2006). *Disease control priorities in developing countries*. Oxford University Press, Oxford.

Jenkins, M., and Curtis, V. (2005). Achieving the 'good life': why some people want latrines in rural Benin. *Social Science and Medicine*, **61**, 2446–59.

Judah G., Aunger, R., Schmidt, W.P., Michie, S., Granger, S., and Curtis, V. (2009a). Experimental pretesting of hand-washing interventions in a natural setting. *American Journal of Public Health*, **99**, S405–411.

Judah, G., Donachie, P., Cobb, E., Schmidt, W., Holland, M., and Curtis, V. (2009b). Dirty hands: bacteria of faecal origin on commuters' hands. *Epidemiology and Infection*, **138**, 409–14.

Kalivas, P., and Volkow, N. (2005). The neural basis of addiction: a pathology of motivation and choice. *The American Journal of Psychiatry*, **162**, 1403–13.

Kermode, M. (2004). Unsafe injections in low-income country health settings: need for injection safety promotion to prevent the spread of blood-borne viruses. *Health Promotion International*, **19**, 95–103.

Kessler, D. (2009). *The end of overeating*. Rodale Press, New York.

Kurzban, R., Dukes, A., and Weeden, J. (2010). Sex, drugs and moral goals: reproductive strategies and views about recreational drugs. *Proceedings of the Royal Society B*, **277**, 3501–8.

Quinn, T., Mann, J., Curran, J., and Piot, P. (1986). AIDS in Africa: an epidemiologic paradigm. *Science*, **234**, 955–63.

Leigh, J., Macaskill, P., Kuosma, E., and Mandryk, J. (1999). Global burden of disease and injury due to occupational factors. *Epidemiology and Infection*, **10**, 626–31.

Miller, G. (2009). *Spent: sex, evolution and consumer behaviour*. Viking Press, London.

Miller, M., Humphrey, J., Johnson, E., Marinda, E., Brookmeyer, R., and Katz, J. (2002). Why do children become Vitamin A deficient? *Journal of Nutrition*, **132**, 2867S–2880S.

Milton, K. (2000). Back to basics: why foods of wild primates have relevance for modern human health. *Nutrition*, **16**, 481–3.

Milton, K. (2002). Hunter-gatherer diets: wild foods signal relief from diseases of affluence. In: P. Ungar and M. Teaford (eds), *Human diet: its origins and evolution*, pp. 111–22. Bergin and Garvey, Westport, CT.

Nesse, R.M. (2004). Natural selection and the elusiveness of happiness. *Philosophical Transactions of the Royal Society B*, **359**, 1333–47.

Nesse, R.M., Bergstrom, C.T., Ellison, P.T., *et al.* (2010). Making evolutionary biology a basic science for medicine. *Proceedings of the National Academy of Sciences of the USA*, **107**, 1800–7.

Nesse, R.M. and Berridge, K.C. (1997). Psychoactive drug use in evolutionary perspective. *Science*, **278**, 63–6.

Nesse, R.M. and Williams, G.C. (1995). *Evolution and healing*. Weidenfeld and Nicolson, London.

Nesse, R.M. and Williams, G.C. (1998). Evolution and the origin of disease. *Scientific American*, **29**, 86–93.

Oeppen, J. and Vaupel, J. (2002). Broken limits to life expectancy. *Science*, **296**, 1029–31.

Perkins, H.W. (2004). *The social norms approach to prevention*. Jossey Bass, San Francisco, CA.

Pomerleau, C.S. (1997). Co-factors for smoking and evolutionary psychobiology. *Addiction*, **92**, 397–408.

Popkin, B. (2003). The nutrition transition in the developing world. *Development Policy Review*, **21**, 581–97.

Porzig-Drummond, R., Stevenson, R., Case, T., and Oaten, M. (2009). Can the emotion of disgust be harnessed to promote hand hygiene? Experimental and field-based tests. *Social Science and Medicine*, **68**, 1006–12.

Power, M.L. and Schulkin, J. (2009). *The evolution of obesity*. Johns Hopkins University Press, Baltimore, MD.

Prahalad, C.K. (2005). *The fortune at the bottom of the pyramid: eradicating poverty through profit.* Wharton School Publishing, Upper Saddle River, NJ.

Prentice, A.M. (2006). The emerging epidemic of obesity in developing countries. *International Journal of Epidemiology*, **35**, 93–9.

Rabie, T. and Curtis, V. (2006). Handwashing and risk of respiratory infections: a quantitative systematic review. *Tropical Medicine and International Health*, **11**, 269–78.

Richerson, P. and Boyd, R. (2005). *Not by genes alone: how culture transformed human evolution*. Chicago University Press, Chicago, IL.

Ridley, M. (2010). *The rational optimist: how prosperity evolves*. Fourth Estate, London.

Scott, B., Schmidt, W., Aunger, R., Garbrah-Aidoo, N., and Animashaun, R. (2007). Marketing hygiene behaviours: the impact of different communications channels on reported handwashing behaviour of women in Ghana. *Health Education Research*, **22**, 225–33.

Smith, V.S. (2007). *Clean: a history of personal hygiene and purity*. Oxford University Press, New York.

Stoltzfus, R., Chwaya, H., Tielsch, J., Schulze, K., Albonico, M., and Savioli, L. (1997). Epidemiology of iron deficiency anemia in Zanzibari school children: the importance of hookworms. *The American Journal of Clinical Nutrition*, **65**, 153–9.

Udoh, I.A., Mantell, J., Sandfort, T., and Eighmy, M. (2009). Potential pathways to HIV/AIDS transmission in the Niger Delta of Nigeria: poverty, migration and commercial sex. *AIDS Care*, **21**, 567–74.

Valdiserri, R., Arena, V., Proctor, D., and Bonati, F. (1989). The relationship between women's attitudes about condoms and their use: implications for condom promotion programs. *American Journal of Public Health*, **79**, 499–501.

Williams, G. and Nesse, R. (1991). The dawn of Darwinian medicine. *Quarterly Review of Biology*, **66**, 1–22.

Wilson, C. (1954). *The history of Unilever: a study in economic growth and social change*. Cassel, London.

World Health Organization (2009). *Global health risks: mortality and burden of disease attributable to selected major risks*. World Health Organization, Geneva.

Chapter 17

Mental health and well-being: clinical applications of Darwinian psychiatry

Alfonso Troisi

Introduction

I was recently invited to give a lecture on Darwinian psychiatry at a major meeting of clinical psychiatrists. The audience was quite large and mostly included clinicians who were engaged daily in diagnosing and treating patients with mental disorders. At the end of my talk, I asked two questions of the audience, receiving individual replies via televoting. The first question was 'Does the evolutionary approach make an important theoretical contribution to clinical psychiatry?'. The great majority (87%) of the participants replied 'Yes'. The second question was 'Is Darwinian psychiatry relevant for your clinical practice?'. Only 8% of the participants gave an affirmative response.

The results of this snapshot survey do not reflect the idiosyncratic opinion of a group of conformist clinicians. Rather, they accurately illustrate the general attitude toward the evolutionary approach in the medical and psychiatric literature (McCrone 2003). For example, Samuel Guze, one of the most influential educators and researchers in contemporary clinical psychiatry, wrote: 'Unless evolutionary explanations are tied directly to genetic and physiological knowledge of why some people get sick in certain ways while others do not, they are too vague and general to be useful in medicine' (1992, p. 92). More recently, Randolph Nesse, one of the founders of the emerging field of evolutionary medicine, has acknowledged that at present there are no evolutionary-based treatments for mental disorders and has concluded that evolutionary biology's main contribution to psychiatry is that of offering a new conceptual framework for integrating findings from the many disciplines which study the human mind and behaviour (Nesse 2005).

Sceptics have sound arguments against the clinical utility of Darwinian psychiatry. As a branch of medicine, psychiatry has as its main objectives the diagnosis, treatment, and prevention of mental disorders. Yet these clinical aspects have not been a major focus of research and discussion in evolutionary psychiatry. Almost certainly, this is the most important reason that can explain why the evolutionary approach has had so little impact on clinical psychiatry (Troisi and McGuire 2006). When evolutionary psychiatrists suggest ultimate explanations of vulnerability to psychopathology based on phenomena such as evolutionary trade-offs (different traits are adaptive in different environments), genome lag (evolved traits may be out of step with modern environments), and historical constraints (how a trait has evolved may result in particular susceptibility to disease), they certainly offer a deeper understanding of the origin of psychiatric disorders. However, this is not enough for clinical psychiatrists. When people experiencing mental distress arrive at the hospital or psychiatrist's office, what they need is a proper diagnosis and an effective treatment; evolutionary hypotheses and explanations, as currently presented by most articles and books on Darwinian psychiatry, do not seem to be of great help to address patients' needs.

In this chapter, contrary to the belief of sceptics, I will try to convince the reader that Darwinian psychiatry has great potential in terms of clinical utility. The term 'clinical utility' refers to a variety of functions that help clinicians to conceptualize diagnostic entities, communicate medical information to patients and their families, choose effective interventions, and predict treatment response, complications, and course of illness (First et al. 2004). My ambition is not to demonstrate that the evolutionary approach can match the practical applications of those disciplines that are revolutionizing contemporary psychiatry, such as, for example, brain imaging or psychopharmacology. This is not only untrue for the present but also highly improbable for the future. More modestly, I suggest that the inclusion of evolutionary theories and methods within the clinician's know-how can improve, to some extent, the diagnosis, treatment, and prevention of mental disorders.

Diagnosis

Diagnosis is the first step of the clinical process. In medicine, diagnostic reasoning consists of collecting and analysing all the information necessary for identifying a disease. Compared to other medical disciplines, psychiatry has devoted much attention to the issue of diagnostic assessment, and the question of how to improve the reliability and validity of psychiatric diagnoses continues to be debated in the research and clinical literature. A major focus of the current debate is on the necessity to include empirical data from genetics and neuroscience into the diagnostic criteria of psychiatric disorders (Insel et al. 2010). A neuroscience-based approach to psychiatric diagnosis will surely be useful to refine classification schemes, improve prognostic predictions, and develop better treatment strategies. In other fields of medicine, laboratory tests and data related to pathophysiological mechanisms are routinely used to reach a diagnosis. By contrast, psychiatric diagnoses essentially reflect information collected through clinical observation and are based solely upon clinical manifestations (i.e. presenting signs and symptoms.)

Clinical observation suffers from inevitable limitations in terms of diagnostic accuracy. However, it is also true that psychiatrists have not made full use of its potential. When examined from an evolutionary perspective, clinical observation of people with mental disorders should be revised with regard to what information the clinician should collect, how information should be sought, and where information should be gathered. The changes suggested by an evolutionary approach derive from the application of the concepts and methods of ethology and behavioural ecology to diagnostic reasoning. Ethology and behavioural ecology are the disciplines that, since the second half of the 20th century, have applied evolutionary principles to the study of animal and human behaviour, a scientific field that was previously the exclusive domain of psychology. Unlike psychologists, ethologists and behavioural ecologists emphasize the importance of direct observation and measurement of behaviour in natural settings. In addition, they call attention to the functional consequences of behaviour for biological adaptation. Let us take a look at the changes that the application of these ideas would cause in psychiatric diagnosis. Table 17.1 lists the 'instructions' to put these changes into practice.

What to collect?

Current diagnostic criteria for psychiatric disorders are based on signs (e.g. psychomotor retardation, bizarre behaviour, or pseudoseizures) and symptoms (e.g. depressed mood, delusions, hallucinations, or obsessive ideas). Such an emphasis on symptom profiles makes sense because signs and symptoms are the most proximal indicators of a disorder. However, from an evolutionary viewpoint, a clinical assessment that focuses exclusively on signs and symptoms is limited, only partially explaining features of mental disorders. Like mentally healthy people, individuals with

Table 17.1 Darwinian psychiatry provides the clinician with a list of instructions to revise current methods of collecting diagnostic information. The acronym GOAL may serve as a mnemonic for remembering the list.

Give less weight to symptoms	G
Observe and measure behaviour	O
Assess functional capacities	A
Leave your office	L

mental disorders act to optimize the achievement of short-term goals and their behaviour reflects interactions between their strategies, their functional capacities, and environmental contingencies. With few exceptions, the motivations and goals of persons with and without mental disorders do not appear to differ. Rather, the difference between the two groups lies in the capacity of enacting efficient strategies to achieve biological goals. Whatever else they are, mental disorders are conditions of failed functions. It follows that an accurate assessment of functional capacities is essential to the development of precise diagnostic evaluation and outcome measurement (McGuire and Troisi 1998).

According to Darwinian psychiatry, clinical assessment should focus primarily on functional capacities and person–environment interactions. At this point, a brief clarification of how Darwinian psychiatry conceptualizes functional capacities is needed. Humans have been designed by natural selection to strive for the achievement of specific short-term goals, such as acquiring resources, making friends, developing social support networks, having high status, attracting a mate, and establishing intimate relationships. In ancestral environments, the achievement of these short-term goals correlated consistently with a gene-transmitting advantage, the ultimate goal of any evolved strategy (Troisi and McGuire 1998). In many respects, humans no longer live in an environment for which they were adapted. Because the modern world is so different from ancestral environments, the ancestral-fitness consequences of evolved strategies may no longer be realized. Nevertheless, the capacity to achieve short-term biological goals remains a valid measure of mental health because it is an indication that the individual possesses those optimal functional capacities that, in ancestral environments, promoted biological adaptation.

In theory, contemporary psychiatry is not blind to the importance of assessing functional capacities, and its 'diagnostic bible', the DSM (*Diagnostic and Statistical Manual of Mental Disorders*; American Psychiatric Association 2000), includes a separate axis for measuring the patient's overall level of functioning. In practice, psychiatrists rarely use measures of functional disability during routine clinical assessment, and there has been a recent proposal that 'no functioning or disability should appear as part of the threshold of diagnosis' for any disorder (Ustun and Kennedy 2009, p. 82). According to this view, symptoms alone are sufficient to diagnose a mental disorder. Underestimation of the diagnostic importance of functional assessment may in part derive from the conviction that symptom criteria already include social role impairment. If symptoms and disability were strongly correlated and if functional impairments were always secondary to symptoms, then the inclusion of functional assessment into the diagnostic process would be redundant. Yet, the redundancy hypothesis is questioned by abundant evidence showing that functioning and symptoms hold a tenuous relationship and that functional impairments may either cause or be caused by symptoms (McKnight and Kashdan 2009).

For example, treatment outcome studies of depressed patients have shown that change in social functioning lags behind change in depression symptoms. Patients who present with symptoms and functional impairment before treatment often show clinically meaningful change in symptoms but relatively little change in social functioning afterwards (Hirschfeld et al. 2002). In some studies, social disability persisted up to 4 years longer than depressive symptoms (Bothwell and Weissman 1977). Similar findings have been reported for patients with schizophrenia (Wunderink et al. 2009). Schizophrenic patients have pronounced deficits in social competence; these impairments in social functioning are largely independent of the signs and symptoms currently used to diagnose schizophrenia (Horan et al. 2010). By directly observing schizophrenic patients' behaviour, investigators have found only weak correlations between psychotic symptom ratings and social skills required for effective functioning (Gilbert et al. 2000). Other studies have shown that deficits in social competence manifest many years before the onset of psychiatric symptoms among adolescents at risk for schizophrenia (Schiffman et al. 2004).

The findings reported above (and others omitted for brevity) show that the relationship between symptoms and functioning is complicated and that the severity of psychiatric symptoms is often a poor predictor of the degree of functional impairment. The Darwinian concept of mental disorder (Troisi and McGuire 2002) encourages clinicians to consider re-prioritizing their selection of diagnostic criteria to ensure that the focus shifts away from mere symptom profiles and toward a comprehensive data collection programme that includes functional capacities.

How to collect

Psychiatric diagnosis takes place within the context of psychiatric interview, a conversation conducted by the clinician in which facts and statements are elicited from the patient. The principles that inspire psychiatric interview largely derive from the psychological tradition. The clinician aims at gaining knowledge of a patient's thoughts, emotions, and feelings. The main sources of information for arriving at a diagnostic hypothesis are patient reports concerning subjective experiences. The neuroscience revolution also holds the promise of enriching information about the psychological state of the patient with data obtained through genetic tests, brain imaging, or neurophysiological records. However, even if this will be possible in the next few years, an important component of psychiatric phenotypes will remain largely neglected: the patient's behaviour.

Ultimately, the ingredients of what we call 'psychiatric disorders' belong to three different domains: the psychological domain (what the patients thinks and feels), the physiological domain (the bodily changes associated with the disorder), and the behavioural domain (what the patient does in everyday life). Until recently, ignorance about the organic (mostly neural) changes that underlie mental disorders has brought psychological aspects to the centre of the diagnostic stage. Now that neuroscience is gaining a dominant role in psychiatric research, clinicians are turning to physiological correlates of mental disorders with increasing interest. But who cares about behaviour? Contemporary psychiatry seems to be squeezed between mind and brain, with no room for a rigorous analysis of behavioural aspects.

From an evolutionary point of view, neglecting behaviour is a major methodological error. Behaviour is crucial for biological adaptation, and many different evolutionary disciplines including sociobiology, behavioural ecology, Darwinian anthropology, and human ethology focus on behaviour as the primary object of study. While some evolutionary psychologists have argued that behaviour itself does not contribute to fitness and that the psychological mechanisms that produce behaviour should be the focus of evolutionary analysis (e.g. Symons 1989), and although such an emphasis on psychological processes to the detriment of behavioural analysis is in line with prevailing psychiatric thinking, I consider such a concordance potentially dangerous because it could immunize psychiatry against the evolutionary critiques that stress its weakness in assessing the actual behaviour of persons with mental disorders.

Someone could object that my claim that psychiatrists do not base their diagnoses on patients' behaviour is unfounded. Expert clinicians look at the patient's appearance, posture, tone of voice, language, and way of dressing to get information about the patient. And the use of rating scales to obtain quantitative data on patients' behaviour is widespread in clinical practice. These objections miss a crucial point: to observe and measure behaviour with the methods of evolutionary disciplines is not as straightforward as the application of everyday knowledge or the use of scales that translate subjective impressions into crude behavioural ratings, sometimes of the type 'better/worse' or 'much/less'.

Human ethology has developed distinctive and powerful methods for observing and measuring non-verbal behaviour during clinical interviews. Non-verbal communication is important in all human communication and indeed also in clinical practice. During the clinical interview, non-verbal information is often sent, perceived, and interpreted unconsciously. For the detection and interpretation of this information, clinicians draw on everyday knowledge and their own past experiences with people's responses in and to the clinical setting. By contrast, the ethological approach involves direct observation, coding, and analysis of behaviour patterns. Rigorous description is combined with a theoretical emphasis on those behaviours that are most closely related to adaptive functioning (Troisi 1999). A brief and selective review of the findings of the ethological studies of non-verbal communication during interviews shows the clinical utility of such an alternative method of data collection.

Non-verbal behaviour of depressed patients has been studied to determine whether their behaviour predicts response to antidepressant drugs and recurrence of depression. In one study, prior to drug treatment, responders and non-responders did not differ with respect to sex, age, education, diagnostic subtype, and severity of depression. In contrast, ethological profiles at baseline differed, with drug non-responders showing significantly more assertive and affiliative behaviours (Troisi et al. 1989). Another study (Bos et al. 2007) showed that problems in non-verbal communication are related to higher subsequent exposure to stressful life events, and, via this route, to higher risk of recurrence of depression. Lack of congruence between the levels of non-verbal involvement of remitted depressed outpatients and their conversation partners was predictive of the subsequent occurrence of stressful life events, particularly of events that were interpersonal in nature. Stressful interpersonal events in turn were predictive of recurrence of depression. Interesting findings have also emerged from ethological studies of persons with schizophrenia. In these patients, facial expressivity during clinical interview is a better predictor of work and social functioning in the real world than negative symptoms (i.e. those symptoms that are currently considered the best clinical predictor of functional disability) (Troisi et al. 2007).

These findings have several implications. First, they show that the ethological recording of patients' behaviour may yield different results from those obtained using psychiatric rating scales or clinical observation. Second, they suggest that, when there is a discrepancy between the verbal and the non-verbal message, the message conveyed by the latter should take precedence. While verbal communication contains opinions and facts, the non-verbal component more strongly communicates feelings and social motivations. Because the verbal information may be censored by the patient, the non-verbal communication may be more 'true' than the verbal component. Finally, these findings cast doubt on the validity of routine psychiatric assessments and suggest caution in basing important clinical decisions (e.g. when to discharge a patient, whether to increase drug dosage) exclusively on patients' reports of their symptoms.

Where to collect?

A basic principle of the evolutionary study of behaviour is that the assessment of functional capacities cannot be properly made without consideration of the environment in which the individual lives. Observing people in natural settings allows an understanding of the relationship of specific behaviours to the capacity of dealing with real-life situations. The majority of functions that are of interest to psychiatry are carried out in the social arena. Because features of the social environment change, carrying out the same function often requires the use of different strategies and capacities. Thus, behaviour and its outcomes need to be assessed on a moment-to-moment basis (Shiffman et al. 2008). For example, the efficiency of a social signal is defined in part by the response of the person receiving the signal. In short, Darwinian psychiatry draws clinicians' attention to the issue of ecological validity of information collected during routine diagnostic assessment.

The artificial settings where mental health assessments generally occur (the hospital ward, the consulting room, or the psychiatrist's office) may give a distorted picture of what happens to patients in their natural environment. For example, depressed individuals can engender negative mood and rejection in those with whom they interact. Roommates of depressed college students like their roommates less and tend to reject them more than do roommates of non-depressed subjects, and spouses of depressed partners report feeling more depressed, hostile, and critical following interactions with their partners. However, studies focusing on interactions of depressed people with strangers have failed to replicate these findings. Thus, the negative quality of depressed persons' interactions will fully emerge only in specific social environments, namely, those in which there is a close relationship among participants (Hale et al. 1997; Nezlek et al. 2000).

Ideally, to take advantage of the full potential of the ecological approach to diagnosis, diagnostic assessment should include observations carried out in patients' natural habitats. Even though such an approach has been employed in ethological studies of healthy people, this is admittedly difficult to do with psychiatric patients for ethical and practical reasons. Nevertheless, it is important to be aware that patients' behaviour in the hospital or the psychiatrist's office is not necessarily representative of their behaviour in other contexts. It is encouraging that clinicians from other fields of medicine have arrived at the same conclusion and are striving to develop new technologies allowing collection of diagnostic data in the real world (e.g. Boriani et al. 2008; Lin et al. 2010).

Therapy

Delineating the approach

To be clinically relevant, any new theory or hypothesis applied to medicine must have therapeutic implications. When patients consult a psychiatrist, they want the doctor to tell them the cause of their illness, talk to them about their condition, and let them know how long their illness will last, but, above all, they want the doctor to cure their symptoms and make the final decision about their treatment plan (Channa and Siddiqui 2008). In my introduction, I made reference to the admission of the evolutionary psychiatrist Randolph Nesse that, unfortunately, there are no evolutionary-based treatments for mental disorders. Thus, the question arises 'Can the evolutionary approach contribute to improve treatment of mental disorders?'. The answer depends on how therapy is conceptualized.

The prevailing metaphor of medicine is that of the body as a machine that the doctor is called upon to fix when it breaks. The doctor's role is that of an engineer who uses technology (i.e. therapeutic tools) to reverse the pathways leading to machine malfunctioning (i.e. the pathogenic

Table 17.2 How Darwinian psychiatry rethinks therapy: summary points

Adaptive symptoms should be treated only if their suppression does not increase the probability of unfavourable outcomes
Therapy should aim at improving the patient's chances of achieving short-term biological goals
Persons with intact functional capacities who experience dysfunctional states should be helped to identify the environmental and personal constraints that interfere with achieving short-term goals
Persons with suboptimal functional capacities due to trait variation should be helped to refine suboptimal functional capacities
Persons with suboptimal functional capacities should be helped to develop the use of alternative capacities that improve the likelihood of achieving high-priority goals
Persons with suboptimal functional capacities should be advised to actively search for environments in which they are most likely to achieve high-priority goals
Clinical manifestations reflecting evolved adaptive strategies should be distinguished from those that are caused by compromised functional capacities. Treatment of adaptive strategies should be avoided if they do not cause discomfort to the individual or those in the individual's environment

mechanisms of disease) (Childs 1999). Within such a theoretical framework, I see no role for an evolutionary approach to therapy of mental disorder. It is highly improbable that evolutionary studies will lead to the discovery of a new drug to eliminate psychotic symptoms of schizophrenia, a prenatal kit to switch off the susceptibility genes for autism, or a new psychotherapy technique to reverse the cognitive distortions underlying depressive personality. Yet, if we accept a broader concept of the aims of therapeutic interventions, the answer to the question above is more optimistic. According to an alternative view of medicine, the aim of therapy is not only to reverse the pathogenesis (i.e. proximate mechanisms) of illness but also to restore the congruence between a patient's individuality and the conditions of the environment (Childs 1999). If therapy is conceived of in these terms, the therapeutic relevance of Darwinian psychiatry emerges clearly (McGuire and Troisi 1998). Darwinian psychiatry can contribute to the strategic aspects of therapy, not to its tactical execution (i.e. the actual means used to gain the therapeutic objectives). Table 17.2 summarizes the basic points of psychiatric therapy as conceived by an evolutionary perspective. A detailed explanation of each of these points can be found elsewhere (McGuire and Troisi 1998; Troisi and McGuire 1998).

Symptomatic therapy

Based on the metaphor of the broken machine, the signs and symptoms of disease are just epiphenomena that reflect the underlying pathological process. For example, the symptoms of type I diabetes are direct consequences (epiphenomena) of an insufficient production of insulin caused by autoimmune destruction of insulin-producing beta cells of the pancreas. There is nothing good in the symptoms of diabetes; frequent urination, weight loss, and increased thirst are just manifestations of a disrupted mechanism.

Yet, not all medical symptoms are the same: many manifestations of disease are sophisticated adaptations, not just epiphenomena of a broken machine (Nesse and Williams 1994). When classified from an evolutionary perspective, symptoms can be divided into two broad categories: symptoms as defects in the body's mechanisms, and symptoms as useful defences. For example, seizures, jaundice, coma, and paralysis have apparently no adaptive function and arise from defects in the organism. But many other manifestations of disease are defences. Vomiting eliminates toxins from the stomach. The low iron levels associated with chronic infection limit the growth of pathogens. Coughing clears foreign matter from the respiratory tract.

The distinction between defects and defences is of paramount importance for therapeutic decision-making. The suppression of adaptive symptoms may worsen the course of illness and lead to unfavourable outcomes. The risk is highest with those treatments that alleviate symptoms without reversing the pathogenic mechanisms causing the disease (the so-called symptomatic or palliative treatments). Empirical evidence for the adaptive function of symptoms and the negative effects of their suppression through symptomatic therapy comes mostly from studies of infectious diseases (Blumenthal 1997; Oppenheimer 2001). Similar evidence for psychiatric symptoms is sparse, but preliminary data suggest that the distinction between defects and defences is also important in psychiatry.

Keller and Nesse (2006) have recently introduced and tested a new framework for understanding the evolutionary origin of depressive symptoms. Their hypothesis (called the 'situation-symptom congruence (SCC)' hypothesis) predicts that, if different depressive symptoms serve different functions, then different events that precipitate a depressive episode should give rise to different symptom patterns that increase the ability to cope with the adaptive challenges specific to each situation. Thus, for example, crying should be especially prominent when social bonds are threatened, lacking, or lost; decreased ability to experience positive emotions should be prominent when the environment is unpropitious; and pessimism should arise when future efforts are unlikely to succeed. These predictions are based on the different adaptive functions that each of these depressive symptoms serves in the regulation of behaviour and psychological processes. Crying elicits comforting behaviours and strengthens social bonds; positive emotions facilitate approach behaviour and increase risk-taking; pessimism dissuades the individual from pursuing current and potential goals. The SSC hypothesis was tested by asking 445 participants to identify depressive symptoms that followed a recent adverse situation. Guilt, rumination, fatigue, and pessimism were prominent following failed efforts; crying, sadness, and desire for social support were prominent following social losses. These significant differences were replicated in an experiment in which 113 students were randomly assigned to visualize a major failure or the death of a loved one.

These findings raise the question of whether drug-induced suppression of symptoms in some forms of depression or in some patients may have potentially deleterious effects. In effect, there is an ongoing debate in the psychiatric literature about the correctness of treating persons who have been recently bereaved (Zisook and Kendler 2007). The *British National Formulary* (British Medical Association 2000) and the Committee on Safety of Medicines (1998) advise against using benzodiazepines after bereavement and suggest that these anxiolytic compounds may inhibit the grieving process. However, this advice appears to be based on anecdote. The only controlled trial found no evidence of a positive or negative effect of benzodiazepines on the course of bereavement, except that those subjects who received drug treatment appeared less likely to have resolution of sleep problems in the weeks after bereavement (Warner et al. 2001). Stronger evidence for the potential risks of symptomatic therapy comes from studies of post-traumatic stress disorder (PTSD).

PTSD is an anxiety disorder that can develop after exposure to a terrifying event or ordeal in which grave physical harm occurred or was threatened. Traumatic events that may trigger PTSD include violent personal assaults, natural or human-caused disasters, accidents, or military combat. Since the distressing acute symptoms following traumatic experiences deserve effective and rapid alleviation, pharmacological interventions in the post-stress acute phase are common. However, some prospective studies (Gelpin et al. 1996; Mellman et al. 2002) have shown that there may be an increased incidence of PTSD in individuals treated with anxiolytics immediately after exposure to trauma. The deleterious effects of acute phase treatments are likely to be caused by their interference with the acute stress response, a set of physiological and psychological mechanisms that evolved to cope with traumatic events in the natural environment. Matar et al. (2009) tested this hypothesis in a rodent model, finding that a brief course with alprazolam (an anxiolytic drug) in the immediate aftermath of stress-exposure is associated with less favourable responses to additional, subsequent stress-exposure. Drug treatment was associated with a significant attenuation of corticosterone levels, suggesting a possible link between disruption of the initial hypothalamic–pituitary–adrenal axis response and subsequent unfavourable outcomes.

The preceding discussion of the potential risks of palliative therapy of psychiatric symptoms invites two conclusive reflections. First, the evolutionary warning against palliative therapy should not be interpreted as a medical version of the naturalistic fallacy (i.e. what is natural and adaptive is inherently good) (Troisi 2005). If suppression of adaptive symptoms does not increase the probability of unfavourable outcomes, then symptomatic therapy is not only appropriate but also mandatory because one of the basic aims of medicine is to alleviate suffering. Second, the scarcity of empirical evidence to distinguish whether or not a given psychiatric symptom is an adaptive defence should persuade evolutionary psychiatrists to spend more time and energy in testing hypotheses in clinical samples rather than indulging in post hoc invention of adaptive explanations for virtually every disorder (Troisi 2006).

Prevention

Prevention is the holy grail of medicine and psychiatry. Diagnosis and therapy are important for helping patients and fulfilling their expectations but both take place after the disease process has already struck the individual. Prevention is much more; it holds the promise of reducing or eliminating the risk of getting sick through the application of a body of knowledge concerning the causal factors that set in motion the pathogenetic process. In particular, primary prevention aims at avoiding the development of a disease by removing modifiable risk factors involved in its aetiology.

Contemporary medicine puts a major emphasis on primary prevention, and the success of population-based campaigns against infectious diseases in the past (Anonymous 1937) and against lung cancer and coronary heart disease more recently (Day 2010), supports the optimism of those who view the multi-factorial aetiology of most human pathologies as the only limit to the triumph of preventive medicine.

Compared to other fields of medicine, psychiatry seems to be a step back, as attested by minor changes over time in the prevalence rates of mental disorders (Kessler et al. 2005). However, epidemiological studies of major psychiatric disorders have identified a number of modifiable risk factors that could be minimized by focused preventive interventions (Tandon et al. 2010; Gottlieb et al. 2011). Given this state of play, what has evolutionary psychiatry to offer to primary prevention of mental disorders?

Unlike diagnosis and therapy, prevention can be discussed from an evolutionary perspective without provoking perplexed reactions from clinicians who are unfamiliar with evolutionary thinking. Many of the evolutionary insights on preventive interventions are easily understandable, or even already known, to practising psychiatrists. Studying physiology during their medical training, doctors learn that our bodies are designed to function under specific environmental conditions, and that we have a limited capacity to adjust to altered environments. Psychiatrists have no difficulty in applying these principles to mental health. For example, the finding that we need satisfying social relationships for good mental and physical health (Heinrich and Gullone 2006; Holt-Lunstad et al. 2010) can be easily accommodated within the same theoretical framework that explains why we need adequate vitamin intake. Of course, there are ultimate causes that account for our necessity for both vitamins and social relationships, but a detailed knowledge of our evolutionary history is not required for implementing programmes to prevent scurvy or depression caused by loneliness.

An innovative and original contribution that Darwinian psychiatry can offer to the prevention of mental disorders is the identification of those modifiable risk factors that were absent in ancestral environments but common in modern environments. This is the mismatch hypothesis (see also Chapter 16, this volume), and, based on feedback from colleagues, it is the evolutionary proposition that enjoys highest popularity among practising psychiatrists. Among medical disorders, a well-known example is obesity. In ancestral environments, high-calorie foods were rarely available and the selective power of famine favoured the evolution of eating habits appropriate to starvation; that is, seeking out and devouring any available high-calorie food. In addition, carriers of genes encoding efficient mechanisms of energy extraction and storage from dietary sources had an advantage during times of food shortage. In contemporary affluent societies, individuals carrying these genes are now being exposed to sedentary lifestyles and fat-rich diets. Therefore, obesity can be viewed as the maladaptive outcome of a lack of fit of 'Stone Age genes' with new nutritional patterns and lifestyle changes (Eaton and Eaton 2000). Among psychiatric disorders, substance abuse and addiction are good examples of maladaptive conditions resulting from the mismatch between the modern environment and evolved psychological traits (Troisi 2001b). Under natural conditions, brain reward systems were activated when (and only when) the individual was pursuing or achieving a goal relevant to biological adaptation. In contrast, drugs of abuse (a novelty in evolutionary terms) directly interact with specific receptors in the brain that normally help mediate feelings of satisfaction and pleasure associated with the execution of adaptive behaviours. Direct chemical stimulation of these receptors creates a signal in the brain that indicates, falsely, the achievement of biological goals (Nesse and Berridge 1997).

Obesity and drug abuse are straightforward examples of the mismatch hypothesis but do not permit full appreciation of its implications for prevention. These implications become clear when one understands that evolutionary insights into modifiable risk factors may sometimes depart considerably from common sense and conventional clinical wisdom, and that their application can literally revolutionize primary prevention, as in the case of the impact of attachment theory on paediatric hospitalization and prevention of childhood emotional disorders.

Until the late 1950s, professional child care workers maintained institutional practices that assigned little importance to a child's relationship with the mother or primary caregiver (Kobak and Madsen 2008). If a young child needed to be hospitalized, it was standard to prevent or severely restrict parental visitation. In the United Kingdom, during the 1940s and 1950s, parents were allowed to visit their sick children in the hospital for only one hour per week (Karen 1994). The physiological needs of the children, particularly for food and warmth, were placed ahead of their need for an affectionate relationship with their mothers. The prevailing view was that if a child was fed by a variety of caregivers, the relationship with mother would hold no special significance for the child. Yet, the reality was dramatically different. James Robertson used films to demonstrate the devastating impact of then-current hospital policy on the emotional well-being of young children (Robertson 1953). The films documented how the disruption of children's bonds with their parents resulted in severe emotional distress and desperate efforts to find the missing parents. These early efforts to document young children's reaction to hospitalization were initially met with a great deal of disbelief and hostility from professional audiences. To many child care workers, young children's apparent distress at being separated from their parents in a well-managed hospital setting could easily be dismissed as unrealistic and immature. In absence of a theory explaining the origin and complexity of the mother–child bond, professionals stressed the physical needs of children and paid little attention to the emotional effects of separation.

John Bowlby (1969/1982) developed such a theory. In search for a paradigm alternative to psychoanalysis and learning theory, he discovered the field of ethology, with its roots in naturalistic observation and evolutionary biology. Inspired by the evolutionary perspective, Bowlby showed that the attachment system was genetically 'wired' into human nature through intense directional selection during evolutionary history. The reason why the attachment system evolved and remains so deeply ingrained in the human brain and behaviour is that it provided a good solution to one of the most daunting adaptive problems our ancestors faced: how to survive through the most perilous years of social and physical development (Simpson and Belsky 2008). Guided by the incorrect assumptions of psychoanalysis and learning theory, professionals and childcare workers downplayed the significance of mother–child separation and ignored the emotional needs of hospitalized children. Growing acceptance of the attachment theory in the 1960s eventually altered hospital practice and signalled, with unmistakable clarity, the mismatch between professionals' prejudices and the real needs of young children. Thanks to Bowlby's work, successful prevention of emotional disorders related to paediatric hospitalization became feasible.

The example of attachment theory illustrates a general principle that guides the application of evolutionary psychology to prevention of mental disorders. Effective prevention requires a detailed understanding of psychological stressors (i.e. those life events that elicit negative emotions and can precipitate the onset of a psychiatric disorder). Yet, mental health professionals lack a body of knowledge about normal emotional functioning comparable to the understanding physiology offers to general medicine (Nesse 2005). Often, cultural prejudices and evolutionary analysis weigh up the congruence between a patient's individuality and the conditions of the environment in a very different way. Based on evolutionary principles such as sexual selection and life history theory, Darwinian psychiatry sheds new light on individual values, goals, and reactivity to life situations (Troisi 2008). When a life event interferes with achieving a biological goal, its harmful impact will depend primarily on the importance of the goal to an individual, and the importance attached to the same biological goals may differ according to age and sex. This can explain why, for example, women are more vulnerable than men to the stressful impact of infertility and interpersonal negative events (Troisi 2001a).

Conclusions

Reflecting on the reasons that have delayed the incorporation of evolutionary knowledge into the fields of mental health and clinical psychiatry, and on the ways to speed up the process, Nesse (2005, p. 919) wrote: 'All it would take is discovery of a single cure. Even discovery of the definitive cause for a single illness would do'. I agree with Nesse that it is unlikely that clinicians will embrace the evolutionary approach until Darwinian psychiatry can prove its practical utility on a vast scale and refocus its attention on clinical decision-making. However, even now, there are many clinical applications of the evolutionary approach that testify its practical utility. This chapter has dealt with those activities that clinicians exert in their everyday practice, which those suffering from mental disorders expect to progress at a fast pace: diagnosis, therapy, and prevention. Each of these clinical tasks can already benefit from the contributions of evolutionary theory. This is not surprising if one agrees with Kurt Lewin (1951, p. 169) that 'there is nothing so practical as a good theory'. And evolutionary theory is something more than a good theory.

References

American Psychiatric Association. (2000). *Diagnostic and statistical manual of mental disorders (4th ed., text rev.)*. APA, Washington, DC.

Anonymous. (1937). A triumph of preventive medicine. *American Journal of Public Health and the Nation's Health*, **27**, 182–3.

Blumenthal, I. (1997). Fever–concepts old and new. *Journal of the Royal Society of Medicine*, **90**, 391–4.

Boriani, G., Diemberger, I., Martignani, C., et al. (2008). Telecardiology and remote monitoring of implanted electrical devices: the potential for fresh clinical care perspectives. *Journal of General and Internal Medicine*, **23**(Suppl), 73–7.

Bos, E.H., Bouhuys, A.L., Geerts, E., van Os, T.W., and Ormel, J. (2007). Stressful life events as a link between problems in nonverbal communication and recurrence of depression. *Journal of Affective Disorders*, **97**, 161–9.

Bothwell, S. and Weissman, M.M. (1977). Social impairments four years after an acute depressive episode. *The American Journal of Orthopsychiatry*, **47**, 231–7.

Bowlby, J. (1969/1982). *Attachment and loss, Vol. 1., Attachment.* Basic Books, New York.

British Medical Association and Royal Pharmaceutical Society of Great Britain. (2000). *British National Formulary* (September). BMJ Books and Pharmaceutical Press, London and Wallingford.

Channa, R. and Siddiqui, M.N. (2008). What do patients want from their psychiatrists? A cross-sectional questionnaire based exploratory study from Karachi. *BMC Psychiatry*, **8**, 14.

Childs, B. (1999). *Genetic medicine. A logic of disease.* The John Hopkins University Press, Baltimore, MD.

Committee on Safety of Medicines. (1998). *Current problems* (21). *Guidance on benzodiazepines*. London, Committee on Safety of Medicines and the Medicines and Healthcare Products Regulatory Agency.

Day, M. (2010). Lifestyle change helped to halve number of Americans dying from heart disease. *British Medical Journal*, **341**, c4198.

Eaton, S.B. and Eaton, S.B. III. (2000). Paleolithic vs. modern diets: selected pathophysiological implications. *European Journal of Nutrition*, **39**, 67–70.

First, M.B., Pincus, H.A., Levine, J.B., Williams, J.B., Ustun, B., and Peele, R. (2004). Clinical utility as a criterion for revising psychiatric diagnoses. *American Journal of Psychiatry*, **161**, 946–54.

Gelpin, E., Bonne, O., Peri, T., Brandes, D., and Shalev, A.Y. (1996). Treatment of recent trauma survivors with benzodiazepines: a prospective study. *Journal of Clinical Psychiatry*, **57**, 390–4.

Gilbert, E.A., Liberman, R.P., Ventura, J., et al. (2000). Concurrent validity of negative symptom assessments in treatment refractory schizophrenia: relationship between interview-based ratings and inpatient ward observations. *Journal of Psychiatric Research*, **34**, 443–7.

Gottlieb, L., Waitzkin, H., and Miranda, J. (2011). Depressive symptoms and their social contexts: a qualitative systematic literature review of contextual interventions. *International Journal of Social Psychiatry*, **57**, 402–417.

Guze, S.B. (1992). *Why psychiatry is a branch of medicine.* Oxford University Press, New York.

Hale, W.W. III., Jansen, J.H., Bouhuys, A.L., Jenner, J.A., and van den Hoofdakker, R.H. (1997). Non-verbal behavioral interactions of depressed patients with partners and strangers: the role of behavioral social support and involvement in depression persistence. *Journal of Affective Disorders*, **44**, 111–22.

Heinrich, L.M. and Gullone, E. (2006). The clinical significance of loneliness: a literature review. *Clinical Psychology Review*, **26**, 695–718.

Hirschfeld, R., Dunner, D.L, Keitner, G., et al. (2002). Does psychosocial functioning improve independent of depressive symptoms? A comparison of nefazodone, psychotherapy, and their combination. *Biological Psychiatry*, **51**, 123–33.

Holt-Lunstad, J., Smith, T.B., and Layton, J.B. (2010). Social relationships and mortality risk: a meta-analytic review. *PLoS Medicine*, **7**, e1000316.

Horan, W.P., Rassovsky, Y., Kern, R.S, Lee, J., Wynn, J.K., and Green, M.F. (2010). Further support for the role of dysfunctional attitudes in models of real-world functioning in schizophrenia. *Journal of Psychiatry*, **44**, 499–505.

Insel, T., Cuthbert, B., Garvey, M., *et al.* (2010). Research domain criteria (RDoC): toward a new classification framework for research on mental disorders. *American Journal of Psychiatry*, **167**, 748–51.

Karen, R. (1994). *Becoming attached*. Warner, New York.

Keller, M.C. and Nesse, R.M. (2006). The evolutionary significance of depressive symptoms: Different adverse situations lead to different depressive symptom patterns. *Journal of Personality and Social Psychology*, **91**, 316–30.

Kessler, R.C., Demler, O., Frank, R.G., *et al.* (2005). Prevalence and treatment of mental disorders, 1990 to 2003. *New England Journal of Medicine*, **352**, 2515–23.

Kobak, R. and Madsen, S. (2008). Disruptions in attachment bonds. Implications for theory, research, and clinical intervention. In: J. Cassidy and P.R. Shaver (eds), *Handbook of attachment. Theory, research, and clinical applications 2nd Edition*, pp. 23–47. The Guilford Press, New York.

Lewin, K. (1951). *Field theory in social science; selected theoretical papers*. D. Cartwright (ed.), Harper and Row, New York.

Lin, C.T., Ko, L.W., Chang, M.H., *et al.* (2010). Review of wireless and wearable electroencephalogram systems and brain-computer interfaces – a mini-review. *Gerontology*, **56**, 112–19.

Matar, M.A., Zohar, J., Kaplan, Z., and Cohen, H. (2009). Alprazolam treatment immediately after stress exposure interferes with the normal HPA-stress response and increase vulnerability to subsequent stress in an animal model of PTSD. *European Neuropharmacology*, **19**, 283–95.

McCrone, J. (2003). Darwinian medicine. *Lancet Neurology*, **2**, 516.

McGuire, M.T. and Troisi, A. (1998). *Darwinian psychiatry*. Oxford University Press, New York.

McKnight, P.E. and Kashdan, T.B. (2009). The importance of functional impairment to mental health outcomes: a case for reassessing our goals in depression treatment research. *Clinical Psychology Review*, **29**, 243–59.

Mellman, T.A., Bustanmante, V., David, D., and Fins, A. (2002). Hypnotic medication in the aftermath of trauma. *Journal of Clinical Psychiatry*, **63**, 1183–4.

Nesse, R.M. (2005). Evolutionary psychology and mental health. In: D. Buss (ed.), *Handbook of evolutionary psychology*, pp. 903–27. Wiley, Hoboken, NJ

Nesse, R.M. and Berridge, K.C. (1997). Psychoactive drug use in evolutionary perspective. *Science*, **278**, 63–6.

Nesse, R.M. and Williams, G.C. (1994). *Why we get sick: the new science of Darwinian medicine*. Random House, New York.

Nezlek, J.B., Hampton, C.P., and Shean, G.D. (2000). Clinical depression and day-to-day social interaction in a community sample. *Journal of Abnormal Psychology*, **109**, 11–19.

Oppenheimer, S. (2001). Iron and its relation to immunity and infectious disease. *The Journal of Nutrition*, **131**, 616S–635S.

Robertson, J. (1953). *A two-year-old goes to hospital: a scientific film record*. Concord Film Council, Nacton, UK.

Schiffman, J., Walker, E., Ekstrom, M., Schulsinger, F., Sorensen, H., and Mednick, S. (2004). Childhood videotaped social and neuromotor precursors of schizophrenia: a prospective investigation. *American Journal of Psychiatry*, **161**, 2021–7.

Shiffman, S., Stone, A.A., and Hufford, M.R. (2008). Ecological momentary assessment. *Annual Review of Clinical Psychology*, **4**, 1–32.

Simpson, J.A. and Belsky, J. (2008). Attachment theory within a modern evolutionary framework. In: J. Cassidy and P.R. Shaver (eds), *Handbook of attachment. Theory, research, and clinical applications 2nd Edition*, pp. 131–57. The Guilford Press, New York.

Symons, D. (1989). A critique of Darwinian anthropology. *Ethology and Sociobiology*, **10**, 131–44.

Tandon, R., Nasrallah, H.A., and Keshavan, M.S. (2010). Schizophrenia, 'Just the Facts' 5. Treatment and prevention. Past, present, and future. *Schizophrenia Research*, **122**, 1–23

Troisi, A. (1999). Ethological research in clinical psychiatry: the study of nonverbal behavior during interviews. *Neuroscience and Biobehavioral Review*, **23**, 905–13.

Troisi, A. (2001a). Gender differences in vulnerability to social stress: a Darwinian perspective. *Physiology and Behavior*, **73**, 443–9.

Troisi, A. (2001b). Harmful effects of substance abuse: a Darwinian perspective. *Functional Neurology*, **16**, 237–43.

Troisi, A. (2005). The concept of alternative strategies and its relevance to psychiatry and clinical psychology. *Neuroscience and Biobehavioral Review*, **29**, 159–68.

Troisi, A. (2006). Adaptationism and medicalization: The Scylla and Charybdis of Darwinian psychiatry. *Behavioral and Brain Sciences*, **29**, 422–3.

Troisi, A. (2008). Psychopathology and mental illness. In: C. Crawford and D. Krebs (eds), *Foundations of evolutionary psychology*, pp. 453–74. Lawrence Erlbaum., New York

Troisi, A. and McGuire, M.T. (1998). Evolution and mental health. In: H.S. Friedman (ed.), *Encyclopedia of mental health*, vol. 2, pp. 173–81. Academic Press, San Diego, CA.

Troisi, A. and McGuire, M.T. (2002). Darwinian psychiatry and the concept of mental disorder. *Neuroendocrinology Letters*, **23**(suppl 4), 31–8.

Troisi, A. and McGuire, M.T. (2006). Darwinian psychiatry: it's time to focus on clinical questions. *Clinical Neuropsychiatry*, **3**, 85–6.

Troisi, A., Pasini, A., Bersani, G., Grispini, A., and Ciani, N. (1989). Ethological predictors of amitriptyline response in depressed outpatients. *Journal of Affective Disorders*, **17**, 129–36.

Troisi, A., Pompili, E., Binello, L., and Sterpone, A. (2007). Facial expressivity during the clinical interview as a predictor of functional disability in schizophrenia. A pilot study. *Progress in Neuropsychopharmacolology and Biological Psychiatry*, **31**, 475–81.

Ustun, B. and Kennedy, C. (2009). What is 'functional impairment'? Disentangling disability from clinical significance. *World Psychiatry*, **8**, 82–5.

Warner, J., Metcalfe, C., and King, M. (2001). Evaluating the use of benzodiazepines following recent bereavement. *British Journal of Psychiatry*, **178**, 36–41.

Wunderink, L., Sytema, S., Nienhuis, F.J., and Wiersma, D. (2009). Clinical recovery in first-episode psychosis. *Schizophrenia Bulletin*, **35**, 362–9.

Zisook, S. and Kendler, K.S. (2007). Is bereavement-related depression different than non-bereavement-related depression? *Psychological Medicine*, **37**, 779–94.

Chapter 18

Evolutionary perspectives on sport and competition

Diana Wiedemann, Robert A. Barton, and Russell A. Hill

Introduction

Sports are ubiquitous in human cultures and are valued as a form of physical competition (Chick and Loy 2001; Llaurens et al. 2009a). At the same time, watching sport is a pastime that captivates viewers across the world, with major sporting events such as the FIFA World Cup or Olympic Games attracting audiences of hundreds of millions, both as live spectators, and as television, Internet, and radio viewers or listeners. The worldwide appeal of sports such as football has led economists and psychologists to devote an increasing body of research to this global phenomenon (Kocher and Sutter 2010). Recently, there has also been a dramatic increase in research examining sport from an evolutionary perspective.

This chapter describes research into evolutionary aspects of human behaviour in sports, illustrating how evolutionary insights have been applied and have enriched our understanding of human sports and competition, and points to promising ideas for further study. We focus on four issues that have received recent attention. First, we describe the relationship between testosterone levels and sporting outcomes and how they play an important role in the phenomenon of home advantage. Second, we explore the issue of home advantage and its psychological and hormonal mediators in more detail. Third, we elucidate possible explanations for why left-handedness may be advantageous in physical combats in westernized but also traditional societies. Finally, we describe and assess influences of colour on human and animal agonistic behaviour.

Testosterone and human competition

Androgens are hormones involved in human male competition, antisocial norm breaking, dominant, and aggressive behaviour (Elias 1981; Susman et al. 1987; Mazur and Booth 1998). The androgen testosterone is a steroid hormone responsible for the development and maintenance of masculine features, but it is also found, in lesser amounts (in most species), in females. Studies of non-human primates have shown that a male's status and testosterone levels are linked: elevated testosterone levels are recorded when males achieve high status, but decline again when status is lost (Eberhardt et al. 1980; McGuire et al. 1986; Setchell et al. 2008). Similar effects can be found in human competitions; sporting winners gain more status than losers (Mazur and Booth 1998) and males achieving high status are likely to have increased testosterone levels as a result (Mazur and Lamb 1980). Higher testosterone levels have been associated with offensive behaviour such as attacks, fights, and threats in male judo competitors (Salvador et al. 1999) and winning tennis players have also been reported to have increased testosterone levels (Mazur and Lamb 1980; Booth et al. 1989). Male basketball players were found to have higher testosterone levels the more their individual contribution to the team outcome (González-Bono et al. 1999) suggesting that this is a highly consistent relationship.

Although the relationship between human aggression and testosterone levels is still debated (Archer 1991), Mazur and Booth (1998) concluded that competitiveness and dominance in humans appear to be linked to testosterone and that testosterone encourages 'behavior apparently intended to dominate – to enhance one's status over – other people'. Intriguingly, if a sportsman anticipates a competitive event as being a real challenge, his testosterone levels rise directly before competition; if he wins, his testosterone rises after competition relative to the loser (Mazur and Lamb 1980; Elias 1981). The function of such changes remains unclear, but Mazur and Booth (1998) suggested that high testosterone in winners may prepare them to engage in subsequent competitions and that the decrease among losers may prevent them from injury as they withdraw from further challenges. Mehta and Josephs (2006) showed that a male's decision on whether to go back into a game, having lost a one-on-one competition against another man, can be predicted by the direction of change in his testosterone level. Although this was not a competitive sports setting, the same is likely to apply to physical competition. Indeed, Archer (2006) illustrated that sports competitions lead to larger increases in testosterone than contrived competitions.

Studies on hormones and competition in humans have mostly focused on male athletes. Limited evidence suggests that women are generally unresponsive to the effect of competition on testosterone (Mazur et al. 1997). However, Bateup et al. (2002) indicated that women's testosterone levels are as responsive to competition as those of men. DeBoer (2004) argues that the motivation for competition differs with gender: women are motivated to express themselves and men are motivated to prove themselves. More evidence is required, but provisionally it appears that the relationship between testosterone and aggression is different in men and women (Archer 2006).

Other factors, such as a competitor's mood (Booth et al. 1989), coping strategies, and state and trait psychological factors (Filaire et al. 2001), may also be important and should not be neglected in studies of endocrine responses of competitors. Similarly, behaviour influences hormones, as well as vice versa, so causal inferences cannot easily be made from correlations (Kivlighan et al. 2005). Studying hormones in humans is also challenging as ethical limitations make it difficult to manipulate hormones and measure aggressive behaviour. Overall, therefore, while it is clear that hormonal responses have an important role to play in sporting contests, the situation is complex, and psychological, biological, and anthropological elements must all be taken into account when assessing the role of endocrine responses in competition.

Home advantage

It is common knowledge that teams have a greater chance of winning whenever the match takes place at their home venue, and this general perception is supported by statistical analyses (Pollard and Gomez 2009). The received wisdom suggests that the home fans are tantamount to an extra player on the field, with the visitors having to cope with the home fans' hostility and inimical shouting and chanting. The visiting team may also have made a long journey to reach the venue, possibly with a night in an unfamiliar hotel, while the hosts remain at home and follow their familiar routines. The combined effect is thought to give the home team the edge in the game, a well-documented phenomenon referred to as the 'home advantage' (Neave and Wolfson 2003; Pollard and Gomez 2009). Despite being a robust effect, home advantage varies season by season, from region to region, across divisions, and from sport to sport. Indeed, levels of home advantage appear highest in the early years of each league's existence in all sports (Pollard and Pollard 2005b;

Pollard and Gomez 2009). Interestingly, declines in home advantage have been reported over the last two decades in ice-hockey and basketball, as well as a large drop from 67% to 60% in English football after World War II (Pollard and Pollard 2005b).

Crowd support, in both team (Schwartz and Barsky 1977; Nevill et al. 1996) and individual sports (Balmer et al. 2005), travel fatigue and geographical distance (Clarke and Norman 1995; Pollard 2006b), familiarity with the pitch (Pollard 2002), and referee bias (Nevill et al. 2002; Dawson et al. 2007) are clearly all fruitful explanations for the robust home advantage phenomenon. However, none of them has been proved to have a very strong effect alone. Indeed, in their review of home advantage in football, Pollard and Pollard (2005a) proposed a model demonstrating that the interacting effects of various psychological factors and tactics, rather than any single cause, led to home advantage. They noted that levels of home advantage were variable across European domestic football leagues, with Balkan nations showing a much higher home advantage (79%) than elsewhere. The authors suggested that the territoriality principle of Neave and Wolfson (2003) was likely to be a good reason for high home advantage numbers in regions like the Balkans, where considerable conflict is part of each country's recent history.

Neave and Wolfson (2003) explored the relationship between testosterone, dominance, and territoriality in human competitive encounters. The authors define territoriality as 'the protective response to an invasion of one's perceived territory', which is also prevalent in various animal species giving residents 'home advantage' in territorial disputes (Alcock 1998). Intruders trigger territorial aggression and a rise in circulating levels of testosterone in residents (Wingfield and Wada 1989). Intriguingly, similar effects are found in humans: a footballer's salivary testosterone concentration is significantly higher before playing games at home than away, and this effect is further increased when the opponent was an 'extreme' rival (Figure 18.1) (Neave and Wolfson 2003). Strikers (specialist goal-scorers) generally had the highest levels; goalkeepers had the highest concentrations when playing against a bitter rival but the lowest levels in training sessions. Even though sample sizes were small and further investigation is required, Neave and Wolfson (2003) argued that goalkeepers, as the last defending line in a team, might be particularly inclined to testosterone changes when facing an important opponent. They concluded that their results suggest that testosterone, a hormone associated with aggression and territoriality in humans (Mazur and Booth 1998) and non-human animals (Nelson 2001), plays a mediating role in the

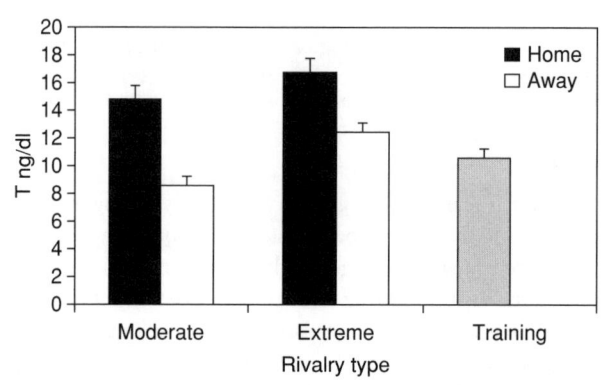

Fig. 18.1 Mean testosterone level in 19 soccer players playing against moderate and extreme rivals before home and away games, with mean level before neutral training sessions included for comparison. Reprinted from Physiology & Behavior, 78 (2), Nick Neave and Sandy Wolfson, Testosterone, territoriality, and the 'home advantage', pp. 269–275, copyright (2003), with permission from Elsevier.

differential performances of teams when playing away or at home. Nevertheless, the ways in which testosterone is able to improve home player performance remain obscure, although likely explanations relate to increases in motivation, confidence, reaction times, information processing, and physiological potential (Neave and O'Connor 2009).

Knowledge of the potential hormonal basis of home advantage offers up a series of avenues for future research. Pollard (2006a) states that the direct analysis of hormone concentrations could deliver fruitful insights into the high levels of home advantage in players from Balkan countries. Neave and O'Connor (2009) highlight the need to determine 'how individual differences in testosterone might relate to performance during a game, and whether it is possible to (legally) manipulate testosterone to improve team performance when playing away'. Additionally, it would be interesting to test for 'leader effects', since a team's captain or coach may also show additional testosterone changes; indeed winning has been shown to lead to increased testosterone levels even in fans (Bernhardt et al. 1998). Conversely, it might be possible to discover ways of mitigating loser effects mediated by testosterone reductions. Cortisol levels (which measure stress responses: Sapolsky et al. 2000) could also be a promising avenue for further research on territoriality in both male and female athletes. Overall, examining the endocrine responses associated with home advantage has opened up an exciting range of opportunities for future research that could greatly enhance our understanding of the complex inter-relationships of factors explaining home advantage and success in sports.

While the precise hormonally-mediated mechanisms remain elusive, home advantage is nevertheless a robust phenomenon. Knowing that home advantage can have such a major influence on sports outcomes, it is crucial for sports teams, their coaches, and psychologists to prepare for, minimize, and even counteract these effects when walking into an unfamiliar stadium. Wolfson and Neave (2004) discuss potential strategies, emphasizing the importance of discipline, concentration, mental preparation, and the establishment of rituals. At the same time, future research may identify 'legal' ways of raising testosterone levels through behavioural means (Neave and O'Connor 2009). As a consequence, while home advantage remains a robust and widespread phenomenon, its effects seem far from insurmountable. Recent decades have seen a decline in the magnitude of home advantage across a range of sports (Pollard and Pollard 2005b) and future research has the potential to increase the speed of this downwards trajectory.

Handedness

The list of famous left-handers extends to all walks of life, including leading scientists such as Albert Einstein, Marie Curie, and Isaac Newton; artists such as Michelangelo and Leonardo da Vinci; and politicians such as David Cameron, Winston Churchill, Barack Obama, Bill Clinton, and Benjamin Franklin (indeed five of the last seven US presidents are reported to be left-handed). It is amongst sportswomen and sportsmen, however, that left-handedness appears especially common, with a high representation amongst athletes considered amongst the greatest ever in their sport: Babe Ruth (baseball), Garfield Sobers (cricket), Pelé, Diego Maradona, and Johan Cruyff (soccer), Mark Spitz (swimming), and John McEnroe and Martina Navratilova (tennis) were all left-handed. To what extent is such anecdotal evidence supported by more systematic analysis?

Though frequencies of sinistrality (left-handedness) vary across human cultures and can range between 3.3% and 26.9% (Faurie et al. 2005), a consistent minority (10–13%) of individuals in all human populations is left-handed (Raymond et al. 1996; Raymond and Pontier 2004). Such frequencies have a long history, and have existed since at least the Upper Palaeolithic (Faurie and Raymond 2004); similar patterns are also reported for chimpanzees (Hopkins and

Morris 1993, McGrew and Marchant 1999). It has therefore been suggested that some sort of evolutionary mechanism must be involved in the persistence of this polymorphism (Llaurens et al. 2009b).

While left-handers are overrepresented in interactive forms of sports such as fencing (50%), table tennis (32%), badminton (23%), cricket (18.5%), and tennis (15%) (Bisiacchi et al. 1985; Aggleton and Wood 1990; Coren 1993; Goldstein and Young 1996; Raymond et al. 1996), sports without dual confrontations or direct opponents, such as swimming (Raymond et al. 1996), skiing, cycling, or gymnastics (Grouios et al. 2000) show no such bias (Figure 18.2). Furthermore, the smaller the physical distance between opponents (such as in combat sports like karate and judo), the higher the frequency of left-handed individuals (Grouios et al. 2000). For many sports, the advantages of left-handedness have been long recognized (Hagemann 2009; Harris 2010), and in some sports left-handed players are thought to hit with greater power (cricket: Brooks et al. 2004; baseball: Grondin et al. 1999). These effects are not confined to elite athletes; expert, intermediate and novice tennis players all find it more difficult to predict the direction of strokes by left-handers than those of non-left-handers (Hagemann 2009). Being left-handed, therefore, appears to be advantageous in interactive sports.

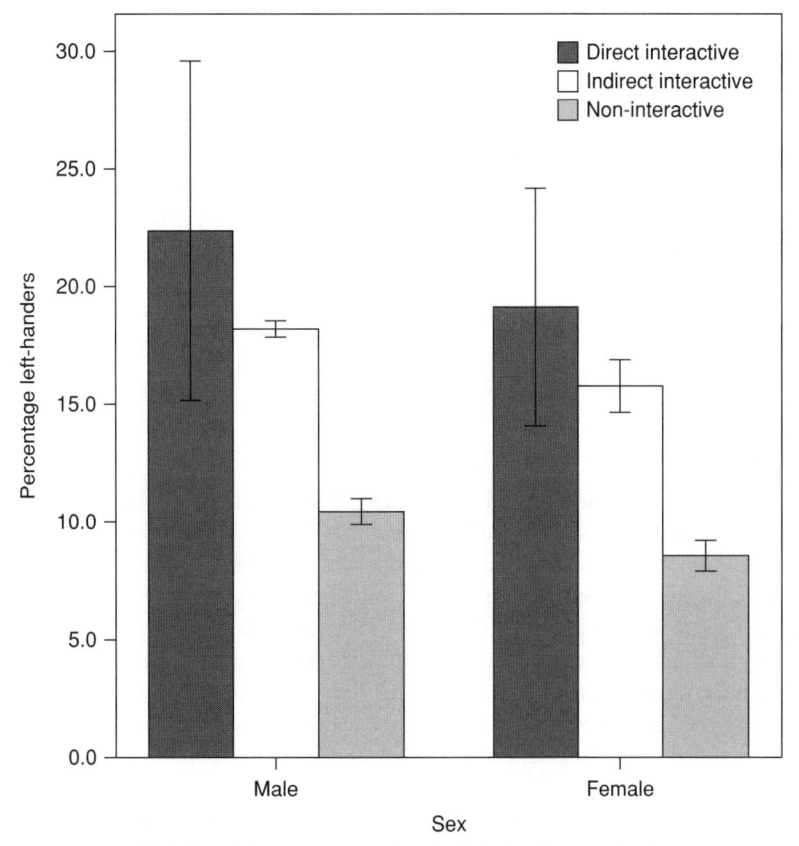

Fig. 18.2 Frequency of left-handedness based on level of physical interaction between competitors in sport. Reproduced with permission of author(s) and publisher from: Grouios, G., Tsorbatzoudis, H., Alexandris, K., and Barkoukis, V., Do left-handed competitors have an innate superiority in sports? Perceptual and Motor Skills, 90, pp. 1273–1282 © 2000.

The role of left-handedness and laterality in sports has attracted interest from researchers from sports psychology, neuropsychology, kinesiology, as well as evolutionary psychology, biology, and anthropology (Annett 1985; Porac and Coren 1981; Aggleton and Wood 1990; Faurie et al. 2005). In the context of sport, a simple reason why the minority of left-handers have an advantage has been termed the 'surprise effect' (Coren 1993; Faurie and Raymond 2005). Since the majority of sportspeople are right-handed, right-handers are more accustomed to competing against opponents favouring the same side. As a consequence, when they encounter a left-handed opponent, who already has the advantage of being practised to compete in a mostly right-handed world, right-handers find themselves at a disadvantage. Left-handers thus have a frequency-dependant advantage (Goldstein and Young 1996; Brooks et al. 2004).

Such effects are not restricted to ritualised Western sports, and left-handers are more frequent in most warlike and violent societies (Raymond et al. 1996; Faurie and Raymond 2005). Indeed, the 'fighting hypothesis' (Raymond et al. 1996; Faurie and Raymond 2005) suggests that left-handers thrive in traditional societies with high levels of violence because of the inherent advantages of left-handedness in these aggressive interactions. Across a range of societies, the frequency of left-handedness correlates positively with homicide rates (Figure 18.3). For instance, in the Dioula population of Burkina Faso, murder rates are as low as 0.013 murders per 1000 residents per year, and there are only 3.4% left-handers within the population. In contrast, 27% of the Eipo of Irian Jaya are left-handed and three out of 1000 inhabitants are murdered each year in this society. One possible interpretation is that, in confrontations where death and injuries are likely, left-handedness confers a competitive advantage. It is not, however, currently possible to exclude alternative hypotheses, such as a causal relationship between androgens and both handedness (e.g. Mathews et al. 2004; Sperling et al. 2010) and homicide rates (Dabbs et al. 2001).

Various environmental and developmental factors have been identified as playing a role in hand preference determinism (reviewed by Llaurens et al. 2009b), while genetics may also play a

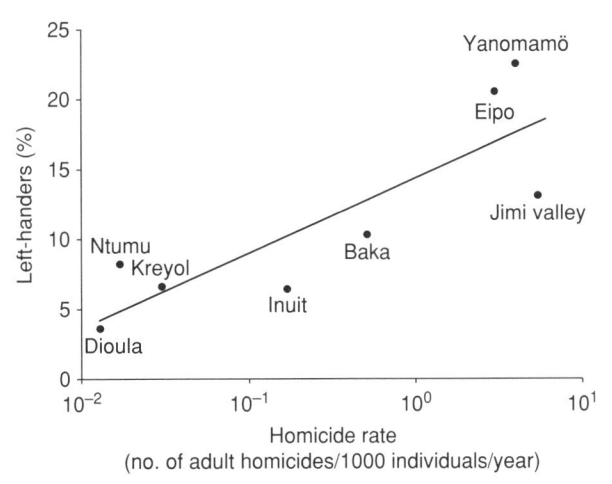

Fig. 18.3 Percentage of left-handers and homicide rate in traditional and sub-traditional cultures. Handedness refers to hand preference for machete use (Kreyol, Ntumu, Dioula, Baka), knife use (Inuit, Eipo), and tool use (Yanomamö, Jimi Valley). Homicide rate was computed as the number of homicides per 1000 individuals per year. Faurie, C. and Raymond, M., Handedness, homicide, and negative frequency-dependent selection. Proceedings of the Royal Society of London B, 272, 25–28. Copyright (2005) The Royal Society.

substantial role (Annett 1985; McManus 1991; Sicotte et al. 1999; Francks et al. 2002). Left-handedness may be triggered by various factors associated with the *in utero* environment, prenatal androgens, maternal age, birth weight, birth stress, or birth trauma (see review in Llaurens et al. 2009b).

In order to understand left-handedness, it is sensible to investigate why a predominance of right-handedness may have evolved in humans. Various evolutionary hypotheses have been put forward, such as the development of language in the left hemisphere (Annett 1985). It has been suggested that an 'impairment' of the right hemisphere (thus, a possible negative influence on skilled performances such as fast reactions, fine control with both hands, and visuospatial thinking) may in some cases result in the left hemisphere language specialization that can be found in nearly all right-handers (Annett 1985). Right-handedness is also suggested to have evolved in association with one-handed throwing ability and its associated cognitive, motor, and postural demands as a possible pre-adaptation for the emergence of left hemisphere specialization in language and motor skills (Calvin 1982; Hopkins et al. 2005). Another idea is that infant handling on the left side permitted a caregiver to move freely on the right side for other purposes (Hopkins and Morris 1993). However, Raymond et al. (1996) and Grouios et al. (2000) emphasize that none of these hypotheses has been advanced to explain the ongoing existence and persistence of left-handers.

Outside of sports and competition, left-handedness is often reported to correlate with advantages in cognitive tasks and socioeconomic status (see review in Faurie et al. 2008), creativity (Newland 1981; Coren 1995), or in the existence of a larger *corpus callosum* (Witelson 1985). Better spatial and visual skills due to relatively larger right hemispheric brain regions are another suggestion as to why left-handers might have an advantage (Geschwind and Galaburda 1985). Cherubin and Brinkman (2006) asked participants to spot matching letters across their left and right visual fields, finding that extreme left-handed subjects were up to 43 milliseconds faster. It seems that left-handers are more bicerebral and that the transfer time between hemispheres is faster in left-handed than in right-handed persons (Cherubin and Brinkman 2006).

If left-handers have a fitness advantage in a variety of situations, why are they not more common? What are the costs that maintain left-handedness at low frequencies within most populations? Left-handedness is associated with several fitness costs including a lower height in adulthood, lower weight, higher age at puberty, lower life expectancy, diverse immune and neurological disorders, and an elevated accident risk (Olivier 1978; Coren 1989; Coren and Halpern 1991; McManus and Bryden 1991; Aggleton et al. 1993; Gangestad and Yeo 1997), though some contradictory evidence exists especially for reduced longevity (e.g. Wood 1988). Smaller size and weight (Olivier 1978) have obvious potential fitness costs for left-handers, particularly in relation to a male's reproductive value (Pawłowski et al. 2000; Mueller and Mazur 2001). Right-handers also have advantages for some life-history traits, such as number of offspring, which is lower in left-handers (Gangestad and Yeo 1994; Faurie et al. 2006). The fitness costs of left-handedness thus appear to limit its frequency in most societies, and provide strong support for adopting an evolutionary approach in addressing the role of laterality in sporting contexts.

Evolutionary accounts also help to explain the higher prevalence rates of left-handers found amongst men (Annett 1985). Following from the fighting hypothesis (Raymond et al. 1996; Faurie and Raymond 2005), the importance that survival in violent fights has on men's reproductive value and social status (e.g. Chagnon 1988) may directly increase the winners' own fecundity. The Yanomamö offer one such example as warfare involves kidnapping females (Gibbons 1993) and left-handers may gain an advantage as their surviving offspring would pass on their genes. Interestingly, however, a child is more likely to become left-handed when its mother is left-handed (Porac and Coren 1981; McManus 1991). One explanation for the maintenance of left-handed women in human populations is that they may benefit indirectly from their left-handed sons (Billard et al. 2005). Indeed, both male and female student athletes have been found to have more sexual partners than their non-sportive counterparts, and this effect is further elevated in high performance athletes (Faurie et al. 2004). Given the disproportionate prevalence of left-handed athletes, their sexual success may help to maintain the handedness polymorphism in contemporary societies, although this idea remains to be tested.

Left-handers are more frequent in traditional societies with high levels of violence and warfare due to the inherent advantages of left-handedness in these aggressive interactions. Similar arguments provide compelling evidence for the advantage of left-handed athletes in interactive sports, with the frequency-dependent benefits of sinistrality leading to an elevated frequency of left-handers in many sports. Although quantifying the costs and benefits of laterality remains challenging, it is clear that an evolutionary perspective offers unique insights into the role of handedness in sporting competition. Nevertheless, the evolutionary perspective offers few solutions to right-handed athletes; as long as left-handers remain comparatively rare, they will retain their frequency-dependant advantage in competitive and sporting interactions.

Colour influences in sports

Having won 15 German Cups and 22 German Championships, F.C. Bayern Munich is by far the most successful German football club. Liverpool and Manchester United dominate the roll call of English domestic honours, while Ajax is the Netherlands' most decorated team. These successful teams are also united by a second characteristic; all have red as the primary colour on their signature football strip. Although this could easily be construed as pure coincidence, recent evidence suggests that clothing colour may play an important role in deciding sporting contests.

Two decades ago, Frank and Gilovich (1988) analysed penalty records of the National Hockey League and the National Football League and showed a bias in the referee's judgement of black sports attire. In both sports, teams playing in non-black uniforms were penalized significantly less often than opponents wearing black. The data also revealed that a team's switch from non-black to black sportswear was followed by a rise in penalties; whereas the change of a team's colour from, for example, blue-and-gold to red-and-green uniforms, did not have the same effect. Although the authors could not find any difference in the likelihood of winning or losing a game when playing in black versus non-black uniforms, the study did indicate that the colour of attire can influence sporting contests.

The first evidence that colour might influence the outcome of sporting events was provided by Hill and Barton (2005) based on an analysis of four combat sports (boxing, taekwondo, freestyle wrestling, and Greco-Roman wrestling) at the 2004 Olympic Games in Athens, Greece. During this competition, red or blue uniforms were randomly assigned to competitors, providing a natural experiment. If colour had no effect on outcomes, an equal number of red and blue winners would be anticipated. Instead, Hill and Barton (2005) found that wearing red was associated with a significantly increased probability of winning, with 55% of all bouts won by competitors in red (Figure 18.4a). Wearing red appeared to tip the balance between winning and losing in close contests: significant effects were found when competitors were closely matched (60% red winners), but not in more asymmetric encounters (Figure 18.4b). Interestingly, no such winning bias due to colour effects was found in judo contests when opponents were dressed in either blue or white (Dijkstra and Preenen 2008). This suggests that while factors such as skill and ability will inevitably have the greatest say in determining sporting outcomes, the subtle effects of red colouration may decide contests where competitors are evenly matched.

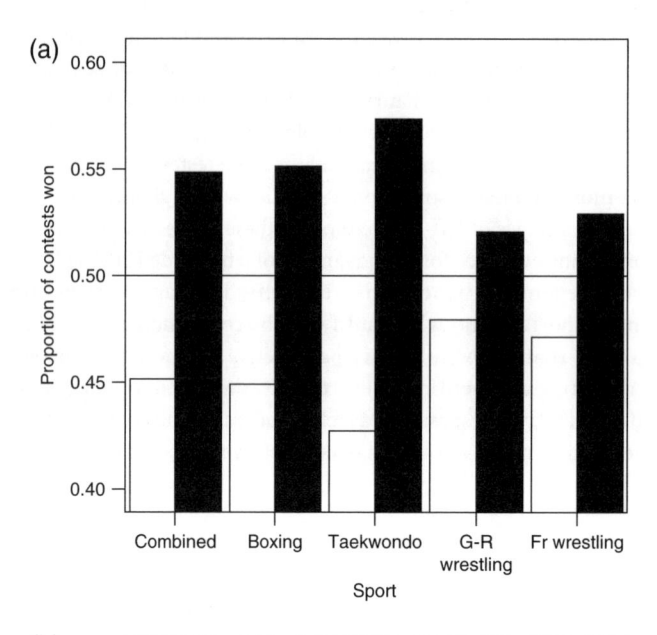

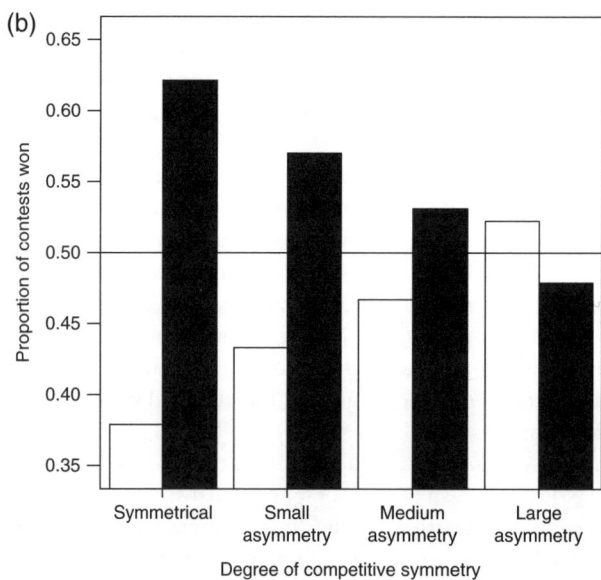

Fig. 18.4 Influence of colour of sporting attire on the outcome of competitive sports. a) Proportion of contests in Olympic combat sports won by competitors wearing red (solid bars) or blue (open bars) outfits for all sports combined and for the individual sports of boxing, tae kwon do, Greco-Roman (G–R) wrestling and freestyle (Fr) wrestling. b) Proportion of contests won by competitors wearing red (solid bars) or blue (open bars) given different degrees of relative ability (asymmetry) in the two competitors in each bout. Black lines at 0.5 indicate the expected proportion of wins by red or blue under the null hypothesis that colour has no effect on contest outcomes. Reprinted by permission from Macmillan Publishers Ltd: Nature, 450 (7040), Russell A. Hill and Robert A. Barton, Psychology: Red enhances human performance in contests, copyright 2005.

Recently, a series of studies have provided substantial support for the role of colour in determining sporting outcomes. A review of English football data found an association between teams wearing red shirts and long-term success (Attrill et al. 2008). Since World War II, red teams have provided more champions and averaged higher finishing league positions than teams in other colours. Most significantly, within cities with more than one team, red teams have significantly outperformed their non-red neighbours over the 55-year period (Attrill et al. 2008). Similar findings by Hill and Barton (2005) showed that teams at the Euro 2004 tournament had better results when playing in red. Greenlees et al. (2010) found that penalty-takers were least successful when they were opposed to a goalkeeper wearing red, supporting Greenlees et al.'s (2008) earlier finding that penalty-takers in red shirts are perceived to possess character traits such as confidence, assertiveness, and composure (compared with those in white). Such effects also appear to operate in virtual contests: Ilie et al. (2008) found similar patterns among experienced players in a first-person-shooter online computer game. Within the game, individuals joined either the red or blue team, and despite better players showing no preference for either colour, red teams triumphed in significantly more (55%) matches. A growing body of evidence thus suggests that the colour red may play a significant role in deciding sporting contests, although these relationships may be confined to male competitors (Barton and Hill 2005).

Why should wearing red enhance performance in human contests? One possibility relates to the biological and cultural associations between red and anger, aggression, and danger. On one hand, red's signalling presence in traffic lights and stop signs reminds us to halt. On the other hand, red hearts and lingerie stand for romantic, passionate attraction and can even signal sexual opportunities, such as in red-light districts (Mahnke 1996; Kaya and Epps 2004). It is thus not surprising that women regard men as being sexually desirable and more attractive when presented in red clothing or on a red background (Elliot et al. 2010a, see also Roberts et al. 2010). Emotional states associated with red identify it as the colour of aggression, passion, anger, energy, and danger (Greenfield 2005). Encountering red in various ways causes ambiguous feelings in humans as positive and negative emotions are associated with the colour red. As humans, we have impressions such as red being intense and bloody, but also warm and passionate (Kaya and Epps 2004). Compared to that, black seems consistently negative, relating to death and evil in almost all cultures (Williams and McMurty 1970). Still, one would assume that different colours have various meanings in different societies and cultures. Surprisingly, despite cross-cultural differences in colour meanings, similarities in emotional states associated to colours were found. Adams and Osgood (1973) carried out a survey in 23 cultures and postulated an affective salience of the colour red. Mostly, red is perceived as active and black as passive. However, both colours are considered to be strong. This could explain why athletes' performances are impaired as an opponent dressed in red may be perceived as particularly active. It could, however, also mean that just the simple fact of being dressed in red creates some sort of higher motivation/more activity in the wearer.

Darwin (1876) noticed that primate males use colouration for attracting females and displaying dominance. Subsequent studies of primates provide support for the significance of red. Recent evidence has found an adaptive link between primate trichromatic vision and their ability to distinguish mood differences due to redness in skin flushing (Changizi et al. 2006). The intensity of red colouration in rhesus macaques (*Macaca mulatta*) and mandrills (*Mandrillus sphinx*) offers a cue to male quality and status (Waitt et al. 2003; Setchell and Dixson 2001) and female rhesus macaques looked longer at images of redder males (Waitt et al. 2003). However, it is in male–male competition that the role of red is most pronounced. In mandrills, male red colouration on rump, face, and genitalia depends on their testosterone levels (Setchell and Dixson 2001) and Setchell and Wickings (2005) concluded that mandrills assess a rival's fighting ability and dominance rank based on the brightness of his red colouration. In fact, red colouration is associated with dominance and aggression in a wide range of taxa and may be a signal of intimidation to rivals (Tinbergen 1955; Pryke 2009).

In humans, skin redness is not such a pronounced signal. However, males tend to be redder than females (Edwards and Duntley 1939, in Ioan et al. 2007), and facial redness correlates with testosterone levels (Edwards et al. 1941), so may indicate high status and dominance. Redness due to oxygenated (redder) blood is a signal of increased aerobic fitness (Stephen et al. 2009a,b), whereas paler skin (reduced redness) can signal anaemia (Jeghers 1944) and can be caused by lower testosterone levels (Ferrucci et al. 2006). While blushing is a sign of social discomfort in situations of shame or embarrassment (Drummond and Quah 2001; Montoya et al. 2005), red skin colouration is also associated with anger and dominance (Drummond and Quah 2001) and blanching is associated with submissiveness, fright, and fear (Montoya et al. 2005; Changizi et al. 2006). Indeed, humans interpret skin blood colouration as an honest cue to underlying health (Stephen 2009a, in both Caucasians and black South Africans) and a 'ruddy' face is often associated with healthiness. Hence, there is evidence from humans and non-human species that red colouration signals both biological traits (health, testosterone, dominance) and emotional states (anger, arousal). It is also known that, in animals, artificial colours can exploit innate responses to natural stimuli. Cuthill et al. (1997) found that, when red and green leg bands were placed on the legs of male zebra finches (a species in which beak redness signals condition), red-banded males dominated green-banded males. Possibly, then, the effects of clothing colour in sporting contests in humans reflects a similar response based on the biological significance of natural skin redness.

What are the underlying mechanisms of these effects? Barton and Hill (2005) suggested the impact of colour might operate through hormonal influences such that wearing red may elevate testosterone levels in the wearer and/or reduce them in the opponent. So far only a small pilot study has examined this possibility and, while the sample size was too small for rigorous statistical analysis, there was qualitative support for athletes in red experiencing elevated testosterone levels (Hackney 2006—but note that Hackney interpreted the non-significant result as contradicting the hypothesis rather than due to lower statistical power). Further research is clearly required to assess the psychological and hormonal effects of wearing red, although a growing body of evidence is showing pronounced effects on the receivers of the stimulus.

Negative effects of perceiving (as opposed to wearing) red are now well documented. In an ingenious experiment, Hagemann et al. (2008) noted a perceptual bias in decisions taken by professional taekwondo referees in response to competitor colour. Hagemann et al. (2008) digitally manipulated video footage to reverse the competitor colours and found that the referees evaluated the identical performances of the taekwondo competitors differentially according to whether they were wearing blue or red protective gear. In the original video footage, the red competitors scored more points, but with the colours reversed, the original blue competitor, now shown in red, was scored more highly. The results provide compelling support for the association between red and dominance influencing perceptions of sporting success.

Sportsmen could thus be distracted by a 'threatening stimulus' such as the colour red, even if they are not aware of this fact (Elliot and Niesta 2008). Ioan et al. (2007) found that 'seeing red' was a particular distractor for men in competitive situations; specifically, men seemed to experience more interferences than women when they were asked to name the colour red in the Stroop test (e.g. reading the word 'blue' written in red). These findings support Hill and Barton's (2005) view that red colour may act as a distractor for men through a psychological effect that evolved in response to sexual competition. Similar results were reported by Little and Hill (2007), who showed that in a study of perceptions of abstract shapes, red dominated the colour blue but the social perception of these colours again differed with gender. Mehta and Zhu (2009) also reported a significant difference in avoidance motivation of red compared with blue conditions in cognitive tasks but, at the same time, red enhanced performance in detail-oriented tasks. Elliot et al. (2007) and others (Kwallek et al. 1997; Stone and English 1998) showed that longer wavelengths in colours such as red and orange degraded task performance and appeared to arouse, while short presentations of red have been found to significantly reduce motivation (Elliot et al. 2010b). This reduced motivation can even go so far that participants would knock fewer times on a door when briefly shown a red cover sheet for a test relative to participants shown a green one (Elliot et al. 2010b).

Colour thus appears to play a small but significant role in deciding sporting contests, with an increasing number of studies reporting the red advantage across a range of datasets. The results appear to reflect the evolutionary and cultural associations of red with dominance and aggression, and raise a number of significant implications. In particular, one repercussion of these findings is that colour represents a significant obstacle to ensuring a level playing field in sport, and suggests that governing bodies should play close attention to the colour of sporting attire. Recent analyses (Dijkstra and Preenen 2008) have shown that in judo, where competitors are assigned either blue or white outfits for each bout, there are no systematic colour biases in sporting outcomes. These results provide an easy solution to the influence of colour in many combat sports, where a simple change of colours could reinstate the level playing field. The situation is more challenging in team sports, however, since signature shirt colours often have a long history as well as enormous commercial value. Nevertheless, provided sporting associations are aware of the potential significance of colour in determining sporting success, individual teams can make their own informed assessments of the relative importance of sporting and commercial success. While this may inevitably bias results to those teams already using red as their signature colour, the opportunity at least exists for other teams to exploit this advantage.

Conclusion

Evolutionary approaches have significantly increased our understanding of the factors influencing the outcome of human sporting competitions. Across a range of sports, an evolutionary approach helps to explain phenomena such as home advantage, the high frequency of left-handedness, and the 'red advantage' in human competitive interactions. The evolutionary approach has also opened up new questions and areas of research, which, alongside traditional studies of sporting behaviour, may offer novel avenues for improving sporting performance.

The significance of testosterone levels has been a recurrent theme throughout the chapter, as endocrine responses appear to play a mediating role in territoriality, aggressive, and dominance behaviour. Clearly, humans and animals differ in the way they express aggressive or dominant behaviour, and yet the relationships between rank, aggression, and testosterone levels appear remarkably consistent. Further research is still clearly required, not least into gender differences in endocrine responses, but endocrine responses appear to underlie widespread phenomena such as home advantage, as well as recent investigations into the more subtle effects of colour stimuli. Given the significance of these effects for determining sporting outcomes, such evolutionary explanations increase the potential for further sporting enhancements, not least through investigations into how testosterone levels might be 'legally' manipulated before sporting encounters. Such investigations could adapt existing approaches or build on recent developments in evolutionary approaches to colour psychology, but either way the next few years offer exciting opportunities for research into factors underpinning sporting success in humans.

References

Adams, F.M. and Osgood, C.E. (1973). Cross-cultural study of affective meanings of color. *Journal of Cross-Cultural Psychology*, **4**, 135–56.

Aggleton, J.P., Kentridge, R.W., and Neave, N.J. (1993). Evidence for longevity differences between left handed and right handed men: an archival study of cricketers. *Journal of Epidemiology and Community Health*, **47**, 206–9.

Aggleton, J.P. and Wood, C.J. (1990). Is there a left-handed advantage in 'ballistic' sports? *International Journal of Sport Psychology*, **21**, 46–57.

Alcock, J. (1998). *Animal behaviour: an evolutionary approach*. Sinaur Associates, Sunderland, MA.

Annett, M. (1985). *Left, right, hand and brain: the right shift theory*. LEA Publishers, London.

Archer, J. (1991). The influence of testosterone on human aggression. *British Journal of Psychology*, **82**, 1–28.

Archer, J. (2006). Testosterone and human aggression: an evaluation of the challenge hypothesis. *Neuroscience and Biobehavioral Reviews*, **30**, 319–45.

Attrill, M.J., Gresty, K.A., Hill, R.A., and Barton, R.A. (2008). Red shirt colour is associated with long-term team success in English football. *Journal of Sport Sciences*, **26**, 577–82.

Balmer, N.J., Nevill, A.M., and Lane, A.M. (2005). Do judges enhance home advantage in European championship boxing? *Journal of Sports Sciences*, **23**, 409–16.

Barton, R.A. and Hill, R.A. (2005). Hill and Barton reply. *Nature*, **43**, 10–11.

Bateup, H.S., Booth, A., Shirtcliff, E.A., and Granger, D.A. (2002). Testosterone, cortisol, and women's competition. *Evolution and Human Behavior*, **23**, 181–92.

Bernhardt, P.C., Dabbs, J.M. Jr, Fielden, J.A., and Lutter, C.D. (1998). Testosterone changes during vicarious experiences of winning and losing among fans at sporting events. *Physiology and Behavior*, **65**, 59–62.

Billard, S., Faurie, C., and Raymond, M. (2005). Maintenance of handedness polymorphism in humans: a frequency-dependent selection model. *Journal of Theoretical Biology*, **235**, 85–93.

Bisiacchi, P.S., Ripoll, H., Stein, J.F., Simonet, P., and Azemar, G. (1985). Left-handedness in fencers: an attentional advantage? *Perceptual and Motor Skills*, **61**, 507–13.

Booth, A., Shelley, G., Mazur, A., Tharp, G., and Kittok, R. (1989). Testosterone, and winning and losing in human competition. *Hormones and Behavior*, **23**, 556–71.

Brooks, R., Bussière, L.F., Jennions, M.D., and Hunt, J. (2004). Sinister strategies succeed at the cricket World Cup. *Proceedings of the Royal Society B*, **271**, 64–6.

Calvin, W.H. (1982). Did throwing stones shape hominid evolution? *Ethology and Sociobiology*, **3**, 115–24.

Chagnon, N.A. (1988). Life histories, blood revenge, and warfare in a tribal population. *Science*, **239**, 985–92.

Changizi, M.A., Zhang, Q., and Shimojo, S. (2006). Bare skin, blood and the evolution of primate colour vision. *Biology Letters*, **2**, 217–21.

Cherubin, N. and Brinkman, C. (2006). Hemispheric interactions are different in left-handed individuals. *Neuropsychology*, **20**, 700–7.

Chick, G. and Loy, J.W. (2001). Making men of them: male socialization for warfare and combative sports. *World Cultures*, **12**, 2–17.

Clarke, S.R. and Norman, J.M. (1995). Home ground advantage of individual clubs in English soccer. *The Statistician*, **44**, 509–21.

Coren, S. (1989). Southpaws: somewhat scrawnier. *Journal of the American Medical Association*, **17**, 2682–83.

Coren, S. (1993). *Left hander: everything you need to know about lefthandedness*. Murray, London.

Coren, S. (1995). Differences in divergent thinking as a function of handedness and sex. *American Journal of Psychology*, **108**, 311–25.

Coren, S. and Halpern, D.F. (1991). Left-handedness: a marker for decreased survival fitness. *Psychological Bulletin*, **109**, 90–106.

Cuthill, I.C., Hunt, S., Cleary, C., and Clark, C. (1997). Colour bands, dominance, and body mass regulation in male zebra finches (*Taeniopygia guttata*). *Proceedings of the Royal Society B*, **264**, 1093–9.

Dabbs, J.M., Riad, J.K., and Chance, S.E. (2001). Testosterone and ruthless homicide. *Personality and Individual Differences*, **31**, 599–603.

Darwin, C.R. (1876). Sexual selection in relation to monkeys. *Nature*, **15**, 18–19.

Dawson, P., Dobson, S., Goddard, J., and Wilson, J. (2007). Are football referees really biased and inconsistent? Evidence on the incidence of disciplinary sanction in the English Premier League. *Journal of the Royal Statistical Society A*, **170**, 231–50.

DeBoer, K.J. (2004). *Gender and competition: how men and women approach work and play differently*. Coaches Choice, Monteray CA.

Dijkstra, P.D. and Preenen, P.T.Y. (2008). No effect of blue on winning contests in judo. *Proceedings of the Royal Society B*, **275**, 1157–62.

Drummond, P.D. and Quah, S.H. (2001). The effect of expressing anger on cardiovascular reactivity and facial blood low in Chinese and Caucasians. *Psychophysiology*, **38**, 190–6.

Eberhart, J.A., Keverne, E.B., and Meller, R.E. (1980). Social influences on plasma testosterone levels in male talapoin monkeys. *Hormones and Behavior*, **14**, 247–66.

Edwards, E.A., Hamilton, J.B., Duntley, S.Q., and Hubert, G. (1941). Cutaneous vascular and pigmentary changes in castrate and eunuchoid men. *Endocrinology*, **28**, 119–28.

Elias, M. (1981). Serum cortisol, testosterone, and testosterone-binding globulin responses to competitive fighting in human males. *Aggressive Behavior*, **7**, 215–24.

Elliot, A. and Niesta, D. (2008). The effect of red on men's attraction to women. *Journal of Personality and Social Psychology*, **95**, 1150–64.

Elliot, A.J., Maier, M.A., Moller, A.C., Friedman, R., and Meinhardt, J. (2007). Color and Psychological functioning: the effect of red on performance attainment. *Journal of Experimental Psychology: General*, **136**, 154–68.

Elliot, A.J., Niesta Kayser, D., Greitemeyer, T., et al. (2010a). Red, rank, and romance in women viewing men. *Journal of Experimental Psychology: General*, **139**, 399–417.

Elliot, A.J., Maier, M.A., Binser, M.J., Friedman, R., and Pekrun, R. (2010b). The effect of red on avoidance behavior in achievement contexts. *Personality and Social Psychology Bulletin*, **25**, 365–75.

Faurie, C. and Raymond, M. (2004). Handedness frequency over more than 10000 years. *Proceedings of the Royal Society B*, **271**, 43–5.

Faurie, C. and Raymond, M. (2005). Handedness, homicide and negative frequency-dependent selection. *Proceedings of the Royal Society B*, **272**, 25–8.

Faurie, C., Pontier, D., and Raymond, M. (2004). Student athletes claim to have more sexual partners than other students. *Evolution and Human Behavior*, **25**, 1–8.

Faurie, C., Schiefenhövel, W., Le Bomin, S., Billiard, S., and Raymond, M. (2005). Variation in the frequency of lefthandedness in traditional societies. *Current Anthropology*, **46**, 142–7.

Faurie, C., Alvergne, A., Bonenfant, S., et al. (2006). Handedness and reproductive success in two large cohorts of French adults. *Evolution and Human Bevavior*, **27**, 457–72.

Faurie, C., Goldberg, M., Hercberg, S., Zins, M., and Raymond, M. (2008). Socio-economic status and handedness in two large cohorts of French adults. *British Journal of Psychology*, **99**, 533–54.

Ferrucci, L., Maggio, M., Bandinelli, S., et al. (2006). Low testosterone levels and the risk of anemia in older men and women. *Archives of Internal Medicine*, **166**, 1380–90.

Filaire, E., Maso, F., Sagnol, M., Ferrand, C., and Lac, G. (2001). Anxiety, hormonal responses, and coping during a judo competition. *Aggressive Behavior*, **27**, 55–63.

Francks, C., Fisher, S.E., MacPhie, I.L., et al. (2002). A genomewide linkage screen for relative hand skill in sibling pairs. *American Journal of Human Genetics*, **70**, 800–5.

Frank, M.G. and Gilovich, T. (1988). The dark side of self- and social perception: black uniforms and aggression in professional sports. *Journal of Personality and Social Psychology*, **54**, 74–85.

Gangestad, S.W. and Yeo, R.A. (1994). Parental handedness and relative hand skill: a test of the developmental instability hypothesis. *Neuropsychology*, **8**, 572–8.

Gangestad, S.W. and Yeo, R.A. (1997). Behavioral genetic variation, adaptation and maladaptation: an evolutionary perspective. *Trends in Cognitive Sciences*, **1**, 103–8.

Geschwind, N. and Galaburda, A.M. (1985). Cerebral lateralization: biological mechanisms, associations and pathology. I. A hypothesis and a program for research. *Archives of Neurology*, **42**, 428–59.

Gibbons, A. (1993). Evolutionists take the long view on sex and violence. *Science*, **261**, 987–8.

Goldstein, S.R. and Young, C.A. (1996). 'Evolutionary' stable strategy of handedness in major league baseball. *Journal of Comparative Psychology*, **110**, 164–9.

González-Bono, E., Salvador, A., Serrano, M.A., and Ricarte, J. (1999). Testosterone, cortisol, and mood in a sports team competition. *Hormones and Behavior*, **35**, 55–62.

Greenfield, A.B. (2005). *A perfect red: empire, espionage, and the quest for the color of desire.* Harper Collins, New York.

Greenlees, I., Leyland, A., Thelwell, R., and Filby, W. (2008). Soccer penalty takers' uniform colour and pre-penalty kick gaze affect the impressions formed of them by opposing goalkeepers. *Journal of Sports Sciences*, **26**, 569–76.

Greenlees, I.A., Eynon, M., and Thelwell, R.C. (2010). 'The effects of goalkeepers' clothing colour on penalty kick performance.' Paper presented at the Annual Conference of the British Psychological Society, Stratford-Upon-Avon.

Grondin, S., Guiard, Y., Ivry, R.B., and Koren, S. (1999). Manual laterality and hitting performance in major league baseball. *Journal of Experimental Psychology: Human Perception and Performance*, **25**, 747–54.

Grouios, G., Tsorbatzoudis, H., Alexandris, K., and Barkoukis, V. (2000). Do left-handed competitors have an innate superiority in sports? *Perceptual and Motor Skills*, **90**, 1273–82.

Hackney, A.C. (2006). Testosterone and human performance: influence of the color red. *European Journal of Applied Physiology*, **96**, 330–3.

Hagemann, N. (2009). The advantage of being left-handed in interactive sports. *Attention, Perception and Psychophysics*, **71**, 1641–8.

Hagemann, N., Strauß, B., and Leißing, J. (2008). When the referee sees red... *Psychological Science*, **19**, 769–71.

Harris, L.J. (2010). In fencing, what gives left-handers the edge? Views from the present and the distant past. *Laterality: Asymmetries of Body, Brain and Cognition*, **15**, 15–55.

Hill, R.A. and Barton, R.A. (2005). Red enhances human performance in contests. *Nature*, **435**, 293.

Hopkins, W.D. and Morris, R.D. (1993). Handedness in great apes: a review of findings. *International Journal of Primatology*, **14**, 1–25.

Hopkins, W.D., Russell, J., Cantalupo, C., Freeman, H., and Schapiro, S. (2005). Factors influencing the prevalence and handedness for throwing in captive chimpanzees (*Pan troglodytes*). *Journal of Comparative Psychology*, **119**, 363–70.

Ilie A., Ioan, S., Zagrean, L., and Moldovan, M. (2008). Better to be red than blue in virtual competition. *Cyberpsychology and Behaviour*, **11**, 375–7.

Ioan, S., Sandulachea, M., Avramescub, S., et al. (2007). Red is a distractor for men in competition. *Evolution and Human Behaviour*, **28**, 285–93.

Jeghers, H. (1944). Pigmentation of the skin. *New England Journal of Medicine*, **231**, 88–100.

Kaya, N. and Epps, H. H. (2004). Relationship between color and emotion: a study of college students. *College Student Journal*, **38**, 396.

Kivlighan, K.T., Granger, D.A., and Booth, A. (2005). Gender differences in testosterone and cortisol response to competition. *Psychoneuroendocrinology*, **30**, 58–71.

Kocher, M.G. and Sutter, M. (2010). Introduction to the special issue 'The Economics and Psychology of Football'. *Journal of Economic Psychology*, **31**, 155–7.

Kwallek, N., Woodson, H., Lewis, C.M., and Sales, C. (1997). Impact of three interior color schemes on worker mood and performance relative to individual environmental sensitivity. *Color Research and Application*, **22**, 121–32.

Little, A.C. and Hill, R.A. (2007). Attribution to red suggests special role in dominance signalling. *Journal of Evolutionary Psychology*, **5**, 161–8.

Llaurens, V., Raymond, M., and Faurie, C. (2009a). Ritual fights and male reproductive success in a human population. *Journal of Evolutionary Biology*, **22**, 1854–9.

Llaurens, V., Raymond, M., and Faurie, C. (2009b). Why are some people left-handed? An evolutionary perspective. *Philosophical Transactions of the Royal Society B*, **364**, 881–94.

Mahnke, F.H. (1996). *Color, environment, and human response.* Van Nostrand Reinhold, New York.

Mathews, G.A., Fane, B.A., Pasterski, V.L., Conway, G.S., Brook, C., and Hines, M. (2004). Androgenic influences on neural asymmetry: handedness and language lateralization in individuals with congenital adrenal hyperplasia. *Psychoneuroendocrinology*, **29**, 810–22.

Mazur, A. and Booth, A. (1998). Testosterone and dominance in men. *Behavioral and Brain Sciences*, **21**, 353–97.

Mazur, A. and Lamb, T.A. (1980). Testosterone, status and mood in human males. *Hormones and Behavior*, **14**, 236–46.

Mazur, A., Susman, E., and Edelbrock, S. (1997). Sex difference in testosterone response to a video game contest. *Evolution and Human Behavior*, **18**, 317–26.

McGrew, W.C. and Marchant, L.F. (1999). Laterality of hand use pays off in foraging success for wild chimpanzees. *Primates*, **40**, 509–13.

McGuire, M.T., Brammer, G.L., and Raleigh, M.J. (1986). Resting cortisol levels and the emergence of dominant status among male vervet monkeys. *Hormones and Behavior*, **20**, 106–17.

McManus, I.C. (1991). The inheritance of left-handedness. In: G.R. Bock and J. Marsh (eds), *Biological asymmetry and handedness, Ciba Foundation Symposium*, pp. 251–81. Wiley, London.

McManus, I.C. and Bryden, M.P. (1991). Geschwind's theory of cerebral lateralization: developing a formal, causal model. *Psychological Bulletin*, **110**, 237–53.

Mehta, P.H. and Josephs, R.A. (2006). Testosterone change after losing predicts the decision to compete again. *Hormones and Behavior*, **50**, 684–92.

Mehta, R. and Zhu, R. (2009). Blue or red? Exploring the effect of color on cognitive task performances. *Science*, **323**, 1226–9.

Montoya, P., Campos, J.J., and Schandry, R. (2005). See red? Turn pale? Unveiling emotions through cardiovascular and hemodynamic changes. *Spanish Journal of Psychology*, **8**, 79–85.

Mueller, U. and Mazur, A. (2001). Evidence of unconstrained directional selection for male tallness. *Behavioral Ecology and Sociobiology*, **50**, 302–11.

Neave, N. and O'Connor, D.B. (2009). Testosterone and male behaviours. *The Psychologist*, **22**, 28–31.

Neave, N. and Wolfson, S. (2003). Testosterone, territoriality and the 'home advantage'. *Physiology and Behavior*, **78**, 269–75.

Nelson, R.J. (2001). *An introduction to behavioural endocrinology.* Sinauer, Sunderland, MA.

Nevill, A.M., Newell, S.M., and Gale, S. (1996). Factors associated with home advantage in English and Scottish soccer matches. *Journal of Sports Sciences*, **14**, 181–6.

Nevill, A.M., Balmer, N.J., and Williams A.M. (2002). The influence of crowd noise and experience upon refereeing decisions in football. *Psychology of Sport and Exercise*, **3**, 261–72.

Newland, G.A. (1981). Differences between left- and righthanders on a measure of creativity. *Perceptual and Motor Skills*, **53**, 787–92.

Olivier, G. (1978). Anthropometric data on left-handed. *Biometrie Humaine*, **13**, 13–22.

Pawłowski, B., Dunbar, R.I.M., and Lipowicz, A. (2000). Tall men have more reproductive success. *Nature*, **403**, 156.

Pollard, R. (2002). Evidence of a reduced home advantage when a team moves to a new stadium. *Journal of Sports Sciences*, **20**, 969–73.

Pollard, R. (2006a). Home advantage in soccer: variations in its magnitude and a literature review of the inter-related factors associated with its existence. *Journal of Sport Behavior*, **29**, 169–89.

Pollard, R. (2006b). Worldwide regional variations in home advantage in association football. *Journal of Sports Sciences*, **24**, 231–40.

Pollard, R. and Gomez, M.A. (2009). Home advantage in football in South-West Europe: long-term trends, regional variation, and team differences. *European Journal of Sport Science*, **9**, 341–52.

Pollard, R. and Pollard, G. (2005a). Home advantage in soccer: a review of its existence and causes. *International Journal of Soccer and Science*, **3**, 25–33.

Pollard, R. and Pollard, G. (2005b). Long-term trends in home advantage in professional team sports in North America and England (1876–2003). *Journal of Sports Sciences*, **23**, 337–50.

Porac, C. and Coren, S. (1981). *Lateral preferences and human behavior*. Springer-Verlag, New York.

Pryke, S.R. (2009). Is red an innate or learned signal of aggression and intimidation? *Animal Behaviour*, **78**, 393–8.

Raymond, M. and Pontier, D. (2004). Is there geographical variation in human handedness? *Laterality*, **9**, 35–52.

Raymond, M., Pontier, D., Dufour, A.B., and Møller, A.P. (1996). Frequency-dependent maintenance of left handedness in humans. *Proceedings of the Royal Society B*, **263**, 1627–33.

Roberts, S.C., Owen, R.C., and Havlic̆ek, J. (2010). Distinguishing between perceiver and wearer effects in clothing color-associated attributions. *Evolutionary Psychology*, **8**, 350–64.

Salvador, A., Suay, F., Martinez-Sanchis, S., Simón, V.M., and Brain, P.F. (1999). Correlating testosterone and fighting in male participants in judo contests. *Physiology and Behavior*, **68**, 205–9.

Sapolsky, R.M., Romero, L.M., and Munck, A.U. (2000). How do glucocorticoids influence stress responses? Integrating permissive, suppressive, stimulatory, and preparative actions. *Endocrine Reviews*, **21**, 55–89.

Schwartz, B. and Barsky, S.F. (1977). The home advantage. *Social Forces*, **55**, 641–61.

Setchell, J.M. and Dixson, A.F. (2001). Changes in the secondary sexual adornments of male mandrills (*Mandrillus sphinx*) are associated with gain and loss of alpha status. *Hormones and Behavior*, **39**, 177–84.

Setchell, J.M., Smith, T., Wickings, E.J., and Knapp, L. A. (2008). Social correlates of testosterone and ornamentation in male mandrills. *Hormones and Behavior*, **54**, 365–72.

Setchell, J.M. and Wickings, E.J. (2005). Dominance, status signals and coloration in mandrills (*Mandrillus sphinx*). *Ethology*, **111**, 25–50.

Sicotte, N.L., Woods, R.P., and Mazziotta, J.P. (1999). Handedness in twins: a meta-analysis. *Laterality*, **4**, 265–86.

Sperling, W., Biermann, T., Bleich, S. et al. (2010). Non-right-handedness and free serum testosterone levels in detoxified patients with alcohol dependence. *Alcohol and Alcoholism*, **45**, 237–40.

Stephen, I.D., Coetzee, V., Law Smith, M., and Perrett, D.I. (2009a). Skin blood perfusion and oxygenation colour affect perceived human health. *PLoS One*, **4**, e5083.

Stephen, I.D., Law Smith, M.J., Stirrat, M.R., and Perrett, D.I. (2009b). Facial skin colouration affects perceived health of human faces. *International Journal of Primatology*, **30**, 845–57.

Stone, N.J. and English, A.J. (1998). Task type, posters, and workspace color on mood, satisfaction, and performance. *Journal of Environmental Psychology*, **1**, 175–85.

Susman, E.J, Inoff-Germain, G., Nottelmann, E.D., Loriaux, D.L., Cutler, G.B., and Chrousos, G.P. (1987). Hormones, emotional dispositions, and aggressive attributes in young adolescents. *Child Development*, **58**, 1114–34.

Tinbergen, N. (1955). *The study of instinct*. Oxford University Press, Oxford.

Waitt, C., Little, A.C., Wolfensohn, S., *et al.* (2003). Evidence from rhesus macaques suggests that male coloration plays a role in female primate mate choice. *Proceedings of the Royal Society B*, **270**, 144–6.

Williams, J.E. and McMurty, C.A. (1970). Color connotations among Caucasian 7th graders and college students. *Perceptual and Motor Skills*, **30**, 701–13.

Wingfield, J.C. and Wada, M. (1989). Changes in plasma levels of testosterone during male-male interactions in the song sparrow, *Melospiza meldoia*: time course and specificity of response. *Journal of Comparative Physiology, A*, **166**, 189–94.

Witelson, S.F. (1985). The brain connection: the corpus callosum is larger in left-handers. *Science*, **229**, 665–8.

Wolfson, S. and Neave, N. (2004). Preparing for home and away matches. *Insight*, **8**, 43–6.

Wood, E.K. (1988). Less sinister statistics from baseball records. *Nature*, **335**, 212.

Section 5

Marketing and communication

Chapter 19

Why we buy: evolution, marketing, and consumer behaviour

Vladas Griskevicius, Joshua M. Ackerman, and Joseph P. Redden

Because advertising, effective advertising, is an appeal to human fundamental needs, desires, and motivations, it is an appeal to basic human nature. People the world over have... the same ambitions, the same egotism, the same temptations. The setting changes, the climate, the culture, the idiom, but the basic human nature is the same everywhere.
Leo (1964, pp. 181–182)

Introduction

Researchers have been examining consumer behaviour, marketing, and advertising for nearly a century. Yet almost none of this work has considered why we buy the things we buy from an evolutionary perspective. After all, our ancestors did not shop at the supermarket giant Wal-Mart; people do not have genes for preferring the soft drink Coke over Pepsi; and shoppers rarely deliberate how a purchase will help propagate their genes.

In this chapter we discuss how a closer inspection of our ancestral roots can provide important insight into the underlying motives driving our purchases. Scholars and marketers are just beginning to consider evolutionary influences on modern marketing and consumer behaviour (e.g. Saad 2007; Miller 2009; Griskevicius et al. 2009a). Because *Homo sapiens* is an ultra-social species, an evolutionary perspective suggests that people interact with the present-day world using brains that evolved to confront recurring social challenges faced in ancestral environments. For example, our ancestors had to solve the social challenges of attracting and retaining a mate, affiliating with others, protecting themselves from danger, gaining status, and caring for family. As we discuss later, these deep-seated evolutionary motives continue to drive modern consumer behaviour, albeit rarely in obvious or conscious ways.

Consider that the majority of people's monthly income is spent on clothing, housing, and transportation.[1] What motivates our consumption choices? Fundamentally, the answer is straightforward: we buy clothing to stay warm, transportation because we need to get from point A to

---
[1] From www.visualeconomics.com/how-the-average-us-consumer-spends-their-paycheck/

point B, and housing because we need shelter from the elements. Yet people typically do not purchase the first product they encounter; rather, consumers often decide on a purchase while facing multiple options. Thus, the relevant question for marketers is why someone chooses to buy one car, house or pair of shoes over another. The answer is rarely that the purchased car is more effective at getting a person from point A to point B, that the house provides superior shelter, or that the shoes are better at protecting one from the elements. Indeed, although a pair of shoes may cost £15 to produce, some consumers have been persuaded to pay £150 for such leather coverings, even when they are fully aware that the shoes are functionally equivalent to another pair half its price. What drives these types of decisions?

Our goal in this chapter is to show how an evolutionary perspective on motivation relates to consumer behaviour. This approach can encourage marketers and advertisers to better meet the evolutionary needs of consumers, and help consumers make more informed decisions. First, we provide an overview to a modern evolutionary approach to consumer behaviour. Next, we present the evolutionary social motives that continue to drive modern behaviour, highlighting emerging findings relevant to marketing. Finally, we discuss implications of evolutionary motivations for marketers, including how they can be applied to market segmentation, consumer targeting, and brand positioning.

A modern evolutionary approach to consumer behaviour

A modern evolutionary approach is based on the seminal work of Charles Darwin (1859, 1871). This approach suggests that, just as the forces of natural selection can shape morphological features, so they can shape psychological and behavioural tendencies. An evolutionary approach maintains that human and non-human animals inherit brains and bodies equipped to behave in ways that are *adaptive*—that are fitted to the demands of the environments within which their ancestors evolved. Just as human morphological features have been shaped by evolutionary pressures, our brains are designed to solve recurrent problems in the ancestral world (Tooby and Cosmides 1992; Buss 1995; Barrett et al. 2002). Below we review two key distinguishing features of a modern evolutionary approach to consumer behaviour.

Distinguishing proximate and ultimate levels of explanation

An evolutionary approach asks the following question about a behaviour: What could be its adaptive (ultimate) function? That is, how might it have helped our ancestors survive, reproduce, or solve another ancestral challenge? This approach is concerned with a particular type of 'why' question. For example, when asking *why* children prefer doughnuts to spinach, one answer is that doughnuts taste better and produce more pleasure than spinach. An evolutionary approach, however, would also ask why doughnuts taste better than spinach in the first place. In this case, the reason is because humans have inherited a preference for fatty and sweet foods (originally meat and ripe fruit, etc.) which provided our ancestors with much-needed calories in a food-scarce environment. An evolutionary perspective thus draws an important distinction (see Tinbergen 1963) between the more mechanistic *proximate* explanations (the first answer) and strategic *ultimate* explanations (the second answer).

Marketers, like most social scientists, have typically been concerned with proximate levels of explanations. Proximate explanations for purchasing behaviour include factors such as preferences, learning, attitudes, values, presentation of information, emotions, and personality. For example, in the popular book *Why We Buy* (Underhill 2000), the reasons given for why we buy are all proximate reasons, such as a product's location in a store and its specific placement on a shelf. However, an evolutionary perspective contends that it is also important to consider the ultimate reasons for why people buy what they do.

Proximate and ultimate explanations are not competitors; they are complementary perspectives. For instance, it is misguided to ask if a man bought an expensive sports car because it makes him feel good (a proximate reason) *or* because it enhances his reproductive opportunities (an ultimate reason). Both explanations can be correct, each explaining the same behaviour at a different level of analysis. The important point is that neglecting ultimate reasons for behaviour can limit or hinder marketing strategies. For example, if an organization desired to curb conspicuous consumption by dissuading people from purchasing pricey sports cars, ignoring ultimate reasons for such purchases might lead to a marketing strategy that tries to persuade people that expensive sports cars shouldn't make them feel good. This kind of a strategy is likely to be fighting an uphill battle.

Finally, an evolutionary approach does not presume that people will be conscious of the ultimate reasons for their behaviour; indeed, research has shown that such motives often guide behaviours in an automatic and non-conscious manner (e.g. Barrett and Kurzban 2006; Kenrick et al. 2010a). However it does highlight that people usually have multiple motives for a given behaviour.

The mind evolved to solve adaptive challenges

An evolutionary approach also does not assume that humans or other organisms inherit some capacity to determine in advance which behaviours will enhance reproductive fitness. Instead, many organisms, including humans, have evolved motivations that increase the probability of solving a set of *recurrent adaptive challenges* confronted during the ancestral past.

Converging evidence suggests that our minds are adapted to life in hunter-gatherer bands of about 100–150 individuals, in an environment similar to present-day African savannah (Dunbar 2004). As we discuss later, our ancestors needed to solve a set of recurring social problems, such as mating, protecting themselves from danger, making friends, gaining status, and taking care of offspring (Kenrick et al. 2010a). Individuals in the past who did not solve these problems did not pass their genes on to us.

The notion that the mind comes equipped with motivational systems for solving recurring adaptive problems has important implications. Foremost, this suggests that the human mind is not a blank slate (Kenrick et al. 2010b). A blank slate approach posits that there is no human nature, meaning that preferences and desires are determined solely by exposure to culture or immediate context, which itself is often believed to be arbitrarily determined (e.g. Ariely et al. 2006). Such an approach implies that marketers can be as effective in persuading people to behave in one way as another. Advertising campaigns, for example, might be equally capable in persuading people to like spinach or ice cream, to want friends or be a hermit, or to desire high or low status. Although blank slate approaches might seem outdated to some readers, it is important to realize that such approaches dominated social science research throughout the twentieth century, and continue to be prevalent in economics, business, anthropology, sociology, and psychology today (Pinker 2002; Tooby and Cosmides 1992).

In stark contrast to blank-slate approaches, an evolutionary perspective maintains that there is a human nature: that humans have a set of evolved motives and dispositions reflecting adaptive psychologies that helped solve recurring challenges confronted during the ancestral past. According to this perspective, human preferences, desires, and inclinations are not infinitely malleable. Instead, people across cultures are more likely to learn one thing than another, and to care about and desire some things more than others (Hsee et al. 2009). An evolutionary approach suggests that marketers can be more effective at changing people's behaviours and persuading them to adopt new practices through strategic communication that considers the evolutionary motives behind why we buy.

Evolutionary social motives and consumer behaviour

To best understand the evolved motivational structure of the mind, we need to identify the recurrent adaptive problems that ancestral humans faced. Although there are an innumerable number of goals that people pursue throughout their lives, there is remarkable commonality in the ultimate purposes these goals serve. Like all other animals, for example, at a base level our ancestors needed nourishment and shelter. Because humans are intensely social animals, however, we also faced significant and recurrent social problems whose solutions have important implications for modern consumer behaviour. Indeed, researchers reviewing cognitive, behavioural, neurophysiological, cross-cultural, and cross-species evidence have suggested a number of ways of organizing fundamental social motives that are critically relevant to humans (Fiske 1992; Buss 1999; Bugental 2000; Kenrick et al. 2003, 2010; Ackerman and Kenrick 2008). These frameworks converge on a core set of ultimate social motives that include attracting and retaining mates, protecting oneself from interpersonal dangers, making friends, gaining status and power, and caring for offspring.

All individuals possess this core set of ultimate motives and associated motivational systems inherited from our ancestors. Each motive can have strong effects on behaviour depending on its degree of activation within an individual. Although the power of a given motive can be temporarily increased by features of the situation (e.g. a job opening may activate status concerns), the relative enduring strength of each motive can be conceptualized as an aspect of an individual's personality. This means that for a given individual, some of these evolutionary motives are much more likely to be chronically active (e.g. some people are persistently focused on status pursuit), while others are less influential. Below, we discuss the workings of each of these evolutionary social motives, highlighting emerging findings relevant to marketing and consumer behaviour.

Romantic partner attraction

The motive to attract romantic partners lies at the heart of an evolutionary approach, whereby maximizing reproduction is the name of the game. Attracting a mate is not merely about sex. Instead, it involves a two-sided, yin-and-yang set of challenges and opportunities: choosing between potential mates and being chosen by them. Thus, a mate-attraction motive drives both evaluation and display. For instance, people attempt to discern a potential partner's status and resources, physical attractiveness, and attributes such as kindness and intelligence (e.g. Buss 1989; Li et al. 2002; Li and Kenrick 2006). Because such qualities are valued by perceivers interested in finding mates, such qualities are also put on heightened display (for example, in the products and brands we choose to wear, drive, and consume). Although the motive to attract a mate is universal, individuals differ in the extent to which they are chronically motivated to repeatedly pursue this goal (Simpson and Gangestad 1991).

The motive to attract a mate operates on many types of behaviours. For example, men motivated to attract mates exhibit more creative thinking, heroic helping, gift-giving, impulsive decision-making, and desire to purchase conspicuous goods (Griskevicius et al. 2006; Griskevicius et al. 2007; Saad and Gill 2003; Wilson and Daly 2004; Sundie et al. 2011). This motive also has specific influences on men's financial decisions, such as erasing loss aversion for men (Li et al. 2010). These types of behavioural changes lead men to be more appealing as mates, and encourage them to act quickly when mating opportunities present themselves. Women are also affected by a mate-attraction motive, for example by becoming more benevolent (Griskevicius et al. 2007). Women are also influenced by relevant internal factors such as hormonal fluctuations associated with the monthly ovulatory cycle. During the peak fertility phase, women's mate-attraction motive tends to be more chronically active, leading women to non-consciously prefer products that will make them more alluring at this time (Durante et al. 2011).

Romantic partner retention

Issues involved in maintaining partnerships are qualitatively different from those involved in acquiring mates. For example, a man who expends a great deal of resources to throw a big party may impress a woman he is interested in dating, but the same behaviour may indicate a lack of commitment once the two are married. A mate-retention motive involves both maintenance of current relationship bonds and managing potential romantic competitors (Campbell and Ellis 2005). Emotionally, whereas a mate-attraction motive is linked to feelings of lust, a mate-retention motive is linked to feelings of romantic love and jealousy (Kenrick and Shiota 2008).

As with the mate attraction motive, individuals differ in the extent to which they are motivated to retain a mate (Maner et al. 2007a). Individuals high in this motive are more likely to show increased levels of love and care for their current partner, of physical and verbal possession signals, and of elevated resource display and appearance enhancement (Buss and Shackelford 1997). Even simply imagining one's partner displaying romantic interest in another person can increase vigilance, including increasing attention to and memory for potential competitors (Maner et al. 2007a, 2009). As with a mate-attraction motive, one method of retaining a mate is through gifts (Saad and Gill 2003). Indeed, men's mate retention motives and purchase of gifts for their partner increase when their romantic partner is ovulating (Pillsworth and Haselton 2006), suggesting that men may not always be fully conscious of the reasons for these purchases.

Protecting oneself from physical harm

Before one can attract or retain a mate, one must first survive long enough to be able to reproduce. Thus, our ancestors had to be successful at avoiding physical dangers and disease that stem from interpersonal contact. Because self-protection is imperative from an evolutionary perspective, the mind is designed to be hyper-sensitive to threats, even if such threats are ambiguous or hypothetical, similar to how a sensitive smoke detector detects and acts upon the tiniest bit of evidence indicating potential danger at the cost of occasional false alarms (Nesse 2005; Haselton and Nettle 2006). The self-protection motive is facilitated by the emotions of fear and disgust (Kenrick and Shiota 2008). Although everyone is at least somewhat motivated to protect themselves from physical harm, there are individual differences in this motive, whereby some individuals are chronically worried about danger or disease and others take risks with abandon (Altemeyer 1988; Duncan et al. 2009; Tybur et al. 2009).

Self-protection principally involves the avoidance of threats (e.g. violence, disease) from other people. To aid this goal, a self-protection motive increases people's sensitivity to certain features of a person or object, including unfamiliarity, physical abnormality, and membership in other religions, races, or other out-groups (e.g. Cottrell and Faulkner et al. 2004; Neuberg 2005; Ackerman et al. 2007; Ackerman et al. 2009a). Such cues may have signalled an elevated potential for harm in ancestral times, meaning that a self-protection motive promotes biases designed to improve protection against heuristically dangerous others (e.g. Shapiro et al. 2009).

A motive for self-protection has been shown to affect a variety of behaviours relevant for consumer behaviour. For example, it leads people to take less risk in financial decisions and to become more loss-averse (Lerner and Keltner 2001; Li et al. 2010). A self-protection motive also promotes a 'strength in numbers' response, leading people to become more conforming and band together with similar others (e.g. Kugihara 2005; Griskevicius et al. 2006b). For example, advertising messages appealing to the behaviours of the masses (e.g. 'this product has been purchased by over a million people') are more persuasive when a self-protection motive is active (Griskevicius et al. 2009a). Accordingly, the motive for self-protection is associated with increased preferences for, and defence of, the status quo (e.g. Jost and Hunyady 2005), and people chronically concerned with self-protection (e.g. disease vulnerability) are also quicker to show avoidant physical behaviours (Mortensen et al. 2010).

Forming friendships

People who were hermits and loners are unlikely to have passed on their genes. Our ancestors lived within social groups much as we do now, and the motivation to form alliances with non-kin would have been highly valuable (e.g. Baumeister and Leary 1995). Friendship is an important way of managing potential dangers, receiving support, and exerting control over the world. Friendships often begin as reciprocal relationships, with favours being equally exchanged, but over time they can become more communal and trusting (Clark et al. 1986; Lydon et al. 1997). Individual differences exist in this universal motive, whereby some people are more motivated to pursue and maintain friendships and are likewise more sensitive to social rejection (Maner et al. 2007c).

An affiliation motive can have powerful influences on behaviour. It encourages psychological connection with others, which can lead one person to treat another person's actions as their own. For example, people take on the personality characteristics of others (Goldstein and Cialdini 2007), become vicariously depleted by the self-control actions of others (Ackerman et al. 2009b), and even feel the pain of other people (Jackson et al. 2006). A friendship motive is especially strong when social relationships are threatened. For instance, social rejection promotes the reinforcement of existing friendships, and motivates behaviours to make new friends (Maner et al. 2007c). This affiliative response can lead people to spend more money on relationship-affirming products and even on items that a friend enjoys but the purchaser does not (Mead et al. 2011). People's relationships with brands can also mimic those they have with friends, and people feel the sting of brand transgressions much like they do the slights of friends (Aaker et al. 2004).

Gaining status

Humans also desire power and prestige in their groups. Because 'success' in evolution is always relative to the fate of competitors, an evolutionary perspective highlights the importance of a motivation to attain relative status, which can result in greater interpersonal influence (Miller et al. 1995), material resources (Cummins 1998), self-esteem (Tesser 1988), and health (Marmot 2004). Because a desire for status is associated with both climbing a social hierarchy and protecting one's position in that hierarchy, this motive is related to both feelings of pride and anger (Kenrick and Shiota 2008). Again, there are individual differences in this universal motive (e.g. McClelland 1975), whereby some people are more motivated to pursue status and are more sensitive to losses in status. For example, although status is important for both sexes, status tends to afford more benefits to men. This is because men's value as a mate is often related to their social and economic status (Buss 1989; Li et al. 2002), leading men to be relatively more sensitive to status losses (Daly and Wilson 1988; Gutierres et al. 1999).

An evolutionary approach highlights two main routes by which to attain status (Henrich and Gil-White 2001): through dominance (e.g. overpowering others and forcing deference) or through prestige (freely conferred deference). Although status motives can lead to more dominant and aggressive behaviour (Griskevicius et al. 2009b), cultural prescriptions often encourage humans to seek status through prestige rather than dominance. For example, self-sacrifice for a group has been shown to increase the self-sacrificer's status, including the likelihood that the person will be selected as a leader (Hardy and Van Vugt 2006). Once status has been obtained, people exhibit a number of biases designed to preserve their status. For example, threats to one's position in a hierarchy may lead to more conservative behaviour patterns (Maner et al. 2007b) and to sacrifice group goals for personal goals (Maner and Mead 2010). A status motive is also directly related to the acquisition of products that signal elevated status and power. For example, status-seeking individuals will pay more for luxury and reputational goods (Rucker and Galinsky 2008), and status motives can lead people to choose 'green' products over counterpart non-green products because doing so enhances their reputations (Griskevicius et al. 2010). When status-signalling goods are unattainable, status motives can even lead to purchase of counterfeit products (Wilcox et al. 2009).

Kin care

Because human offspring are altricial, the motive to care for offspring is of central importance to evolutionary success. The care and support people provide to their children and close relatives is driven by a fundamentally different evolutionary process than when benefits are provided to unrelated individuals. Relatives' overlapping genetic structure allows for cooperation to emerge in the absence of reciprocal norms or laws, because help given by one person to a biological relative provides a net benefit to the genes shared between those individuals (inclusive fitness: Hamilton 1964). Indeed, people give higher value gifts to closer kin (Saad and Gill 2003); after death, inheritance bequests follow the same principle (Smith et al. 1987).

The motive for kin care is related to feelings of nurturance (Kenrick and Shiota 2008), whereby a person motivated by kin care is more likely to be pro-social and relatively unconcerned with having their support returned in an equitable fashion. Note that a kin care motive is not the same as a desire to have children. Instead, the kin care system evolved to motivate investment into assisting helpless others that are nearby, typically one's offspring but in the modern world the same system can motivate giving to helpless strangers (e.g. starving children in Africa) or even taking care of puppies and kittens. Once more, although all people possess a motive for kin care, individual differences exist (e.g. Oyserman et al. 2002), whereby some people are more motivated to care for others and more sensitive to pain felt by close others.

Surprisingly little research has examined how a kin care motive influences behaviour. One interesting set of findings comes from research in organizational behaviour. Family businesses have always made up a substantial portion of the corporate world, with estimates as high as 90% for all businesses in the US, including 37% of the Fortune 500 companies.[2] As Fritz (1997, p. 51) notes: 'A Chinese proverb says – with less whimsy and more hard-nosed sense than most – "You can only trust close relatives"'. Research on firms has uncovered a puzzling occurrence: family-run firms tend to perform better and operate more efficiently, yet the nepotism they engender can lead to free-riding and worsening performance in subsequent generations (McConaughy et al. 2001; Perez-Gonzalez 2006; see also Chapter 4, this volume). Despite these potential downsides, organizations often try to capitalize on the fitness gains likely to accompany genetic relatedness by providing messages that amplify feelings of psychological kinship (Bailey 1988), including reminders of similarity, common goals, and even using fictive kinship terminology.

---

[2] Data from the University of Tulsa College of Business Administration (2000). www.galstar.com/~persson/fobi2/

Marketing applications

In our introduction, we noted that one of the most important questions for marketers is why people prefer one product over another. Consideration of the six evolutionary social motives reviewed in the 'Evolutionary social motives and consumer behaviour' section can help marketers answer this question more successfully by providing insight into the ultimate reasons why consumers buy. In fact, an evolutionary perspective of consumer motivation can add value to nearly every aspect of marketing strategy, helping firms more effectively segment consumers, target consumers, and position brands—three activities that are often cited as the key components of developing a successful marketing strategy (Kotler 1999). Below we discuss in more detail how an evolutionary approach to consumer motivation can be applied to these and other aspects of marketing, including advertising and persuasion.

Customer segmentation

Marketers would like to be able to customize each product and message for each individual consumer. However, such a resource-intensive approach is often impractical. Marketers also realize that a 'one size fits all' approach is usually largely ineffective. To balance these needs, marketers have long realized the importance of segmenting their customers (Hotelling 1929; Smith 1956). Marketers try to group consumers into a limited number of segments, hoping that those within a specific segment will like and want the same things, while people in different segments will like and want different things (Wedel and Kamakura 2000). The process of segmentation is like playing a game of 20 questions, whereby one person is trying to figure out what another is thinking. Just as in the game, marketers attempt to read the mind of consumers, trying to extract what a consumer wants. Unlike the game, however, marketers get only a couple of questions to figure this out. Nevertheless, making marketing decisions at the segment level simplifies marketing planning and communications, allowing marketers to customize the strategy and message to each segment.

The critical question for a marketer is what basis to use for segmenting consumers. Perhaps the most commonly used methods are demographics such as gender, age, and income. For example, when deciding which specific offers to send to which customers, many direct marketers rely on a database called PRIZM. This database uses zip codes to segment consumers, the assumption being that this indicates some information about that person's demographics, which should relate to what they want and need. Unfortunately, segmenting consumers in this way often produces little useful information to predict what people want beyond broad generalities (Yankelovich and Meer 2006). Similarly disappointing results have been found when marketers attempt to segment based on psychographic variables such as lifestyle, interests, and aspirations. For example, one popular psychographic method is the VALS (Values and Lifestyles) approach, which categorizes people based on their level of innovation and access to resources. VALS segments include 'Innovators' on the leading edge, 'Thinkers' who are motivated by ideals, and 'Strivers' focused on achievement. The idea underlying psychographic approaches is that what someone values is related to the products and brands they want to buy. Unfortunately, much like demographic-based approaches, these methods are often poor predictors of actual consumer behaviour (Bucklin et al. 1995; Rossi et al. 1996; Fennell et al. 2003).

An evolutionary approach suggests that marketers should segment based on consumer needs (i.e. the desires and motives that drive consumer purchases). Although marketers understand that consumer needs are important (e.g. most successful marketers run focus groups and conduct interviews and surveys), such efforts at segmentation have been plagued by three critical issues. First, most efforts have attempted to segment based on proximate needs. Because consumers have an innumerable number of proximate needs, however, this makes it difficult to identify the specific needs to examine, to hone in on a few needs that work best, and to define a theoretically-based taxonomy. Second, proximate needs by their very nature are typically specific to a particular product category, often even to a particular product model. Thus, identifying the specific proximate needs to create a segmentation scheme for one product typically has little use for marketing other products. Because the set of proximate needs is constantly changing from product to product (and between product categories), this lack of consistency makes it difficult to use such approaches to manage customer relationships or develop an optimal portfolio of products. Third, the methodologies marketers use in hope of uncovering consumer needs (e.g. focus groups, surveys) often makes it impossible that the ultimate needs driving purchases will be revealed. Given that people often have no idea why they did something (Nisbett and Wilson 1977; Nolan et al. 2008), and given that this is especially true when trying to uncover the evolutionary motives for a behaviour (Barrett and Kurzban 2006; Kenrick et al. 2010a), typical marketing methods of introspection and verbal descriptions will rarely be able to uncover the ultimate need driving a behaviour.

An evolutionary framework of ultimate needs addresses all of these problems. By focusing on a single motivation level directly connected to adaptive fitness, the number of operational needs can be reduced to a small, manageable set of evolutionary social motives—motives that operate as the backseat drivers of consumer behaviour, often steering conscious decisions in unconscious ways. Of course, although the research reviewed here suggests that the six social motives are likely to capture much of what consumers prefer across product categories, future research is needed to examine the extent to which these motives do indeed predict consumer preferences and purchases. Given that current segmentation approaches (VALS, PRIZM, etc.) are often not very successful, any improvement in market segmentation would be considered substantial.

Marketers should consider systematically asking about the six evolutionary motives in their ongoing surveys. This information could be used to produce a person-specific 'motive profile' for consumers, whereby this profile of evolutionary needs could be used by different firms and for different product categories. We can easily imagine that such consumer-specific motive profiles could be added to consumer databases, whereby marketers could purchase this information to better understand their customers' needs and group them in meaningful ways. It may even be possible to develop consumers' motive profiles not solely through direct survey research, but based on the products that individuals purchase.

Consumer targeting

Once a marketer has placed consumers into a limited number of segments, the next step in strategy is deciding which segments to target. Targeting requires developing a plan to reach a target segment and delivering the value of the segment to the firm. Targeting helps firms focus their marketing efforts on a few select segments where they can be successful and avoids wasting effort on segments that will likely never buy the product.

When choosing which targets to pursue, a firm takes into consideration factors such as the size, growth rate, and profitability of each segment (Kotler 1991; Wedel and Kamakura 2000). However, an often-overlooked consideration is the ability to reach a target in an economical fashion. As with market segmentation, a key challenge includes being either too general or too specific in targeting efforts. For example, one common way to target is by considering the demographics of a target group. In television advertising, for instance, it is important to take into account that one programme might be watched more by younger female viewers, whereas another might interest older male viewers. By considering viewer demographics, a firm marketing a low-cost cell phone plan with unlimited text messaging might target younger viewers by advertising on programmes that tend to appeal to youth. Unfortunately, this kind of demographic approach is often not very economical because only a small percentage of even target viewers may have any interest in this particular offering, yet the firm has to pay advertising costs for reaching all viewers watching the programme. In contrast to such general demographic approaches, a firm can also be highly specific in its targeting. For example, makers of the original Hummer sport utility vehicle (SUV) might have targeted readers of select magazines like '*4 Wheel & Off-Road Magazine*'. Although the Hummer might appeal to the small readership of this magazine, this targeting strategy is missing out on potential consumers who are not automotive enthusiasts, but who might nevertheless be interested in purchasing this kind of product.

An evolutionary approach suggests that marketers might be more effective at targeting if they targeted based on evolutionary needs rather than on traditional segments. For example, the low-cost cell phone plan with unlimited texting is likely to be especially appealing to a segment with a motive profile high on friendship needs and low on status-seeking needs (people who desire to keep in constant touch with friends but who don't crave the latest technology). In contrast, a large, flashy SUV is likely to appeal to a segment that is high in status-seeking and self-protection (i.e. people who might see a Hummer as a status symbol as well as providing safety in auto accidents).

The use of ultimate social motives to form the basis of each target segment could also facilitate more economic access to segments. For example, consider two prime-time television dramas: a police crime drama and corporate legal drama. Each programme is watched by similar demographics (middle-aged men and women), meaning that both are essentially identical from a traditional advertising perspective. However, an evolutionary motives approach suggests that viewers for each programme might differ systematically in their motive profiles. Whereas a police crime drama might be watched by individuals high in the need for self-protection, a corporate legal drama might be watched by people desiring to climb the corporate hierarchy and gain status. Thus, a firm targeting a segment high in self-protection should advertise on the crime drama but not on the corporate law drama, whereas a firm targeting a status segment should take the reverse approach.

The evolutionary social motives approach extends well beyond targeting for television advertising. Firms can use this approach to gain clues about where to find prospects in that target group and reach them in an economical fashion. A segment with a strong kin care motive, for example, might spend more time playing at the park, reading the lifestyle section of the newspaper, and surfing parenting websites; whereas those with a strong motive to form friendships might instead meet at malls, read the entertainment section, and use social networking websites. Those with a strong desire for self-protection might be more likely to get their information from familiar sources that are perceived as expert, trustworthy, and sharing common core values (e.g. national TV stations and national newspapers), while those focused on forming friendships might instead read more socially-oriented, local materials (newspapers, magazines, websites). In summary, the social motives of a targeted segment can inform marketers about where these people spend their time, therefore presenting economical avenues for reaching them with a marketing message.

Product development and strategic messaging

Consideration of social motives can also help marketers develop improved product offerings for a targeted segment. For example, if a segment (unconsciously) prefers designer apparel because it raises their perceived status, then the marketer would want to emphasize this aspect for that segment. Perhaps this explains why Ralph Lauren offers shirts with horse logos that cover nearly the entire shirt, and Christian Louboutin adds distinctive red soles to its shoes. These seemingly small design choices are likely to create much value in the eyes of a segment high on status-seeking, but not for a different segment low on status-seeking. Another segment might find value in products that satisfy other motives such as protection from danger. Such a segment might instead prefer the REI shirt with special UV blocking technologies, for example, or the Nike shoes with greater ankle support but minimal branding features.

In addition to improving their products, marketers might also consider how a social motives approach can help them craft more effective messages. Such an approach could prove especially effective in digital media (e.g. retail websites and email), where individual customization techniques are already widely employed (Mulvenna et al. 2000; Mulpuru 2007). For example, online retailers (such as Amazon.com) may not have a single, uniform homepage; instead, the homepage that uploads on a given terminal can be changed by the firm depending on who is accessing the site. This allows firms to strategically decide what products to show on its homepage based on the motive profile of the user.

E-mail and direct marketers could also use social motive profiles as a basis for deciding what version of an advert to send. For example, messages tend to be more persuasive (Cialdini 2009) when they incorporate either appeals to social proof (e.g. 'over a million sold', 'most popular') or appeals to scarcity (e.g. 'limited edition', 'exclusive offer'). However, a social motives approach suggests marketers should be more strategic about when to use one type of appeal over another. Specifically, self-protection motives enhance the effectiveness of social proof appeals, while decreasing the effectiveness of scarcity appeals (Griskevicius et al. 2009a). Conversely, mate-attraction motives enhance the persuasiveness of scarcity appeals, while decreasing persuasiveness of social proof appeals.

Because consumers' individual social motives drive the content and manner in which information is processed, firms should react accordingly. An evolutionary approach suggests that marketing should be less about trying to persuade consumers to take a desired action and more about presenting the desired action as a natural solution to an already-active need. Because consumers have different problems that they desire to solve (problems that map onto the evolutionary social motives), consumers will naturally desire the products that will help them solve the problem that is most pressing. Financial companies, for example, should view themselves as problem solvers, whereby they offer different solutions to different problems. Thus a company could send out two different versions of its financial newsletter based on the motive profile of the recipients. For customers high in self-protection motives, it could emphasize products containing cues to safety, such as investments with guaranteed returns, insurance or inflation protection. In contrast, for consumers high in status motives, it could emphasize products that contain cues to power and prestige, such as exclusive financial products with potential for high returns. Consumers are likely to prefer motive-consistent solutions even when those consumers cannot consciously identify what motives are currently active. In these instances, options that satisfy needs are liable to simply 'feel right', a concept known as regulatory fit (Avnet and Higgins 2006).

Brand positioning

Another key decision in marketing strategy is the positioning of the brand. In branding, marketers want to own a particular space in the minds of consumers (Aaker 1991; Kotler 1999). For instance, when people hear the name of the car manufacturer Volvo, they instantly think of 'safety'. Wal-Mart and the electronics giant Apple similarly invoke the notions of 'low prices' and 'creativity' respectively. All three firms have invested heavily in creating these valuable brand associations. A difficult question for many brands, especially new ones, is what associations they should focus on making salient. Currently, marketers must sift through endless positioning possibilities with little guidance from any underlying structure or theoretical frameworks. For example, the delivery company UPS has spent millions of dollars positioning itself as the brown brand ('What can brown do for you?'), the Budweiser beer brand pours its resources into being known as the 'King of Beers', while many other brands spend large amounts of money trying to be everything to all people (e.g. 'Visa is everywhere you want to be'). Given such disparate types of positioning (that seem to be somewhat successful), determining a brand positioning can be an overwhelming task.

An evolutionary social motives approach offers some recommendations for brand positioning. From this perspective, a brand can successfully own one of six social niches that map onto underlying social motives. This means that a brand can be associated with status (e.g. luxury), family, safety (e.g. health), mate attraction (e.g. sex appeal), friendship, or relationship maintenance. For example, Wal-Mart is known for low prices and its brand image and advertising is strongly linked to family, but will appeal less to those particularly motivated by status. Similarly for Apple, although it is associated with creativity, it owns the high-status niche for computer electronics, whereby its products are more expensive, flashier, and more likely to be used by celebrities. In fact, Apple products are intentionally overrepresented in television shows and Hollywood films because Apple spends a great deal of money to ensure that their products are seen being used by celebrities.

An important implication of an evolutionary perspective is that it will be difficult for one brand to own multiple social motive niches. That is, because each of the six recurrent adaptive problems is qualitatively different and requires qualitatively different solutions, it is likely to be difficult to persuade consumers that a single brand can be highly effective at solving multiple adaptive problems. For example, although Volvo branding emphatically emphasized safety throughout the 1970s, 80s, and 90s, its brand positioning began to change after it was acquired by Ford in 1999. Instead of being *the* safe car, Volvos began to also be positioned as high-status, thus spanning two different evolutionary niches. Perhaps not coincidentally, sales of Volvos and its market share plummeted in the last decade. Even brands that have positioned their identity to attracting people in all walks of life have found that appealing to specific motives can increase returns. Visa, which sells a product that should be equally useful to people in any purchase situation, has found success marketing a Black Card to select customers that is billed as the 'world's most prestigious credit card'. Here, this additional sub-brand helps Visa simultaneously speak to groups with different social motives.

The fact that a single brand is unlikely to be successful at occupying multiple evolutionary niches has important implications for how the social motives framework can serve as a basis for building an effective brand portfolio. For example, a firm producing beverages might consider having different brands for beverages that serve different social evolutionary functions: *the* drink to celebrate achievements (status), *the* family drink, *the* drink for hanging out with friends, *the* sexy drink, *the* healthy drink, and *the* drink that keeps one's marriage going. Coca-Cola, for example, does not label its healthy drinks with a Coca-Cola brand; instead, this segment is branded as Dasani and Vitamin Water.

Decisions regarding brand portfolios often have more to do with business than with psychology. For example, General Motors had one of the most well-known brand portfolios after acquiring a variety of different car manufacturers in the early 20th century. It tried to position some of these brands based on socioeconomic status segments: Chevrolet was the mainstream brand for middle-America; Buick was the entry-level luxury brand; Cadillac was the premier luxury brand. Still other brands such as Oldsmobile were positioned as being everything to all people, while Pontiac focused on high-performance engines. Although innovative at the time, the General

Motors approach has not fared well over time, with many of the brands being discontinued or marginalized. Although poor business practices apparently contributed to the demise of some of these brands, an evolutionary approach suggests that General Motors would have been wiser to position these brands along social motives such as *the* family brand, *the* status brand, and *the* sexy brand. For example, one of the most lasting and successful GM models is the Corvette, which is considered to be one of the sexiest and flashiest American cars. Indeed, most people consider Corvette to be its own brand (rather than a Chevrolet model) because it occupies a distinctly different psychological niche (the sexy car niche) than Chevrolet. By considering the ultimate motives driving consumer preferences, firms can position their brands to more directly speak to what consumers really want, even if those preferences are latent and not identifiable through traditional, conscious response methodologies.

Situational influences on social motives

Thus far we have concentrated on evolutionary motives as reflected by individual differences, whereby it is useful for marketers to consider that people have different motive profiles. However, it is also worthwhile for marketers to consider how specific evolutionary motives can be heightened by specific contexts. For example, a self-protection motive will become temporarily more influential after a person reads a news story about a murder; a mate-attraction motive will be heightened after seeing an attractive opposite-sex individual; a mate-retention motive by remembering an approaching wedding anniversary; a status motive by hearing about a promotion; an affiliation motive after social rejection; and a kin care motive after seeing a photo of one's child. The key implication is that consumers will be more easily persuaded to want products that are consistent with the motive that is currently most influential. After watching a clip from a scary film, people are more easily persuaded to want a product that is owned by many other people and helps them blend into the crowd, rather than a product that is unique and distinctive (Griskevicius, Goldstein et al. 2009). Conversely, after watching a clip from a sexy film, people are more easily persuaded to want a product that that is unique and will help them stand out from the crowd, rather than one that is already owned by many consumers.

The notion that situational cues can be used to heighten the influence of specific motives can be used strategically by marketers. Consider advertising during local news programmes. The first segment is often filled with reports of frightful incidents such as murders, rapes, abductions, wars, or fires: content that is likely to heighten a self-protection motive. Yet the products and services that are often advertised during the news, such as those that improve one's appearance or luxury cars, often have no relation to self-protection. Instead, advertising after the first news segment is likely to be more successful if it focuses on products relevant to self-protection (which, of course, may involve more than goods explicitly referencing protection, such as goods that highlight maintenance of the status quo). In contrast to the first news segment, one of the last news segments tends to be about sports. Sport features reports about competitions that might serve to temporarily heighten a motive for status; viewers might thus be more easily persuaded to purchase status-related products by adverts immediately following the end of the programme.

Consideration of the situational influences of evolutionary motives suggests that marketers can be more strategic in how and when they use contexts to heighten specific motives. For example, a particular motive may be heightened when a person encounters a particular background on a website (Mandel and Johnson 2002), through the store environment and atmospherics (Kaltcheva and Weitz 2006), or even via particular scents or music (Zhu and Meyers-Levy 2005; Bosmans 2006). Consider the influence of a pungent scent that is likely to temporarily heighten a self-protection (disease avoidance) motive. In such a situation, people would be more likely to avoid products that are strange and unfamiliar, instead desiring things that are safe, clean, and familiar (Argo et al. 2006). Such cues may even increase people's desire for domestically manufactured products, reducing desire for things considered to be foreign. A marketer's sensitivity to the influence of the context suggests that firm partnerships might be an effective means of increasing joint returns. For example, consider a car dealer who is planning to place a print ad featuring a high-status vehicle. Joining forces with a company that is interested in advertising executive job advancement training by placing car and job ads in close proximity to each other may elevate the chance that a status motive is heightened for the reader, driving them to prefer both *that* car and *that* training class.

Conclusion

Evolutionary approaches are only beginning to make inroads into our understanding of marketing, advertising, and consumer behaviour. Although an examination of current mainstream research in marketing would suggest that evolution has little to do with modern consumer behaviour, a closer inspection of our ancestral roots can provide much insight into why we buy. An evolutionary perspective suggests that we interact with our present-day world using brains that evolved to solve a recurring and limited set of ancestral social challenges. Solving the challenges of making friends, gaining status, attracting a mate, keeping a mate, protecting ourselves from danger, and taking care of offspring requires the application of specific motives that play a role not only in life on the ancestral savannah but also in modern consumer environments. These deep-seated motives can shape all stages of the consumer judgement and decision-making process, and thus represent especially important tools for firms to leverage to better meet consumer preferences. Several specific areas of marketing that might be aided by an evolutionary approach include market segmentation, consumers targeting, brand positioning, and advertising. Although future research is needed to examine the effectiveness of this approach, an evolutionary perspective holds much promise for unlocking the fundamental mysteries of consumer behaviour.

References

Aaker, D.A. (1991). *Managing brand equity*. Free Press Publishing, New York.

Aaker, J., Fournier, S., and Brasel, S.A. (2004). When good brands do bad. *Journal of Consumer Research*, **31**, 1–16.

Ackerman, J. and Kenrick, D.T. (2008). The costs of benefits: help-refusals highlight key trade-offs of social life. *Personality and Social Psychology Review*, **12**, 2, 118–14.

Ackerman, J., Kenrick, D.T. and Schaller, M. (2007). Is friendship akin to kinship? *Evolution and Human Behavior*, **28**, 365–74.

Ackerman, J.M., Becker, D.V., Mortensen, C.R., Sasaki, T., Neuberg, S.L., and Kenrick, D.T. (2009a). A pox on the mind: disjunction of attention and memory in the processing of physical disfigurement. *Journal of Experimental Social Psychology*, **45**, 478–85.

Ackerman, J.M., Goldstein, N.J., Shapiro, J.R., and Bargh, J.A. (2009b). You wear me out: the vicarious depletion of self-control. *Psychological Science*, **20**, 326–32.

Altemeyer, B. (1988). *Enemies of freedom*. Jossey-Bass, San Francisco, CA.

Argo, J.J., Dahl, D.W., and Morales, A.C. (2006). Consumer contamination: how consumers react to products touched by others. *Journal of Marketing*, **70**, 81–94.

Ariely, D., Loewenstein, G., and Prelec, D. (2006). Tom Sawyer and the construction of value. *Journal of Economic Behavior and Organization*, **60**, 1–10.

Avnet, T. and Higgins, E.T. (2006). How regulatory fit affects value in consumer choices and opinions. *Journal of Marketing Research*, **43**, 1–10.

Bailey, K.G. (1988). Psychological kinship: implications for the helping professions. *Psychotherapy: Theory, Research, Practice, Training*, **25**, 132–41.

Barrett, H.C. and Kurzban, R. (2006). Modularity in cognition: framing the debate. *Psychological Review*, **113**, 628–47.

Barrett, L., Dunbar, R.I.M., and Lycett, J.E. (2002). *Human evolutionary psychology*. Palgrave, London.

Baumeister, R.R. and Leary, M.R. (1995). The need to belong: desire for interpersonal attachments as a fundamental human motivation. *Psychological Bulletin*, **117**, 497–529.

Bosmans, A.M.M. (2006). Scents and sensibility: When do (in)congruent ambient scents influence product evaluations? *Journal of Marketing*, **70**, 32–43.

Bucklin, R.E., Gupta, S., and Han, S. (1995). A brand's eye view of response segmentation in consumer brand choice behavior. *Journal of Marketing Research*, **32**, 66–74.

Bugental, D.B. (2000). Acquisition of the algorithms of social life: a domain-based approach. *Psychological Bulletin*, **126**, 187–219.

Buss, D.M. (1989). Sex differences in human mate preferences: evolutionary hypotheses tested in 37 cultures. *Behavioral and Brain Sciences*, **12**, 1–49.

Buss, D.M. (1995). Psychological sex differences: origins through sexual selection. *American Psychologist*, **50**, 164–71.

Buss, D.M. (1999). Adaptive individual differences revisited. *Journal of Personality*, **67**, 259–64.

Buss, D.M. and Shackelford, T.K. (1997). From vigilance to violence: mate retention tactics in married couples. *Journal of Personality and Social Psychology*, **72**, 346–61.

Campbell, L. and Ellis, B.J. (2005). Commitment, love, and mate retention. In: D.M. Buss (ed.), *The handbook of evolutionary psychology*, pp. 419–42. John Wiley and Sons, Hoboken, NJ.

Cialdini, R.B. (2009). *Influence: science and practice* (5th Edition). Allyn and Bacon, Boston, MA.

Clark, M.S., Mills, J., and Powell, M.C. (1986). Keeping track of needs in communal and exchange relationships. *Journal of Personality and Social Psychology*, **51**, 333–38.

Cottrell, C.A. and Neuberg, S.L. (2005). Different emotional reactions to different groups: a sociofunctional threat-based approach to 'prejudice'. *Journal of Personality and Social Psychology*, **88**, 770–89.

Cummins, R.A. (1998). The second approximation to an international standard for life satisfaction. *Social Indicators Research*, **43**, 307–34.

Daly, M. and Wilson, M. (1988). *Homicide*. Aldine de Gruyter, New York.

Darwin, C. (1859). *On the origin of species by means of natural selection*. John Murray, London.

Darwin, C. (1871). *The descent of man, and sexual selection in relation to sex*. John Murray, London.

Dunbar, R.I.M. (2004). *The human story*. Faber and Faber, London.

Duncan, L.A., Schaller, M., and Park, J.H. (2009). Perceived vulnerability to disease: development and validation of a 15-item self-report instrument. *Personality and Individual Differences*, **47**, 541–6.

Durante, K.M., Griskevicius, V., Hill, S.E., Perilloux, C., and Li, N.P. (2011). Ovulation, female competition, and product choice: hormonal influences on consumer behavior. *Journal of Consumer Research*, 37, 921–34.

Faulkner, J., Schaller, M., Park, J.H., and Duncan, L.A. (2004). Evolved disease avoidance mechanisms and contemporary xenophobic attitudes. *Group Processes and Intergroup Relations*, **7**, 333–53.

Fennell, G., Allenby, G.M., Yang, S., and Edwards, Y. (2003). The effectiveness of demographic and psychographic variables for explaining brand and product category use. *Quantitative Marketing and Economics*, **1**, 223–44.

Fiske, A.P. (1992). The four elementary forms of sociality: framework for a unified theory of social relations. *Psychological Review*, **99**, 689–723.

Fritz, R. (1997). *Wars of succession: the blessings, curses and lessons that family owned firms offer anyone in business*. Merritt Publishing, Santa Monica, CA.

Goldstein, N.J. and Cialdini, R.B. (2007). The spyglass self: a model of vicarious self-perception. *Journal of Personality and Social Psychology*, **92**, 402–17.

Griskevicius, V., Cialdini, R.B., and Kenrick, D.T. (2006a). Peacocks, Picasso, and parental investment: the effects of romantic motives on creativity. *Journal of Personality and Social Psychology*, **91**, 63–76.

Griskevicius, V., Goldstein, N.J., Mortensen, C.R., Cialdini, R.B., and Kenrick, D.T. (2006b). Going along versus going alone: when fundamental motives facilitate strategic (non)conformity. *Journal of Personality and Social Psychology*, **91**, 281–94.

Griskevicius, V., Tybur, J.M., Sundie, J.M., Cialdini, R.B., Miller, G.F. and Kenrick, D.T. (2007). Blatant benevolence and conspicuous consumption: when romantic motives elicit costly displays. *Journal of Personality and Social Psychology*, **93**, 85–102.

Griskevicius, V., Goldstein, N.J., Mortensen, C.R., Sundie, J.M., Cialdini, R.B., and Kenrick, D.T. (2009a). Fear and loving in Las Vegas: evolution, emotion, and persuasion. *Journal of Marketing Research*, **46**, 385–95.

Griskevicius, V., Tybur, J.M., Gangestad, S.W., Perea, E.F., Shapiro, J.R., and Kenrick, D.T. (2009b). Aggress to impress: hostility as an evolved context-dependent strategy. *Journal of Personality and Social Psychology*, **96**, 980–94.

Griskevicius, V., Tybur, J.M., and Van den Bergh, B. (2010). Going green to be seen: status, reputation, and conspicuous conservation. *Journal of Personality and Social Psychology*, **98**, 392–404.

Gutierres, S.E., Kenrick, D.T., and Partch, J.J. (1999). Beauty, dominance, and the mating game: contrast effects in self-assessment reflect gender differences in mate selection. *Personality and Social Psychology Bulletin*, **25**, 1126–34.

Hamilton, W.D. (1964). The genetical evolution of social behavior: I and II. *Journal of Theoretical Biology*, **7**, 1–52.

Hardy, C.L. and Van Vugt, M. (2006). Nice guys finish first: the competitive altruism hypothesis. *Personality and Social Psychology Bulletin*, **32**, 1402–13.

Haselton, M.G. and Nettle, D. (2006). The paranoid optimist: an integrative evolutionary model of cognitive biases. *Personality and Social Psychology Review*, **10**, 47–66.

Henrich, J. and Gil-White F.J. (2001). The evolution of prestige: freely conferred status as a mechanism for enhancing the benefits of cultural transmission. *Evolution and Human Behavior*, **22**, 165–96.

Hotelling, H. (1929). Stability in competition. *The Economic Journal*, **39**, 41–57.

Hsee, C.K., Yang, Y., Li, N., and Shen, L. (2009). Wealth, warmth and wellbeing: whether happiness is relative or absolute depends on whether it is about money, acquisition, or consumption. *Journal of Marketing Research*, **46**, 396–409.

Jackson, P.L., Brunet, E., Meltzoff, A.N., and Decety, J. (2006). Empathy examined through the neural mechanisms involved in imagining how I feel versus how you feel pain. *Neuropsychologia*, **44**, 752–61.

Jost, J.T. and Hunyady, O. (2005). Antecedents and consequences of system-justifying ideologies. *Current Directions in Psychological Science*, **14**, 260–5.

Kaltcheva, V. and Weitz, B. (2006). When should a retailer create an exciting store environment? *Journal of Marketing*, **70**, 107–18.

Kenrick, D.T. and Shiota, M.N. (2008). Approach and avoidance motivation(s): an evolutionary perspective. In A.J. Elliot (ed.), *Handbook of approach and avoidance motivation*, pp. 271–85. Psychology Press, New York.

Kenrick, D.T., Li, N.L., and Butner, J. (2003). Dynamical evolutionary psychology: individual decision rules and emergent social norms. *Psychological Review*, **110**, 3–28.

Kenrick, D.T., Griskevicius, V., Neuberg, S.L., and Schaller, M. (2010a). Renovating the pyramid of needs: contemporary extensions built upon ancient foundations. *Perspectives on Psychological Science*, **5**, 292–314.

Kenrick, D.T., Nieuweboer, S., and Buunk, A.P. (2010b). Universal mechanisms and cultural diversity: replacing the blank slate with a coloring book. In: M. Schaller, A. Norenzayan, S. Heine, T. Yamagishi, and T. Kameda (eds), *Evolution, culture, and the human mind*, pp. 257–71. Lawrence Erlbaum Associates, Mahwah, NJ.

Kotler, P. (1991). *Marketing management: analysis, planning, implementation and control.* Prentice-Hall, Englewood Cliffs, NJ.

Kotler, P. (1999). *Kotler on marketing.* Free Press Publishing, New York.

Kugihara, N. (2005). Effects of physical threat and collective identity on prosocial behaviors in an emergency. In: J.P. Morgan (ed.), *Psychology of aggression,* pp. 45–67. Nova Science, Hauppauge, NY.

Leo, N.B. (1964). Creative strategy for international advertising. In: S.W. Dunn (ed.), *International handbook of advertising*, pp. 181–82. McGraw-Hill, New York.

Lerner, J.S. and Keltner, D. (2001). Fear, anger, and risk. *Journal of Personality and Social Psychology*, **81**, 146–59.

Li, N.P. and Kenrick, D.T. (2006). Sex similarities and differences in preferences for short-term mates: what, whether, and why. *Journal of Personality and Social Psychology*, **90**, 468–89.

Li, N.P., Bailey, J.M., Kenrick, D.T., and Linsenmeier, J.A. (2002). The necessities and luxuries of mate preferences: testing the trade-offs. *Journal of Personality and Social Psychology*, **82**, 947–55.

Li, Y.J., Kenrick, D.T., Griskevicius, V., and Neuberg, S.L. (2010). 'The evolutionary roots of decision biases: erasing and exacerbating loss aversion.' Paper presented at the Human Behavior and Evolution Society meeting, Eugene, OR.

Lydon, J.E., Jamieson, D.W., and Holmes, J.G. (1997). The meaning of social interactions in the transition from acquaintanceship to friendship. *Journal of Personality and Social Psychology*, **73**, 536–48.

Mandel, N. and Johnson, E.J. (2002). When web pages influence choice: effects of visual primes on experts and novices. *Journal of Consumer Research*, **29**, 235–45.

Maner, J.K., Gailliot, M.T., Rouby, D.A., and Miller, S.L. (2007a). Can't take my eyes off you: attentional adhesion to mates and rivals. *Journal of Personality and Social Psychology*, **93**, 389–401.

Maner, J.K., DeWall, C.N., Baumeister, R.F., and Schaller, M. (2007c). Does social exclusion motivate interpersonal reconnection? Resolving the 'Porcupine Problem'. *Journal of Personality and Social Psychology*, **92**, 42–55.

Maner, J.K., Gailliot, M.T., Butz, D., and Peruche, B.M. (2007b). Power, risk, and the status quo: does power promote riskier or more conservative decision-making? *Personality and Social Psychology Bulletin*, **33**, 451–62.

Maner, J.K., Miller, S.L., Rouby, D.A., and Gailliot, M.T. (2009). Intrasexual vigilance: the implicit cognition of romantic rivalry. *Journal of Personality and Social Psychology*, **97**, 74–87.

Maner, J.K. and Mead, N. (2010). The essential tension between leadership and power: when leaders sacrifice group goals for the sake of self-interest. *Journal of Personality and Social Psychology*, **99**, 482–97.

Marmot, M. (2004). *Status syndrome: how your social standing directly affects your health and life expectancy.* Bloomsbury, London.

Mead, N.L., Baumeister, R.F., Stillman, T.F., Rawn, C.D., and Vohs, K.D. (2011). Social exclusion causes people to spend and consume in the service of affiliation. *Journal of Consumer Research*, **37**, 902–19.

McClelland, D.C. (1975). *Power: the inner experience.* Irvington, New York.

McConaughy, D.L., Matthews, C.H., and Fialko, A.S. (2001). Founding family controlled firms: performance, risk and value. *Journal of Small Business Management*, **39**, 31–49.

Miller, A.G., Collins, B.E., and Brief, D.E. (1995). Perspectives on obedience to authority: the legacy of the Milgram experiments. *Journal of Social Issues*, **51**, 1–19.

Miller, G.F. (2009). *Spent: sex, evolution, and consumer behavior.* Penguin/Putnam, New York.

Mortensen, C.R., Becker, D.V., Ackerman, J.M., Neuberg, S.L., and Kenrick, D.T. (2010). Infection breeds reticence: the effects of disease salience on self-perceptions of personality and behavioral avoidance tendencies. *Psychological Science*, **21**, 440–7.

Mulpuru, S. (2007). *Which personalization tools work for e-commerce—and why*. Forrester Research, Cambridge, CA.

Mulvenna, M.D., Anand, S.S., and Buchner, A.G. (2000). Personalization on the Net using Web mining: introduction. *Communications of the ACM*, **43**, 122–5.

Nesse, R.M. (2005). Natural selection and the regulation of defenses: A signal detection analysis of the smoke detector principle. *Evolution and Human Behavior*, **26**, 88–105.

Nisbett, R. and Wilson, T. (1977). Telling more than we can know: verbal reports on mental processes. *Psychological Review*, **84**, 231–59.

Nolan, J.P., Schultz, P.W., Cialdini, R.B., Goldstein, N.J., and Griskevicius, V. (2008). Normative social influence is underdetected. *Personality and Social Psychology Bulletin*, **34**, 913–23.

Oyserman, D., Coon, H., and Kemmelmeier, M. (2002). Rethinking individualism and collectivism: evaluation of theoretical assumptions and meta-analyses. *Psychological Bulletin*, **128**, 3–73.

Perez-Gonzalez, F. (2006). Inherited control and firm performance. *American Economic Review*, **96**, 1559–88.

Pillsworth, E.G. and Haselton, M.G. (2006). Women's sexual strategies: the evolution of long-term bonds and extra-pair sex. *Annual Review of Sex Research*, **17**, 59–100.

Pinker, S. (2002). *The blank slate: the modern denial of human nature*. Viking-Penguin, New York.

Rossi, P.E., McCulloch, R.E., and Allenby, G.M. (1996). The value of purchase history data in target marketing. *Marketing Science*, **15**, 321–40.

Rucker, D.D. and Galinsky, A.D. (2008). Desire to acquire: powerlessness and compensatory consumption. *Journal of Consumer Research*, **35**, 257–67.

Saad, G. (2007). *The evolutionary bases of consumption*. Lawrence Erlbaum Associates Publishers, Mahwah, NJ.

Saad, G. and Gill, T. (2003). An evolutionary psychology perspective on gift giving among young adults. *Psychology and Marketing*, **20**, 765–84.

Shapiro, J.R., Ackerman, J.M., Neuberg, S.L., Maner, J.K., Becker, D.V., and Kenrick, D.T. (2009). Following in the wake of anger: when not discriminating is discriminating. *Personality and Social Psychology Bulletin*, **35**, 1356–67.

Simpson, J.A. and Gangestad, S.W. (1991). Individual differences in sociosexuality: evidence for convergent and discriminant validity. *Journal of Personality and Social Psychology*, **60**, 870–83.

Smith, M.S., Kish, B.J., and Crawford, C.B. (1987). Inheritance of wealth as human kin investment. *Ethology and Sociobiology*, **8**, 171–82.

Smith, W.R. (1956). Product differentiation and market segmentation as alternative marketing strategies. *The Journal of Marketing*, **21**, 3–8.

Sundie, J.M., Kenrick D.T., Griskevicius, V., Tybur, J.T., Vohs, K.D., and Beal, D.J. (2011). Peacocks, Porsches, and Thorstein Veblen: conspicuous consumption as a sexual signaling system. *Journal of Personality and Social Psychology*, **100**, 664–80.

Tesser, A. (1988). Toward a self-evaluation maintenance model of social behavior. In: LL. Berkowitz (ed.), *Advances in experimental social psychology (Vol. 21)*, pp. 181–228. Academic Press, San Diego, CA.

Tinbergen, N. (1963). On the aims and methods of ethology. *Zeitschrift für Tierpsychologie*, **20**, 410–33.

Tooby, J. and Cosmides, L. (1992). The psychological foundations of culture. In: J. Barkow, L. Cosmides, and J. Tooby (eds), *The adapted mind: evolutionary psychology and the generation of culture*, pp. 19–136. Oxford University Press, New York.

Tybur, J.M., Lieberman, D., and Griskevicius, V. (2009). Microbes, mating, and morality: individual differences in three functional domains of disgust. *Journal of Personality and Social Psychology*, **97**, 103–22.

Underhill, P. (2000). *Why we buy: the science of shopping*. Simon and Schuster, New York.

Wedel, M. and Kamakura, W.A. (2000). *Market segmentation: conceptual and methodological foundations.* Kluwer Academic Publishers, Boston, MA.

Wilcox, K., Kim, H., and Sen, S. (2009). Why do consumers buy counterfeit luxury brands? *Journal of Marketing Research*, **46**, 247–59.

Wilson, M. and Daly, M. (2004). Do pretty women inspire men to discount the future? *Proceedings of the Royal Society B*, **271**, 177–9.

Yankelovich, D. and Meer, D. (2006). Rediscovering market segmentation. *Harvard Business Review*, **84**, 122–31.

Zhu, R. and Meyers-Levy, J. (2005). Distinguishing between the meanings of music: when background music affects product perceptions. *Journal of Marketing Research*, **43**, 333–45.

Chapter 20

Evolutionary psychology and perfume design

S. Craig Roberts and Jan Havlicek

Human olfaction: the neglected sense?

In most mammalian species, smell is an integral component of successful intraspecific communication. From aardvark to zebra, olfactory information is signalled via urinary or faecal deposits, or through emanations of specialized secretory structures evolved for this purpose. Olfaction mediates social behaviour—from the effusive sniff-greeting ceremonies of group-living species like wolves to the ritualized naso-anal and naso-genital inspections of neighbouring territorial male antelopes (Gosling and Roberts 2001). In many cases, selection has shaped such signals to have specific properties, particularly in terms of potency, robustness, and longevity, so that signals can withstand degradation in extreme climates and persist to allow unambiguous interpretation by a signal receiver, often after a considerable time (e.g. Gosling and Roberts 2001). Indeed, some of these same properties have been recognized and exploited by perfumers, so that musk, castoreum and civetone, components of the sexual and territorial signals of the musk deer (*Moschus* spp.), the beaver (*Castor canadensis* and *C. fiber*), and the African civet (*Civettictis civetta*), respectively, have also been used for many centuries as ingredients in human fragrances (Stoddart 1990).

Mammalian olfaction underpins the two arms of sexual selection—via the assessment of competitors and mates and subsequent modulation of social interactions, it regulates the processes of intrasexual competition and mate choice (e.g. Gosling and Roberts 2001). Furthermore, in some cases, specific compounds within olfactory signals, pheromones, can interfere with the physiology and behaviour of other individuals in particular ways. The boar pheromone androstenone, for example, induces lordosis in females, a characteristic mating posture (Signoret and du Mesnil du Buisson 1961). More insidiously, major histocompatibility complex (MHC) class-1 peptides present in male urinary scent marks are responsible for mediating pregnancy block in rodents (Bruce 1959; Leinders-Zufall et al. 2004), while other compounds manipulate female oestrus condition to the male's advantage (Whitten 1956).

Against this background, humans have appeared to be rather an exception, a primarily visual species in which olfactory abilities have largely been lost. Indeed, there are only about 10 million olfactory receptor cells in the human olfactory epithelium, compared to about 230 million in dogs (Schaal and Porter 1991), such reductions being partly due to changing facial architecture and reduction in size of the snout through the positioning of the eyes for depth perception, and through other changes associated with bipedalism (Jones et al. 1992). In terms of olfactory receptor genes, humans possess approximately 1000, of which two-thirds are non-functional or pseudogenes (Glusman et al. 2001; Menashe et al. 2003), compared to mice in which a much larger fraction (around 1100 of 1300 genes) are functional (Young et al. 2002).

However, the assertion that humans are especially microsmatic (that is, relying little on sense of smell) has been increasingly questioned in recent years (e.g. Schaal and Porter 1991, Shepherd 2004) on the basis that humans exhibit remarkable sensitivity to some odours (e.g. Li et al. 2007), and that odours regulate human reproductive physiology and behaviour to a greater extent than previously realized. One way in which this might occur, despite fewer receptors and functional receptor gene repertoire, is by increased cognitive power in the processing of odours through proliferation of brain areas associated with olfactory perception compared to other species (Shepherd 2004).

Furthermore, Stoddart (1990) reviews the enormous interest we invest in adorning our bodies with artificial fragrances, and recounts extensive ethnographic and anthropological evidence that demonstrates a much more prevalent role for odour in many non-industrial cultures than in modern urban societies. Here we pick just two examples: in the Ongee tribe of the Andaman Islands, odour is seen as the source of one's personality, and seasons are described according to the fragrance of the flowers that happen to be in bloom at that time (Pandya 1993). Meanwhile in Elizabethan times, it was the custom of young ladies to impregnate slices of apples with armpit odour and then to dispense these as gifts to potential suitors. Stoddart also pointed out that the human body possesses more odour-producing sources, especially apocrine glands, than those of closely-related primate species. Furthermore, the extensive development of axillary hair in humans, compared with other primates, coupled with the potential for communication via axillae as a result of acquired bipedality, led Montagna and Parakkal (1974) to muse that axillae appear to be excellent odour-producing organs. In light of all this evidence, Stoddart referred to humans as 'the scented ape'. Building on this idea, research over the past two decades has extended our understanding of the extent to which body odour is perceived by others, and particularly the kinds of information about an individual that we are able to glean (often subconsciously) from their body odour.

Perfume use in human society

Like other animals, we make several inter-personal attributions using odour, or at least have the potential to do so, and these may influence biological processes that are visible to selection, particularly our choice of mate. Why then should we invest so much time, effort and money in attempting to get rid of our natural odour by masking it under a myriad of other fragrances?

Origins of perfume use

A large part of the answer lies in cultural phenomena such as the practices of clothes-wearing and bathing. Historically, clothes were worn for weeks or months at a time, becoming impregnated with sweat and odour, as well as extraneous smells from cooking, fires and so on. Furthermore, the habit of whole-body bathing was very uncommon in European culture; indeed it was considered unhealthy (Corbin 1988) and people at most would wash hands and faces. Until around the 1850s in Europe and America, for example, bathing remained the privilege of the upper classes, promulgating the association between odour and low social status (Largey and Watson 1972; Hyde 1997). This association was already evident in ancient times: in his *Natural History*, the Roman naturalist and philosopher Pliny the Elder wrote:

> Perfume ought by right to be accredited to the Persian race: they soak themselves in it, and quench the odour produced from dirt by its adventitious attraction. The first case that I am able to discover was when a chest of perfumes was captured by Alexander among the rest of the property of King Darius when his camp was taken. Afterwards the pleasure of perfume was also admitted by our fellow-countrymen as well among the most elegant and also most honourable enjoyments of life.

Pliny went on to note the use of perfume as appropriate tributes to the dead and to the gods, and subsequently other authors have noted the use of fragrances as a symbol of religious sanctity (e.g. Classen et al. 1994). Many traditional societies, such as the Ilahita of Papua New Guinea, also have deeply-embedded cultural associations between odour and morality (Tuzin 2006). Against this background, we see the widespread use of perfumes as a testament to the fundamental human desire to be accepted within one's society and to be viewed positively by one's peers and potential mates.

Perfume diversity

The choice of perfumes is extraordinarily diverse—far more so than strictly necessary for any psychological or social function alone. One reason for this is that perfume choice follows the vagaries of fashion, as was the case even in the ancient world. As Pliny noted,

> ... the first thing proper to know about them is that their importance changes, quite often their fame having passed away. The perfume most highly praised in the old days was made on the island of Delos, but later that from the Egyptian town of Mendes ranked the highest... the iris perfume of Corinth was extremely popular for a long time, but afterwards that of Cyzicus, and similarly the attar of roses made at Phasehs...

At one level, at least historically, the arrival of new scents was constrained by the availability of naturally occurring ingredients in the local area or within trading networks. Here is Pliny once more, describing the acquisition of the key and tremendously valuable ingredient myrrh:

> It is bought up all over the district from the common people and packed into leather bags; and our perfumiers have no difficulty in distinguishing the different sorts by the evidence of the scent and consistency.

Today, however, the design of new fragrances is largely unshackled from availability. In a global society, anything is available at a price. Many ingredients that were previously harvested, extracted and purified have now been replaced with artificial compounds which can be manufactured at relatively low cost.

A second reason for the huge diversity of perfumes is the way in which the modern perfumer searches for novel combinations of compounds and even entirely novel compounds produced for the first time in the laboratory (for an excellent account of this fascinating industry, see Pybus and Sell 1999). It is largely a subjective process, new candidates being evaluated by a 'nose', a virtuoso in the art of fragrance perception, someone trained to be adept at distinguishing individual constituents in a complex odour mixture and to label these using a dedicated vocabulary to a degree well beyond the layperson. Once a pleasing fragrance is found, it is further evaluated by panels and focus groups, always with a view to testing how pleasing it smells. Thus, in the typical design process, other than customary concerns regarding compound stability and pricing, the two main benchmarks of a marketable new product lie in its hedonic perception (simply, does it smell good?) and its novelty.

Psychological effects

Today, we are more than ever aware of the deep psychological reach of odours. Certain fragrances, for example, are known to influence emotion and mood (Herz et al. 2004; Warrenburg 2005; Weber and Heuberger 2008), and performance in cognitive tasks (Herz et al. 2004; Herz 2009; Zucco et al. 2009) including memory (Smith et al. 1992; Herz and Schooler 2002). Unsurprisingly, the growing recognition of the power of odour to elicit deep-seated and often subconscious psychological effects has led to a plethora of applications, one of the best known being in retail marketing. For example, use of ambient odours increases product evaluations and brand name recall, particularly for unfamiliar brands, apparently through inducing longer periods of attention to the products in the presence of the odour (Morrin and Ratneshwar 2000; Morrin and Ratneshwar 2003). To some extent, congruity between the odour and the product can further increase marketing success (Spangenberg et al. 2006; Bosmans 2006).

Effects on the wearer

In Western culture, almost everyone can afford to use some product aimed at controlling body odour. And most do, even though most also bathe regularly: amongst a sample of 176 women and 71 men in the UK (Roberts et al. 2010), 79% of women and 60% of men reported using a deodorant every day (only 4% and 8%, respectively, never used a deodorant), while 44% of women attested to use of a perfume on an everyday basis (with only 3% reporting they never did so).

Recently, research has turned to how the use of personal fragrances influences behaviour within a social context, specifically by action on the perfume wearer. In one study (Higuchi et al. 2005), women were interviewed by a female confederate while being filmed. Halfway through the interview, half the women applied a perfume while half did not. Independent and naïve participants subsequently scored silent video clips of the interviews for rates of non-verbal behaviours and for impressions of self-confidence. Women in the perfume group were found to display fewer non-symbolic movements (e.g. touching their hair or nose) than those in the controls, became more relaxed and more dominant according to subjective rating scales, and were judged by female observers to be more self-confident.

In a second study, 35 heterosexual men were randomly allocated to a group given a commercially available deodorant or a placebo deodorant (containing alcohol spray but no active ingredients), both of which were presented in unmarked white spray cans of different shape to the product's characteristic form (Roberts et al. 2009). The men provided subjective self-ratings of confidence and attractiveness before use, and then on two subsequent occasions over 4 days while using the provided deodorants within their daily hygiene regimen. The men were also photographed and videotaped after having been asked to imagine introducing themselves to an attractive woman. The study (see also Figure 20.1) found that men in the placebo group progressively rated themselves as less attractive and self-confident compared to men in the deodorant group, although there was no initial difference. There was no difference in the perceived attractiveness of the men in the two groups as judged from their photographs by a panel of independent female judges. However, the same judges watching the muted video clips scored men in the deodorant group as significantly more attractive than men in the placebo group. Furthermore, the more the men reported liking the allocated deodorant, the more attractive they were judged in the video than expected based on their photo-rating (i.e. suggesting relatively favourable behaviour or body language). These results show how personal fragrance appeared to have an effect on men's non-verbal behaviour in such a way that it influenced how they were perceived by others, even in the absence of odour cues.

Results such as this demonstrate the potential for incorporating evolutionary psychological studies on wearer effects of personal fragrances, allowing real evaluations of important innovations in design, beyond simple provision of a convenient marketing angle. What we would like to argue in this chapter is that the process of perfume design would be further enhanced if another critical benchmark was introduced: how does the fragrance influence the communicative value of the underlying body odour of the individual who uses it? Before we address this, however, let us first examine the nature of the biological cues available in body odour and how these might influence inter-personal attributions.

Body odour as a biological signal

Animal and human work points to five main categories of information that are transferred using body odour. These are: 1) individual recognition cueing and kin-related behaviour; 2) cues of current state; 3) mediation of female reproductive physiology; 4) cues of underlying good-genes; and 5) cues of complementary genes in partner choice.

Individual recognition

The similarity of body odour among twins (Roberts et al. 2005) and other relatives (Porter and Moore 1981) indicates that body odour has, to some significant extent, a genetic basis. This, coupled with ultra-high variability in odour chemistry, has led some scholars to coin the term 'body odour signature' (Porter et al. 1985) and indicates that body odour can be involved in individual recognition. A solid body of evidence suggests that adults are able to recognize their own odour (Russell 1976; Hold and Schleidt 1977; Schleidt 1980; Schleidt et al. 1981; Platek et al. 2001), the odour of their partner (Schleidt 1980; Schleidt et al. 1981) and that of relatives (Porter and Moore 1981; Porter et al. 1985, 1986; Weisfeld et al. 2003). It should be pointed out, however, that success rates in recognition vary extensively, mostly due to the experimental paradigm, and are typically far lower than facial recognition. The ability to recognize odours develops soon after birth, as newborns prefer their mother's breast and axillary odour within several days of delivery. Similarly, mothers quickly learn the odour of their offspring (Russell et al. 1983) and it is thought that smell plays a significant role in development of mother–infant attachment. This is supported by studies which show that mothers who recognized the odour of their babies behaved more affectionately toward them a month later (Fleming et al. 1995; Steiner et al. 1996).

Furthermore, odour recognition of relatives within a family, and hedonic attributions towards such odours, may be an important component of incest avoidance. Mutual odour aversion occurs between father and daughter, and between brothers and sisters (Weisfeld et al. 2003). Hedonic shifts in odour preferences away from kin and towards non-kin is expected to develop during puberty when mate choice becomes relevant. However, a recent study by (Ferdenzi et al. 2010) did not find clear support for this idea, perhaps due to the relatively small sample size, and we await more data to resolve this issue.

Cues of current state

Body odour variability reflects not only genetic make-up but also some environmental factors which may provide cues to an individual's current state. Examples of such cues include: 1) reproductive state, 2) diet, 3) health status, and 4) affective state.

Although human ovulation is traditionally regarded as concealed, results of several recent studies show changes in hedonic ratings of female axillary odour across the menstrual cycle, peaking around ovulation (Havlicek et al. 2006; Singh and Bronstad 2001). In contrast, such shifts were not observed in women using hormonal contraception (Kuukasjarvi et al. 2004). The apocrine glands of the axillary region, which are primarily responsible for production of chemicals involved in development of body odour, initiate activity only in puberty and it is thought they are sensitive to fluctuations in steroid hormone concentration. Direct empirical evidence is scarce, but we can expect changes in body odour related to puberty, and perhaps also to menopause.

Studies on rodents show that diet affects body odours, and subsequent odour preferences among conspecifics (Beauchamp 1976; Ferkin et al. 1997). Odours of mice which had been temporarily starved were not attractive to opposite-sex individuals, but attractiveness was restored within 24h of feeding recommencing (Pierce and Ferkin 2005). In contrast to rodents, most evidence on dietary effects on human body odours is only anecdotal or indirect. For instance, communities that traditionally eat high amounts of fish are said to have a characteristic odour. Analytical studies suggest that some aromatic foods, for example plants of the *Alliaceae* or *Brassicaceae* families which are a common part of European cuisine, could affect human axillary odour (Buttery et al. 1976). However, the only direct evidence comes from a study in which consumption of red meat negatively affected subjectively perceived quality of the body odour (Havlicek and Lenochova 2006).

In addition, body odour may reflect underlying health, although similarly to dietary effects, we currently have only limited experimental evidence about this issue. However, numerous reports from the clinical literature provide strong support for this view (Havlicek and Lenochova 2008). It is known that individuals affected by some metabolic disorders such as diabetes have a characteristic smell of acetone on the breath, while the rarer maple syrup syndrome causes a distinctive odour (Monastiri et al. 1997). More commonly, changes in body odour are caused by infectious agents. For example, some skin inflammations or ulcers have a specific and foul odour (Finlay et al. 1996). It is noteworthy that rodent studies show that the smell of individuals infected by parasitic worms is avoided by their conspecifics (Willis and Poulin 2000). There is no evidence of a similar phenomenon in humans; however, in view of the significant role of parasitic diseases in human evolution, this issue is no doubt worth exploring.

Last, results of several studies point to the fact that humans are sensitive to changes in body odour caused by affective tuning (Ackerl et al. 2002; Chen and Haviland-Jones 2000). These studies evoked specific affective states in participants by showing them films of either a fearful or happy nature while they were wearing experimental tee-shirts. Subsequently, raters were asked to judge how the odour donors might have felt, based solely on their odour. Raters, women in particular, judged the odours in the fearful condition (and to some extent also the happy one) correctly at levels above chance. Further, several recent studies suggest that human odour collected under conditions of anxiety cause increases in the startle reflex and risk-taking behaviour, and induces empathy and anxiety in smellers (e.g. Pause et al. 2009; Albrecht et al. 2011). Anxiety-related body odours also enhance decoding of emotionally ambiguous human faces (Zhou and Chen 2009). The significance of these findings is further supported by brain imaging studies which show that these chemosensory cues activate neural segments, for instance the amygdala, involved in processing emotionally-laden stimuli (Mujica-Parodi et al. 2009).

In summary, body odour provides various cues about individual current state. Although this field started to expand only very recently, the reviewed evidence clearly support the notion that communication of affective states is partly mediated by olfaction.

Mediation of reproductive physiology

More than 30 years ago, McClintock (1971) noted in her classical paper that cohabitating female students tend to synchronize the onset of their menstrual cycle. This early observation gave rise to several dozen studies on highly variable samples of cohabiting women, such as among mothers and daughters (Weller and Weller 1993a), Bedouin families (Weller and Weller 1997), kibbutz shared households (Weller and Weller 1993b) or lesbian couples (Weller and Weller 1992). A majority, but not all (Trevathan et al. 1993; Yang and Schank 2006) of the studies showed similar shifts toward menstrual synchrony. Menstrual synchrony was also found in some samples of women spending protracted amounts of time together (e.g. in women sharing offices) and this effect is modulated by closeness between workers (Weller and Weller 1995b). On the other hand, a similar effect was not shown in other samples, for instance basketball team players (Weller and Weller 1995a). It should also be pointed out that the methodology of a majority of these studies

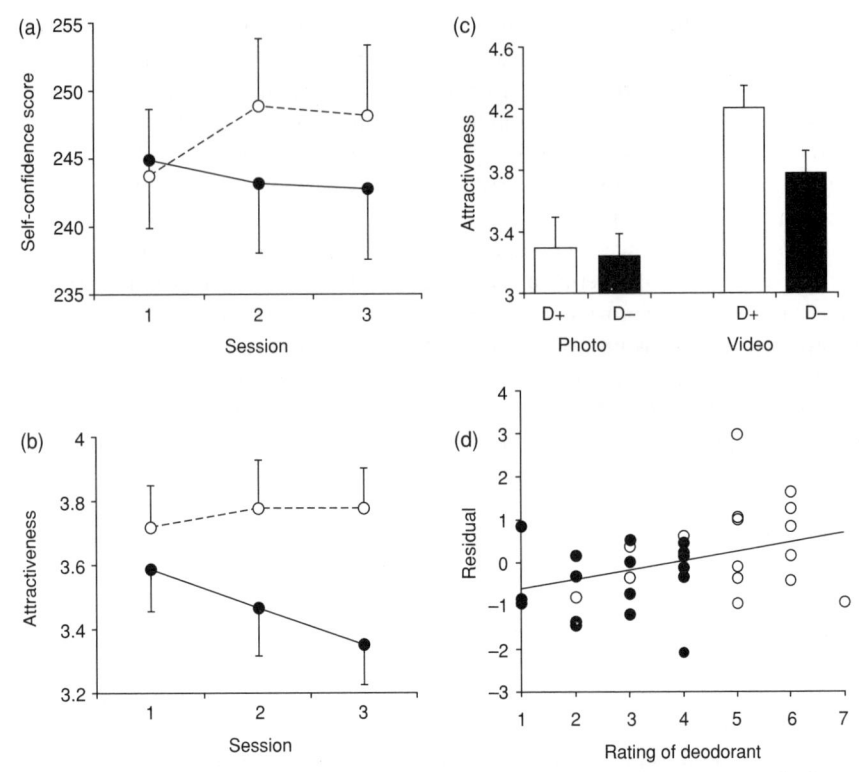

Fig. 20.1 Effects of perfume use on wearer's behaviour and perceived attractiveness of others. (a) Mean (± standard error) self-confidence scores collected from participants using an allocated full deodorant formulation (open circles or D+) or a placebo deodorant (closed circles or D–), assessed across three sessions spanning 72 hours. (b) Self-rated attractiveness scores during the same experiment. (c) Mean (± standard error) ratings of male facial attractiveness by female judges based on either digital photos or video clips taken at the end of the experiment. (d) Standardized residuals of video-rated over photograph-rated attractiveness and male-assessed odour pleasantness, for the D+ group (open circles) and D– group (closed circles). The figure shows that men who expressed liking for the allocated deodorant were more likely judged attractive in video-ratings than expected based on their photograph-rated attractiveness.

has been questioned (Wilson 1992) and some authors even suggest that the whole effect is an artefact of convergence of irregular cycles (Schank 2001, 2006).

In her original paper, McClintock suggested that menstrual synchrony could be mediated by odour. More recently, her team carried out an experiment in which a group of women was exposed to axillary extracts from other women (Stern and McClintock 1998). The odour recipients tended to shift the onset of their cycle towards the odour donors, suggesting a priming pheromonal effect. The chemicals responsible for this phenomenon are currently unknown, but it was shown that women sensitive to androstenol (a steroidal compound found in axillary odour) have a higher tendency to synchronize their cycles (Morofushi et al. 2000). Although menstrual synchrony is often regarded as the best examined example of human pheromones, its potential function is rather unclear. It is also possible that its function was more significant in cooperatively breeding primate ancestors of our species, or in polygynous social settings (Burley 1979), and simply remain as an evolutionary vestige in modern humans.

Several studies have also found that regular dyadic sexual activities, but not autosexual ones, have an effect on female menstrual functions. Women with weekly sexual activity had more regular cycles and a higher frequency of ovulatory cycles (Cutler et al. 1985; Burleson et al. 1991). Again it is thought that this effect may be mediated by male odour, supported by a study showing shortenings of length between individual LH (luteinizing hormone) pulses in women smelling male axillary extracts (Preti et al. 2003). Levels of LH are involved in hormonal regulation of ovulation.

Lastly, it is known that male partners of pregnant women show changes in hormonal profiles of testosterone and prolactin, which might have an impact on their nurturing behaviour (Storey et al. 2000). The mechanism underlying these shifts is currently unclear; however, it is possible that it is again due to some chemicals produced by pregnant women. This idea is supported by an analytical study showing that pregnant women emit different chemical substances compared with non-pregnant women (Vaglio et al. 2009).

Indicators of genetic quality

It is thought that the main role of odour communication is in mate choice and the regulation of romantic relationships. Current models of sexual selection predict that at least some cues which are involved in mate choice reflect the quality of the individual (Kokko et al. 2002). We can thus expect this would also be the case for body odour. One of the indicators which has been extensively studied in the last two decades is fluctuating asymmetry. In bilaterally symmetric organisms (including humans), we can see small non-directional imperfections in symmetry, termed fluctuating asymmetry. These fluctuations are thought to reflect the ability of the organism to cope with environmental stresses like toxins and infections (Møller et al. 1999). A number of studies on both facial and body attractiveness have shown that more symmetrical individuals tend to be perceived as attractive (Thornhill and Gangestad 1999a). It has also been shown that the level of symmetry is related to the attractiveness of body odour in both men and women (Gangestad and Thornhill 1998; Rikowski and Grammer 1999; Thornhill and Gangestad 1999b; Thornhill et al. 2003), and that women are particularly sensitive to these cues around the time of ovulation (i.e. when conception is most likely). Furthermore, there is evidence that periovulatory women prefer men with high scores of psychological dominance, which may in turn reflect their biological quality (Havlicek et al. 2005).

Cues of complementary genes

Whilst it is expected that the above-mentioned preferences to markers of genetic quality will show a rather uniform preferential pattern across individuals, the same does not apply to cues of genetic compatibility, which follow more individually specific patterns. One of the model systems in which this mechanism is studied are the highly variable genes of the major histocompatibility complex (MHC). Products of these genes play a key role in immune system functioning by presenting peptides of foreign origin on the cellular surface. Each of the genes in the complex have several tens of different known alleles, suggesting that the region is under strong balancing selection (Apanius et al. 1997). It has been shown in various vertebrate species that females prefer the odours of potential partners with MHC genes that are dissimilar to their own (Penn and Potts 1999). Several human studies have also found that naturally cycling women prefer the odour of MHC-dissimilar (Wedekind et al. 1995; Wedekind and Füri 1997; Santos et al. 2005) or moderately dissimilar partners (Jacob et al. 2002). When men judge female odours, the results are more ambivalent, with one study (Thornhill et al. 2003) showing a preference for dissimilarity while another failed to find a significant link (Santos et al. 2005). Moreover, a pioneering paper by

Wedekind et al. (1995) suggested disruption of these preferences in women using hormonal contraception. More recently, Roberts et al. (2008) replicated this study using a more sensitive within-subject design. Although they found a shift in preferences in women who started to use contraception, they were not able to replicate preferences for the odour of dissimilar individuals (see also review in Havlicek and Roberts 2009).

Towards biologically-informed design?

As we have shown, odour carries several important cues to biological quality. Since these cues have been shaped by selection over evolutionary time, and play a role in coordination of key social interactions, incorporation of this knowledge into perfume design could potentially provide a springboard for transforming the success of specific perfumes. Current understanding of body odour-associated effects on inter-individual perception and attribution lead us to several promising possible avenues to take this process forward.

Constant fertility enhancement

The finding that there is a periovulatory peak in attractiveness of women's axillary odour, as perceived by men (Singh and Bronstad 2001; Havlicek et al. 2006), clearly suggests a change in the chemical composition of women's axillary odour associated with ovulation. This is not unsurprising, since similar fertility-associated odour cues are well-known in other species including mice (Schwende and Novotny 1982; Andreolini et al. 1987), elephants (Dehnhard et al. 2001), cattle (Dehnhard and Claus 1996), sheep (Blissitt et al. 1994), and horses (Ma and Klemm 1997).

These changes influence attraction to the odour by males (Ferkin and Johnston 1995) and also modulate their intrasexual behaviour such that aggression between males is higher when exposed to the scent of a fertile female than a non-fertile one (Dixon and Mackintosh 1975). The potential for a similar effect in humans was recently demonstrated in an experiment in which exposure of men to the odour of women at the fertile stage of the cycle elicited higher levels of testosterone after testing than men exposed to odours of non-fertile women (Miller and Maner 2010).

Given this, if the compound or group of compounds that provide cues to ovulation were discovered, it might then be possible to incorporate this into a perfume in such a way as to raise the perceived attractiveness of the wearer, throughout the menstrual cycle, to the level normally experienced only at ovulation. Even if this were only a subtle effect, it could provide a unique selling point and one that would likely prove irresistible to perfume manufacturers and marketers.

Amplifying quality perception

In a similar fashion, perfumes that incorporate the constituents of body odour associated with any measure of biological quality should improve the attractiveness attributions of the wearer, as well as (again) providing a gift to the product's advertisement campaign. For example, Havlicek et al.'s (2005) discovery of a correlation between perceived pleasantness of axillary odour and men's relative psychometric dominance indicates a chemical basis which could be exploited in male perfume. Once again, this finding reflects similar phenomena in other species, particularly rodents, where odours of dominant males are preferred over that of subordinates (White et al. 1986; Mossman and Drickamer 1996; Kruczek 1997; for reviews see Gosling and Roberts 2001; Roberts 2007).

In mice, unlike in humans, the compounds responsible for dominance signalling are known. Within seven days of establishment of dominant or subordinate relationships, gas chromatography reveals quantitative differences in at least sixteen compounds found in male urine (Harvey et al. 1989).

Subordinate male urine shows a decrease in concentrations of ketones, dihydrofurans, and acetates, while dominant male urine increases in levels of 2,3 dihydro-*exo*-brevicomin, 2-*sec*-butyl-4,5-dihydro-thiazole and two sesquiterpenic compounds, alpha- and beta-farnesene. These compounds are androgen-dependent and are absent from the urine of castrated males, but are restored by administration of testosterone (Harvey et al. 1989). Importantly, these compounds have been synthesized and, when added to urine of castrated males, they invoke levels of aggression (Novotny et al. 1985) and restore levels of interest and attractiveness to females (Jemiolo et al. 1985) that are equivalent to those seen in intact males.

Cueing complementary genes

There has been much recent interest in how odour might reveal genetic compatibility between potential mates. As before, the research in humans has been preceded by work in other species, notably mice, where genotype at the MHC is revealed in urinary odour and, when given the opportunity to choose, females prefer to mate disassortatively, that is, with males who are relatively dissimilar at the MHC (Yamazaki et al. 1976; Potts et al. 1991). Although the task of characterizing the chemical basis underlying MHC-correlated odours is a great deal more complex than that of determining the basis of social dominance, it appears that the cues to relative dissimilarity in mouse urine are carried by a series of volatile carboxylic acids which vary in relative concentration according to MHC type (Singer et al. 1997; Yamaguchi et al. 1981). Although the extent to which MHC-correlated odour influence human behaviour is still a topic of debate and research interest, accumulating evidence suggests the real possibility for a significant role in affecting mating preferences (see also reviews in Roberts and Little 2008; Havlicek and Roberts 2009).

The potential for interactions between perfumers and evolutionary psychologists has been neatly demonstrated in an experiment by Milinski and Wedekind (2001), in which they correlated perfume preferences with MHC genotype. They asked 137 men and women to score 36 individual perfume ingredients using the item 'Would you like to smell like that yourself?'. Two years later, a subset of 18 ingredients was scored for self (the same item as before) and also as potential ingredients for a partner's perfume, using the question 'Would you like your partner to smell like that?'. Where results could be compared across the 2 years, a high degree of repeatability in ratings indicated consistency in odour perception. Furthermore, there were significant correlations between a rater's genotype and the extent to which they liked or disliked individual perfume components as ingredients for their own use. However, the same correlation did not exist between genotype and ratings for their partner's perfume, indicating that the results could not be explained simply by generalized odour preferences. These important results raise the possibility that perfume choice might not be arbitrary or random and simply mask underlying body odour as is generally assumed, but more intriguingly, that the choice might serve to complement one's own body odour. In other words, the use of perfume might not interfere with the underlying biological signal contained in body odour as much as previously thought.

In a study led by our colleague Pavlina Lenochova, and with several other researchers, we recently tested between these ideas in a series of experiments in Prague and Vienna. We hypothesized that if perfumes act by simply masking body odour, we should find uniformly higher ratings of perfumed axillary samples compared with non-perfumed samples from the same individuals, as well as reduced individual variability in odour pleasantness across individuals. Although we found in two experiments that naïve smellers rated perfumed samples more positively than non-perfumed samples, significant interactions with perfume wearers indicated that the perfume acted differently on different individuals (Figure 20.2a). This suggests that the application of perfumes to existing body odour produces an odour mixture with an emergent

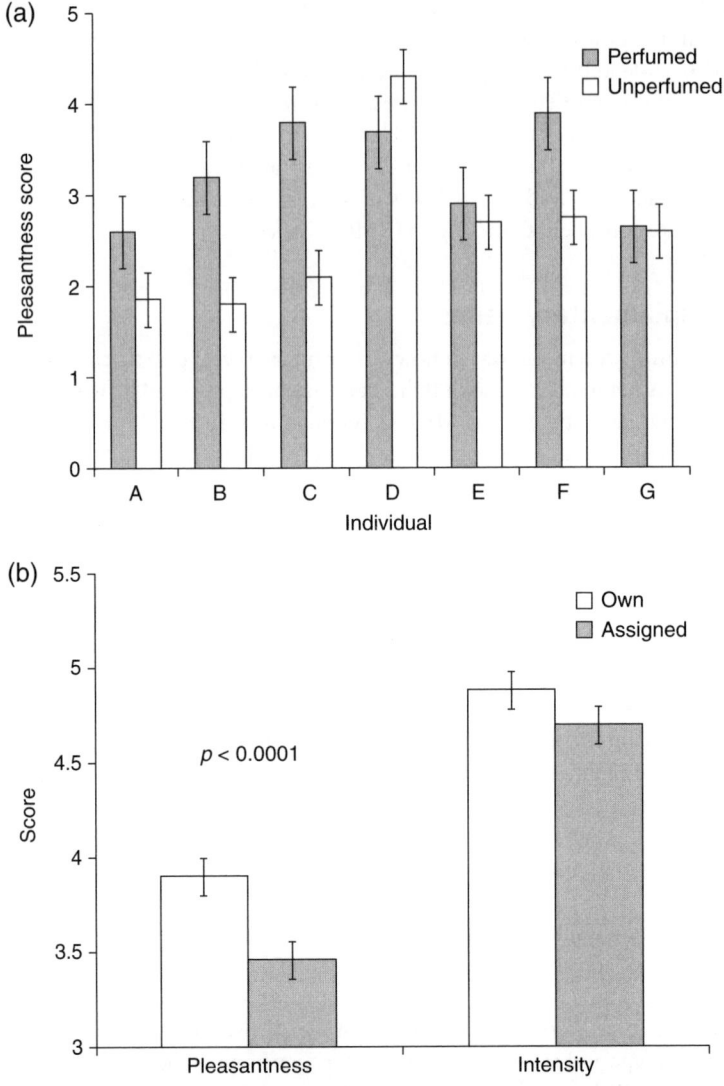

Fig. 20.2 Effect of perfume on perception of body odour. (a) Mean (± standard error) pleasantness scores of body odour from perfumed and non-perfumed axillae of the same individuals (A–G). Perfumed odours are perceived as more pleasant on average, but not in all cases and there is a significant odour donor × perfume condition interaction. (b) Mean (± standard error) pleasantness and intensity scores of body odour and perfume blends. Participants applied their own perfume to one axilla, and one assigned by the experimenters to the other. The body odour/perfume blend is more pleasant when composed of the individual's preferred perfume.

quality that is perceptually different from either constituent. We took this further in a third experiment, in which we compared ratings of axillary samples collected when the odour donors were wearing either their own perfume or another perfume assigned by the experimenters. In subsequent rating sessions, smellers perceived the body odour-perfume blend to be more pleasant when the donors wore their own perfume, compared with the body odour blend with the assigned perfume (Figure 20.2b). These results challenge the conventional view that perfume use acts to simply mask underlying odour. On the contrary, they support the view of Milinski and Wedekind (2001) and suggest that perfumes blend with body odour in a complementary manner, working with rather than against the underlying odour. The results show that perfume use can thus potentially enhance the communicatory significance of body odour, as well as providing an explanation for the highly individual nature of perfume choice.

The implications of these results for perfumers are potentially revolutionary. First, it is equally relevant to both males and females; in contrast, the constant fertility enhancement is restricted to the female market and amplifying quality perception is likely to be mainly male-oriented. Second, it emphasizes the individuality of perfume preferences, providing a firm scientific basis for further development of a niche market in customized designer products. Third, it carries implications for the ways in which perfumes are purchased. Perfumes are often bought as presents for others. However, the research suggests that one is more likely to choose a perfume that suits oneself than the intended beneficiary. Possibly this effect would be ameliorated by buying for relatives, but, according to this study at least, one should avoid buying a perfume for an unrelated sexual partner.

Some caveats

Although there appears to be ways in which insights from evolutionary psychology can inform perfume design and marketing, there are also some important considerations which urge for a cautious approach, and it is to these that we now turn.

The product works too well

There are at least two possible undesirable side effects on the behaviour of the perfume wearer and those with whom the wearer comes into contact. First, imagine a perfume which contained some presently-unknown compound with attractant properties akin to bombykol and other classic pheromones of the insect world. We think this very unlikely but, according to marketing hype, this is the dream offered to young men by at least one well-known commercial product, where one spray results in a stampede of beautiful women converging from all directions towards the lucky consumer. Undoubtedly an amusing idea and excellent marketing device, in reality, such a product would generate undesirable levels of attention. Even if we assume a product existed which exerted a much more subtle effect, a user would have no control over who approached, or when. In general terms, it might be expected that it would be more popular with men than women (according to views of sex-typical evolved strategies) but even amongst young men, it might eventually become too much.

Second, one could also imagine a design which built upon the association between odour chemistry and male dominance, or one that boosted a wearer's self-confidence. While this might attract attention from potential mates, the intended receivers of the signal, it could also attract unwanted attention from others, notably other males. In the mouse studies described above, addition of synthetic compounds associated with dominance to the odour of target males increased levels of aggression from other males. Such a fragrance might also elicit higher levels of aggressive behaviour in wearers than would be normally experienced.

Disruption of function

Another issue with a perfume that included analogues of active body odour components is that it might often disrupt the underlying signalling interaction on which it was based. The most obvious way in which this could occur is simply if detection of the signal is precluded by the overpowering nature of the fragrance. Although Milinski and Wedekind's (2001) study suggests that the masking of the biological signal by artificial fragrances (at least those made from ingredients not occurring in body odour—their study did not examine compounds of human origin) may be less problematic than previously thought, any such masking that did occur would be likely to disrupt evolutionarily stable patterns of mate selection. A more targeted manipulation of a perfume by manipulating or mimicking elements of the underlying biological signal would likely exacerbate this disruption. For example, addition of compounds associated with attractiveness, dominance, or some other trait associated with high quality, would disproportionately favour relatively poor-quality individuals and this might influence their chances of being selected by a higher quality opposite-sex partner than would otherwise be the case. This may be to the benefit of the perfume user, but could come at a cost to the partner since s/he would potentially have been deceived about the partner's quality.

A further problem could arise in the use of compounds that mimicked a woman's periovulatory state in a perfume aimed at promoting a constantly high perception of attractiveness. While at first sight this might seem to be an effective design solution to the problem of cyclical variation in odour perception by others, careful attention would need to be paid to possible disruptive effects on the user's (or others') own menstrual cycle.

Ethical concerns

Some of the potential design improvements we have outlined also carry certain philosophical concerns surrounding the manipulation of the perception of another person through use of a biologically-manipulated perfume. In essence, the issue boils down to this: is it acceptable to subliminally manipulate Y's perception of X in order to make Y fall in love with X? How would Y respond to discovering s/he has been chemically duped?

Our current view is that this is likely to be of trivial concern, and is no more problematic than use of, say, a facial cosmetic that enhanced eyelid or lip colour to alter perception of a woman's attractiveness, a foundation make-up that disguised poor underlying skin condition, or even cosmetic surgery (e.g. breast augmentation). Sooner or later, these minor deceptions tend to come to light within a developing relationship, and may or may not play a part in its continuance.

However, we raise the point mainly because it could become relevant should there be in the future any discovery and use of a compound that has the property, in the sense of a classical insect pheromone, of insidiously triggering a specific behaviour which is beyond the control of the signal receiver. No such compound (or behavioural response) has yet been found, however (nor do we think personally one will be found, but as we have said, this is common advertisement hype). The closest we do come to this scenario are the androgen steroids which occur in the axillary sweat of both men and women and which are already known to alter physiology and behaviour, although these effects appear to be dependent on context and experience, and modulate existing behaviour in subtle ways rather than altering or dictating behavioural responses. Compounds such as androstadienone and androstenol, for example, induce changes in hormonal levels of exposed participants, along with induced moods and affective states (reviewed in Havlicek et al. 2010) and can even modulate attractiveness attributions in certain situations (Saxton et al. 2008a,b). If a still more potent compound does come to light, perhaps its use would come to be treated as seriously as mind-altering drugs like flunitrazepam (rohypnol).

Conclusions

Although we have highlighted some cautionary issues, we believe there is at least potential for a beneficial relationship between evolutionary psychologists and designers and marketeers in the perfume industry. Evolutionary psychology has already and will continue to produce insights into the informative capacity of body odour in human perception and its role in social interactions, particularly in partner preferences and mate selection. These insights could increasingly be harnessed to increase both the potency of perfume function and the strategies employed in its marketing.

We think one of the most interesting approaches would be to address in more detail the relationship between an individual's perfume preference and their genotype, particularly at the MHC. The possibility that the two are linked has already been demonstrated, although this deserves further work and replication. If true, the connection provides a further clue for explaining the diversity of perfume preferences and choice, since MHC genotype is so variable. For the perfume industry, it has the potential to open up a new realm of individually-customized perfumes—uniquely tailored and bespoke, and thus profitable.

More work is needed particularly to understand the ramifications of perfume choices on social relationships. The past two decades have demonstrated how body odour plays a more critical role in intimate social decision-making than ever before realized. Yet issues such as how individuals choose specific perfumes, how perfumes interact with body odour, and how perfume-body odour blends are perceived, have scarcely been addressed. What is most needed, for both evolutionary psychologists and perfumers, is greater understanding of the chemistry of body odour and the chemical basis for specific biological cues. With this knowledge, some of the possibilities we have described become truly real.

References

Ackerl, K., Atzmueller, M., and Grammer, K. (2002). The scent of fear. *Neuroendocrinology Letters*, **23**, 79–84.

Albrecht, J., Demmel, M., Schopf, V., et al. (2011). Smelling chemosensory signals of males in anxious versus nonanxious condition increases state anxiety of female subjects. *Chemical Senses*, **36**, 19–27.

Andreolini, F., Jemiolo, B., and Novotny, M. (1987). Dynamics of excretion of urinary chemosignals in the house mouse (*Mus musculus*) during the natural estrus cycle. *Experientia*, **43**, 998–1002.

Apanius, V., Penn, D., Slev, P.R., Ruff, L.R., and Potts, W.K. (1997). The nature of selection on the major histocompatibility complex. *Critical Reviews in Immunology*, **17**, 179–224.

Beauchamp, G.K. (1976). Diet influences attractiveness of urine in guinea-pigs. *Nature*, **263**, 587–8.

Blissitt, M.J., Bland, K.P., and Cottrell, D.F. (1994). Detection of estrous-related odor in ewe urine by rams. *Journal of Reproduction and Fertility*, **101**, 189–91.

Bosmans, A. (2006). Scents and sensibility: when do (in)congruent ambient scents influence product evaluations? *Journal of Marketing*, **70**, 32–43.

Bruce, H.M. (1959). Exteroceptive block to pregnancy in the mouse. *Nature*, **184**, 105.

Burleson, M.H., Gregory, W.L., and Trevathan, W.R. (1991). Heterosexual activity and cycle length variability: effect of gynecological maturity. *Physiology and Behavior*, **50**, 863–6.

Burley, N. (1979). Evolution of concealed ovulation. *American Naturalist*, **114**, 835–58.

Buttery, R.G., Guadagni, D.G., Ling, L.C., Seifert, R.M., and Lipton, W. (1976). Additional volatile components of cabbage, broccoli, and cauliflower. *Journal of Agriculture and Food Chemistry*, **24**, 829–32.

Chen, D. and Haviland-Jones, J. (2000). Human olfactory communication of emotion. *Perceptual and Motor Skills*, **91**, 771–81.

Classen, C., Howes, D., and Synnot, A. (1994). *Aroma—the cultural history of smell*. Routledge, London.

Corbin, A. (1988). *The foul and the fragrant: odor and the French social imagination*. Harvard University Press, Harvard, MA.

Cutler, W.B., Preti, G., Huggins, G.R., Erickson, B., and Garcia, R. (1985). Sexual behavior frequency and biphasic ovulatory type menstrual cycles. *Physiology and Behavior*, **34**, 805–10.

Dehnhard, M. and Claus, R. (1996). Attempts to purify and characterize the estrus-signalling pheromone from cow urine. *Theriogenology*, **46**, 13–22.

Dehnhard, M., Heistermann, M., Goritz, F., Hermes, R., Hildebrandt, T., and Haber, H. (2001). Demonstration of 2-unsaturated C-19-steroids in the urine of female Asian elephants, Elephas maximus, and their dependence on ovarian activity. *Reproduction*, **121**, 475–84.

Dixon, A.K. and Mackintosh, J.M. (1975). The relationship between the physiological condition of female mice and the effects of their urine on the social behaviour of adult males. *Animal Behaviour*, **23**, 513–20.

Ferdenzi, C., Schaal, B., and Roberts, S.C. (2010). Family scents: developmental changes in the perception of kin body odor? *Journal of Chemical Ecology*, **36**, 847–54.

Ferkin, M.H. and Johnston, R.E. (1995). Effects of pregnancy, lactation and postpartum estrus on odor signals and the attraction to odors in female meadow voles, *Microtus pennsylvanicus*. *Animal Behaviour*, **49**, 1211–17.

Ferkin, M.H., Sorokin, E.S., Johnston, R.E., and Lee, C.J. (1997). Attractiveness of scents varies with protein content of the diet in meadow voles. *Animal Behaviour*, **53**, 133–41.

Finlay, I.G., Bowszyc, J., Ramlau, C., and Gwiezdzinski, Z. (1996). The effect of topical 0.75% metronidazole gel on malodorous cutaneous ulcers. *Journal of Pain and Symptom Management*, **11**, 158–62.

Fleming, A., Corter, C., Surbey, M.K., Franks, P., and Steiner, M. (1995). Postpartum factors related mother's recognition of newborn infant odours. *Journal of Reproductive and Infant Psychology*, **13**, 197–210.

Gangestad, S.W. and Thornhill, R. (1998). Menstrual cycle variation in women's preferences for the scent of symmetrical men. *Proceedings of the Royal Society B*, **265**, 927–33.

Glusman, G., Yanai, I., Rubin, I., and Lancet, D. (2001). The complete human olfactory subgenome. *Genome Research*, **11**, 685–702.

Gosling, L.M. and Roberts, S.C. (2001). Scent-marking by male mammals: cheat-proof signals to competitors and mates. *Advances in the Study of Behavior*, **30**, 169–217.

Harvey, S., Jemiolo, B., and Novotny, M. (1989). Pattern of volatile compounds in dominant and subordinate male mouse urine. *Journal of Chemical Ecology*, **15**, 2061–72.

Havlicek, J. and Lenochova, P. (2006). The effect of meat consumption on body odor attractiveness. *Chemical Senses*, **31**, 747–52.

Havlicek, J. and Lenochova, P. (2008). Environmental effects on human body odour. In: J.L. Hurst, R.J. Beynon, S.C. Roberts, and T.D. Wyatt (eds), *Chemical Signals in Vertebrates 11*, pp. 199–212. Springer, New York.

Havlicek, J. and Roberts, S.C. (2009). MHC-correlated mate choice in humans: a review. *Psychoneuroendocrinology*, **34**, 497–512.

Havlicek, J., Roberts, S.C., and Flegr, J. (2005). Women's preference for dominant male odour: effects of menstrual cycle and relationship status. *Biology Letters*, **1**, 256–9.

Havlicek, J., Dvorakova, R., Bartos, L., and Flegr, J. (2006). Non-advertized does not mean concealed: Body odour changes across the human menstrual cycle. *Ethology*, **112**, 81–90.

Havlicek, J., Murray, A.K., Saxton, T.K., and Roberts, S.C. (2010). Current issues in the study of androstenes in human chemosignaling. *Vitamins and Hormones: Pheromones*, **83**, 47–81.

Herz, R.S. (2009). Aromatherapy facts and fictions: a scientific analysis of olfactory effects on mood, physiology and behavior. *International Journal of Neuroscience*, **119**, 263–90.

Herz, R.S. and Schooler, J.W. (2002). A naturalistic study of autobiographical memories evoked by olfactory and visual cues: testing the Proustian hypothesis. *American Journal of Psychology*, **115**, 21–32.

Herz, R.S., Schankler, C., and Beland, S. (2004). Olfaction, emotion and associative learning: effects on motivated behavior. *Motivation and Emotion*, **28**, 363–83.

Higuchi, T., Shoji, K., Taguchi, S., and Hatayama, T. (2005). Improvement of nonverbal behaviour in Japanese female perfume-wearers. *International Journal of Psychology*, **40**, 90–9.

Hold, B. and Schleidt, M. (1977). Personal odor and nonverbal communication. *Zeitschrift für Tierpsychologie*, **43**, 225–38.

Hyde, A. (1997). *Bodies of law*. Princeton University Press, Princeton, NJ.

Jacob, S., McClintock, M.K., Zelano, B., and Ober, C. (2002). Paternally inherited HLA alleles are associated with women's choice of male odor. *Nature Genetics*, **30**, 175–9.

Jemiolo, B., Alberts, J., Sochinski-Wiggins, S., Harvey, S., and Novotny, M. (1985). Behavioural and endocrine responses of female mice to synthetic analogues of volatile compounds in male urine. *Animal Behaviour*, **33**, 1114–18.

Jones, S., Martin, R., and Pilbeam, D. (1992). *The Cambridge encyclopedia of human evolution*. Cambridge University Press, Cambridge.

Kokko, H., Brooks, R., McNamara, J.M., and Houston, A.I. (2002). The sexual selection continuum. *Proceedings of the Royal Society B*, **269**, 1331–40.

Kruczek, M. (1997). Male rank and female choice in the bank vole, *Clethrionomys glareolus*. *Behavioral Processes*, **40**, 171–6.

Kuukasjarvi, S., Eriksson, C.J.P., Koskela, E., Mappes, T., Nissinen, K., and Rantala, M.J. (2004). Attractiveness of women's body odors over the menstrual cycle: the role of oral contraceptives and receiver sex. *Behavioral Ecology*, **15**, 579–84.

Largey, G.P. and Watson, D.R. (1972). The sociology of odors. *American Journal of Sociology*, **77**, 1021–34.

Leinders-Zufall, T., Brennan, P., Widmayer, P., et al. (2004). MHC class I peptides as chemosensory signals in the vomeronasal organ. *Science*, **306**, 1033–7.

Li, W., Moallem, I., Paller, K.A., and Gottfried, J.A. (2007). Subliminal smells can guide social preferences. *Psychological Science*, **18**, 1044–9.

Ma, W. and Klemm, W.R. (1997). Variations of equine urinary volatile compounds during the oestrous cycle. *Veterinary Research Communications*, **21**, 437–46.

McClintock, M.K. (1971). Menstrual synchrony and suppression. *Nature*, **229**, 244–5.

Menashe, I., Man, O., Lancet, D., and Gilad, Y. (2003). Different noses for different people. *Nature Genetics*, **34**, 143–4.

Milinski, M. and Wedekind, C. (2001). Evidence for MHC-correlated perfume preferences in humans. *Behavioral Ecology*, **12**, 140–9.

Miller, S.L. and Maner, J.K. (2010). Scent of a woman: men's testosterone responses to olfactory ovulation cues. *Psychological Science*, **21**, 276–83.

Møller, A.P., Gangestad, S.W., and Thornhill, R. (1999). Nonlinearity and the importance of fluctuating asymmetry as a predictor of fitness. *Oikos*, **86**, 366–8.

Monastiri, K., Limame, K., Kaabachi, N., et al. (1997). Fenugreek odour in maple syrup urine disease. *Journal of Inherited Metabolic Disease*, **20**, 614–15.

Montagna, W. and Parakkal, P.F. (1974). *The structure and function of skin*. Academic Press, New York.

Morofushi, M., Shinohara, K., Funabashi, T., and Kimura, F. (2000). Positive relationship between menstrual synchrony and ability to smell 5α-androst-16-en-3α-ol. *Chemical Senses*, **25**, 407–11.

Morrin, M. and Ratneshwar, S. (2000). The impact of ambient scent on evaluation, attention, and memory for familiar and unfamiliar brands. *Journal of Business Research*, **49**, 157–65.

Morrin, M. and Ratneshwar, S. (2003). Does it make sense to use scents to enhance brand memory? *Journal of Marketing Research*, **40**, 10–25.

Mossman, C.A. and Drickamer, L.C. (1996). Odor preferences of female house mice (*Mus domesticus*) in seminatural enclosures. *Journal of Comparative Psychology*, **110**, 131–8.

Mujica-Parodi, L.R., Strey, H.H., Frederick, B., et al. (2009). Chemosensory cues to conspecific emotional stress activate amygdala in humans. *Plos One,* **4**, e6415.

Novotny, M., Harvey, S., Jemiolo, B., and Alberts, J. (1985). Synthetic pheromones that promote inter-male aggression in mice. *Proceedings of the National Academy of Sciences of the USA,* **82**, 2059–61.

Pandya, V. (1993). *Above the forest: a study of Andamanese ethnoanemology, cosmology and the power of ritual.* Oxford University Press, Delhi.

Pause, B.M., Adolph, D., Prehn-Kristensen, A., and Ferstl, R. (2009). Startle response potentiation to chemosensory anxiety signals in socially anxious individuals. *International Journal of Psychophysiology,* **74**, 88–92.

Penn, D.J. and Potts, W.K. (1999). The evolution of mating preferences and major histocompatibility complex genes. *American Naturalist,* **153**, 146–63.

Pierce, A.A. and Ferkin, M.H. (2005). Re-feeding and the restoration of odor attractivity, odor preference, and sexual receptivity in food-deprived female meadow voles. *Physiology and Behavior,* **84**, 553–61.

Platek, S.M., Burch, R.L., and Gallup, G.G. (2001). Sex differences in olfactory self-recognition. *Physiology and Behavior,* **73**, 635–40.

Porter, R.H. and Moore, J.D. (1981). Human kin recognition by olfactory cues. *Physiology and Behavior,* **27**, 493–95.

Porter, R.H., Cernoch, J.M., and Balogh, R.D. (1985). Odor signatures and kin recognition. *Physiology and Behavior,* **34**, 445–48.

Porter, R.H., Balogh, R.D., Cernoch, J.M., and Franchi, C. (1986). Recognition of kin through characteristic body odors. *Chemical Senses,* **11**, 389–95.

Potts, W.K., Manning, C.J., and Wakeland, E.K. (1991). Mating patterns in seminatural populations of mice influenced by MHC genotype. *Nature,* **352**, 619–21.

Preti, G., Wysocki, C.J., Barnhart, K.T., Sondheimer, S.J., and Leyden, J.J. (2003). Male axillary extracts contain pheromones that affect pulsatile secretion of luteinizing hormone and mood in women recipients. *Biology of Reproduction,* **68**, 2107–113.

Pybus, D. and Sell, C. (1999). *The chemistry of fragrances.* RSC Paperbacks, Cambridge.

Rikowski, A. and Grammer, K. (1999). Human body odour, symmetry and attractiveness. *Proceedings of the Royal Society B,* **266**, 869–74.

Roberts, S.C. (2007). Scent-marking. In: J.O. Wolff and P.W. Sherman (eds), *Rodent Societies,* pp. 255–66. University of Chicago Press, Chicago, IL.

Roberts, S.C. and Little, A.C. (2008). Good genes, complementary genes and human mate choice. *Genetica,* **132**, 309–21.

Roberts, S.C., Gosling, L.M., Spector, T.D., Miller, P., Penn, D.J., and Petrie, M. (2005). Body odor similarity in noncohabiting twins. *Chemical Senses,* **30**, 651–6.

Roberts, S.C., Gosling, L.M., Carter, V., and Petrie, M. (2008). MHC-correlated odour preferences in humans and the use of oral contraceptives. *Proceedings of the Royal Society B,* **275**, 2715–22.

Roberts, S.C., Little, A.C., Lyndon, A., Roberts, J., Havlicek, J., and Wright, R.L. (2009). Manipulation of body odor alters men's self-confidence and judgements of their visual attractiveness by women. *International Journal of Cosmetic Science,* **31**, 47–54.

Roberts, S.C., Miner, E.J., and Shackelford, T.K. (2010). The future of an applied evolutionary psychology for human partnerships. *Review of General Psychology,* **14**, 318–29.

Russell, M.J. (1976). Human olfactory communication. *Nature,* **260**, 520–2.

Russell, M.J., Mendelson, T., and Peeke, H.V.S. (1983). Mother's identification of their infant's odors. *Ethology and Sociobiology,* **4**, 29–31.

Santos, P.S.C., Schinemann, J.A., Gabardo, J., and Bicalho, M.D. (2005). New evidence that the MHC influences odor perception in humans: a study with 58 Southern Brazilian students. *Hormones and Behavior,* **47**, 384–8.

Saxton, T.K., Little, A.C., and Roberts, S.C. (2008a) Ecological validity in the study of human pheromones. In: J.L. Hurst, R.J. Beynon, S.C. Roberts, and T.D. Wyatt (eds), *Chemical Signals in Vertebrates 11*, pp. 111–20. Springer, New York.

Saxton, T.K., Lyndon, A., Little, A.C., and Roberts, S.C. (2008b). Evidence that androstadienone, a putative human chemosignal, modulates women's attributions of men's attractiveness. *Hormones and Behavior*, **54**, 597–601.

Schaal, B. and Porter, R.H. (1991). Microsmatic humans revisited—the generation and perception of chemical signals. *Advances in the Study of Behavior*, **20**, 135–99.

Schank, J.C. (2001). Menstrual-cycle synchrony: problems and new directions for research. *Journal of Comparative Psychology*, **115**, 3–15.

Schank, J.C. (2006). Do human menstrual-cycle pheromones exist? *Human Nature*, **17**, 448–70.

Schleidt, M. (1980). Personal odor and nonverbal-communication. *Ethology and Sociobiology*, **1**, 225–31.

Schleidt, M., Hold, B., and Attili, G. (1981). A cross-cultural-study on the attitude towards personal odors. *Journal of Chemical Ecology*, **7**, 19–31.

Schwende, F.J. and Novotny, M. (1982). Volatile compounds associated with estrus in mouse urine: potential pheromones. *Experientia*, **40**, 213–15.

Shepherd, G.M. (2004). The human sense of smell: are we better than we think? *PLoS Biology*, **2**, 572–5.

Signoret, J.P. and du Mesnil du Buisson, F. (1961). Étude du comportement de la truie en oestrus. *Proceedings of the 4th International Congress on Animal Reproduction*, 171–5.

Singer, A.G., Beauchamp, G.K., and Yamazaki, K. (1997). Volatile signals of the major histocompatibility complex in male mouse urine. *Proceedings of the National Academy of Sciences of the USA*, **94**, 2210–14.

Singh, D. and Bronstad, P.M. (2001). Female body odour is a potential cue to ovulation. *Proceedings of the Royal Society B*, **268**, 797–801.

Smith, D.G., Standing, L., and Deman, A. (1992). Verbal memory elicited by ambient odor. *Perceptual and Motor Skills*, **74**, 339–43.

Spangenberg, E.R., Sprott, D.E., Grohmann, B., and Tracy, D.L. (2006). Gender-congruent ambient scent influences on approach and avoidance behaviors in a retail store. *Journal of Business Research*, **59**, 1281–7.

Steiner, M., Coote, M., Tran, A., Corter, C., and Fleming, A. (1996). Cortisol and maternal attraction to infant odour. *European Neuropsychopharmacology*, **6**, 58.

Stern, K. and McClintock, M.K. (1998). Regulation of ovulation by human pheromones. *Nature*, **392**, 177–9.

Stoddart, M. (1990). *The scented ape*. Cambridge University Press, Cambridge.

Storey, A.E., Walsh, C.J., Quinton, R.L., and Wynne-Edwards, K. E. (2000). Hormonal correlates of paternal responsiveness in new and expectant fathers. *Evolution and Human Behavior*, **21**, 79–95.

Thornhill, R. and Gangestad, S.W. (1999a). Facial attractiveness. *Trends in Cognitive Sciences*, **12**, 452–60.

Thornhill, R. and Gangestad, S.W. (1999b). The scent of symmetry: a human sex pheromone that signals fitness? *Evolution and Human Behavior*, **20**, 175–201.

Thornhill, R., Gangestad, S.W., Miller, R., Scheyd, G., McCollough, J.K., and Franklin, M. (2003). Major histocompatibility complex genes, symmetry, and body scent attractiveness in men and women. *Behavioral Ecology*, **14**, 668–78.

Trevathan, W.R., Burleson, M.H., and Gregory, W.L. (1993). No evidence for menstrual synchrony in lesbian couples. *Psychoneuroendocrinology*, **18**, 425–35.

Tuzin, D. (2006). Base notes—odor, breath and moral contagion in Ilahita. In: J. Drobnick (ed.), *The smell culture reader*, pp. 59–67. Berg, New York.

Vaglio, S., Minicozzi, P., Bonometti, E., Mello, G., and Chiarelli, B. (2009). Volatile signals during pregnancy: a possible chemical basis for mother-infant recognition. *Journal of Chemical Ecology*, **35**, 131–9.

Warrenburg, S. (2005). Effects of fragrance on emotions: moods and physiology. *Chemical Senses*, **30**(Suppl. 1), i248–9.

Weber, S.T. and Heuberger, E. (2008). The impact of natural odors on affective states in humans. *Chemical Senses*, **33**, 441–7.

Wedekind, C. and Füri, S. (1997). Body odour preferences in men and women: do they aim for specific MHC combinations or simply heterozygosity? *Proceedings of the Royal Society B*, **264**, 1471–9.

Wedekind, C., Seebeck, T., Bettens, F., and Paepke, A.J. (1995). MHC-dependent mate preference in humans. *Proceedings of the Royal Society B*, **260**, 245–9.

Weisfeld, G.E., Czilli, T., Phillips, K.A., Gall, J.A., and Lichtman, C.M. (2003). Possible olfaction-based mechanisms in human kin recognition and inbreeding avoidance. *Journal of Experimental Child Psychology*, **85**, 279–95.

Weller, A. and Weller, L. (1992). Menstrual synchrony in female couples. *Psychoneuroendocrinology*, **17**, 171–7.

Weller, A. and Weller, L. (1993a). Menstrual synchrony between mothers and daughters and between roommates. *Physiology and Behavior*, **53**, 943–9.

Weller, L. and Weller, A. (1993b). Multiple influences on menstrual synchrony—kibbutz roommates, their best friends, and their mothers. *American Journal of Human Biology*, **5**, 173–9.

Weller, A. and Weller, L. (1995a). Examination of menstrual synchrony among women basketball players. *Psychoneuroendocrinology*, **20**, 613–22.

Weller, A. and Weller, L. (1995b). The impact of social interaction factors on menstrual synchrony in the workplace. *Psychoneuroendocrinology*, **20**, 21–31.

Weller, A. and Weller, L. (1997). Menstrual synchrony under optimal conditions: Bedouin families. *Journal of Comparative Psychology*, **111**, 143–51.

White, P.J., Fischer, R.B., and Meunier, G.F. (1986). Female discrimination of male dominance by urine odor cues in hamsters. *Physiology and Behavior*, **37**, 273–7.

Whitten, W.K. (1956). Modification of the oestrous cycle of the mouse by external stimuli associated with the male. *Journal of Endocrinology*, **13**, 399–404.

Willis, C. and Poulin, R. (2000). Preference of female rats for the odours of non-parasitised males: the smell of good genes? *Folia Parasitologica*, **47**, 6–10.

Wilson, H.C. (1992). A critical review of menstrual synchrony research. *Psychoneuroendocrinology*, **17**, 565–91.

Yamaguchi, M., Yamazaki, K., Beauchamp, G.K., Bard, J., Thomas, L., and Boyse, E.A. (1981). Distinctive urinary odors governed by the major histocompatibility locus of the mouse. *Proceedings of the National Academy of Sciences of the USA*, **78**, 5817–20.

Yamazaki, K., Boyse, E.A., Miké, V., et al. (1976). Control of mating preferences in mice by genes in the major histocompatibility complex. *Journal of Experimental Medicine*, **144**, 1324–35.

Yang, Z.W. and Schank, J.C. (2006). Women do not synchronize their menstrual cycles. *Human Nature*, **17**, 433–47.

Young, J.M., Friedman, C., Williams, E.M., Ross, J.A., Tonnes-Priddy, L., and Trask, B.J. (2002). Different evolutionary processes shaped the mouse and human olfactory receptor gene families. *Human Molecular Genetics*, **11**, 535–46.

Zhou, W. and Chen, D. (2009). Fear-related chemosignals modulate recognition of fear in ambiguous facial expressions. *Psychological Science*, **20**, 177–83.

Zucco, G.M., Paolini, M., and Schaal, B. (2009). Unconscious odour conditioning 25 years later: revisiting and extending 'Kirk-Smith, Van Toller and Dodd'. *Learning and Motivation*, **40**, 364–75.

Chapter 21

Television programming and the audience

Charlotte De Backer

In every industrialized society, every night after work the primary form of recreation is to immerse oneself in the broadcast or projected world of fictionalized lives and events (indeed, the appetite for recreation itself requires evolutionary explanation).
Tooby and Cosmides (2001, p. 8)

Introduction

In the last two decades, many public broadcasting companies across the world have been struggling to survive against the rise of commercial television. Viewers across the world seem to prefer the private lives of some unknown fictive characters over documentaries and educational programmes about world history, nature, life in different cultures, learning a new language, and so on. Despite the benefits of such non-fictional programmes, our minds are drawn to the first kind: we want to know why Maggy cheated on Greg with his best friend Marc, and how Greg will respond when he finds out. Entertainment is big business, and when we do switch to a documentary or a more educational programme, it is likely about crime in our neighbourhood or the sex lives of other people. Although the frequency of violence on screen seems to be slightly decreasing in the past few years, sex and crime are still by far the most popular topics in prime-time television (Hetsroni 2007a, b).

Journalists are aware of the audience's preference for such stories, and over the two decades, non-fiction television programmes have also shifted from a more neutral, serious format towards a more sensational one (Slattery et al. 2001). Although some see this tabloidization as a negative trend, we must recognize that journalists are catering to our demands. Human interest stories have always been present in any media format, and have always been popular with many people (Nordin 1979; Davis and McLeod 2003). The question of whether the presence and increase of certain kinds of entertainment products is a good or bad thing is a moral question, which I do not discuss here. Rather, I focus on why we are so very preoccupied with sex and crime, and with the private lives of people that appear on screen. Our biology, not morality, can help us answer this question. I will outline how evolutionary theory can help us explain why we consume so much entertainment, and how our drive for sensationalism has forced television programmers to cut down on educational programmes such as documentaries that deal with the reality we actually live in.

Since we mainly consume stories about *people,* whom we 'encounter' via the visual world of media formats such as television, I will focus on characters on-screen and will present three different approaches that explain a great deal—but not yet all—of our popular media consumption about the private lives of others. I will explain how we use media stories as a means for vicarious learning, how characters become part of our extended network, and lastly how media stories enter our real-life social network as a kind of social glue. But I start with a discussion on how media characters enter our life and trigger our emotional feelings of friendship.

The people in our television

When we think about the people we see on screen, we most often think about glamorous people, like celebrities; especially nowadays, in an age where not only movie stars, but also sportspeople, politicians, and Average Joe have entered this glamorous world. But what are 'celebrities'? Researchers have defined celebrity in different ways, but one of the most cited definitions is Boorstin's (1961, p. 58): 'the celebrity is a person who is well-known for their well-knownness'. Although being somewhat circular, this definition precisely embodies the core idea for defining a 'celebrity': celebrities, or 'stars', are people who are 'known' to an average person because they appear frequently in the media, and are therefore recognized by many. Lists of famous celebrities, such as the Forbes list, even use the criterion of 'media hits' as one of their main discriminators. It is not so much what a celebrity does that matters, but the fact that (s)he attracts attention. Although it might seem as if their prestige is due to their high salaries and expensive assets, these financial aspects alone do not determine celebrity status. This is because prestige, as Henrich and Gil-White (2001) say, equates with merit in the eyes of others and promotes a desire for proximity and sustained observation: the key to becoming prestigious is simply recognition. The more people recognize you, and talk about you, the more prestigious you become in the eyes of others. As Schely-Newman (2004, p. 482) says: 'Appearance in gossip columns has a value of sorts – it provides visibility and recognition, which in turn may be transferred to a permanent position on the social scene'. This translates in a marketing rule of thumb; there is no such thing as bad publicity. Several years ago, the British top model Kate Moss confirmed this, after receiving extensive, scandalous press coverage because she was caught using drugs. Many thought this was the end of her career that had been laying low for a while already; on the contrary, the media stir boosted her career to a level she had not previously reached; she got more famous (and successful) than ever.

But why is mere recognition such a powerful strategy for becoming famous? To understand this, I turn to Goldstein and Gigerenzer's (1999) theory of recognition heuristics, by which we make fast and frugal decisions—under uncertainty—that generally solve the problem at hand. They offered the classical example that, when asked which of two cities is larger, subjects tend to pick the one they recognize, and they are usually correct. 'The recognition heuristic for such tasks is simply stated: *If one of two objects is recognized and the other is not, then infer that the recognized object has the higher value*' (Goldstein and Gigerenzer 1999, p. 41). From this point of view, some celebrities are attributed prestige simply because they are widely recognized due to their frequent media appearances. Without this, more talented and deserving individuals miss out.

If this is true, then no special skills, except the power to attract attention from others, is required to become a celebrity. Anyone can aim for celebrity status, and reality television has provided us with the tools to do it. Andy Warhol once said that in the future every individual would get his or her fifteen minutes of fame; we now live in an age where this has become reality; an age of 'public unknowns', or what Rojek (2001) calls 'celetoids'; unknown people who achieve celebrity status for a brief period. Reality television made the line between reality and television, between ordinary and famous, very thin (Turner 2006).

In this respect, no media format has greater power than audiovisual media, where we see (moving) visual representations of celebrities. But why are visual media so powerful?

Eye witness, I witness

Visual information increases credibility of a message. The media have always been aware of this; before photography was introduced in 1839, engravings were used to illustrate written texts in newspapers. From the mid-19th century, picture papers were present and popular across Europe and the Americas (Goldberg 1993). In addition to increasing accessibility to illiterate people, pictures add credibility to a text as the audience feels as if they witnessed the situation. According to Bird (1992), this is the reason why, today, tabloids and gossip magazines are especially picture prone: they are not regarded as the most trusted formats of journalism, so gain credibility by adding pictures, by presenting articles in interview-style ('we got our information first-hand'), or adding that news comes from a 'very close source'. Pictures seem to hold more truth than words, because our brain processes visual information in a similar sense as real-life information. We experience what we see 'as if I witnessed the situation'. We could say that eye-witness observations translate in our minds to I-witness emotional feedback, even under circumstances where what we witness is not what we experienced, or even does not represent any reality at all.

Framing this in evolutionary psychological terms, this 'eye-witness, I-witness' confusion is not that strange. Many elements of this visual (virtual) environment that we see in the media resemble elements from the historical past that shaped our minds. They might therefore trigger our psychological mechanisms to generate a behaviour strategy designed to solve a problem existing in that historical past. Such *mismatches* between psychological adaptations that evolved in the past and cues we encounter in the current environment often explain behaviours that would otherwise make no sense, in evolutionary terms. A classical example of this is the way that many people experience a fear reaction when encountering a plastic snake, often intended to evoke this reaction as a joke. This seemingly irrational reaction becomes less irrational when we consider the past environment that shaped our psychological mechanisms to fear snakes and other predators that were a real and common threat to the survival and reproduction of our ancestors. Our minds are designed to react with a fight-or-flight reaction to potential threats (LeDoux 1998), or to situations that come across as being threatening; things that look like snakes and other predators also trigger the same responses. This is because the cost of expressing our defence to potential threats is low compared to the potential harm it protects against (the smoke detector effect: Nesse 2001).

I believe we can even extend this principle to other, more positive situations too; mass mediated visuals trigger a range of evolved psychological mechanisms that were designed to deal with real-life visual input. Just like a plastic snake looks as if it were a real snake, mass mediated visuals look as if they were real, and it can be no surprise that eye-witnessed information feels like I-witnessed situations. You observe a murder scene and feel frightened, or you watch a seduction scene and you feel aroused, or maybe slightly jealous. Pictures boost the credibility of information, and in addition they make us feel as if who we *see* is who we *meet*: an individual part of our social network. As Barkow (1992, pp. 629–630) put it:

> We see them in our bedrooms, we hear their voices when we dine: If this hypothesis is correct, how are we not to perceive them as our kin, our friends, perhaps even our rivals? As a result, we automatically seek information about their physical health, about changes in their relative standing, and above all about their sexual relationships.

In what follows I will explain how and why the people we see on screen, such as celebrities, start to function as role-models, friends, and mutual acquaintances in daily life, and how this leads to television programmers delivering us even more of the same.

Television: the school of life

When people get hooked to a soap opera or other television programme, it is often because, while switching channels, they became drawn into a scene where someone is murdered, raped, betrayed, or saved a life. Television characters that are quite unknown to the audience are most likely to attract attention when they are caught up in dramatic or heroic events. This applies to other media formats too. Gossip columns or tabloid magazines, for instance, will cover both the glamour of celebrities and the unusual, bizarre events that happen to Average Joe's. In fact, tabloid journalism used to *only* cover these stories at the start of their success story:

> For the first decade after tabloids infiltrated the chain stores in significant numbers, celebrity coverage was an important, but still relatively small, part of their editorial mix. They devoted far more space and attached much higher priority to stories that embodied the 'gee-whiz' factor – unexplained phenomena (space aliens, UFOs, psychics, telekinesis, out-of-body experiences, life after death), unsung heroes, rags to riches, wacky inventions, weird pastimes, people overcoming handicaps, and the like. Even medical breakthroughs, government waste and skullduggery, burning social issues, and self-help articles were often played above celebrity pieces.
>
> (Sloan 2001, p. 95)

Our lust for sensationalism did not emerge with the rise of these early tabloids, of course; humans have been interested in astounding news long before the first newspapers appeared in the 17th century (e.g. Davis and McLeod 2003). Sensational stories have always been part of our human history, under the form of urban legends and other formats of oral history. Our interest in these stories could easily be explained because of the potential 'schadenfreude' effect: the misfortunes of other give us a relative feeling of being lucky, or having reasons to be happy. But this 'your pain is my gain' explanation does not explain our interest in the heroic stories, like 'miracle tea cured woman from deadly disease' or 'man who won lottery finds love of his life'.

Framing our interest in these stories in an evolutionary perspective, the reason why we are drawn both to the bad and good things happening to others, is because both types of messages contain fitness-relevant information and are an efficient way to learn strategies needed to succeed in daily life. As human beings, we have always relied upon the experience of others to update our knowledge about what we can do best if we want to succeed. As Bandura (1977) puts it:

> Learning would be exceedingly laborious, not to mention hazardous, if people had to rely solely on the effects of their own actions to inform them what to do. Fortunately, most human behaviour is learned observationally through modelling: from observing others one forms an idea of how new behaviours are performed, and on later occasions this coded information serves as a guide for action.
>
> (Bandura 1977, p. 22)

Others have confirmed the importance of Bandura's theory of social learning in human history (Richerson and Boyd 1992; Sugiyama 1996, 2001; Henrich and McElreath 2003). Evolution programmed our ancestors to gain experience by observing others' actions, because personal experiences can be too costly Furthermore, imitation of others can be regarded as a fast and frugal heuristic (Gigerenzer and Todd 1999, pp. 31–32), because it enables individuals to make decisions with limited time and energy.

When information about (un)successful actions is shared orally (rather than by direct observation), the costs of gaining this form of experience become even lower, as larger audiences can be reached and receivers do not even need to be within proximity of the person talked about. And because the cost/benefit ratio is so positive for acquisition of second-hand information, storytelling was probably very popular among our ancestors: 'A hunter-gatherer band might contain scores or even hundreds of lifetimes' worth of experience whose summary can be tapped into if it can be communicated. So, vicarious experience of especially interesting events, communicated from others, should be aesthetically rewarding' (Tooby and Cosmides 2001, p. 23). Studies such as Kaplan and Hill's (1992) analysis of food acquisition among the Aché have confirmed this. Other proof of the fact that we have an innate preference for stories that contain potential fitness-relevant information comes from Sperber's (1985) experiment, who illustrated how we much more easily remember complex narrative stories about the lives of others than (relatively easier to remember) information about, for instance, Stock Exchange figures. We seem to be hardwired to pay attention to narratives that contain fitness-relevant information.

Before finishing this section, I want to add two final remarks. First of all, we are much more drawn to negative than to positive or neutral information (Taylor 1991) and it is more accurately retained (Ito et al. 1998). People more easily believe information about the immoral (negative) behaviour of others than about moral (positive) behaviour. Moreover, information about immoral behaviour is regarded to have greater news value (Lupfer et al. 2000). To explain this 'negativity bias', Rozin and Royzman (2001) have argued that humans and animals possess innate predispositions to learn faster how to escape danger than to improve well-being. Learning what is beneficial is, of course, good, but missing such positive information does not decrease one's fitness. In contrast, missing opportunities to learn how to avoid extremely dangerous situations can be extremely costly, even fatal. This explains our extra interest in negative sensational news; one is more likely to be drawn into a programme where violent murder has just been committed compared to one where someone has just been promoted to a function that will improve his quality of life.

Second, we are particularly prone to adopting celebrities as role models. This is because well-known individuals on screen have a greater impact on our emotional and behavioural responses. We are interested in a lot more about them: we do not only want to know whether an A-list celebrity survived the plane crash, or is again the best earning actor in Hollywood; we also like to know if he is still happily married, if his wife is pregnant or not, where he does his grocery shopping and where he is heading for his next holiday. This is because of the prestige attributed to celebrities and based on the simple principle that in daily life we tend to pay more attention to people who are slightly more successful than we are (prettier, smarter, richer, and so on). People are selective in their copying behaviour (Richerson and Boyd 1992), and this starts with being selective in who to take on as a role model. On average, we are most attracted to higher status others, because mimicking their behaviour has greater potential to lead to an increase in one's own status (Boyd and Richerson 1985; Henrich and Gil-White 2001; Henrich et al. 2001). Success may be guaranteed if we copy the attribute that caused our chosen role model to be so prestigious.

However, this is not always straightforward. As we have discussed, many contemporary celebrities do not seem to have one particular skill that brought them to their level of fame and it leaves the audience confused: if we do not know what exactly made them famous, then what should we be copying? This dilemma has probably been around throughout evolutionary history; more successful people are known but the key to their success was unclear. Boyd and Richerson (1985) therefore argue that evolution most probably has shaped human psychology for a *general copying bias* rather than a specific copying bias. It is less costly to mimic the general behavioural pattern of an individual than to analyse precisely which behavioural combinations lead to success. This general copying bias can therefore explain why we often do appear to mimic higher-status individuals, and why celebrities become popular endorsers of brands. We buy the way they look, smell, behave, what they eat and drink, we go where they spend their holidays, and we donate money to causes they believe are important.

To summarize the above, one reason for keeping up with what happens in the lives of the people that live in our television (or other screens) can be seen as a mismatch outcome from the interaction of our innate desire to gain experience via vicarious learning, and the fact that our modern world provides us with detailed information of (fictive) people. At first glance, the outcome of this mismatch seems quite innocent and not as maladaptive as the mismatch may be between, say, our desire for fatty foods and the abundance of fast-food in today's society. However, there might be reasons to believe that taking television characters, and especially celebrities, as role models can do more harm than good to the fitness of some. One specific example is the effects of images of on-screen women representing a rather thin body ideal to impressionable teenage girls, which can be detrimental for their self-perception/-esteem, in some cases negatively impacting their eating behaviour (e.g. Groesz et al. 2002; Grabe et al. 2008).

From friends on screen to friends in daily life

A second explanation for our addiction to what happens on screen is fear of loneliness, and desire to be surrounded by others. As Hartup and Stevens (1997) have pointed out, friendship correlates with well-being; friends foster our self-esteem and can therefore be considered as important cognitive and affective resources. Friendships have developmental advantages and lead to good outcomes. The ultimate reason why friendship is so important can be explained with the Banker's paradox: 'The harsh irony of the Banker's paradox is this: just when individuals need money most desperately, they are also the poorest credit risks and, therefore the least likely to be selected to receive a loan' (Tooby and Cosmides 1996, p. 131). Translating this to human social life, those who need help the most are the most costly to others, and most at risk of being left alone. Friendships can resolve this problem since true friends do not count costs and benefits of their transactions (Silk 2003). Friends support one another, and more importantly, tolerate a lot more from one another than in interactions between other non-related individuals. In many respects friends, behave as if they were family, and Cords (1997) assumes friendship emerged and evolved because it returns so many adaptive benefits to individuals (including joint access to food patches; support, acquisition, and maintenance of dominance; tolerance and protection; and potential mates, since friends sometimes become lovers). Friends are so important that people who lack them not only feel lonely, but also become more depressed.

Friends are the people we meet on regular basis or, in modern times, with whom we are in touch on a regular basis over the phone, via email, and so on. Our ancestors however, only had one way of communicating with friends: face-to-face. Friends for them were the people they met most often. No surprise that today, the impact of *seeing* your friends is still stronger compared to hearing or reading something from them; out-of-sight is out-of-mind. Managers of modern companies working with people in different locations are very well aware of this, and try to bridge the distances by implementing technologies to bring the 'visual encounter' back to modern life; we use pictures and webcams, and immediately feel more connected to our distant colleagues (Armstrong and Cole 2002). Visual representations seem to work just as well as actual face-to-face encounters.

This brings us to the second reason why watching television can lead to addiction of entertainment; we meet people on a regular basis and start to consider them as friends. This is especially true for television fiction, like soap operas where the same people reappear on a weekly or often daily basis. Inspired by this idea, Kanazawa (2002) assumed that people who watch more television have a more expanded social network, and tested the hypothesis that those who view more TV have higher friendship satisfaction. He found that people who watched a lot of television reported greater overall friendship satisfaction than people who did not, an effect that he attributed to the fact that they simply had more friends to be satisfied from, even if some were just para-social friends. Interestingly, for women, the friendship satisfaction only results from dramas and sitcoms which deal with family issues, whereas men get friendship satisfaction from watching television programmes that depict men at work, such as news programmes and public broadcasting services. According to Kanazawa (2002), this sex difference is due to the fact that women have more family members and close friends in their real-life social networks, while men are more likely to have colleagues in theirs. In other words, a sex difference in real-life social networks translates directly into people's extended social network of television characters. However, Kanazawa's findings were criticized by Freese (2002), who claimed that overall life satisfaction needed to be taken into account. On reanalysing his results, Kanazawa (2003) found only significant effects for female viewers, being somewhat more satisfied about their friendships when heavily consuming television programmes. However, other studies do support the notion that television characters trigger emotions in audience members that correspond to emotional responses we show in real-life interpersonal interactions. We feel closely connected to our favourite television characters, and even feel distressed when we need to break up with one or more of them, for instance, when a television show is being ended or a character dies (Eyal and Cohen 2006).

Perhaps Kanazawa did not take sufficiently into account the degree of perceived loneliness. The late Kurt Vonnegut once quoted that 'Television is providing artificial friends and relatives to lonely people'. It may be that lonely people do get particular friendship satisfaction from consuming entertainment formats where characters reappear on a regular basis. Indeed, it has been shown that lonely people are more drawn to celebrities than those with many real-life friends, and actively seek out information about them. People who score highest in scales such as the Celebrity Worship Scale tend to have an introvert nature and to report lacking meaningful relationships in real life (McCutcheon et al. 2002). Elderly people also experience this through physical immobility and lost part of their social network through bereavement. Saunders (1999), who investigated gossip among elderly people, noticed that they indeed showed a great interest in talk about celebrities or other people they knew from consuming media products. They become their new friends, replacing the (real) ones they have lost.

If television delivers support to lonely people, should we not be unconcerned about the fact that more entertainment is being produced and consumed every day? Ironically, our growing consumption of television is one of the key reasons behind growing social isolation. Putnam (2000) described the decline of social life in western societies in his book *Bowling Alone*. When summing up the key factors that influenced the decline of the social life of Americans, he listed electronic entertainment (especially television), as one of the strongest influences, contributing up to 25% of the overall effect. Television keeps us away from real-life social contacts. Television also gives us the illusion of having friends who are always there for us (we need only switch on our machine), but does not provide the essence of friendship: emotional support. What is special about our interactions with television characters is that while they reveal their (fictive or real) private lives to us, and we show emotions towards them, we do not share our private lives with them and they do not show emotions towards us (personally). Such interactions where reciprocity is lacking are called para-social, or one-way interactions (Horton and Wohl 1956; Rubin et al. 1985). Unfortunately, the tendency to have one para-social friend leads inexorably to more, simply because we tune in more often, continually triggering the emotions associated with such friendships (Brown et al. 2003). This spiralling effect might soon deliver us a whole network of para-social friends, perhaps elevating friendship satisfaction as Kanazawa assumed, but at the cost of losing real-life friendships in which emotional support travels in both directions. And it is exactly the need for this social support that, throughout evolutionary history, shaped our desire to be surrounded with friends, because it is social support that entails the adaptive benefit of friendship.

We can summarize the second explanation for being so interested in the lives of on-screen characters as a mismatch outcome of our innate desire to establish intimate bonds with people we regularly encounter, and the way that modern entertainment devices trigger these emotions. It seems that on-screen characters positively affect our perceived friendship gratification, but the results of such studies are mixed. On the other hand, it may be that by pursuing such relationships, we may be losing exactly that what we innately desire: real social connections. Our television addiction seems maladaptive, rather than adaptive or harmless, from this point of view.

The social power of television talk

A final explanation for our interest in television entertainment and its characters stems from our desire to talk about others, to gossip. Among other functions, gossip is essential in the construction and maintenance of social bonds (e.g. Gluckman 1963, 1968; Dunbar 1997). Gossip helps in social bonding because it deals with people who are mutually known to the gossipers, creating a sphere of intimacy; people who gossip share connections and trust each other (Bergmann 1993; Nevo et al. 1993; Ayim 1994). Gossip is often associated with themes of 'us versus them' (Gottman and Mettetal 1986; Gelles 1989; Leaper and Holliday 1995), and the use of a little negative gossip about a mutually-known third party can help cement an early-stage friendship (Bosson et al. 2006). The social bonding effect can also be due to the shared values people exchange through gossip (Liang 1993; Riegel 1996). Outsiders simply cannot understand the gossip conversation of a group sharing a social history and culture. In a similar sense, gossip can be compared to humour. Laughing for the same, shared reason presumes that the jokers (and gossipers) share similar attitudes (Liang 1993; Morreall 1994). Lastly, Gluckman (1963, 1968) also explained how gossip reinforces group bonds because it is used as means to scandalize those who violate group norms (Wilson et al. 2000; Young 2001). In this perspective, gossip unites because it enforces conformity on group members.

However in modern societies, having mutual acquaintances to talk about is not always a given. We are often faced with a dilemma: we have people to talk *to*, but we don't have mutual acquaintances to talk *about*. What John Locke (1998) calls 'social de-voicing' refers to this stunning fact; in Western societies, people talk less to each other due to urbanization, relocation, television, individualism, economic success, and disappearance of social programmes. We now live in larger cities, but conversation across the fence with our neighbours has disappeared; in fact, people often do not know who their neighbours are.

Caughey (1984) pointed out that media figures could become shared interests to talk about with real social contacts, thus facilitating real-life social contacts. And this is exactly what they have become: 'Discussions about soap operas takes us into this [. . .] type of gossip, as it is somehow "personal" to discuss the lives of characters on soap operas as if they were real people' (Riegel 1996, p. 204). Gossip between soap opera viewers unites a group of people who otherwise may have nothing in common. Celebrities are known by many and can function as mutual acquaintances to gossip about, acting as social ties in our scattered modern world. And while in some studies (e.g. Brown and Barwick 1987; Riegel 1996) the focus lies mainly on female use of this behaviour, men seem to engage in it too. Analysing the increasing popularity of men's magazines, Benwell (2001) looked at readers' gossip in the letters pages. She noted that men gossip mainly about celebrities, and public knowledge, but without engaging in very private or emotional talk, which is mostly the subject in women's magazines. However we remain sufficiently aware that we are talking about fictive people or real people we will never encounter in reality. Benwell (2001, p. 23) writes: '. . . gossip about unknown celebrities is sufficiently removed from real experience so as not to constitute a threat'. Gossip about television characters is safe; there is no threat of hurting the person gossiped about, and it is food for casual talk, without getting too personal or intimate with the people we are talking to. Celebrity gossip therefore is a safe means, for example, to test whether our beliefs and values are similar to those of people we meet, without having to dish up the private lives of people we really know, or even worse, having to dig up past personal experiences that would make the conversation too personal. Celebrity gossip is small talk that bridges distance between people, while maintaining sufficient emotional distance to still feel comfortable in casual close connections.

To sum up, from this perspective, our interest in the lives of people on screen stems from a mismatch between our innate desire to gossip about mutual acquaintances with real life friends, and the fact that modern entertainment provides us a group of mutual acquaintances that can cross large social distances. The outcome is not maladaptive, and could even be seen as more positive than negative. If talking about the lives of (fictive) television characters reinforces the social bonds we uphold in real life, the reason *why* we feel connected might not be real, but the relationship we are establishing with it *is* real.

Conclusion

'When I was a kid we did not have phones, televisions, and all other stuff people now have. We had a radio, and in the evening we would all sit together, our parents and my sisters and brothers, and we would listen to the radio. And once a week I would go to the movie theatre with my father, to watch the news. It was a real treat, and afterwards we would talk about what we had seen. Then, as I started my own family, televisions became more popular. Some people in our street owned one, and we would gather a few nights a month to watch television at their place. Those were the days; we would have a laugh together and discuss what we saw on screen. Not like your family. You are all so serious when you watch television, and you don't talk with each other anymore. You see my girl, a lot has changed in just a few generations' time. But what they show on television is often still the same; in *Coronation Street* people still fall in love and fight over nothing. Which reminds me that my show is about to start. I'll call you next week, same time ok?'

Except for the fact that she doesn't watch *Coronation Street*, but some Flemish soap opera spoken in Dutch (with very similar content), the above could have been a conversation with my grandmother. So, perhaps, might it have been one with yours. Entertainment for our grandparents' generation was a true social activity, while nowadays it has become an individual addiction. The one thing that is unchanged is our desire to be entertained; we love to hear stories about other people and we love to gossip about them. We seem to be hardwired to be entertained, and the modern age we live in gives us exactly what we want: around the clock opportunities to consume the lives of others.

In this chapter I have tried to explain what underlies this drive for entertainment. One important factor is vicarious learning: paying attention to what happens to others can be of tremendous value to ourselves, we take on life lessons. But that only explains part of our media consumption. Our addiction to soaps and our obsession with celebrities is more often driven by emotional responses that evolved to function in an interpersonal face-to-face context. The reason our emotions respond to media environments can be explained as a mismatch reaction to modern artefacts that look a lot like the environmental features that in our evolutionary past that shaped our human minds. Cooperation, forming long-term alliances, finding mates, and competing with rivals have always acted as evolutionary pressures, as adaptive problems. Our ancestors were the ones whose behaviour was most adapted to face these problems, and we have inherited their way of feeling, living, and dealing with these daily problems. As a result we express towards virtual worlds and virtual people the emotions that evolved to function in face-to-face interactions.

A key factor in this process that I highlighted in this chapter is the use of visual information: pictures and moving images. Our minds somehow process the information we see with our eyes as if it were a true, real experience; an eye-witness event triggers an I-witness emotional response. As a result we become emotionally absorbed in a world that looks a lot like reality, but that also lacks important features, such as emotional support. Further research on this matter is necessary to fully understand the impact, and potential negative influence, of our ever-increasing entertainment consumption.

References

Armstrong, D.J. and Cole, P. (2002). Managing distances and differences in geographically distributed work groups. In: P. Hinds and S. Kiesler (eds), *Distributed work*, pp. 167–86. MIT Press, Cambridge, MA.

Ayim, M. (1994). Knowledge through the grapevine: gossip as inquiry. In: R.F. Goodman and A. Ben-Ze'ev (eds), *Good gossip*, pp. 85–99. The University Press of Kansas, Lawrence, KS.

Bandura, A. (1977). *Social learning theory*. General Learning Press, New York.

Barkow, J.H. (1992). Beneath new culture is old psychology: gossip and social stratification. In: J. Barkow, L. Cosmides, and J. Tooby (eds), *The adapted mind: evolutionary psychology and the generation of culture*, pp. 627–37. Oxford Univeristy Press, New York.

Benwell, B. (2001). Male gossip and language play in the letters pages of men's lifestyle magazines. *The Journal of Popular Culture*, **34**, 19–33.

Bergmann, J.R. (1993). *Discreet indiscretions: the social organization of gossip*. Aldine de Gruyter, New York.

Bird, E.S. (1992). *For enquiring minds: a cultural study of supermarket tabloids*. University of Tennessee Press, Knoxville, TN.

Boorstin, D.J. (1961). *The image or what happened to the American dream*. Weidenfeld and Nicolson, New York.

Bosson, J.K., Johnson, A.B., Niederhoffer, K., and Swann, W.B. (2006). Interpersonal chemistry through negativity: bonding by sharing negative attitudes about others. *Personal Relationships*, **13**, 135–50.

Boyd, R. and Richerson, P.J. (1985). *Culture and the evolutionary process*. University of Chicago Press, Chicago. IL.

Brown, M.E. and Barwick, L. (1987). Fables and endless genealogies: soap opera and women's culture. *Continuum: The Australian Journal of Media and Culture*, **1**, 71–82.

Brown, W.J., Basil, M.D., and Bocarnea, M.C. (2003). Social influence of an international celebrity: responses to the death of Princess Diana. *Journal of Communication*, **53**, 587–605.

Caughey, J.L. (1984). *Imaginary social worlds: a cultural approach*. University of Nebraska Press, Lincoln, NE.

Cords, M. (1997). Friendship, alliances, reciprocity and repair. In: A. Whiten and R.W. Byrne (eds), *Machiavellian intelligence II*, pp. 24–49. Cambridge University Press, Cambridge.

Davis, H. and McLeod, S.L. (2003). Why humans value sensational news: an evolutionary perspective. *Evolution and Human Behavior*, **24**, 208–16.

Dunbar, R.I.M. (1997). Groups, gossip, and the evolution of language. In: A. Schmitt (ed.), *New aspects of human ethology*, pp. 77–89. Plenum Press, New York.

Eyal, K. and Cohen, J. (2006). When good friends say goodbye: a parasocial breakup study. *Journal of Broadcasting and Electronic Media*, **50**, 502–23.

Freese, J. (2002). Imaginary imaginary friends: television viewing and satisfaction with friendships. *Evolution and Human Behavior*, **24**, 65–9.

Gelles, E.B. (1989). Gossip: an eighteenth-century case. *Journal of Social History*, **22**, 667–83.

Gigerenzer, G. and Todd, P.M. (1999). *Simple heuristics that make us smart*. Oxford University Press, New York.

Gluckman, M. (1963). Papers in honor of Melville J. Herskovits: gossip and scandal. *Current Anthropology*, **4**, 307–316.

Gluckman, M. (1968). Psychological, sociological and anthropological explanations of witchcraft and gossip: a clarification. *Man*, **3**, 20–34.

Goldberg, V. (1993). *The power of photography: how photography changed our lives*. Abbeville Press, New York.

Goldstein, D.G. and Gigerenzer, G. (1999). The recognition heuristic: how ignorance makes us smart. In: G. Gigerenzer and P.M. Todd (eds), *Simple heuristics that make us smart*, pp. 37–58. Oxford University Press, New York.

Gottman, J.M. and Mettetal, G. (1986). Speculations about social and affective development: friendship and acquaintanceship through adolescence. In: J.M. Gottman and J.G. Parker (eds), *Conversations of friends: speculations on affective development*, pp. 192–240. Cambridge University Press, Cambridge.

Grabe, S., Ward, L.M., and Hyde, J.S. (2008). The role of the media in body image concerns among women: a meta-analysis of experimental and correlational studies. *Psychological Bulletin*, **134**, 460–76.

Groesz, L.M., Levine, M.P., and Murnen, S.K. (2002). The effect of experimental presentation of thin media images on body satisfaction: a meta-analytic review. *International Journal of Eating Disorders*, **31**, 1–16.

Hartup, W.W. and Stevens, N. (1997). Friendships and adaptation in the life course. *Psychological Bulletin*, **121**, 355–70.

Henrich, J., Albers, W., Boyd, R., *et al.* (2001). What is the role of culture in bounded rationality? In: G. Gigerenzer and R. Selten. (eds), *Bounded rationality*, pp. 343–59. MIT Press, Cambridge, MA.

Henrich, J. and Gil-White, F.J. (2001). The evolution of prestige: freely conferred deference as a mechanism for enhancing the benefits of cultural transmission. *Evolution and Human Behavior*, **22**, 165–96.

Henrich, J. and McElreath, R. (2003). The evolution of cultural evolution. *Evolutionary Anthropology: Issues, News, and Reviews*, **12**, 123–35.

Hetsroni, A. (2007a). Three decades of sexual content on prime-time network programming: a longitudinal meta-analytic review. *Journal of Communication*, **57**, 318–48.

Hetsroni, A. (2007b). Four decades of violent content on prime-time network programming: a longitudinal meta-analytic review. *Journal of Communication*, **57**, 759–84.

Horton, D. and Wohl, R.R. (1956). Mass communication and para-social interaction. *Psychiatry*, **19**, 215–29.

Ito, T.A., Larsen, J.T., Smith, N.K., and Cacioppo, J.T (1998). Negative information weighs more heavily on the brain: the negativity bias in evaluative categorizations. *Journal of Personality and Social Psychology*, **75**, 887–900.

Kanazawa, S. (2002). Bowling with our imaginary friends. *Evolution and Human Behavior*, **23**, 167–71.

Kanazawa, S. (2003). The relativity of relative satisfaction. *Evolution and Human Behavior*, **24**, 71–3.

Kaplan, H. and Hill, K. (1992). The evolutionary ecology of food acquisition. In: E.A. Smith and B. Winderhalder (eds), *Evolutionary ecology and human behaviour*, pp. 167–202. Aldine de Gruyter, New York.

Leaper, C. and Holliday, H. (1995). Gossip in same-gender and cross-gender friends' conversations. *Personal Relationships*, **2**, 237–46.

LeDoux, J.E. (1998). *The emotional brain: the mysterious underpinnings of emotional life*. Touchstone, New York.

Liang, M. (1993). Gossip: does it play a role in the socialization of nurses? *Image – the Journal of Nursing Scholarship*, **25**, 37–43.

Locke, J.L. (1998). *The de-voicing of society: why we don't talk to each other anymore*. Simon and Schuster, New York.

Lupfer, M.B., Weeks, M., and Dupuis, S. (2000). How pervasive is the negativity bias in judgments based on character appraisal? *Personality and Social Psychology Bulletin*, **26**, 1353–66.

McCutcheon, L.E., Lange, R., and Houran, J. (2002). Conceptualization and measurement of celebrity worship. *British Journal of Psychology*, **93**, 67–87.

Morreall, J. (1994). Gossip and humor. In: R.F. Goodman and A. Ben-Ze'ev (eds), *Good gossip*, pp. 56–64. The University Press of Kansas, Lawrence, KS.

Nesse, R.M. (2001). The smoke detector principle. *Annals of the New York Academy of Sciences*, **935**, 75–85.

Nevo, O., Nevo, B., and Zehavi, A.D. (1993). Gossip and counselling: the tendency to gossip and its relation to vocational interests. *Counselling Psychology Quarterly*, **6**, 229–38.

Nordin, K.D. (1979). The entertaining press: sensationalism in eighteenth-century Boston newspapers. *Communication Research*, **6**, 295–320.

Putnam, R. (2000). *Bowling alone: the collapse and revival of American community*. Simon and Schuster, New York.

Richerson, P.J. and Boyd, R. (1992). Cultural inheritance and evolutionary ecology. In: E.A. Smith and B. Winterhalder (eds), *Evolutionary ecology and human behaviour*, pp. 61–92. Aldine de Gruyter, New York.

Riegel, H. (1996). Soap operas and gossip. *Journal of Popular Culture*, **29**, 201–9.

Rojek, C. (2001). *Celebrity*. Reaktion Books, London.

Rozin, P. and Royzman, E.B. (2001). Negativity bias, negativity dominance, and contagion. *Personality and Social Psychology Review*, **5**, 296–320.

Rubin, A. M., Perse, E.M., and Powell, R.A. (1985). Loneliness, parasocial interaction, and local television news viewing. *Human Communication Research*, **12**, 155–80.

Saunders, P.A. (1999). Gossip in an older women's support group: a linguistic analysis. In: H.E. Hamilton (ed.), *Language and communication in old age: multidisciplinary perspectives*, pp. 267–93. Garland, New York.

Schely-Newman, E. (2004). Mock intimacy: strategies of engagement in Israeli gossip columns. *Discourse Studies*, **6**, 471–88.

Silk, J.B. (2003). Cooperation without counting: the puzzle of friendship. In: P. Hammerstein (ed.), *The genetic and cultural evolution of cooperation*, pp. 37–54. MIT Press, Cambridge, MA.

Slattery, K.L., Doremus, M., and Marcus, L. (2001). Shifts in public affairs reporting on the network evening news: a move toward the sensational. *Journal of Broadcasting and Electronic Media*, **45**, 290–302.

Sloan, B. (2001). *I watched a wild hog eat my baby!: a colorful history of tabloids and their cultural impact*. Prometheus Books, Amherst, NY.

Sperber, D. (1985). Anthropology and psychology: towards an epidemiology of representations. *Man*, **20**, 73–89.

Sugiyama, M.S. (1996). On the origins of narrative. *Human Nature*, **7**, 403–25.

Sugiyama, M.S. (2001). Food, foragers, and folklore: the role of narrative in human subsistence. *Evolution and Human Behavior*, **22**, 221–40.

Taylor, S.E. (1991). Asymmetrical effects of positive and negative events: the mobilization-minimization hypothesis. *Psychological Bulletin*, **110**, 67–85.

Tooby, J. and Cosmides, L. (1996). Friendship and the banker's paradox: other pathways to the evolution of adaptation for altruism. *Proceedings of the British Academy*, **88**, 119–43.

Tooby, J. and Cosmides, L. (2001). Does beauty build adapted minds? Toward an evolutionary theory of aesthetics, fiction and the arts. *Substance*, **94/95**, 6–27.

Turner, G. (2006). The mass production of celebrity: 'Celetoids', reality TV and the 'demotic turn'. *International Journal of Cultural Studies*, **9**, 153–65.

Wilson, D.S., Wilczynski, C., Wells, A., and Weiser, L. (2000). Gossip and other aspects of language as group-level adaptations. In: C.M. Heyes and L. Huber (eds), *The evolution of cognition*, pp. 347–65. MIT Press, Cambridge, MA.

Young, R. (2001). 'There is nothing idle about it: deference and dominance in gossip as a function of role, personality, and social context.' Thesis (PhD), University of California, Berkeley, 2001.

Chapter 22

News as reality-inducing, survival-relevant, and gender-specific stimuli

Maria Elizabeth Grabe

Introduction

Media scholars are strikingly uncomfortable with evolutionary psychology. I have graduated from making trite excuses for employing it in research to deriving amusement from flabbergasted facial displays when I invoke it during a presentation. Perhaps this audience response reflects the intellectual contours of a young discipline, one that is still trying to find comfort in the pluralism of its origins in both social science and the humanities. Drop biology into this mix and intellectual indigestion is a likely outcome.[1] Indeed, my discipline treats the possibility that biological predispositions explain-in some part-how we interact with media as stoppage to its celebrations of the self-determining individual and her mastery of personalized meaning-making. Interestingly, the heavy sledgehammer of *social* determinism forms the basis for many venerated media theories.[2] By my diagnosis then, the dominant ontological pathology of my discipline favours social over biological forces in explaining mediated interaction (that is, when the high-fiving about individual agency subsides long enough for structural explanations to be heard).

There are those who have cleared paths through the brush for a slow-growing group of scholars to walk the evolutionary psychology road. This chapter will pay homage to the pioneers, quibble a bit with them, tackle the place of media in the life history of *Homo sapiens*, and focus on journalism as a specific area of media research in desperate need of an evolutionary psychology plunge.

The media equation, qualified

One ground-breaking insight that has cleared the way for applying evolutionary psychology perspectives in media research is the so-called media equation, or *the media-equal-real-life* idea (Reeves and Nass 1996). This notion, growing in popularity since the mid-1980s, simply proposes that the human perceptual system does not appropriately respond to mediated stimuli as originating from the mediated world. Instead, media are treated as if they are real. Reeves and Nass

---

[1] A personal favorite example comes from a letter (dated 7 January 2002) from the editor of *Critical Studies of Media Communication*, Celeste Condit, when she summarized her stance on how I should proceed with a manuscript revision: 'Many of your readers are going to go ballistic at an evolutionary psychology perspective. I myself accept evolutionary perspectives, but don't think your version is a BALANCED evolutionary perspective. So you are going a) to alienate too many readers with this, and b) can't take the space to defend the argument, and c) most importantly, don't need it!'
[2] Agenda Setting, Cultivation, Framing, Priming, The Knowledge Gap, Spiral of Silence, to name a few.

(1996, p. 12) claim, 'There is no switch in the brain that can be thrown to distinguish the real and mediated worlds'. Or as Pinker (1997, p. 29) puts it: 'Even in a lifelong couch potato, the visual system never "learns" that television is a pane of glowing phosphor dots'. This explains why humans have an automatic response to threatening mediated messages as if they represent bona fide danger—a scenario which media producers eagerly take advantage of. The technological nature of the message (e.g. moving specks of light on a screen) is realized retrospectively, if at all. Thus, during the initial seconds of exposure to a negatively-compelling media message, the brain treats it as real and prepares the body for an approach or avoidance response—even when higher-order cognitive processes are at work discounting the message as representational in nature.

The media equation has been observed in humans across demographic and individual differences. Reeves and Nass (1996) also insist that all media have the ability to elicit the media equation response, from entertainment to news and from text messages on a cell phone to wall-sized audiovisual HDTV displays. Central to their explanation for the sweeping nature of the media equation response is the idea of an old human brain unequipped to suspend physical world functions from operating when we enter the contemporary media environment. To the old brain that Reeves and Nass (1996) describe, it is all the same.

Tracking the life history of the human brain produces insights that prompt qualifications to the media equation idea. First, an old brain might not mean an outmoded one. In fact, in evolutionary terms old generally means well adapted, not outmoded. Could the human brain's treatment of media and physical worlds as similar be state-of-the-art adaptation instead of the evolutionary dawdle that the media equation assumes? Second, it is important to acknowledge that the old wiring is for visual processing rather than for text-readiness. Notwithstanding the importance of language in contemporary society, the human brain has had a remarkably short amount of time to adapt and specialize to oral and written forms of communication. Despite much controversy in dating particular evolutionary adaptations, the sequence of eyesight pre-dating speech is uncontested and serves as a reminder of why images are consequential to the media equation idea. After all, *Homo sapiens* employed images long before it wrote. If the old brain is the reason for the media equation response, the oldest part of that brain can be expected to bear the brunt of that response.

At the risk of setting an epic context, let us consider (with aid from Figure 22.1) the grand sequence of: 1) a natural history that positions the emergence and subsequent ramifications of sight in an evolutionary perspective, and 2) a history of human artefact or media making.

The old brain

Understanding the history of vision illuminates the comparatively recent adaptations of language use and the even shorter lifespan of the written word. But even such time-line length comparisons fall short of unveiling the evolutionary havoc that the emergence of the first eye wrought some 543 million years ago when it dramatically hastened biodiversification and the development of species. To no surprise then, from that period forward the eye was ever-present, shaping specialized anatomical and cognitive functions that eventually brought *Homo sapiens* into existence.

Though debated, the formation of the Earth is widely dated to around 4.55 billion years ago (Dalrymple 1991). Life emerged sometime thereafter, perhaps following a flurry of meteorite bombardments (Parker 2003), but the precise dating of the first life-form is a contested topic. Across debates and theories, there is some scientific consensus that abiogenesis (the origin of life from nonlife) occurred between 4.4 and 2.7 billion years ago (Brasier et al. 2002; Schopf et al. 2002; Wilde et al. 2001).

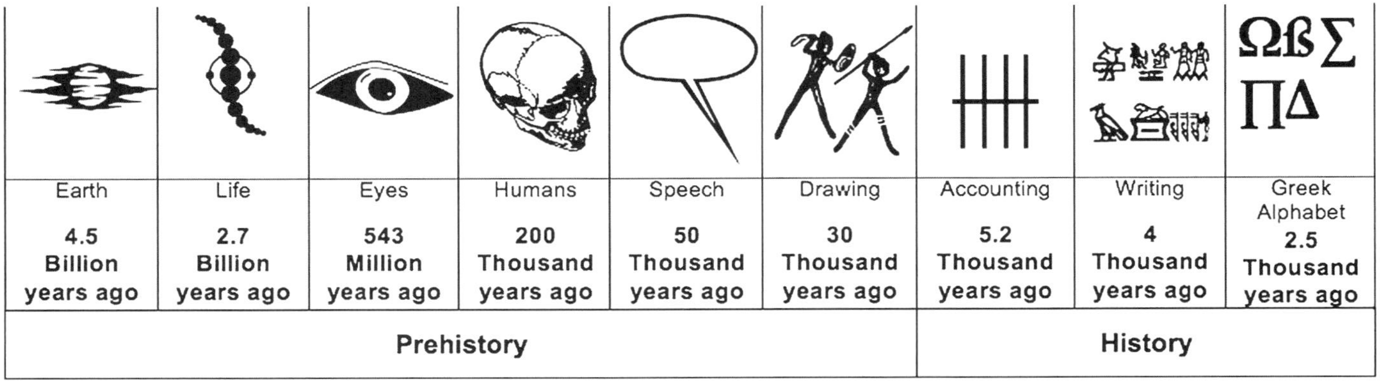

Earth	Life	Eyes	Humans	Speech	Drawing	Accounting	Writing	Greek Alphabet
4.5 Billion years ago	2.7 Billion years ago	543 Million years ago	200 Thousand years ago	50 Thousand years ago	30 Thousand years ago	5.2 Thousand years ago	4 Thousand years ago	2.5 Thousand years ago
			Prehistory				History	

Figure 22.1 Grand sequence of emergence and subsequent ramifications of sight and of human artefact or media making.

Life was slow to diversify. It took more than 2 billion years for three different animal phyla to evolve until the so-called Cambrian explosion, a period approximately 543 million years ago (Morris 1997, 2006; Parker 2003). Spectacularly, over the course of only 5 million years life forms diversified from three to 38 animal phyla, about the same number we have today. Parker (2003, p. 24) describes this degree and tempo of biodiversification as an evolutionary event 'that can be matched in significance only by the beginning of life itself'.

Through close analysis of the fossil record, Parker (2003) has identified development of vision as the accelerant for the Cambrian explosion. The first eye belonged to the trilobite, a small, hard-shelled marine arthropod that appeared about 543 million years ago. The birth of vision changed the rules of survival and accelerated adaptation. Most notably, the development of eyesight gave rise to active predation and, over time, grouped different classes of species into predators and prey—a distinction that did not exist before vision. 'All animals needed to evolve to be adapted to vision before they were eaten, or before they were outwitted by their prey. The early Cambrian thus became a race for adaptation to vision' (Parker 2003, p. 279). Animal body shapes, pigmentation, muscle development, physical movement, and behavioural tendencies, not to mention the positioning and type of eye, are all outcomes of sight. Indeed, the introduction of the eye effectively tore up the existing laws of life and gave rise to a frantic, chaotic period where new rules and shapes of existence prevailed. Following this momentous occasion in natural history, or evolutionary 'chaos' as Parker (2003) describes it, biodiversification settled back into a slower adaptive pace.

Modern humans did not emerge for another half-billion years. Estimates vary, but there is some level of consensus that *Homo sapiens* appeared approximately 200,000, and perhaps as long as 500,000, years ago (Diamond 2006; Goodman et al. 1990). A mutation of the vocal tract, between 40,000 and 60,000 years ago endowed *Homo sapiens* with finer control over sound production (Diamond 2006). Other views on the adaption for speech put it somewhere between 2.4 million and 50,000 years ago. The inexactness of this estimate reveals the scientific controversy surrounding the history of speech.[3] The scarcity of concrete fossil evidence, due to the tenuousness of muscles and ligaments, fuels this debate. Thus, in the relatively wide time span in which speech might have surfaced, a number of potential clues have been identified. Stone tools, dating back some 2.4 million years, are viewed as an indication of linguistic capacity. Two million years ago, the hominid brain experienced rapid expansion, perhaps a sign of language adaptation. Recent identification of a 'speech gene', thought to have originated between 100,000 and 200,000 years ago, offers another clue (Balter 2002). Cultural anthropologists conservatively estimate that humans were, 50,000 years ago, 'creating art, burying their dead, [and performing other] symbolic behaviours that point unequivocally to language use' (Holden 2004, p. 1316).

The will to mediate

Some time after speech evolved, the cultural inclinations of *Homo sapiens* inspired recordings of lived experience in visual form, manifested by what we now call painting and drawing. Artefacts scrawled on cave walls have survived to puzzle contemporary archaeologists in their attempts to date the origin of these drawings. If the new accelerator mass spectrometric techniques and the dates they point to can be trusted, the oldest discovered cave paintings are about 30,000 years old

---

[3] The history of speech became so hotly debated in the 1860s that the British Academy in London and Société de Linguistique in Paris discouraged members from discussing and studying the origin of language and speech, as it was then seen as a speculative and futile theoretical pursuit (Hocckett 1960; Holden 2004). Even today, there is little agreement.

(Valladas 2003; Valladas et al. 2001). Evidence of writing systems, slightly easier to pinpoint but still an inexact science, is traditionally taken as the beginning of history—about 5200 years ago—while the estimated 200,000-year existence of *Homo sapiens* which preceded that point is referred to as prehistory (Clarke 1977). The first writing systems were pictographs and ideographs, visual depictions of ideas dating back some 7000–12,000 years (Hooker 1990). Others regard these artefacts as forms of proto-writing and therefore as evidence of prehistory. There is however, more willingness to view accounting, inspired by economic necessity, as the beginning of the written word and the beginning of human history. The first known form of accounting, cuneiform script, surfaced about 5200 years ago among the Sumerians, who used clay tablets to keep track of commodities and time units spent on labour (Mitchell 1999; Schmandt-Besserat 1996). The use of hieroglyphs by the Egyptians some 4000 years ago marks the origin of the written word, as there is general agreement that hieroglyphs constitute a formal writing system (Houston 2004; Mitchell 1999).

When viewed over time, verbal language thus has a relatively short history, perhaps as short as 40,000 or 50,000 years, and the written word even less (just 5200 years). Compare this to the neurological capacity for visual perception, which evolved over millions of years as the hominoid brain increased in size and specialization. As important as the development of verbal language was in steering the course of (recent) human history, it pales in comparison to the ramifications that the development of sight had on life forms that preceded *Homo sapiens*. In the long history of this vital sense lies an important reminder about the critical role of vision in processing information and guiding behaviour.

The eye 'send[s] more data more quickly and efficiently through the nervous system than any other sense' (Barry 2005, p. 48). Neuro- and cognitive scientists have built the case for visual primacy over other forms of communication and meaning making (see Grabe and Bucy 2009) by documenting: 1) specialized centres in the human brain dedicated to visual processing (Kanwisher et al. 1997; Parker 2003);[4] 2) the speed with which visuals are processed (Barry 2005; Newell 1990; Watanabe et al. 1999); 3) the efficiency with which visual stimuli are classified and stored in memory (Broadbent 1977; Marcus et al. 2000; Paivio 1971); and 4) the way the visual system engages both emotional and 'thinking' centres in the brain (Damasio 1994).

Instrumental as the media equation (Reeves and Nass 1996) idea is as a catalyst for media research grounded in an evolutionary perspective, it is time to consider at least two qualifications. First, given the history of human vision and the evidence for visual primacy emerging from multiple disciplines, all media might not be equal in facilitating the media equation. It is urgently necessary to parse out in a more coherent and comprehensive way how human responses vary across text, audio, visual, and audiovisual messages. The hypothesis on the table is that more lifelike audiovisual media would be the most potent conduits for *media-equal-real-life* responses. In other words, not all media equal real life equally. The encoding of mediated visual information, with its uncanny analogy to the physical world, is considered to be largely 'automatic', a neurologically hardwired process that occurs without conscious control (Barry 2005). By contrast, text-based media do not present an analogical dilemma to the human brain because there is no life-like quality to text and there can thus be no confusion about its purely representational nature. Gazzaniga (1998) goes so far as to say that 'Brains were not built to read' and proceeds to explain that it might be 'why many people have trouble with the process' (p. 56). As international

---

[4] The substantial cranial real estate dedicated to vision is not unique to humans. A large portion of neurophysiology is devoted to vision in animals generally, particularly in diurnal primates, in which up to 50% of neocortex volume is constituted by the primary visual cortex (Joffe and Dunbar 1997; Parker 2003).

demographic studies have found, one in five adults worldwide, two-thirds of them women, are illiterate.[5] In the United States, an estimated 20–25% of the adult population is functionally illiterate (Gee 1989).

The second point of qualification that might serve the original media equation formulation relates to the explanation for the phenomenon as an old brain that lags behind in adaptation—one that is border-line malfunctioning in *mistaking* media for reality. It might be time to strip the media equation from this appraisal. Deeming our brains as old and outmoded implies that we are over-due for an upgrade that would enable *Homo sapiens* to keep the media and physical worlds neatly separated. Early signs suggest that that adaption trajectory is unlikely. Instead, humans display increased ability to seamlessly incorporate media into everyday life, rendering media ubiquitous, and the separation between mediated and physical worlds practically unimportant to lived experience (Deuze 2012). Moreover, it is the human brain that designed and produced, on purpose, reality-like media that are increasingly mobile, screen-based, and stripped from reminders that they are media (consider for example touchscreen interfaces, wireless connectivity, and increasingly lifelike high definition/three-dimensional visualization). Only a human brain could make media (hardware, software, and content) that would advance the media/reality blur in fellow humans. If we find that media users are struggling to separate media from physical reality, is that a sign of an old and outmoded human brain—or the high-functioning innovative brains of the makers of those media messages? More importantly, what existing or anticipated environmental need might compel human adaptation for the ability to separate media from physical reality? At this point, we know of none. In media equation terms then, is it fair to say that the visual brain of contemporary *Homo sapiens*—some 500 million years in the making—serves us well in navigating seamlessly between physical reality and the audiovisual media world we built? We are living in that mediated world at an average of 12 hours per day, often exposed to multiple media at the same time.[6] There is no other human activity—not even sleep—that we do more of. Why should our brains discriminate between media and physical worlds if they are converging? Why would we need to separate specialized brain centres to respond to media versus physical world encounters?

For sure though, the dissolving media-reality distinction inspires plenty of debate—from high-minded philosophical waxing to echo chambers where parents, pundits and politicians belabour the damaging influence of media on children. Loud as they are, none of these musings rise to survival-relevant adaptive thrusts that would deliver a new brain capable of separating media from physical existence.

It is noteworthy that news, as a type of media content, is often excused from concern about the blurring line between the mediated and physical worlds. After all, news is venerated for its informational value in democratic societies precisely because it claims to bring the physical world to citizens in an accurate and fair way—thanks to an elaborate professional code of conduct that supposedly guarantees objectivity. Unlike with entertainment fare or social media, the underlying but unarticulated assumption is that nothing odd or hazardous would come about when citizens treat mediated news about politics and economics as real. Whereas entertainment media produce moral panics about the thawing media-reality distinction, news media ignite head-shaking and sermonizing about the high volume and editorial emphasis on titillating negative events. Evolutionary psychology, arm-in-arm with a qualified media equation perspective, offers a

---

[5] UNESCO. *Literacy.* http://www.unesco.org/en/literacy/
[6] Data from the Kaiser Family Foundation. Generation M2: Media in the lives of 8–18 year olds. www.kff.org/entmedia/mh012010pkg.cfm

powerfully plausible explanation for why the news business thrives on playing up the negative. Liberating as it might be, this explanation requires reconsideration of what journalism is, in ways that strip the profession from the cultural majesty it procured during the past four centuries.

Journalism's grand promise

One cornerstone of democracy is informed citizens who perform their civic duties and exercise their rights (Delli Carpini and Keeter 1996). In response to this ideal, the journalism profession has been revered for its self-appointed mission to deliver informational nutrients to citizens who are, in theory, eager to absorb facts that deepen their understanding and participation in democracy. The noted editor Samuel Bowles, articulating the grand promise, asserted that the 'brilliant mission of the newspaper... is to be the high priest of history, the vitalizer of society, the world's great informer, the earth's high censor, the medium of public opinion and the circulating life blood of the whole human mind' (in Harris 1999, p. 3).

In this view of journalism, information largely means news about politics and economics, delivered through print media. Stories that highlight the death and devastation of everyday life, especially in the spectacularity of life-like audiovisual form, are deemed corrosive and distracting to the making and maintenance of a healthy democracy. A large vocabulary has been developed in aid of defining what proper news is, and terms like sensationalism, tabloid, soft, or human interest news, and infotainment, all serve as slights to signal what news ought not to be. The characterization of broadcast news by Pulitzer Prize–winning journalist Carl Bernstein as servant to an 'idiot culture' offers a more blunt articulation of this sentiment. In his own words, 'We have been moving away from real journalism toward the creation of a sleazoid info-tainment culture... In this new culture of journalistic titillation, we teach our readers and our viewers that the trivial is significant and that the lurid and loopy are more important than real news' (Bernstein 1992, p. 23).

At the heart of this outlook is the worn-out separation of rationality from emotion. Proper news is commended for its assumed ability to enhance the political and social knowledge of the audience by appealing to reason over emotion (Adams 1978; Carroll 1989; Davie and Lee 1995; Scott and Gobetz 1992; Slattery and Hakanen 1994) while the unseemly kind amuses, titillates, and entertains. Stories dealing with negatively-compelling events such as crime, disasters, accidents, and other public fears have consistently been labelled as the lesser topics for informed citizenship (Davie and Lee 1995; Ehrlich 1996; Hofstetter and Dozier 1986; Juergens 1966; Knight 1989; Shaw and Slater 1985; Stevens 1985). Beyond topical categorization, the most distinguished characteristic of improper news is its anticipated effect on consumers. Apparently it stimulates 'unwholesome emotional responses' (Mott 1962, p. 442), shocks and thrills our moral and aesthetic sensibilities (Tannenbaum and Lynch 1960), and emphasizes 'emotion for emotion's sake' (Emery and Emery 1978). Daniels (cited in Tannenbaum and Lynch 1960, p. 382) concludes that these news stories are 'underdistanced'—that is, they violate psychological distance between audience and the physical world by provoking emotional reactions that society has deemed improper. In many ways this sounds like an early articulation of the media equation—in mild moral panic mode.

In other academic circles, rooted in socio-cultural approaches to news analysis, journalism is accepted as a type of dramatic storytelling that portrays an arena of dramatic forces in action (Carey 1975; Epstein 1974, 1975; Hallin 1992; Lule 2001; Tuchman 1978). Less than objective accounts of the world and the potential for provoking audience emotion are not overly problematized in this camp. Moreover, journalism is not judged for topical preferences in accentuating crisis over stability, the present over the past, and scandals involving politicians over political issues (Bennett 1988; Butler 2001; Lule 2001).

In arguing that the dramaturgical properties of journalism serve social maintenance functions, this view of journalism is certainly at odds with the expectation of news media as the distributors of factual information to a public of rational participants in a democracy. The first point of discord is seated in views about objectivity: the ideal of unbiased factual reporting is not in congruence with the view of journalism as value-laden drama-making. Several scholars (e.g. Gans 1979; Lule 2001) have identified measures and presented evidence of the patterned occurrence of values in news. Moreover, the fast-growing body of literature on news framing (e.g. Entman 1991; Liebler and Bendix 1996; Reese et al. 2001) further exemplifies this view of journalism as a driving force in perpetuating social values and reinforcing central myths, rather than imparting weighty information that would spawn knowledgeable citizens engaged in debate, voting, and other activities of idealized participatory democracy. This flows right into the second point of departure between the two views of journalism: assumptions about the news-consuming audience. The democracy-serving view of journalism supposes an active audience who shops the marketplace of news information in preparation for civic participation. The alternative viewpoint is that the defining commonality among news consumers is not the quest for cold hard facts about politics or economics, but rather the reaffirmation of common values and a sense of social cohesion derived from consuming stories about mutual hardship (Lule 2001; Schudson 1978). Human existence involves a chaotic, meaningless, and fragmentary array of experiences. Stories give meaning to life and enable the members of a society to understand each other and to cope with the unknown (Barthes 1972; Cawelti 1975; Himmelstein 1984). News stories are seen as one of many story-telling formats in service of this social integration process.

As much as this sociological view sobers the idealism of journalism's grand promise to society, it falls short of explaining why—beyond society-level functionalism—the high drama of negative news is appealing to citizens to begin with. To call negatively-compelling news a crowd pleaser is an under-statement. Among journalists it is known as the 'if it bleeds, it leads' or 'bodybag journalism' practice to draw news users (Briller 1993; Salerno 1995). As Newhagen (1998) points out, televised newscasts feature 'more images of violence, suffering, and death in half an hour than most people would normally view in a lifetime' (p. 267). Beyond television news, the colourful front pages of tabloid newspapers, and the online availability of full motion video about accidents, disasters, corruption, and conflict are often identified with this professional practice (Slone 2000). Despite the concerns of traditional newsmen like Bernstein, negative compellingness is a cross-cultural standard for newsworthiness (Knobloch-Westerwick et al. 2005). The negative slant in news is also not a contemporary phenomenon or one particular to television—it is as old as the concept of news itself. In fact, European news ballads of the late 1500s epitomize the journalistic sensibility to dwell on negative occurrences (Bird 1992; Shaw and Slater 1985; Stevens 1985).[7]

Strikingly, while compelling negativity is the most persistent news selection principle—over time, across cultures, and despite damning criticism—we have done little to explain it. From a sociological perspective, media scholars conclude that it functions at the macro level to maintain social order. That is a plausible outcome for such a persistent phenomenon. But it is time to entertain explanations for why negative news attracts the attention of society's members in the first place.

---

[7] Interestingly, periods of public outrage about it have become a periodic ritual. In the United States, for example, reactions to the Penny Press of the 1830s, 'yellow journalism' at the end of the 19th century, and the findings of the Hutchins Commission after the Second World War strongly resemble the damning tone of the Bernsteinian sentiment (Altschull 1990; Tannenbaum and Lynch 1960).

Survival-relevancy

The media equation perspective is well-suited for explaining the human propensity for paying attention to negatively-compelling news. The minds and bodies of contemporary humans treat the physical and mediated worlds, in large part, as if they are equally real. Thus, full motion video of an approaching tornado spells threat—during the initial seconds of exposure—every bit as much as it does in the physical world. In response, the human brain automatically mobilizes cognitive resources to attend to this mediated treat (Newhagen and Reeves 1992).

Research in the information processing area shows support for this argument. It is well documented that the brain has an automatic attentive response to negatively-compelling stimuli—in mediated and non-mediated form (Blake et al. 2001; Brosius 1993; Canli et al. 2002; Plutchik 1984; Wrase et al. 2003; Zald 2003). Several studies have confirmed this point in particular reference to television news (Grabe et al. 2003; Hsu and Price 1993; Newhagen 1998). Impose the *why* question to this evidence of an automatic attention response to negative stimuli and one enters evolutionary psychology proper.

Human attention systems have been calibrated, over centuries, to favour stimuli that have adaptive significance; therefore humans ignore some stimuli while attending to others (Cacioppo and Gardner 1999). Surveying the environment for threats was essential to the survival of our ancestors (Plutchik 1980). The pervasive human tendency to automatically classify most, if not all, incoming stimuli as either good or bad can be seen as a screening mechanism to ensure the organism's self-preservation (Plutchik 1980; Rothermund et al. 2001). This mechanism, also referred to as affective processing, relies on a small number of important and easily detectable stimulus characteristics. It occurs without intentional or conscious effort and does not require processing of the non-affective attributes of stimuli (Moors and Houwer 2006). However, the determination of whether a stimulus is good or bad does not by itself provide the organism with adaptive benefit. It is the action tendency that this realization results in that is of value. Evolution favours organisms that learn to withdraw and approach appropriately and quickly (Cacioppo and Gardener 1999; Green et al. 2004). While humans have a strong propensity for monitoring their environment for potential threats and to avoid them, the level of negativity in a stimulus affects the response. Highly-compelling (arousing) negative stimuli elicit avoidance action; less arousing negative stimuli are often approached with a sense of curiosity (Lang 2000) in the same way that positive stimuli are investigated for opportunity. However, attention to and action readiness for a threat is considered more important for survival than curiosity about opportunities (Knoblock-Westerwick et al. 2005).[8]

News media have assumed a surveillance function over the past few centuries by focusing on negative occurrences and even playing-up mild negativity. Audiences respond to this mediated survival-relevant information based on their biological predisposition to pay attention to potential threats in their environment (Shoemaker 1996). Newell (1990), Newhagen and Reeves (1992), and Newhagen (1998) have explained that negatively-compelling television news resembles a non-mediated survival threat to the degree that the biological systems of viewers respond to it as if it is non-mediated. Remarkably, the attentive response to negatively compelling audiovisual news is so strong that intensely negative images of wars, disasters, or social mayhem can proactively enhance memory for audiovisual material that appears after the images are shown, while retroactively inhibiting memory for information that appears before (Newhagen and Reeves 1992). This automatic preparation for premium performance is further revealed by the increase

---

[8] Unlike negative stimuli, messages containing positive emotion require propositional evaluation and additional time to process (Leventhal and Scherer 1987).

in physical and self-reported emotional arousal and attention associated with negatively-compelling news (Grabe et al. 2000, 2003; Newhagen 1998).

No wonder that news is, almost by definition, negative—and that television's audiovisual presentation of these events has been the most attention-drawing in human history. As early as November 1963, a Roper poll commissioned by the Television Information Office (Roper Organization Inc., 1979) found that television was the main source of news for most Americans—a statistic that holds true even today.[9] Life-like audiovisual renditions of news, such as we find online and on television, also seems to have the most potent short-term and enduring long-term impact on citizens. Beyond the evidence from real-time physiological responses to negatively-compelling audiovisual news and encoded memory tests promptly after exposure, long-term memory assessments confirm the imprint that these mediated survival threats leave on the human mind. Six to seven weeks after exposure, television visuals are recalled more easily than news story topics or narrative content, particularly when stories contain negative compelling images (Newhagen and Reeves 1992). Audiovisual news even have the potential to close knowledge gaps between citizens from different socioeconomic groups—while print/word-only news media exacerbate the division between information haves and have-nots (Grabe et al. 2009; Prior 2002).

In an unapologetic evolutionary psychology perspective, Shoemaker (1996) advanced the idea that humans are hard-wired for negative news and offered this as one possible explanation for the historically persistent media emphasis on bad news and the corresponding audience appetite for it. In contrast to the grand promise of journalism as the conduit of an informed citizenry, Shoemaker (1996) argues that journalists serve a specialized function in human societies by surveying the environment and warning against potential survival threats. In early human societies it was the job of watchmen, sitting in treetops or on hills, to perform this surveillance function in their communities. Around the 16th century this task became known as news-gathering and the audience appetite for this type of information has remained remarkably stable since then. In evolutionary psychology terms, this persistent pattern of audience interest in warnings about death, deviance, and destruction (bad news, as we have come to call it), bears a cue to our biological hard wiring. Taking this evolutionary psychology-inspired insight seriously might even serve in figuring out how journalism can do better in delivering on its great promise of informed citizenship. Yet, there is one quirk in this apparently clear-cut evolutionary psychological explanation that should not be under-estimated.

Gender-specificity

In spite of all the journalistic emphasis on bad news and experimental research confirming that this practice automatically engages news users, a number of surveys show that a sizeable chunk of the audience is dissatisfied with the emphasis on negatively-compelling events to the point that they withdraw from news consumption. Closer inspection reveals gender variance in this discontent.

As noted earlier, evolutionary psychology explains a predisposition for automatic attention to negatively-compelling news. Human evolution has favoured traits that make ignoring responses

---

[9] The precise wording of the Roper question was, 'First, I'd like to ask you where you usually get most of your news about what is going on in the world today—from the newspapers or radio or television or magazines or talking to people or where?' The multibarrelled wording of this question has fueled criticism, analysis, and counter-studies in mass media research circles for decades—even while television was strengthening its position as the most widely used source of information (e.g. Lemert, 1980; Reagan and Ducey 1983; Stempel 1991).

in the face of survival threats impossible for both sexes (Leventhal and Scherer 1987). Thus at high levels, negatively-arousing environmental stimuli create a ceiling effect where no variance between sexes in cognitive awareness (attention) and behavioural avoidance (flee response) is observed (Knight et al. 2002; Zillmann 1983). Yet, moderately negative stimuli produce variation in arousal levels such that subjects respond by avoiding, approaching, or even ignoring stimuli (Leventhal and Scherer 1987), and these responses are likely to produce sex differences. Indeed, women have a stronger avoidance response to these lower-intensity negative stimuli than men (Ahmed and Bigelow 1993; Canli et al. 2002; Knight et al. 2002; Wrase et al. 2003; Zald 2003; Zillmann 1983).[10]

This sex difference in responses to moderately negative stimuli may be adaptive for offspring survival. The male tendency for an approach response serves a defensive survival function, detecting, investigating, and protecting offspring from potential danger (Buss 1991). Because the subsistence of offspring is more closely linked to the survival of mothers than fathers, females are predisposed to an avoidance response to negative stimuli, steering themselves and their young clear of potential danger.

There is evidence that women's avoidance response to negative stimuli manifests in their consumption of news. Women are more likely than men to point at news coverage of wars and violence as the main reason why they do not follow international news,[11] and they report that they would like to see more positive news on television (Gallician 1986). Research also shows that women are emotionally more sensitive to negative news than men. Slone (2000) found that women report greater anxiety than men in response to negative news. While 44% of Americans say they are often depressed by the news, women make up 53% of this figure, men only 34%.

Message valence and arousal levels also create a gender divide in acquiring news information. In an experimental study of moderately negative television news, information was kept constant across conditions, but the tone in which the negative news was presented varied in negativity/positivity (Grabe and Kamhawi 2006; Kamhawi and Grabe 2008). Thus positive versions of the news stories were not wholesale positive, but positive renditions of negative events, while negative versions presented a negative spin on an already negative event. Male viewers were linked to a negativity bias, reporting the highest arousal levels and producing the best recognition memory and comprehension scores for negatively valenced versions of the news stories. Women, in contrast, showed signs of an avoidance response to the negatively framed news, rating the positively valenced versions more arousing, as well as comprehending such messages more effectively than negatively framed messages (Grabe and Kamhawi 2006; see also Hendriks Vettehen et al. 1996).[12]

In the same study (Kamhawi and Grabe 2008), women reported enjoying negatively framed news less, were less able to identify with people featured in such stories, and were less appreciative of the journalistic effort exerted in reporting negatively framed stories. Men demonstrated an opposite pattern of effects: they embraced negatively framed news, reporting more enjoyment and involvement with people in negatively than positively framed news.

---

[10] When it comes to positive valence, research shows that unlike women, men have difficulty engaging in positive emotion and tend to ignore positive stimuli, especially at medium and low arousal levels (Ahmed and Bigelow 1993). Fuijita et al. (1991) reckon that men lead a drab emotional life because they are more inhibited in reacting positively to pleasant things.

[11] Gibbons, S. (2003). *Newspaper execs clueless about what women want*. At: http://womensenews.org/article.cfm/dyn/aid/1524/context/archive

[12] Ambiguously valenced news is not clearly advantageous to either sex, but there are tentative indications that women are better positioned than men to gain information from such messages.

Together, these cognitive and emotional responses offer some explanation for the gender gap in news use and knowledge acquisition.[13] The selection of negative events, combined with emphasis on their negative dimensions, might be the reason why women are less avid news consumers and as a result less informed about public affairs issues than men. Not only do women steer clear of negatively framed news, they also question the credibility and objectivity of such news. Explanations that women generally have less education and insufficient knowledge to understand the news (Hendriks Vettehen et al. 1996), or that news is of less interest and relevance to them (Jensen 1986; Morely 1986) do not sufficiently explain the gender gap in news consumption across the world and indeed seem to validate the subtle exclusion of women from full citizenship.

The evidence of a gender gap in response to the valence of news messages also helps to explain why newsrooms have been over-populated by men and why news is defined by negativity. Specifically, if men have a stronger cognitive propensity to approach negatively-valenced stimuli than women, it is not surprising that journalism has a long tradition of employing men to cover mostly negative news for an audience consisting mostly of other men. Historians remind us that news was developed by men for men (Klein 2003; Schudson 1978; Tuchman 1978). Women remain delegated to the sidelines of the profession's informal hierarchy. In the United States, for example, they make up only one-third of all journalists (Becker et al. 1999; Weaver et al. 2007). Women reporters are also generally confined to news beats and genres that are considered feminine and in line with traditional gender expectations of them (soft news, entertainment, human interest), rather than sent out to cover masculine or so-called hard, 'breaking' news that lead front-pages and newscasts (e.g. Cann and Mohr 2001; Desmond and Danilewicz 2010; Nicholson 2007; Thiel Stern 2007). Interestingly, especially in smaller markets, women reporters are more likely to include positive angles in their stories than their male counterparts (Rodgers and Thorson 2003).

Shoemaker's (1996) insight that we are wired for negative news makes advice to report happy news unsound. Yet, by inserting positive (i.e. problem-solving next to problem-signalling) angles into coverage of a negative event, the female news audience might be enticed to engage with news more often. This is not a novel idea. More than 30 years ago, Gans (1979) identified the 'restoration of order' principle as an enduring news value, even in stories about death and destruction. The news coverage of 9/11 and the Haiti earthquake exemplified this by playing up bravery, goodwill, and community support. Unless the noble goal of journalism to inform the public applies predominantly to male citizens, it looks like a focus on the bright side of negative events might be productive.

Conclusion

Media research, a young field overwhelmed by the task of understanding the 'generation, representation, consumption, and interpretation of texts' (Cappella 1996, p. 6) has generally avoided biological avenues of inquiry. The conception of the discipline offers some insight into why researchers have been so determined to explain human co-existence with media from socio-structural or individual-variance approaches. Interest in understanding media processes and effects surfaced against the social backdrop of deepening moral investment in democratic governance at a time when World Wars were fought, at least in part, through emerging mass media

---

[13] See The Pew Research Center for People and the Press commentary, *The Public Affairs Gender Gap* (25 April 2000) http://people-press.org/commentary/?analysisid=10. The gap is independent of other demographic variables such as education, age and work status.

(Cmiel, 1996; Delia, 1987; Gary, 1996). The early efforts to study media therefore focused on how it affected public opinion, persuasion, and informed citizenship.

Media entertainment research came into its own several decades later, largely liberated from the early tradition, and more receptive to cognitive and biological impetus. It is this intellectual climate that embraced the media equation perspective while journalism research, as a subfield, remains—even today—largely consumed by figuring out how news might serve the individual and contribute to democratic governance. This does not provide an easy entry point for biological explanations. Yet, this field has explained itself into corners applying socio-structural and individual agency thinking, crying out for new theoretical milestones, and contemplating redefinition of what journalism is—all whilst continually lamenting the terminal state of the newspaper (and the printed word in general).

Journalism research is in dire need of a jolt. Given the fanfare of the preceding pages, it might appear as if evolutionary psychology is appointed as the conduit for the job. It is. But a required precursor to such a shake-up will entail painful ontological purging of the normative tradition of journalism enquiry. The focus of journalism research must be unhooked from repetitive administrative assessments of how well news aligns with traditional views of democracy and citizenship. Bold and cross-disciplinary sensibilities will be required to understand how humans use media to form memories, experience emotion, and coexist with others. These discoveries might not be pleasing to reigning idealism about news media and democracy. The sharp edge of fitness for survival will undoubtedly emerge in findings about discriminatory messages, hierarchy maintenance manoeuvres, and other anti-democratic patterns of behaviour in the world of news-making and consumption. The point is to move beyond spotting and then critiquing these practices and into ways of explaining them. The evolutionary psychology vantage point is certainly conducive to such an endeavour.

References

Adams, W.C. (1978). Local public affairs content of TV news. *Journalism Quarterly,* **55**, 690–5.

Ahmed, S.M.S. and Bigelow, B. (1993). Latency period as affected by news content. *The Journal of Social Psychology,* **133**, 361–4.

Altschull, J.H. (1990). *From Milton to McLuhan: the ideas behind American journalism.* Longman, New York.

Balter, M. (2002). Language evolution: 'Speech gene' tied to modern humans. *Science,* **297**, 1105.

Barry, A.M. (2005). Perception theory. In K. Smith, S. Moriarty, G. Barbatsis, and K. Keith (eds), *Handbook of visual communication*, pp. 45–62. Lawrence Erlbaum, Mahwah, NJ.

Barthes, R. (1972). *Mythologies.* Hill and Wang, New York.

Becker, L.B., Lauf, E., and Lowry, W. (1999). Differential employment rates in the journalism and mass communication labor force based on gender, race, and ethnicity: exploring the impact of affirmative action. *Journalism and Mass Communication Quarterly,* **76**, 631–45.

Bennett, W.L. (1988). *The politics of illusion.* Sage, New York and London.

Bernstein, C. (1992). The idiot culture: reflections of post-Watergate journalism. *The New Republic,* 8 June, 22–26.

Bird, S.E. (1992). *For inquiring minds: a cultural study of supermarket tabloids.* University of Tennessee Press, Knoxville, TN.

Blake, T., Varnhagen, C. and Parent, M. (2001). Emotionally arousing pictures increase blood glucose levels and enhance recall. *Neurobiology of Learning and Memory,* **75**, 262–73.

Brasier, M.D., Green, O.R., Jephcoat, A.P., *et al.* (2002). Questioning the evidence for Earth's oldest fossils. *Nature,* **416**, 76–81.

Briller, B. (1993). The Tao of tabloid television. *Television Quarterly,* **26**, 51–61.

Broadbent, D.E. (1977). The hidden preattentive process. *American Psychologist,* **32,** 109–18.

Brosius, H. (1993). The effects of emotional pictures in television news. *Communication Research,* **20,** 105–24.

Buss, D. (1991). Evolutionary personality psychology. *Annual Review of Psychology,* **42,** 459–91.

Butler, J.G. (2001). *Television: critical methods and applications.* Lawrence Erlbaum Associates, Mahweh, NJ.

Cacioppo, J.T. and Gardner, W.L. (1999). Emotion. *Annual Review of Psychology,* **50,** 191–215.

Canli, T., Desmond, J.E., Zhao, Z. and Gabrieli, J.D.E. (2002). Sex differences in the neural basis of emotional memories. *Proceedings of the National Academy of Sciences of the USA,* **99,** 10789–95.

Cann, D. and Mohr, P. (2001). Journalist and source gender in Australian television news. *Journal of Broadcasting and Electronic Media,* **45,** 162–85.

Capella, J.N. (1996). Why biological explanation? *Journal of Communication,* **46,** 4–7.

Carey, J.W. (1975). A cultural approach to communication. *Communication,* **2,** 1–22.

Carroll, R.L. (1989). Market size and TV news values. *Journalism Quarterly,* **66,** 49–56.

Cawelti, J.G. (1975). Myths of violence in American popular culture. *Critical Inquiry,* **1,** 521–41.

Clarke, G. (1977). *World prehistory: a new perspective.* Cambridge University Press, Cambridge.

Cmiel, K. (1996). On cynicism, evil, and the discovery of communication in the 1940s. *Journal of Communication,* **46,** 88–107.

Dalrymple, G.B. (1991). *The age of the Earth.* Stanford University Press, Palo Alto, CA.

Damasio, A.R. (1994). *Descartes' error: emotion, reason and the human brain.* G.P. Putnam's Sons, New York.

Davie, W.R. and Lee, J.S. (1995). Sex, violence, and consonance/differentiation: an analysis of local TV news values. *Journalism and Mass Communication Quarterly,* **72,** 128–38.

Delia, J.G. (1987). Communication research: a history. In C. Berger and S. Chaffee (eds) *Handbook of Communication Science,* pp. 20–98. Sage, Newbury Park, CA.

Delli Carpini, M.X. and Keeter, S. (1994). The public's knowledge of politics. In K. J. David (ed.), *Public opinion, the press, and public policy,* pp. 19–40. Greenwood, Westport, CT.

Desmond, R. and Danilewicz, A. (2010). Women are on, but not in, the news: gender roles in local television news. *Sex Roles,* **62,** 822–9.

Deuze, M. (2012). *Media life.* Polity Press, Cambridge.

Diamond, J. (2006). *The third chimpanzee: the evolution and future of the human animal.* Harper Perennial, New York.

Ehrlich, M.E. (1996). The journalism of outrageousness: tabloid television news vs. investigative news. *Journalism and Mass Communication Monographs,* **155,** 1–20.

Emery, E. and Emery, M. (1978). *The press and America; an interpretative history of the mass media.* Prentice Hall, Englewood Cliffs, NJ.

Entman, R.M. (1991). Framing U.S. coverage of international news: contrasts in narratives of the KAL and Iran air incidents. *Journal of Communication,* **41,** 6–27.

Epstein, E.J. (1974). *News from nowhere.* Vintage Books, New York.

Epstein, E.J. (1975). *Between fact and fiction.* Vintage Books, New York.

Fuijita, F., Diener, E., and Sandvik, E. (1991). Gender differences in negative affect and well-being: the case for emotional intensity. *Journal of Personality and Social Psychology,* **61,** 427–34.

Gallician, M. (1986). Perceptions of good news and bad news on television. *Journalism Quarterly,* **63,** 611–16.

Gans, H. (1979). *Deciding what's news.* Pantheon, New York.

Gary, B. (1996). Communication research, the Rockefeller Foundation, and mobilization for the War on Words, 1938–1944. *Journal of Communication,* **46,** 124–48.

Gazzaniga, M.S. (1998). *The mind's past*. University of California Press, Berkeley, CA.

Gee, J.P. (1989). The legacies of literacy: From Plato to Freire through Harvey Graff. *Journal of Education,* **171**, 147–66.

Goodman, M., Tagle, D.A., Fitch, D.H.A., et al. (1990). Primate evolution at the DNA level and a classification of hominoids. *Journal of Molecular Evolution,* **30**, 260–6.

Grabe, M.E. and Bucy, E.P. (2009). *Image bite politics: news and the visual framing of elections*. Oxford University Press, New York.

Grabe, M.E. and Kamhawi, R. (2006). Hard wired for negative news? Gender differences in processing broadcast news. *Communication Research,* **33**, 346–69.

Grabe, M.E., Kamhawi, R., and Yegiyan, N. (2009). Informing citizens: how people with different levels of education process television, newspapers, and web news. *Journal of Broadcasting and Electronic Media,* **53**, 90–111.

Grabe, M.E., Lang, A., and Zhao. X. (2003). News content and form: implications for memory and audience evaluations. *Communication Research,* **30**, 387–413.

Grabe, M.E., Zhou, S., Lang, A. and Bolls, P. (2000). Packaging television news: the effects of tabloid on information processing and evaluative responses. *Journal of Broadcasting and Electronic Media,* **44**, 581–98.

Green, M., Brock, T., and Kaufman, G. (2004). Understanding media enjoyment: the role of transportation into narrative worlds. *Communication Theory,* **14**, 311–27.

Hallin, D.C. (1992). Sound bite news: television coverage of elections 1968–1988. *Journal of Communication,* **42**, 5–24.

Harris, B. (1999). *Blue and gray in black and white: newspapers in the Civil War*. Brassey's, London.

Hendriks Vettehen, P., Hietbrink, N., and Renckstorf, K. (1996). Differences between men and women in recalling TV news. In: K. Renckstorf, D. McQuail, and N. Jankowski (eds), *Media use as social action: a European approach to audience studies*, pp. 151–62. John Libbey, London.

Himmelstein, H. (1984). *Television myth and the American mind*. Praeger, New York.

Hockett, C.F. (1960). The origin of speech. *Scientific American,* **203**, 88–96.

Hofstetter, C.R. and Dozier, D.M. (1986). Useful news, sensational news: quality, sensationalism and local TV news. *Journalism Quarterly,* **63**, 815–20.

Holden, C. (2004). The origin of speech. *Science,* **303**, 1316–19.

Hooker, J.T. (1990). *Reading the past: ancient writing from cuneiform to the alphabet*. University of California Press, Berkeley, CA.

Houston, S.D. (2004). *The first writing script invention as history and process*. Cambridge University Press, New York.

Hsu, M. and Price, V. (1993). Political expertise and affect: effects on news processing. *Communication Research,* **20**, 671–95.

Jensen, K. (1986). *Making sense of the news*. University Press, Aarhus.

Joffe, T.H. and Dunbar, R.I.M. (1997). Visual and socio-cognitive information processing in primate brain evolution. *Proceedings of the Royal Society B,* **264**, 1303–7.

Juergens, G. (1966). *Joseph Pulitzer and the New York world*. Princeton University Press, Princeton, NJ.

Kamhawi, R. and Grabe, M.E. (2008). Engaging the female audience: an evolutionary psychology perspective on gendered responses to news valence frames. *Journal of Broadcasting and Electronic Media,* **52**, 1–19.

Kanwisher, N., McDermott, J., and Chun, M.M. (1997). The fusiform face area: a module in human extrastriate cortext specialized for face perception. *Journal of Neuroscience,* **17**, 4302–11.

Klein, R. (2003). Audience reactions to local TV news. *American Behavioral Scientist,* **46**, 1661–72.

Knight, G. (1989). Reality effects: tabloid television news. *Queen's Quarterly,* **96**, 94–108.

Knight, G., Guthrie, I.K., Page, M.C., and Fabes, R.A. (2002). Emotional arousal and gender differences in aggression: a meta-analysis. *Aggressive Behavior,* **28**, 366–94.

Knobloch-Westerwick, S., Carpentier, F., Blumhojf, A., and Nickel, N. (2005). Selective exposure effects for positive and negative news: testing the robustness of the informational utility model. *Journalism and Mass Communication Quarterly*, **82**, 181–95.

Lang, A. (2000). The limited capacity model of mediated message processing. *Journal of Communication*, **50**, 46–70.

Lemert, J.B. (1980). News media competition under conditions favorable to newspapers. *Journalism Quarterly*, **47**, 272–80.

Leventhal, H. and Scherer, K. (1987). The relationship of emotion to cognition: a functional approach to a semantic controversy. *Cognition and Emotion*, **1**, 3–28.

Liebler, C.M. and Bendix, J. (1996). Old-growth forests on network news: news sources and the framing of an environmental controversy. *Journalism and Mass Communication Quarterly*, **73**, 53–65.

Lule, J. (2001). *Daily news, eternal stories*. The Guilford Press, New York.

Marcus, G.E., Neuman, W. R., and MacKuen, M. (2000). *Affective intelligence and political judgment*. University of Chicago Press, Chicago, IL.

Mitchell, L. (1999). Earliest Egyptian glyphs. *Archaeology*, **52**, 28–9.

Moors, A. and De Houwer, J. (2006). Automaticity: A theoretical and conceptual analysis. *Psychological Bulletin*, **132**, 297–326.

Morley, D. (1986). *Family television: culture, power and domestic leisure*. Comedia, London.

Morris, S.C. (1997). *The crucible of creation: the Burgess shale and the rise of animals*. Oxford University Press, New York.

Morris, S.C. (2006). Darwin's dilemma: the realities of the Cambrian 'explosion'. *Philosophical Transactions of the Royal Society B*, **36**, 1069–83.

Mott, F.L. (1962). *American journalism*. Macmillan, New York.

Newell, A. (1990). *Unified theories of cognition*. Harvard University Press, Cambridge, MA.

Newhagen, J.E. (1998). TV news images that induce anger, fear, and disgust: effects on approach-avoidance and memory. *Journal of Broadcasting and Electronic Media*, **42**, 265–76.

Newhagen, J.E. and Reeves, B. (1992). The evening's bad news: effects of compelling negative television news images on memory. *Journal of Communication*, **42**, 25–41.

Nicholson, J. (2007). Women in newspaper journalism (since the 1990s). In: P. Creedon and J. Cramer (eds), *Women in mass communication*, pp. 35–47. Sage, Thousand Oaks, CA.

Paivio, A. (1971). *Imagery and verbal processes*. Holt, Rinehart and Winston, New York.

Parker, A. (2003). *In the blink of an eye: how vision sparked the Big Bang of evolution*. Perseus, New York.

Pinker, S. (1997). Words and rules in the brain. *Nature*, **387**, 547–48.

Plutchik, R. (1980). A general psychoevolutionary theory of emotion. In: R. Plutchik and H. Kellerman (eds), *Emotion: theory, research, and experience, Vol. 1, Theories of emotion*, pp. 3–33. Academic, New York.

Plutchik, R. (1984). Emotions: A psychoevolutionary theory. In: K. Scherer and P. Ekman (eds), *Approaches to emotion*, pp. 197–220. Harper and Row, New York.

Prior, M. (2002). 'More than a thousand words? Visual cues and visual knowledge.' Paper presented at the Annual meeting of the American Political Science Association, Boston, MA.

Reagan, J. and Ducey, R.V. (1983). Effects of news measure on selection of state government news sources. *Journalism Quarterly*, **60**, 211–17.

Reese, S.D., R. D., Gandy, O.H. Jr, and Grant, A.E. (2001). *Framing public life: perspectives on media and our understanding of the social world*. Lawrence Erlbaum Associates, Mahwah, NJ.

Reeves, B. and Nass, C. (1996). *The media equation: how people treat computers, television, and new media like real people and places*. Cambridge University Press, New York.

Rodgers, S. and Thorson, E. (2003). A socialization perspective on male and female reporting. *Journal of Communication*, **53**, 658–78.

Roper Organization Inc. (1979). *Public perspectives of television and other mass media: a twenty-year review 1959–1978*. Television Information Office, New York.

Rothermund, K., Wentura, D., and Bak, P. (2001). Automatic attention to stimuli signaling chances and dangers: moderating effects of positive and negative goal and action contexts. *Cognition and Emotion*, **15**, 231–48.

Salerno, S. (1995). Trash by any other name. *The Wall Street Journal*, 21 August, A7.

Schmandt-Besserat, D. (1996). *How writing came about*. University of Texas Press, Austin.

Schopf, J.W., Kudryavtsev, A.B., Agresti, D.G., Wdowiak, T.J., and Czaja, A.D. (2002). Laser–Raman imagery of Earth's earliest fossils. *Nature*, **416**, 73–6.

Schudson, M. (1978). *Discovering the news: a social history of American newspapers*. Basic Books, New York.

Scott, D.K. and Gobetz, R.H. (1992). Hard news/soft news content of the national broadcast networks, 1972–1987. *Journalism Quarterly*, **69**, 406–12.

Shaw, D.L. and Slater, J.W. (1985). In the eye of the beholder? Sensationalism in American press news, 1820–1860. *Journalism History*, **12**, 86–91.

Shoemaker, P. (1996). Hard wired for news: using biological and cultural evolution to explain the surveillance function. *Journal of Communication*, **46**, 32–47.

Slattery, K.L. and Hakanen, E.A. (1994). Sensationalism versus public affairs content of local TV news: Pennsylvania revisited. *Journal of Broadcasting and Electronic Media*, **38**, 205–16.

Slone, M. (2000). Responses to media coverage of terrorism. *The Journal of Conflict Resolution*, **44**, 508–22.

Stempel, G.H. III. (1991). Where people *really* get most of their news. *Newspaper Research Journal*, **12**, 2–9.

Stevens, J.D. (1985). Social utility of sensational news: murder and divorce in the 1920's. *Journalism Quarterly*, **62**, 53–8.

Tannenbaum, P.H. and Lynch, M.D. (1960). Sensationalism: the concept and its measurement. *Journalism Quarterly*, **37**, 381–92.

Thiel Stern, S. (2007). Increased legitimacy, fewer women? Analyzing editorial leadership and gender in online journalism. In: P. Creedon and J. Cramer (eds), *Women in mass communication*, pp. 133–47. Sage, Thousand Oaks, CA.

Tuchman, G. (1978). *Making news: a study in the construction of reality*. Free Press, New York.

Valladas, H. (2003). Direct radiocarbon dating of prehistoric cave paintings by accelerator mass spectrometry. *Measurement Science and Technology*, **14**, 1487–92.

Valladas, H., Clottes, J., Geneste, J., *et al*. (2001). Evolution of prehistoric cave art. *Nature*, **413**, 479.

Watanabe, S., Kakigi, R., Koyama, S., and Kirino, E. (1999). Human face perception traced by magneto- and electro-encephalography. *Cognitive Brain Research*, **8**, 125–42.

Weaver, D., Beam, R., Brownlee, B., Voakes, P., and Wilhoit, C. (2007). *The American journalist in the 21st century. US news people at the dawn of a new millennium*. Lawrence Erlbaum, Mahwah, NJ.

Wilde, S.A., Valley, J.W., Peck, W.H., and Graham, C.M. (2001). Evidence from detrital zircons for the existence of continental crust and oceans on the Earth 4.4 Gyr ago. *Nature*, **409**, 175–8.

Wrase, J., Klein, S., Gruesser, S., *et al*. (2003). Gender differences in the processing of standardized emotional visual stimuli in humans: a functional magnetic resonance imaging study. *Neuroscience Letters*, **348**, 41–6.

Zald, D. (2003). The human amygdala and the emotional evaluation of sensory stimuli. *Brian Research Review*, **41**, 88–123.

Zillmann, D. (1983). Arousal and aggression. In: R.G. Green and E.I. Donnerstein (eds), *Aggression: theoretical and empirical reviews*, pp. 75–101. Academic Press, New York.

Section 6

Technology

Chapter 23

Media naturalness theory: human evolution and behaviour towards electronic communication technologies

Ned Kock

Introduction

The advent of the Internet in the early 1990s, and of the World Wide Web in the mid 1990s, led to an explosion in the number of electronic business-to-consumer interactions. Empirical research on electronic communication (e-communication) behaviour also increased considerably and experienced a significant shift in focus from laboratory experiments to field studies (Kock 1999, 2008). Several theories informed this research, including technology-centric theories, of which the most prominent example is media richness theory (Daft and Lengel 1986). Many researchers have tested media richness theory (Carlson and Zmud 1999; Daft et al. 1987; Fulk et al. 1990; Kinney and Dennis 1994; Lengel and Daft 1988; Markus 1994a; Rice 1992; Trevino et al. 2000), and many others continue doing so, even though the theory was first proposed in the mid 1980s, well before the emergence of the Internet as we know it today.

Media richness theory was built around a central hypothesis, the media richness hypothesis, which states that different communication media possess different degrees of a trait called 'richness' (Carlson and Davis 1998; Daft and Lengel 1986; Lee 1994), that make them more or less effective conduits of information and knowledge. Several studies found general support for the media richness hypothesis (Daft et al. 1987; Rice 1993; Rice and Shook 1990; Sproull and Kiesler 1986; Walther 1996), other studies found weak support (Fulk et al. 1990; Markus 1990), and yet other studies found little or no support at all for the media richness hypothesis (Dennis et al. 1999; Kinney and Dennis 1994; Kinney and Watson 1992).

The main goal of this chapter is to extend prior research on human evolution and behaviour towards technology (Kock 2004, 2005, 2009), and offer a solid theoretical basis on which the mixed findings above can be understood. The chapter provides an alternative to the media richness hypothesis, referred to here as the media naturalness hypothesis. Like the media richness hypothesis, the media naturalness hypothesis has important implications for the selection, use, and deployment of e-communication tools in organizations. However, unlike the media richness hypothesis, it is argued here that the media naturalness hypothesis is compatible with social theories of behaviour towards e-communication tools. The 'e' in 'e-communication' stands for 'electronic', so the term 'e-communication' as used here refers to, essentially, any form of computer-mediated communication plus more traditional forms of electronic communication, such as telephone use.

The media richness hypothesis

The media richness hypothesis is used in this chapter to summarize the main idea proposed by media richness theory, originally proposed by Daft and Lengel (1986). According to that idea, communication media can be classified along a continuum of 'richness', where richness is based on the ability of media to carry non-verbal cues, provide rapid feedback, convey personality traits, and support the use of natural language (Daft and Lengel 1986). Matching media to collaborative tasks is based on the need to reduce 'uncertainty', or the absence of information to perform a task, and 'equivocality', or the absence of a shared understanding of what information means in connection with the task being carried out.

The media richness hypothesis argues that the face-to-face medium is the richest and most effective medium for reducing equivocality, which is assumed to be high in, for example, knowledge-intensive tasks that involve different areas of an organization (Kock 1998; Lengel and Daft 1988; Rice and Shook 1990). Communication media created by e-communication tools are placed somewhere in between the face-to-face medium and paper-based written media, depending on their ability to carry non-verbal cues, and so on (Daft et al. 1987; Kock 1998; Lee 1994). According to the media richness hypothesis, rational and effective users choose media of appropriate richness for tasks that involve communication, and if due to accessibility constraints their choice of communication media is restricted to media of lower than appropriate richness, a decrease in task outcome quality will occur.

Evidence in support of the media richness hypothesis

Daft et al. (1987) found that managers who were media sensitive (i.e. who selected appropriately rich media for collaborative tasks) performed better than managers who were not. Later studies provided evidence that e-communication media are more task-oriented than the face-to-face medium, and that users perceived those e-communication media to be less suitable for personal interactions necessary in business communication as compared to richer media (Rice 1993; Rice and Shook 1990). Other studies suggested that use of e-mail and computer conferencing negatively affected group cohesiveness (Sproull and Kiesler 1986; Walther 1996), and argued that e-communication media reduced 'social context cues', making them impersonal and likely to be avoided for business tasks or, if adopted, likely to lead to lower quality task outcomes than the face-to-face medium (for reviews, see Lee 1994; Markus 1994b).

More recent studies continue to provide empirical support for the media richness hypothesis. Walther et al.'s (2001) study, for example, suggests that even the sharing of facial images of group members has a positive effect on group performance in 'zero-history' groups (i.e. groups whose members have no prior history of collaboration). A quote from another study, comparing face-to-face, teleconferencing, and electronic chat groups, is representative of the findings from the recent literature in connection with the media richness hypothesis:

> An analysis of the recorded group discussions revealed that, although most of the groups in the electronic chat condition selected an integrative tallying procedure, an effective strategy that would likely have resulted in the successful solution of the problem, they experienced difficulties coordinating member inputs and verifying information. This slowed their progress and heightened their level of frustration and mental effort. This supports the notion that electronic chat lacks certain characteristics, present in verbal communication, that are necessary for exchanging and structuring information in synchronous groups.
>
> (Graetz et al. 1998, p. 741)

Evidence against the media richness hypothesis

In spite of the existence of supporting empirical evidence for the media richness hypothesis, research on communication media choice and use behaviour also led to results that contradicted the hypothesis. Fulk et al. (1990) and Markus (1990) discuss studies that found weak support for the media richness hypothesis as well as evidence contradicting the hypothesis. Other studies attempting to test, replicate, or extend the media richness hypothesis found little or no support for it (Dennis et al. 1999; Kinney and Dennis 1994; Kinney and Watson 1992). Some found effects contrary to the notion that a lack of social presence and social context cues is necessarily 'bad' for collaborative tasks, as communication media users can compensate for the lack of richness of the media by adapting their communication behaviour (e.g. Carlson and Zmud 1999; Kock 1998; Weisband 1994; Weisband et al. 1995). Similarly, Markus (1994a) found that managers often used e-mail, a lean medium according to the media richness hypothesis, for complex communication in connection with managerial tasks. Others found evidence that users could have rewarding and perceptually rich interaction in computer-based and asynchronous newsgroups, or other on-line social communities, whose underlying communication media were relatively low in richness according to the media richness hypothesis (Rheingold 1993; Walther 1996).

In addition to the empirical evidence contradicting the media richness hypothesis, one main theoretical refutation has been proposed: the social influence refutation (Markus 1994b). This is based on Fulk et al.'s (1990) social influence model, which argues that social influences can strongly shape individual behaviour towards technology in ways that may be independent of technology features. Examples of social influences are patterns of technology use observed in individuals (Bandura 1986) that are consistent with formal and informal social norms of accepted behaviour within the group to which the individual belongs. Social influences on technology-related individual behaviour have been shown to be moderated by a number of factors, particularly an individual's personal attraction to a group (Fulk 1993). Markus (1994a,b) showed that social influences might shape individual behaviour towards communication media in ways that are inconsistent with the media richness hypothesis by focusing on media choices made by managers at a large risk management services provider. Specifically, Markus (1994a) questioned the accuracy of the media richness scale, which places e-mail behind face-to-face interaction, suggesting that social influences can change some of e-mail's attributes that are assumed under the media richness hypothesis to be static and dependent on media attributes. The key piece of evidence was that senior managers' pressure on other employees to reply quickly to e-mail increased the medium's feedback immediacy, and therefore shifted e-mail up from its relative position on the media richness scale. A plausible conclusion that follows from Markus' argument is that social influences, such as pressure from managers on their subordinates, can make a medium that is seen as 'lean' based on the media richness hypothesis to become 'richer' than face-to-face (e.g. if managers require their subordinates to use only e-mail for communication and to avoid face-to-face communication as much as possible).

However, the social influence refutation focused on showing a fatal flaw in the media richness hypothesis, and did not aim to address one striking fact about many of the empirical studies that contradicted the media richness hypothesis. Even though those studies found evidence against the hypothesized positive link between media richness and media choice or quality of task outcomes, they also often found evidence that pointed to the perceived inadequacy of media of low richness by their users, a perception that was aligned with the media richness hypothesis (Carlson and Zmud 1999; Kock 1998; Markus 1994a,b; Rheingold 1993; Walther 1996). This is true in Markus' (1994a) study as well, where managers and employees also perceived e-mail as a poor medium for communication, in spite of the managers' decision to promote the use of e-mail because of some of its advantageous features, such as the ability to enable distributed and asynchronous interaction. That is, those studies successfully questioned the existence of a link between low media richness and two patterns, namely media avoidance and lower quality of task outcomes than in richer media, but not the perception by users that media that veered too far away from the face-to-face medium were somehow less appropriate than the face-to-face medium to support communication concerning business tasks. This chapter tries to explain this phenomenon by introducing the concept of media naturalness and proposing a new hypothesis, called the media naturalness hypothesis. In addition to explaining the phenomenon, as well as explaining evidence in support of the media richness hypothesis, the media naturalness hypothesis is shown to be compatible with Fulk et al.'s (1990) social influence model. The new hypothesis builds on the modern version of Darwin's theory of evolution by natural selection, which allows us to understand how we developed our current biological communication apparatus.

Human evolution and media naturalness

The relevance of understanding the process that led to the evolution of our biological communication apparatus and the effect that this has on e-communication behaviour, comes from one important principle, which is that humans have been 'engineered' by evolutionary forces to communicate primarily in a co-located and synchronous manner, as well as through facial expressions, body language, and speech.

According to the modern version of Darwin's (1859) theory of evolution, the human species evolved through natural selection, a process in which random mutations are introduced in the genetic makeup of offspring, leading to traits that are selected for based on their usefulness for survival and mating (Darwin 1859; Dawkins 1989; Mayr and Provine 1998). The evolutionary pace set by natural selection is usually very slow (Boaz and Almquist 1997; Dobzhansky 1971; Lorenz 1983). Genetic mutations that enhance an individual's chances of survival and mating, in many cases only slightly (Dobzhansky 1971), slowly accumulate and spread through the members of a species, leading to the development of species-wide physical, behavioural and cognitive traits over long periods of time. These may span thousands or millions of years, and are contingent on breeding speed and mortality rates.

Evidence suggests that during much of our evolutionary past we relied on co-located and synchronous forms of communication through facial expressions, body language, and sounds (including speech, which uses a large variety of sound combinations) to exchange information and knowledge among ourselves (Boaz and Almquist 1997; Cartwright 2000). Humans have developed a complex web of facial muscles (22 on each side of the face; more than any other animal) that allow us to generate over 6000 communicative expressions; very few of these muscles are used for other purposes, such as chewing (Bates and Cleese 2001; McNeill 1998). There is a noticeable evolutionary direction towards the development of a biological communication apparatus that supported ever more sophisticated forms of speech, or increased communication complexity, culminating in the development of complex speech by *Homo sapiens*. The advent of complex speech was enabled by the development of the larynx, located relatively low in the neck, and an enlarged vocal tract—key morphological traits that differentiate modern humans from their early ancestors and that allow modern humans to generate the large variety of sounds required to speak most modern languages (Laitman 1984; Lieberman 1998). The morphology of the human ear also suggests a specialized design to decode speech (Lieberman 1998; Pinker 1994).

Since our biological communication apparatus has been used for co-located and synchronous communication using facial expressions, body language, and sounds over such a long period of time, it stands to reason that it should have been designed for communication interaction modes that present those characteristics. A plausible corollary would be that other communication interaction modes, including e-communication in general, would be matched to different degrees to our biological communication apparatus, depending on the degree to which they approximate face-to-face communication.

It is important to note here that optimal biological design rarely occurs in nature because evolution is a slow process that takes time to catch up with environmental change. As a result, changes in the environment often make previous biological designs sub-optimal. A leading evolutionary psychologist has made this point rather eloquently in the past:

> One constraint on optimal design are evolutionary time lags ... evolution refers to change over time ... Because evolutionary changes occur slowly, requiring thousands of generations of recurrent selection pressure, existing humans are necessarily designed for the previous environments of which they are a product. Stated differently, we carry around a stone-aged brain in a modern environment. A strong desire for fat, adaptive in a past environment of scarce food resources, now leads to clogged arteries and heart attacks.
>
> (Buss 1999, p. 20)

Buss's (1999) conclusion is a general one; of which the main hypothesis proposed in this chapter (i.e. the media naturalness hypothesis) can be seen as a corollary. Essentially, what is argued here is that modern humans' brains are not optimally adapted for current e-communication technologies because these technologies often suppress too many of the elements found in face-to-face communication. That is, our brain has likely been to a large extent 'hardwired' for co-located and synchronous communication employing facial expressions, body language, and speech; or, in other words, our brain is genetically programmed to excel in communication interactions that incorporate those elements. This provides the basis on which the concept of *media naturalness* can be defined—the ability of communication media to support co-located and synchronous communication employing facial expressions, body language, and speech. It follows that using modes of communication that veer away from 'natural' communication is likely to put 'extra burden' on the brain, as our brain has been designed for that type of communication. Essentially, natural communication is equated with face-to-face communication.

A simple analogy can help highlight the importance of the above conclusion. Since we evolved two hands, not only one or as many as three hands, we also evolved brain functions designed for the use of two hands (as well as other connected elements, such as arms) to accomplish a number of basic tasks, such as climbing a tree by holding onto its branches. Therefore, trying to climb a tree using only one hand is likely to be more difficult and frustrating for a human than if both hands were used. That is, the lack of 'naturalness' of the act causes a mismatch between the biological makeup of the individual and the task being accomplished. It is argued here that communicating in ways that are not 'natural' is analogous to this, in that it also leads to a mismatch and related consequences. This mismatch refers to our biological communication apparatus, which comprises the brain functions associated with communication, and the various communication instincts (Pinker 1994, 1997) that have been programmed into our brain by evolutionary forces. This conclusion provides the basis for the development of the media naturalness hypothesis.

However, before we go any further, it is important to stress that e-communication tools exist for a reason, which is that they solve key communication problems that exist today (and that did not exist in our prehistoric past). For example, communication through e-mail, with all its limitations, can take place regardless of time and physical location—that is, it can take place in an asynchronous and distributed manner—making it a convenient alternative to face-to-face communication in a variety of business situations. Moreover, e-mail, with all its limitations, generates a record, and thus can be re-processed by its recipient as many times as needed (as long as proper filing takes place); something that is not possible with face-to-face communication. Therefore, the argument put forth in this chapter should not be interpreted as a call for the use of face-to-face communication only in business, but an alternative explanation as to why we often should make e-communication media as face-to-face-like as possible, while at the same time preserving the advantages that led to the widespread use of communication systems such as e-mail.

The effects of media naturalness on communication attributes

The media naturalness hypothesis is an attempt to derive a general predictive statement linking a few dependent communication-related constructs with one main independent construct, namely the mismatch between our biological communication apparatus and communication media characteristics. The inverse of this mismatch is defined as the 'naturalness' of a communication medium—that is, the higher the mismatch, the lower the naturalness of a communication medium. In this section, I define media naturalness and key constructs affected by it, following that with the formal enunciation of the media naturalness hypothesis.

Media naturalness

As discussed in the previous section, there is strong evidence that human beings have been 'engineered' by evolutionary forces to communicate primarily in a co-located and synchronous manner, as well as through facial expressions, body language, and speech. Thus, it is reasonable to assume that natural communication involves at least five key elements: 1) a high degree of co-location, allowing individuals engaged in an interaction to see and hear each other; 2) a high degree of synchronicity, which would allow the individuals engaged in a communication interaction to quickly exchange communicative stimuli; 3) the ability to convey and observe facial expressions; 4) the ability to convey and observe body language; and 5) the ability to convey and listen to speech.

Given this, we can define the naturalness of the communication medium created by an e-communication technology based on the degree to which the technology selectively incorporates (or suppresses) those five elements. That is, it can be stated that, other things being equal, the degree of 'virtual' incorporation of one of the media naturalness elements correlates the degree of naturalness of an e-communication medium. The term 'virtual' means that none of the five media naturalness elements will be incorporated to the e-communication medium to the same extent to which it is available in actual face-to-face communication. For example, flat representations of facial expressions, such as those provided by desktop video conferencing, are a virtual approximation of the actual three-dimensional experience of seeing live facial expressions (Bryson 1996; Mass and Herzberg 1999).

With the main independent construct of the media naturalness hypothesis defined, we can now focus on the identification of key dependent constructs that are relevant from a business perspective. This is done here as a first step, since identifying a comprehensive set of dependent constructs that are directly affected by media naturalness will require extensive empirical research. Nevertheless, the set of dependent constructs identified below can be seen as sufficient to justify the media naturalness hypothesis as a viable alternative for the media richness hypothesis. The dependent constructs discussed here are *cognitive effort, communication ambiguity,* and *physiological arousal.*

Cognitive effort

There is a large body of evidence pointing at our ability to employ the media naturalness elements rather effortlessly in communication interactions. For example, it has been shown that human beings possess specialized brain circuits that are designed for the recognition of faces and the generation and recognition of facial expressions (Bates and Cleese 2001; Le Grand et al. 2001;

McNeill 1998), which artificial intelligence research suggests require complex computations that are difficult to replicate even in powerful computers (Kurzweil 1999; Russel and Norvig 1995). The same situation is found in connection with speech generation and recognition (Kurzweil 1999; Lieberman 1991, 2000; Russel and Norvig 1995).

Since our brain's circuitry has been configured by evolution to excel in communication employing the five media naturalness elements discussed above, one can reasonably conclude that selectively suppressing those elements in communication media will require the development and use of specialized brain circuits to make up for the absence of those elements and enable effective communication. Those brain circuits are not hardwired into our brain but learned over time, primarily through changes in the brain's neocortex, its outer layer, where most learned circuits are concentrated. Lieberman (1991, 1998, 2000) has shown that, as far as human communication is concerned, learned circuits are unlikely to be as efficient as the hardwired circuits endowed on us by evolution; the former usually relying on more convoluted paths than the latter. As pointed out by Pinker and Bloom (1992, p. 477), the latter, genetically coded circuits are a result of the gradual evolution of '. . . neural mechanisms [that make communication] become increasingly automatic, unconscious, and undistracted by irrelevant aspects of world knowledge'.

Moreover, most learned brain circuits used in communication have to be 'refreshed' (or partially re-learned) from time to time, otherwise they are 'erased' (Pinker 1997; Schacter 2001). Those learned circuits usually differ from individual to individual, which can lead to inefficiencies associated with differences in sender/receiver communication brain circuitry (Kotulak 1997). Thus, it is plausible to conclude that since the use of more convoluted paths requires increased neural activity, decreases in media naturalness will generally lead to increased mental effort, or what we refer here to as *cognitive effort*, in connection with communication interactions.

Cognitive effort is defined here as the amount of 'mental activity', or, from a biological perspective, the 'amount of brain activity' involved in a communication interaction. It can be assessed directly, with the use of techniques such as magnetic resonance imaging. Cognitive effort can also be assessed indirectly, based on perceptions of levels of difficulty associated with communicative tasks (Schacter 2001; Todd and Benbasat 1999), as well as through indirect measures such as that of 'fluency', proposed by Kock (1998). The 'fluency' measure builds on the assumption, previously made in many empirical studies (see, e.g. Leganchuk et al. 1998), that the amount of cognitive effort associated with an intellective task correlates with the amount of time required to complete the task. As such, 'fluency' is defined as the amount of time taken to convey a certain number of words through different communication media, which is assumed to correlate (and serve as a surrogate measure of) the amount of time taken to convey a certain number of ideas through different media (Kock 1998).

Empirical studies conducted by Kock (1998, 1999) of process improvement groups interacting through different communication media are particularly well aligned with the notion that a decrease in media naturalness will generally lead to an increase in cognitive effort. Those studies showed that 'fluency' is, on average, 18 times higher face-to-face than over e-mail in complex group tasks. Even in the case of proficient typists, who can usually type half as fast as they can speak or faster, Kock's (2001a, b, c) research suggests fluency in complex collaborative tasks conducted face-to-face to be about 10 times higher than over e-mail, regardless of other factors such as cultural background and familiarity with collaborators. According to this estimate, if exchanging 600 words face-to-face required about 6 minutes, exchanging the same number of words over e-mail would take approximately 1 hour—these figures are comparable to those found in Kock's (2001a, b, c) studies.

Communication ambiguity

Individuals brought up in different cultural environments usually possess different information processing schemas that they have learned over their lifetimes. Different schemas make individuals interpret information in different ways, particularly when information is expected but not actually provided. Bartlett (1932) has unequivocally demonstrated this phenomenon, perhaps for the first time, in his famous experiments involving the American Indian folk tale 'The War of The Ghosts'.

Essentially, the experiments showed that subjects who held different information processing schemas would interpret the tale, which is filled with strange gaps and bizarre causal sequences, in substantially different ways. In Bartlett's (1932) experiments, individuals were expecting certain pieces of information to be provided to them. When they did not get the information that they were expecting, they 'filled in the gaps' based on their existing information processing schemas and the information that they were given (see also Gardner 1985). This led to significant differences in the way individuals interpreted the tale.

The human brain has a series of hardwired information processing schemas that are designed to solve problems that have occurred recurrently during the millions of years that led to the evolution of the human species (Cosmides and Tooby 1992; Tooby and Cosmides 1992). Several of these problems addressed by evolutionary adaptations are related to the communication process (Pinker and Bloom 1992). Our hardwired schemas involved in the communication process make us search for stimuli that will enable us to obtain enough information to effectively interpret the message being communicated, and several of the stimuli we automatically search for are those present in actual face-to-face communication (Lieberman 2000), such as contextual cues (available in co-located communication), immediate feedback (available in synchronous communication) in the form of facial expressions and body language, and voice intonations. When several of these stimuli are not present, by being selectively suppressed by e-communication technologies, individuals 'fill in the gaps' much like the subjects in Bartlett's (1932) experiments.

The problem is that in the absence of information-giving stimuli, 'filling in the gaps' is likely to lead to a higher proportion of misinterpretations, and thus ambiguity, than if the stimuli were not suppressed—as Bartlett's (1932) and other studies (see, e.g. Gardner 1985; and Pinker 1997) show. While different individuals are likely to look for the same types of communicative stimuli, their interpretation of the message being communicated in the absence of those stimuli will be largely based on their learned schemas, which are likely to differ from those held by other individuals (no two individuals, not even identical twins raised together, live through the exact same experiences during their lives). That is, a decrease in medium naturalness, caused by the selective suppression of media naturalness elements in a communication medium, is likely to lead to an increase in the probability of misinterpretations of communicative cues, and thus an increase in *communication ambiguity*.

The above conclusion is consistent with the empirical observation that certain feedback comments, especially those involving constructive criticism, which are often used effectively in face-to-face interaction together with other non-verbal cues that 'soften' their tone, are interpreted in different (and often negative) ways when provided via e-mail in business-related discussions—sometimes as very critical and blunt, sometimes as implying indifference (Kock 1999). Indeed, e-communication in general is perceived as more 'ambiguous' than face-to-face communication (Carlson and Zmud 1999; Graetz et al. 1998; Kock 1998, 2001b; Rheingold 1993; Walther 1996).

While there are studies that show that individuals can voluntarily or involuntarily compensate for this increase in communication ambiguity by means of constructing better thought-out messages (Kock 1998, 2001b), and by becoming familiar with the medium and their partners (Carlson and Zmud 1999; Walther 1996), to the best of my knowledge there have been no studies suggesting that the suppression of media naturalness elements causes a reduction in communication ambiguity (i.e. the opposite effect to what we hypothesize here). That is, even though there is evidence suggesting that the effects of greater communication ambiguity can be moderated by compensatory adaptation behaviour, the evidence suggesting that lower communication media naturalness leads to greater communication ambiguity is beyond much doubt.

Physiological arousal

There is little doubt that a fully developed biological communication apparatus has been particularly important in terms of survival and mating for our prehistoric ancestors, as it is for us today (Boaz and Almquist 1997; Dunbar 1993; Miller 2000). While there is substantial evidence suggesting that our biological communication apparatus is designed for face-to-face communication, there is also ample evidence that such apparatus (including the neural functional language system) cannot be fully developed without a significant amount of practice (Pinker 1994). Evolution is likely to have led to the development of mechanisms to compel human beings to practice the use of their biological communication apparatus; mechanisms that are similar to those compelling animals to practice those skills that play a key role in connection with survival and mating (Wilson 2000). Among these mechanisms, one of the most important is that of *physiological arousal*, which is often associated with 'excitement' and 'pleasure' (Boaz and Almquist 1997; Miller 2000).

It is a plausible conclusion that engaging in communication interactions, particularly in face-to-face situations, is likely to trigger physiological arousal in human beings. This conclusion underlies the theoretical hypothesis that modern humans possess what Pinker (1994) refers to as a 'language instinct', and can be taken further through the associated conclusion that each face-to-face communication element (e.g. the use of facial expressions to convey thoughts and feelings) contributes to physiological arousal. Indeed, there is evidence that face-to-face communication elements such as certain types of facial expressions, oral utterances, and body language expressions, even when used in isolation, evoke physiological arousal in human beings (Bates and Cleese 2001; McNeill 1998; Zimmer 2001).

It would thus be reasonable to also conclude that communication interactions in which certain elements of 'natural' face-to-face communication are suppressed (e.g. the ability to employ/see facial expressions) involve a corresponding suppression of physiological arousal, and, in turn, a consequent decrease in the perceived 'excitement' in connection with the communication interaction. In other words, suppression of media naturalness elements is likely to make communication interactions 'duller' than if those elements were present.

Obviously, as with other conclusions set out in this chapter, the above conclusion assumes 'other things being equal'. For instance, the topic of a communication interaction and the identity of the other person are factors that may influence physiological arousal more strongly than the communication medium itself, which is a point that is not disputed here and is perfectly compatible with the hypothesis. Having said that, it is interesting to point out that the above conclusion is consistent with, and provides a plausible explanation for the ample evidence suggesting that e-communication systems users consistently perceive computer-mediated communication in general as less 'exciting', 'duller', or less 'emotionally fulfilling' than face-to-face communication (Ellis et al. 1991; Kiesler et al. 1988; Kock 1999; Markus 1994b; Reinig et al. 1995; Sproull and Kiesler 1986; Walther 1996).

Decreases in physiological arousal may influence media choices towards media of high naturalness, but, when choice is limited to low naturalness media, they may arguably influence task outcome quality positively under the appropriate circumstances. A decrease in physiological arousal may induce the members of a group to engage in more focused communication, particularly when an e-communication medium is used to support task-oriented interaction, as opposed to relationship-oriented interaction (see Walther 1996). That is, the lack of excitement resulting from the use of an e-communication medium to support a particular group task may be associated with a higher degree of communication focus on the task at hand, rather than gossip or tangential topics, somehow counteracting the negative effects on task outcome quality associated with increase cognitive effort and communication ambiguity. Kock's (1999) study of process improvement groups provides evidence that supports in part this conjecture.

The media naturalness hypothesis

Now that the three main dependent constructs of cognitive effort, communication ambiguity and physiological arousal have been identified and defined, we can formally enunciate the media naturalness hypothesis, as follows:

> Other things being equal, a decrease in the degree of naturalness of a communication medium leads to the following effects in connection with a communication interaction: (a) an increase in cognitive effort, (b) an increase in communication ambiguity, and (c) a decrease in physiological arousal.

The hypothesis assumes that the face-to-face medium is the most natural of all. As discussed before, the media naturalness construct is made up of five main elements: co-location, synchronicity, and the ability to convey facial expressions, body language, and speech. Thus, two assumptions can be made which are useful for managers who need to decide which features to have on their e-communication systems in the face of limited resources. The first assumption is that, other things being equal, an e-communication medium that incorporates one of the media naturalness elements—i.e. co-location, synchronicity, and the ability to convey facial expressions, body language, and speech—will have a higher degree of naturalness than another e-communication medium that does not incorporate that element. The second assumption is that, other things being equal, an e-communication medium that incorporates one of the five media naturalness elements to a larger degree than another will have the highest degree of naturalness.

The media naturalness hypothesis can be used as a basis for management decisions regarding which new features to add to an e-communication tool depending on resource constraints. For example, let us imagine a Web-based application that allows two individuals to communicate through text-based chat in a business-to-consumer type of interaction, such as that involving a customer service representative of an online broker and one of its customers who needs to learn how to purchase a particular investment instrument. According to the media naturalness hypothesis, if a nearly identical application is developed, where the only difference is the ability to convey facial expressions though streamed video (in addition to the text-based chat feature), this latter application will create a communication medium with a higher degree of naturalness than that of the text-based, chat only application. A likely consequence will be higher perceived quality of the online interaction, due to, for example, the lower cognitive effort required.

The media naturalness hypothesis also provides the basis for management decisions regarding 'partial' incorporation of a naturalness element to an e-communication medium depending on resource constraints. Each of the five naturalness elements can be incorporated into an e-communication medium to varying degrees; *full* incorporation means that the element is identical to what would be available in the face-to-face medium.

Contrasting the hypotheses

If the media naturalness hypothesis is to be considered a viable and useful alternative to the media richness hypothesis, key differences between the hypotheses must be explicitly identified. While the media naturalness hypothesis may seem similar to the media richness hypothesis, at least two key differences exist between the two. The first refers to the main dependent constructs of the hypotheses. The media richness hypothesis has two main dependent constructs, which are hypothesized to vary depending on the degree of richness of the communication medium being used. These constructs are: 1) media choice, which is hypothesized to match the richness requirements of a task in the case of 'effective' workers; and 2) task outcome quality, which is hypothesized to be negatively affected if choice is limited to media that possess a level of richness that is lower than the optimal for the task (Daft et al. 1987; Lee 1994; Lengel and Daft 1988; Markus 1994b).

The media naturalness hypothesis, on the other hand, does not relate low media naturalness with certain types of behaviour or task outcomes, like the media richness hypothesis does, but with high cognitive effort and communication ambiguity, and with low physiological arousal. This in turn may or may not lead to certain types of behaviour or task outcomes.

For example, as mentioned before, empirical studies conducted by Kock (1998, 1999) of process improvement groups interacting through different communication media showed that 'fluency' (the number of words conveyed per unit time) is about 10 times higher face-to-face than over e-mail in complex group tasks, even when the effect that 'typing is slower than speaking' is controlled for. Yet, the studies found that most groups voluntarily chose e-mail to perform their tasks (even though interviews suggested that they consistently perceived e-mail as an 'ambiguous' and 'poor' communication medium), and that the task outcome quality was slightly better for the e-mail groups when compared with the face-to-face groups. Interviews also suggested that e-mail was chosen primarily because of what most perceived as advantages, such as the ability to support distributed and asynchronous communication.

The explanation given for these seemingly paradoxical results was that individuals 'compensated' for the lack of naturalness of the communication medium used by preparing better thought out, more focused, and better structured contributions than in face-to-face meetings (Kock 1998, 1999). The studies provide evidence that while cognitive effort was increased, which is consistent with the media naturalness hypothesis, media choice and task outcome quality were different from predicted by the media richness hypothesis. The media naturalness hypothesis provides a basis on which media choices and task outcomes can be more deeply understood as the result of the interplay of biological, social, and environmental influences.

The other key difference between the two ideas is that the media richness hypothesis assumes that different communication media can be classified according to a continuum of media richness, particularly based on the information-carrying capacity of the media. This opens the door for the conclusion that communication media that incorporate more of those features that increase their richness (e.g. feedback immediacy or synchronicity) will be even 'better' than face-to-face interaction in some circumstances—for instance, they may allow individuals to deal with tasks of extremely high equivocality by supporting parallel communication interactions with several individuals at the same time.

The media naturalness hypothesis, however, argues that the face-to-face medium is, other things being equal (including the communication topic and task), the one likely to lead to the least cognitive effort and communication ambiguity, and the most physiological arousal during communication. The reason is that the face-to-face medium is precisely the medium used for communication during the vast majority of our evolutionary history. This implies that e-communication tools with features that allow group members to synchronously generate and access substantially more information than in face-to-face interactions, will also lead to problems, likely due to information overload and other negative effects.

That is, the media naturalness hypothesis allows us to place the face-to-face medium at the centre of a one-dimensional scale of naturalness where the distance from the centre (either to the 'left' or 'right') could be seen as a measure of decreased naturalness. Anything 'less' or 'more', so to speak, than face-to-face communication would be likely to lead to problems in communication interactions. This conclusion is consistent with previous studies of group decision support systems (Dennis et al. 1996; Reinig et al. 1995). Those systems are typically used to enhance face-to-face communication by allowing individuals in the same room to interact synchronously through computers without having to share 'airtime' with each other, i.e. all individuals can contribute ideas at the same time, which a human facilitator manages for the group with the help of the system. Even if used by pairs of individuals, these systems are generally believed to allow for the exchange of significantly more information than 'pure' face-to-face meetings (Johansen 1988).

Consistent with this conclusion, Reinig et al. (1995) found that the use of group decision support systems make meetings less 'exciting' for the participants. Furthermore, a consistent finding from studies of the impact of group decision support systems on meetings has been that they increase the number of ideas generated, but do not improve the quality of the outcomes produced through the meetings (Dennis et al. 1996), which led Dennis (1996) to conclude that the use of the systems leads to information overload. That is, on the surface, individuals seem to exchange more information, but the information is never used to achieve better task outcomes because the rate of information exchange is higher than the information processing capacity of the individuals.

Implications for researchers

The media naturalness hypothesis cannot fully explain e-communication behaviour. Arguing otherwise would be akin to proposing a modern-day version of biological determinism. Other factors in connection with the use of e-communication media need to be considered, such as social influences (Fulk et al. 1990; Markus 1994a, b). For example, the media naturalness hypothesis could not have been used to fully explain the behaviour of the employees in Markus's (1994a) study, who used e-mail, in spite of perceiving it as a poor communication medium, because of pressure from senior managers. That is, it was primarily a social influence (Fulk et al. 1990) that led them to behave in the way they did in connection with media choice. Moreover, they used e-mail in a relatively effective way for complex communication, in spite of their negative perceptions about the medium, which suggests another phenomenon that is not predicted by the media naturalness hypothesis, namely the phenomenon of compensatory adaptation (Kock 1998, 2001b).

Nevertheless, a key contribution of the media naturalness hypothesis, and one of its most important implications for researchers, is that it provides a missing link that may pave the way for the integration of different e-communication theories. Previous theoretical reviews have categorized e-communication theories in similar ways. Webster and Trevino (1995) grouped them into two categories: theories proposing rational explanations of media choice, and theories proposing social explanations. Carlson and Davis (1998) classified theories into two similar categories: trait theories and social interaction theories.

These categorizations suggest key differences between 'socially-deprived' theories, of which Daft and Lengel's (1986) media richness theory is often seen as the paragon, and theories placing emphasis on social elements as determinants of behaviour towards e-communication media.

This 'theoretical polarization' was later highlighted in another theoretical review, suggesting that these two main types of theories have '. . . often been pitted against each other rather than considered as complementary in more comprehensive studies' (Trevino et al. 2000, p. 163).

In spite of attempts to combine both types of theories (Trevino et al. 2000; Webster and Trevino 1995), other studies provide evidence that the media richness hypothesis cannot be effectively combined with hypotheses espoused by social theories without radical revisions (Lee 1994; Ngwenyama and Lee 1997). The development of the media naturalness hypothesis is a first step in the pursuit of a solution to this problem. Evidence of this is that the media naturalness hypothesis, as discussed earlier, is not only compatible with Fulk et al.'s (1990) social influence model, but also adds to our understanding of e-communication behaviour phenomena that are not completely explained by that model.

The above must be followed by a caveat regarding the limitations of the theoretical perspective taken here. The argument presented in this chapter focuses on a human-to-human communication perspective, which is arguably a narrow perspective of human communication and cognition. Different and possibly broader perspectives exist, such as the systemic perspective proposed by Hutchins (1996), who deems a culturally diverse group of individuals as a key unit of cognition, and the 'metaindexicality' perspective proposed by Henderson (1998), who sees visual representations as cognitive artefacts that allow for rich communication at multiple cognitive levels. One could also take a human-to-object perspective, by looking at interactions between humans and inanimate objects as legitimate instances of communication. The relatively narrow human-to-human communication perspective is adopted here because it is arguably the perspective adopted by the media richness hypothesis, for which this chapter attempts to propose a viable alternative.

Implications for managers

The media naturalness hypothesis leads to predictions that are particularly relevant for communicative tasks brought about by the advent of e-business. The hypothesis leads to the prediction that cognitive effort and communication ambiguity should increase, and physiological arousal decrease, with decreases in e-communication media naturalness. And, in business-to-consumer interactions conducted online, increased cognitive effort and communication ambiguity, and (possibly) decreased physiological arousal (especially in entertainment-related interactions), may lead to lower perceived quality and dissatisfaction from the part of customers.

Since the Internet makes it much easier for customers to change suppliers, who are literally 'a few clicks away', the use of e-communication media of lower naturalness than those provided by the competition can have negative consequences for companies that rely heavily on online interactions with their customers to increase or maintain their revenues. This conclusion is aligned with, and partially explains, the constant calls in the popular business literature for the use of more natural forms of online communications between business and consumers (Metz 2000; Wasserman 2001)[1]. That is, even though a decrease in communication medium naturalness may not have a negative effect on task outcome quality (Kock 1998, 2001b), it will lead to other problems in certain situations—e.g. online business-to-consumer interactions.

The media naturalness hypothesis provides the basis on which managers with limited resources can decide how to maximize the naturalness of their companies' online communications with their customers. One area in which these decisions have to be made in many businesses,

---

[1] See also Mottl, J.N. (2000). A wake-up call for e-retailers, *InternetWeek*, September 25, http://www.internetweek.com/indepth/indepth092500.htm

regardless of type and size, is that of online customer support, where customer support representatives interact with customers electronically.

The widespread availability of generic video players and instant-messaging technologies allow for the selective incorporation of synchronicity and the ability to convey speech and facial expressions to these Internet-based interactions, which according to the media naturalness hypothesis is likely to lead to a decrease in the amount of cognitive effort and communication ambiguity. This is likely to contribute to an increase in perceived customer service quality, and potentially in market share (Macdonald 1995; Walkins 1992), particularly in sectors such as financial services, which relies heavily on e-communication tools to provide customer support through the Internet.

The media naturalness hypothesis also provides the basis on which managers and venture capitalists can predict the likely evolution of e-communication technologies and thus better target their investments in those types of technologies. This chapter argues that this evolution will likely be towards e-communication tools that achieve the maximum naturalness at the lowest cost possible. Although this may not be obvious at first glance, e-mail fits this prediction reasonably well, because e-mail is more natural (e.g. it provides a higher degree of synchronicity) and arguably less costly today than paper-based mail, which is what it was originally meant to replace (Keen 1994; Sproull and Kiesler 1991).

This conclusion also explains the relative commercial success of sophisticated text-based chat tools that add synchronicity to online business-to-consumer interactions, making it easier and more exciting for customers to obtain information about products and services (Eichler and Halperin 2000; Gilbert 1999). Finally, it explains the relative commercial success of virtual news anchors such as 'Ananova' (Cracknell 2000; Orubeondo 2000), whose cost is a fraction of their human counterparts', since many Internet users seem to prefer to listen to news online while looking at a virtual newscaster, rather than the arguably more cognitively demanding and less exciting option of reading them on a Web page.

Acknowledgements

This chapter is based on two articles by the author, one published in 2005 in the journal *IEEE Transactions on Professional Communication*, and the other published in 2009 in the journal *MIS Quarterly*. The author thanks Nap Chagnon, Leda Cosmides, Martin Daly, Allen Lee, Geoffrey Miller, John Nosek, Steven Pinker, Gad Saad, Achim Schutzwohl, John Tooby, Rick Watson, and Margo Wilson, for ideas and enlightening discussions on topics that underlie several of the ideas proposed in this chapter.

References

Bandura, A. (1986). *Social foundations of thought and action*. Prentice Hall, Englewood Cliffs, NJ.

Bartlett, F. (1932). *Remembering: a study in experimental and social psychology*. Cambridge University Press, Cambridge, MA.

Bates, B. and Cleese, J. (2001). *The human face*. DK Publishing, New York.

Boaz, N.T. and Almquist, A.J. (1997). *Biological anthropology: a synthetic approach to human evolution*. Prentice Hall, Upper Saddle River, NJ.

Bryson, S. (1996). Virtual reality in scientific visualization. *Communications of the ACM*, **39**, 62–71.

Buss, D.M. (1999). *Evolutionary psychology: the new science of the mind*. Allyn and Bacon, Needham Heights, MA.

Carlson, P.J. and Davis, G.B. (1998). An investigation of media selection among directors and managers: from 'self' to 'other' orientation. *MIS Quarterly*, **22**, 335–62.

Carlson, J.R. and Zmud, R.W. (1999). Channel expansion theory and the experiential nature of media richness perceptions. *Academy of Management Journal*, **42**, 153–70.

Cartwright, J. (2000). *Evolution and human behavior: Darwinian perspectives on human nature.* MIT Press, Cambridge, MA.

Cosmides, L. and Tooby, J. (1992). Cognitive adaptations for social exchange. In: J.H. Barkow, L. Cosmides, and J. Tooby (eds), *The adapted mind: evolutionary psychology and the generation of culture*, pp. 163–228. Oxford University Press, New York.

Cracknell, M. (2000). A cyberstar is born. *Marketing Week*, **23**, 54.

Daft, R.L. and Lengel, R.H. (1986). Organizational information requirements, media richness and structural design. *Management Science*, **32**, 554–71.

Daft, R.L., Lengel, R.H., and Trevino, L.K. (1987). Message equivocality, media selection, and manager performance: implications for information systems. *MIS Quarterly*, **11**, 355–66.

Darwin, C. (1859). *On the origin of species by means of natural selection.* Harvard University Press, Cambridge, MA.

Dawkins, R. (1989). *The selfish gene.* Oxford University Press, New York.

Dennis, A.R. (1996). Information exchange and use in group decision making: you can lead a group to information, but you can't make it think. *MIS Quarterly*, **20**, 433–55.

Dennis, A.R., Haley, B.J., and Vanderberg, R.J. (1996). A meta-analysis of effectiveness, efficiency, and participant satisfaction in group support systems research. In: J.I. DeGross, S. Jarvenpaa, and A. Srinivasan (eds), *Proceedings of the 17th International Conference on Information Systems*, pp. 278–89. The Association for Computing Machinery, New York.

Dennis, A.R., Kinney, S.T., and Hung, Y.C. (1999). Gender differences and the effects of media richness. *Small Group Research*, **30**, 405–37.

Dobzhansky, T. (1971). *Mankind evolving: the evolution of the human species.* Yale University Press, New Haven, CN.

Dunbar, R.I.M. (1993). Coevolution of neocortical size, group size and language in humans. *Behavioral and Brain Sciences*, **16**, 681–735.

Eichler, L. and Halperin, L. (2000). LivePerson: keeping reference alive and clicking. *Econtent*, **23**, 63–6.

Ellis, C.A., Gibbs, S.J., and Rein, G.L. (1991). Groupware: some issues and experiences. *Communications of ACM*, **34**, 38–58.

Fulk, J. (1993). Social construction of communication technology. *Academy of Management Journal*, **36**, 921–38.

Fulk, J., Schmitz, J., and Steinfield, C.W. (1990). A social influence model of technology use. In: J. Fulk and C. Steinfield (eds), *Organizations and communication technology*, pp. 117–40. Sage, Newbury Park, CA.

Gardner, H. (1985). *The mind's new science.* Basic Books, New York.

Gilbert, J. (1999). LivePerson focuses on the human touch. *Advertising Age*, **70**, 62.

Graetz, K.A., Boyle, E.S., Kimble, C.E., Thompson, P., and Garloch, J.L. (1998). Information sharing in face-to-face, teleconferencing, and electronic chat groups. *Small Group Research*, **29**, 714–43.

Henderson, K. (1998). *On line and on paper: visual representations, visual culture, and computer graphics in design engineering.* MIT Press, Cambridge, MA.

Hutchins, E. (1996). *Cognition in the wild.* MIT Press, Cambridge, MA.

Johansen, R. (1988). *Groupware: computer support for business teams.* The Free Press, New York.

Keen, T. (1994). E-mail: from simplicity to ubiquity. In: P. Lloyd (ed.), *Groupware in the 21st century*, pp. 92–6. Praeger, Westport, CT.

Kiesler, S., Siegel, J., and McGuire, T.W. (1988). Social psychological aspects of computer-mediated communication. In: I. Greif (ed.), *Computer-supported cooperative work: a book of readings*, pp. 657–82. Morgan Kaufmann, San Mateo, CA.

Kinney, S.T. and Dennis, A.R. (1994). Re-evaluating media richness: cues, feedback, and task. In: J.F. Nunamaker and R.H. Sprague (eds), *Proceedings of the Hawaii International Conference on System Sciences, Vol. 4*, pp. 21–30. IEEE Computer Society Press, Hawaii.

Kinney, S.T. and Watson, R.T. (1992). Dyadic communication: the effect of medium and task equivocality on task-related and interactional outcomes. In: J.I. DeGross, J.D. Becker, and J.J. Elam, (eds), *Proceedings of the 13th International Conference on Information Systems*, pp. 107–117. The Association for Computing Machinery, Dallas, TX.

Kock, N. (1998). Can communication medium limitations foster better group outcomes? An action research study. *Information and Management*, **34**, 295–305.

Kock, N. (1999). *Process improvement and organizational learning: the role of collaboration technologies*. Idea Group Publishing, Hershey, PA.

Kock, N. (2001a). Asynchronous and distributed process improvement: the role of collaborative technologies. *Information Systems Journal*, **11**, 87–110.

Kock, N. (2001b). Compensatory adaptation to a lean medium: an action research investigation of electronic communication in process improvement groups. *IEEE Transactions on Professional Communication*, **44**, 267–85.

Kock, N. (2001c). The ape that used email: understanding e-communication behavior through evolution theory. *Communications of the Association for Information Systems*, **5**, 1–29.

Kock, N. (2004). The psychobiological model: towards a new theory of computer-mediated communication based on Darwinian evolution. *Organization Science*, **15**, 327–48.

Kock, N. (2005). Media richness or media naturalness? The evolution of our biological communication apparatus and its influence on our behavior toward e-communication tools. *IEEE Transactions on Professional Communication*, **48**, 117–30.

Kock, N. (ed.) (2008). *Encyclopedia of e-collaboration*. Hershey, PA.

Kock, N. (2009). Information systems theorizing based on evolutionary psychology: an interdisciplinary review and theory integration framework. *MIS Quarterly*, **33**, 395–418.

Kotulak, R. (1997). *Inside the brain: revolutionary discoveries of how the mind works*. Andrews McMeel Publishing, Kansas City, KA.

Kurzweil, R. (1999). *The age of spiritual machines*. Viking, New York.

Laitman, J.T. (1984). The anatomy of human speech. *Natural History*, **20**, 20–7.

Lee, A.S. (1994). Electronic mail as a medium for rich communication: an empirical investigation using hermeneutic interpretation. *MIS Quarterly*, **18**, 143–57.

Leganchuk, A., Zhai S., and Buxton, W. (1998). Manual and cognitive benefits of two-handed input: an experimental study. *ACM Transactions on Computer-Human Interaction*, **5**, 326–59.

Le Grand, R., Mondloch, C.J., Maurer, D., and Brent, H.P. (2001). Early visual experience and face processing. *Nature*, **410**, 890.

Lengel, R.H. and Daft, R.L. (1988). The selection of communication media as an executive skill. *Academy of Management Executive*, **2**, 225–32.

Lieberman, P. (1991). *Uniquely human: the evolution of speech, thought, and selfless behavior*. Harvard University Press, Cambridge, MA.

Lieberman, P. (1998). *Eve spoke: human language and human evolution*. W.W. Norton and Company, New York.

Lieberman, P. (2000). *Human language and our reptilian brain: the subcortical bases of speech, syntax, and thought*. Harvard University Press, Cambridge, MA.

Lorenz, K. (1983). *The waning of humaneness*. Little, Brown and Co, Boston, MA.

Macdonald, J. (1995). Quality and the financial service sector. *Managing Service Quality*, **5**, 43–6.

Markus, M.L. (1990). Toward a critical mass theory of interactive media. In: J. Fulk and C. Steinfield (eds), *Organizations and Communication Technology*, pp. 194–218. Sage, Newbury Park, CA.

Markus, M.L. (1994a). Electronic mail as the medium of managerial choice. *Organization Science*, **5**, 502–27.

Markus, M.L. (1994b). Finding a happy medium: explaining the negative effects of electronic communication on social life at work. *ACM Transactions on Information Systems*, **12**, 119–49.

Mass, Y. and Herzberg, A. (1999). VRCommerce: electronic commerce in virtual reality. *Proceedings of the 1st ACM Conference on Electronic Commerce*, pp. 103–9. The Association for Computing Machinery Press, New York.

Mayr, E. and Provine, W.B. (1998). *The evolutionary synthesis: perspectives on the unification of biology*. Harvard University Press, Cambridge, MA.

McNeill, D. (1998). *The face: a natural history*. Little, Brown and Company, Boston, MA.

Metz, C. (2000). Customer support: service on the fly. *PC Magazine*, **19**, 143–4.

Miller, G.F. (2000). *The mating mind: how sexual choice shaped the evolution of human nature*. Doubleday, New York.

Ngwenyama, O.K. and Lee, A.S. (1997). Communication richness in electronic mail: critical social theory and the contextuality of meaning. *MIS Quarterly*, **21**, 145–67.

Orubeondo, A. (2000). Security and senses enlivened with speech technologies. *InfoWorld*, **22**, 86–7.

Pinker, S. (1994). *The language instinct*. William Morrow and Co., New York.

Pinker, S. (1997). *How the mind works*. W.W. Norton and Co., New York.

Pinker, S. and Bloom, P. (1992). Natural language and natural selection. In: J.H. Barkow, L. Cosmides, and J. Tooby (eds), *The adapted mind: evolutionary psychology and the generation of culture*, pp. 451–93. Oxford University Press, New York.

Reinig, B.A., Briggs, R.O., Shepherd, M.M., Yen, J., and Nunamaker, J.F. (1995). Affective reward and the adoption of group support systems: productivity is not always enough. *Journal of Management Information Systems*, **12**, 171–85.

Rheingold, H. (1993). *The virtual community: homesteading on the electronic frontier*. Addison-Wesley, Reading, MA.

Rice, R.E. (1992). Task analyzability, use of new media, and effectiveness: a multi-site exploration of media richness. *Organization Science*, **3**, 475–500.

Rice, R.E. (1993). Media appropriateness: using social presence theory to compare traditional and new organizational media. *Human Communication Research*, **19**, 451–84.

Rice, R.E. and Shook, D.E. (1990). Relationship of job categories and organizational levels to use of communication channels, including electronic mail: a meta-analysis and extension. *Journal of Management Studies*, **27**, 195–230.

Russel, S. and Norvig, P. (1995). *Artificial intelligence: a modern approach*. Prentice Hall, Upper Saddle River, NJ.

Schacter, D.L. (2001). *The seven sins of memory: how the mind forgets and remembers*. Houghton Mifflin, New York.

Sproull, L. and Kiesler, S. (1986). Reducing social context cues: electronic mail in organizational communication. *Management Science*, **32**, 1492–512.

Sproull, L. and Kiesler, S. (1991). Computers, networks and work. *Scientific American*, **265**, 84–91.

Todd, P. and Benbasat, I. (1999). Evaluating the impact of DSS, cognitive effort, and incentives on strategy selection. *Information Systems Research*, **10**, 356–74.

Tooby, J. and Cosmides, L. (1992). The psychological foundation of culture. In: J.H. Barkow, L. Cosmides, and J. Tooby (eds), *The adapted mind: evolutionary psychology and the generation of culture*, pp. 19–136. Oxford University Press, New York.

Trevino, L.K., Webster, J., and Stein, E.W. (2000). making connections: complementary influences on communication media choices, attitudes, and use. *Organization Science*, **11**, 163–82.

Walkins, J. (1992). Information systems: the UK retail financial services sector. *Marketing Intelligence and Planning*, **10**, 13–18.

Walther, J.B. (1996). Computer-mediated communication: impersonal, interpersonal, and hyperpersonal interaction. *Communication Research*, **23**, 3–43.

Walther, J.B., Slovacek, C., and Tidwell, L.C. (2001). Is a picture worth a thousand words? photographic images in long term and short term virtual teams. *Communication Research*, **28**, 105–34.

Wasserman, L. (2001). Live interaction: what's needed on the Web. *Customer Interaction Solutions*, **19**, 58–61.

Webster, J. and Trevino, L.K. (1995). Rational and social theories as complementary explanations of communication media choices: two policy-capturing studies. *Academy of Management Journal*, **38**, 1544–73.

Weisband, S. (1994). Overcoming social awareness in computer-supported groups: does anonymity really help? *Computer Supported Cooperative Work*, **2**, 285–97.

Weisband, S.P., Schneider, S.K., and Connolly, T. (1995). Computer-mediated communication and social information: status salience and status differences. *Academy of Management Journal*, **38**, 1124–51.

Wilson, E.O. (2000). *Sociobiology: the new synthesis*. Harvard University Press, Cambridge, MA.

Zimmer, C. (2001). *Evolution: the triumph of an idea*. HarperCollins Publishers, New York.

Chapter 24

Evolutionary psychology, demography, and driver safety research: a theoretical synthesis

David L. Wiesenthal and Deanna M. Singhal

Evolutionary psychology and risk-taking behaviours

This chapter will argue that evolutionary psychology offers a theoretical paradigm for the interpretation of a variety of troubling traffic safety issues related to risk-taking by young men. Young men are over-represented as both victims and perpetrators for a variety of antisocial phenomena, ranging from violent criminality to drunken driving. Their risk-taking occurs in a variety of spheres, ranging from criminality to civil wars. Principles from evolutionary models and demographic analysis offer an explanation for both the causation of problematic driving behaviours and changes in their occurrence over time. The utility of evolutionary theory is that it offers a broad, encompassing explanation for diverse social phenomena. This need not imply biological determinism, but the interaction of biological forces with societal variables of stability and resource availability, along with socialization and a variety of social influence processes, may provide a fuller explanation.

Preamble

In the summer of 2002, I (DLW) served as a member of Chris Mesquida's doctoral dissertation committee. He had developed an interesting model that synthesized the strengths of evolutionary psychological theory with a demographic and economic analysis to explain the outbreak of what he termed 'coalitional aggression'. Using his skills as a research librarian, he found a number of databases to test his model. The model held up over varying time periods and across differing regions of the world. I was particularly impressed because his model was not based upon simple biological determinism, but took into consideration the level of economic scarcity and the distribution of resources in a given place and time. Civil and tribal wars could be explained (and predicted) nicely. Areas of the globe (e.g. the Middle East, Africa) that contained a disproportionate number of young males, struggling to support themselves and establish families in the face of severe economic hardships, were prime targets for gang recruiters and those fomenting domestic unrest. Interestingly, this model predicts that when the economic plight of these young men is improved, they will cease hostile activities. In fact, this is the tactic now used by the American occupational forces to reduce terrorism and militias in Iraq, where former Sunni insurgents (termed 'the Sons of Iraq') are paid to police their neighbourhoods. Once on the payroll, hostilities have been considerably reduced (Bruno 2009).

Not long after reading the dissertation, I attended the meeting of the Canadian Multidisciplinary Road Safety Conference in Ottawa. At the meeting, Paul Gutoskie, a researcher at Transport Canada, the federal ministry responsible for all forms of transportation across Canada, described his ministry's initiatives to make Canadian roadways the safest in the world. He proudly described the safety improvements that had been implemented. At the same time, several European countries also were involved with similar programmes that resulted in impressive safety improvements. While I did not doubt that these initiatives were effective (graduated licensing comes to mind as a well-documented policy shift; see Simpson 2003; Ulmer et al. 2000), I was struck by the fact that Canada, along with the other countries discussed, was also characterized by a decline in the number of young men in the population. I recalled hearing about the 'young male syndrome' in a colloquium given at York University by Martin Daly, whose work on vengeance (Daly and Wilson 1988) was particularly relevant to our interest in highway aggression (Hennessy and Wiesenthal 1999, 2001a,b, 2002a, 2004, 2005; Wiesenthal et al. 2000a) and what has come to be known as 'road rage'. Was it not possible that the decrease in the number of young men in each country could be partially responsible for the improvement in road safety? The notion of combining a demographic approach with evolutionary theory, as Mesquida (2002) had done, made me curious to examine what might also be influencing safety on Canadian roadways. Together with Deanna Singhal, who then worked as a statistical analyst at the Traffic Injury Research Foundation (TIRF) in Ottawa, we decided to examine some of the data using national driving data from TIRF databases.

The 'young male syndrome'

The risky behaviour of young men is a well-documented phenomenon (see Byrnes et al.'s (1999) meta-analysis). Wilson and Daly (1985) have argued that males are disproportionately involved in a variety of risky behaviours (e.g. gambling, illicit drug use, homicide, theft, robbery, vehicle crashes, drunk driving). Extreme sports (e.g. hang gliding, parachuting, snowboarding) may also exhibit the same gender differences (Anderson 1999). Wilson and Daly (1985, 1994) labelled these behaviours the 'young male syndrome' and argued that evolutionary psychology represents a reasonable model for relating a variety of antisocial behaviours. They argue that sexual selection theory proposes that risky and/or violent behaviours represent competitive behaviours in groups experiencing intense reproductive competition in both evolutionary history and in present situations where contemporary circumstances may indicate the possibility of reproductive failure (1985, p. 59).

Drawing upon the sociobiological theorizing of Bateman (1948), Trivers (1972), and Williams (1966), Wilson and Daly (1985) concluded that males and females have developed distinct behavioural strategies arising from differing selective forces, such that male competitiveness has developed to deal with the desire to acquire the scarce and valued resource of females. Males can enhance their reproductive success by mating with a variety of females, while female reproductive biology offers little reproductive gain from acquiring several sexual partners. Polygyny is theorized to be the outcome of these evolutionary pressures. Although females may compete (see Hrdy 1981; Wasser 1983), male intrasexual competition is of far greater intensity. This disparity between males and females may be heightened when the competitive stakes are greater, increasing the pressure for engaging in all forms of risky behaviour (Wilson and Daly 1985). Males, more than females, vie for status and economic resources and engage in physical conflict when status and resources are involved. In general, it has been noted that males, at all ages, are at greater risk for death than are females (Kruger and Nesse 2004; Owens 2002).

Homicidal conflicts are an overwhelmingly male activity, with males over-represented as both perpetrators and victims. In their study of homicides occurring in Detroit in 1972, Wilson and Daly (1985) concluded that the vast majority involved forms of status competition. The victims, as well as the murderers, were unemployed and unmarried young men (see also Chapter 13, this volume). Murder often is the outcome of honour-related disputes where aggression follows a perceived threat to a man's virility, competence, or implications of the lack of virtue of one's female family members. Male oriented team sports may also be a similar realm where competition and sexuality meet.

Murder and professional sports are not the only sphere of activity for male risk-taking—gambling and illicit drug use are mostly male activities, especially when higher stakes are involved. Wilson and Daly (1985) further state that the disproportionately high male involvement in motor vehicle accidents may be explained by males' greater engagement in risky driving behaviours such as speeding, tailgating, refusing to yield right-of-way, running amber signals, and responding more aggressively to what male drivers perceive as the misdeeds of other drivers. By analysing American motor vehicle fatalities, estimates on distance driven, and number of licensed drivers (all categorized by age and sex), Wilson and Daly concluded that mere risk exposure (i.e. the greater distance driven by males) fails to provide an adequate explanation for the heightened vulnerability of male motorists.

Driving, for Wilson and Daly, is merely another stage where the drama of male competitiveness is displayed. As an example, they cited Jackson and Gray (1976), who observed that male drivers had shorter waiting times to make a left turn into traffic when accompanied by male passengers than when either alone in their vehicles or with female passengers. Krahé and Fenske (2002), in a study of male drivers in Germany, found that younger drivers, those with powerful vehicles, and those with a 'macho' personality were more likely to report driver aggression. In a second study with female participants, Krahé (2005) found that having a feminine orientation on the Bem Sex Role Inventory was inversely related to driving aggression.

Initiatives to improve road safety

Gutoskie (2004) reviewed a series of governmental initiatives in Canada, Scandinavia, and Europe aimed at improving road safety. These policies involved increased safety education, stronger enforcement of laws mandating the wearing of seat belts and prohibiting drinking and driving, lowering the blood alcohol level (commonly referred to as 'BAC'—blood alcohol content) constituting impaired driving, graduated licensing of new drivers, requiring helmets for bicyclists and motorcycle riders, along with improvements in the design, construction, and maintenance of roads. Environmental initiatives, such as improved removal of snow and ice during the winter months, the use of rumble strips to indicate off-road occurrences, and the employment of guardrails and impact absorption systems for bridge abutments, have also been implemented. Engineering improvements in vehicle design, such as airbags, anti-lock braking systems, stability and traction controls, have also been linked to improvements in road safety. Gutoskie described programmes implemented in Canada and the OECD member states of Finland, Great Britain, the Netherlands, and Switzerland, and reviewed the improvements in road safety as measured by data and projections of decreasing injury and fatality rates. The decrease and projected decline in future years of crashes and road deaths were attributed to these initiatives.

Demography and driver safety

But is this really the case, or are there other mediating factors? To address this issue, we must examine the role of demography in road safety. Specifically, we need to examine population growth and the relative change in the proportion of different age groups by sex in the population to see if there is a relationship with road safety statistics. Mesquida (2002) and Mesquida and Wiener (1996, 1999) have argued that the evolutionary psychology model of the risky male needs to be integrated with demography to explain the intensity or severity of male competition.

By examining the various proportions of young males, either over time in a given population or across different populations, it is possible to predict domestic unrest that may culminate in civil wars, other forms of coalitional aggression, and higher levels of criminality. Mesquida has argued that focusing solely on demographic variables is insufficient because it fails to provide an explanation for why males would engage in hostile, competitive activities. The linking of the evolutionary model of male competition achieved through risk-taking with demography, represents a significant advance in theory development. It not only explains behaviour across diverse cultures and species, but allows for predictions, while accounting for environmental conditions. The model predicts that when valued commodities are scarce, male intrasexual competition will intensify to secure the necessary economic status to attract mates. If a society were to provide ample resources (employment, housing, transportation, etc.), then the severity of the competition would be reduced.

This model has received considerable support in cross-national comparison data and in examining a specific nation over time (e.g. Columbia) to account for domestic violence. Criminology has long related age and gender as predictors of criminal behaviour (Fox 2000). Although age effects may not always be present (Gartner and Parker 1990), age does receive considerable support in the literature (Greenberg 1985, see also Hirschi and Gottfredson 1983). Shifts in the proportion of young males in the population have been related to changes in a variety of criminal behaviours (Cohen and Land 1987).

If evolutionary pressures exist for male competition, why should traffic safety issues such as street racing, burnouts (the practice of drivers spinning their vehicles' tires at high speeds until smoke appears), and drifting (controlled, high-speed skidding) be the arena? How can we explain how these behaviours are among the ones selected for competitive activities? We need to examine societal as well as media influences on youth to show how these values are socially constructed as masculine behaviours (Courtenay 2000). Mast et al. (2008) demonstrated that men who receive primes for masculine-identified words were more likely to engage in speeding (as assessed on a realistic driving simulator) compared with men receiving either neutral or feminine primes.

Media influences and dangerous driving

We are all exposed to a variety of media influences promoting risky driving practices. The Internet (e.g. the popular YouTube website) features thousands of video clips promoting dangerous driving and life-endangering stunts. Advertisements, films, and television often depict dangerous driving, street racing, and speeding vehicles. The high-speed car chase has now become almost standard in crime dramas ever since Steve McQueen drove a high performance Mustang over the streets of San Francisco in the 1968 film *Bullitt*. The increase in street racing may be attributable to the promotion of street racing culture in the media, through such things as car advertisements, video games, films, and Internet sites. Even young children view cartoons such as *Speed Racer*. These influences can increase the potential for 'copycat stunts', for which there are anecdotal reports (Small 2007).

Recent computer games promoting racing have also been a concern to policy-makers. For example, Australian road safety authorities tried to ban a computer game which allowed the player to street race through virtual images of Sydney. The promotional material for the Project Gotham Racing 2 game by Xbox stated, 'Burn up a storm past famous landmarks such as the Opera House and Sydney Harbour' (Dowling 2003).

Racing games have become a very popular genre (Fischer et al. 2007). Within realistic virtual environments, players race on circuits or through urban and suburban traffic. Driving actions in these games often include competitive and reckless driving, speeding, crashing into cars or pedestrians, or performing risky stunts. Beullens et al. (2008) conducted a cross-sectional survey, examining the relationship between the playing of racing games and the intention to engage in risky driving among Belgian adolescents. They found that playing video games was a significant predictor of positive attitudes toward risky driving which, in turn, predicted intention to drive this way in the future. Similarly, Fischer et al. (2007) found that playing racing games was positively associated with self-reported competitive driving, exhibitionistic driving, and collisions. Beullens and Van den Bulck (2008) found that music videos, however, were related to perceptions of lowered risk for speeding and drunk driving, with girls expressing greater caution.

Recent experimental studies were conducted by Fischer and colleagues examining social-cognitive theories. They hypothesized that racing video games can 'prime' not only other cognitions, but also affective and behavioural reactions (Fischer et al. 2007). Those who were randomly assigned to play a racing game subsequently exhibited significantly more cognitions and affect associated with risk-taking than participants who played a neutral game. Furthermore, men who played a racing game subsequently took higher risks in video-simulated critical road traffic situations than males who played a neutral game. Overall, the effects were stronger for men than for women.

Another study examined whether risk-promoting print, film, and video media increased accessibility of risk-promoting cognitions and increased risky driving (Fischer et al. 2008). First, high school students assigned to view high-risk sporting activity pictures exhibited increased accessibility of risk-promoting cognitions, and an underestimation of risks associated with high-risk sports, compared with those who viewed low-risk pictures. Second, they examined the effects of film clips featuring risky sport stunts versus neutral scenes on risky driving in the 'Need for Speed' video game. As expected, risk-promoting film clips increased participants' risk-taking in the game. A third experiment reinforced that exposure to risk-promoting (racing) video games increased risk-related cognitions and risky decision-making. Finally, Fischer et al. (2009) ruled out demand characteristics as a cause of the increased risk, finding that (1) risk increased for active participants (rather than passive observers) of racing games that rewarded risk, and (2) that risk was mediated by perceptions of the self as reckless.

During the last decade, performance, speed, and power have been dominant themes in vehicle advertisements (Burns 1999; Ferguson et al. 2003; Rudin-Brown et al. 2008; Schonfeld et al. 2005; Shin et al. 2005). A magazine advertisement for Acura displays a black car underneath a dark sky with the only wording in the advertisement being 'A civilized way to handle your aggression'. Another magazine advertisement for the over-sized Hummer sport utility vehicle proclaims, 'Teach cabbies some respect', with a drawing of a cringing taxi driver alongside the smiling Hummer driver. Internationally, Mazda has run a much-viewed television advertising campaign featuring speeding cars with a voice-over of 'Zoom zoom!'.

Shin et al. (2005) content analysed 250 automobile television commercials. Forty-five per cent contained an unsafe driving activity, with aggressive driving accounting for 85% of the commercials deemed unsafe. Speed violations accounted for 56% of the aggressive driving category. Male drivers were shown in 81% of all commercials. Young males, in particular, seem susceptible to these themes. A survey of novice drivers from northern Ontario found that 'a car's speed and power' was significantly more important to males than females (Tilleczek 2004).

Demographics and drunk driving

Macdonald (2003) examined the decline in alcohol-related car crashes in Ontario, Canada, over a 25-year period and attributed part of the decrease to population changes. He argued that media campaigns against drunk drivers and legislative changes are often cited as explanations for the decline in alcohol-related crashes, but that demographic changes in the age and sex composition of Ontario over this period may provide a better explanation. The rate of impaired crashes decreased by 78.1% from 1974 to 1999. Over the same period, the average age of the Ontario driver increased from 39.4 to 43.2 years, and the proportion of female drivers increased from 39.6% to 46.8%. Macdonald concluded that the ageing population accounted for an 8.6% decline in impaired crashes, and the increasing proportion of licensed female drivers for a further 9.4%. It should be noted that Macdonald did not place his findings in any theoretical context. Evans (2002) also has recognized the importance of examining age and sex in a population in the determination of driver safety, but similarly has failed to develop a theoretical explanation of why young males should be prone to risky and often fatal motoring behaviours.

While both Macdonald (2003) and Evans (2002) recognize the importance of age and sex distributions, it is too soon to evaluate their possible influence on driver safety scholarship. The important model of the young male syndrome, published over 25 years ago by Wilson and Daly (1985), has been cited more than 200 times. Yet, to date, with the exception of the present authors and colleagues, none of the citing articles were related to driver behaviour, nor were they published in safety-related journals.

Gender differences in traffic injuries and dangerous driving

In addition to the statistics provided by Wilson and Daly (1985), there is ample evidence over a variety of geographic areas (e.g. the United Kingdom, the United States, Europe, and Australia) and time periods that males are at greater risk for traffic injuries and death (Social Issues Research Centre, 2004). The World Health Organization (WHO) reported in 2002 that international statistics indicated that 2.7 times as many males died from injuries sustained in road traffic mishaps than females. In Spain, the male death rate was three times higher than females. The figure for pedestrians killed also reflected this gender imbalance (WHO 2002). This gender difference exists across the age distribution, but is most pronounced for younger males—teenagers and young adults (Abdel-Aty and Abdelwahab 2000; Evans 1991; McKenna et al. 1998; Waller et al. 2001). The vast majority of reckless drivers are young men (Begg and Langley 2001; Harre et al. 2000).

Societal and evolutionary process model

Figure 24.1 shows the original explication of the 'young male syndrome' by Wilson and Daly (1985), expanded to deal with a variety of additional elements that reflect research developments since the original publication of the model. Using the demographic feature of the Mesquida (2002) model, the proportion of young males (under 30 years of age), aids in the understanding of a major factor promoting competition between males to establish themselves and attract mates. The scarcity of resources, also drawn from Mesquida's work, serves to sharpen the competition between teenagers and older men. Scarcity and economic downturns may also be linked with a greater proportion of unemployed young men with time on their hands to engage in criminal activities, or drug and alcohol use (Arkes 2006). The greatest potential gain from risk occurs during the stage when mate competition begins. The level of aggression and risk-taking should be directly related to the proportion of young males in a population. It is important to note that the degree of environmental stability and social learning play a role in determining these outcome behaviours of risk, so we are *not* proposing a model based upon simple biological determinism.

The additional feature of societal and situational influences helps explain what areas become important for males to compete in. Situational and state factors, like the presence of young male

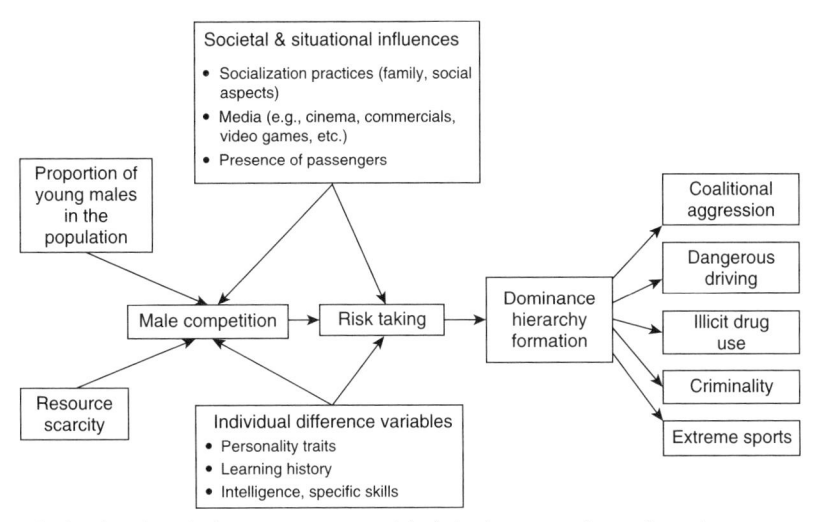

Fig. 24.1 Societal and evolutionary process model of the 'young male syndrome'.

passengers, alcohol consumption, and modelling of parental driving behaviours need to be included. These factors influence both competition and risk-taking. Male competition and risk-taking are also influenced by individual difference variables, such as personality traits, driving skill, reinforcement history for risk, and intelligence, along with the momentary drive states of hunger, thirst, fatigue, and sexual arousal. These all play a role in forming dominance hierarchies.

Male competition existed prior to the development of the automobile in the 19th century and its widespread adoption in the 20th century. The role of the media can assist in explaining why the vehicle and the road are selected as the venue for intense competition, rather than the chess tournament. The struggle for a favourable position in young male hierarchies may be acted out in a number of resultant realms—coalitional aggression, dangerous driving, illicit drug use, and criminality, to name a few. Research is needed to predict which of these, or what specific combination of these behaviours may be likely, in any given period, for an individual teenager or young man.

Canadian driver safety data

An examination of Canadian census data indicates a steady increase in the population since 1971, marked by a decline in the proportion of males aged 15–29 years (Figure 24.2).[1] Both the number of fatalities and personal injuries, due to collision, have also fallen dramatically since the late 1980s (Figure 24.3). The similarities in decline for both of these curves, compared with the declining proportion of young men (Figure 24.2), are quite compelling. All curves show a steady decrease in their respective numbers, suggesting that the decline in young men in the Canadian population could be partly responsible for the decline we see in both collision fatalities and injuries.

Although there is a marked decline in motor vehicle fatality rates between 1990 and 2003, the number of vehicles on Canadian roads has increased during the same period (Figure 24.4).

---

[1] Population pyramids for Canada, constructed by Statistics Canada, are available at http://www12.statcan.ca/english/census01/products/analytic/companion/age/pyramid.cfm

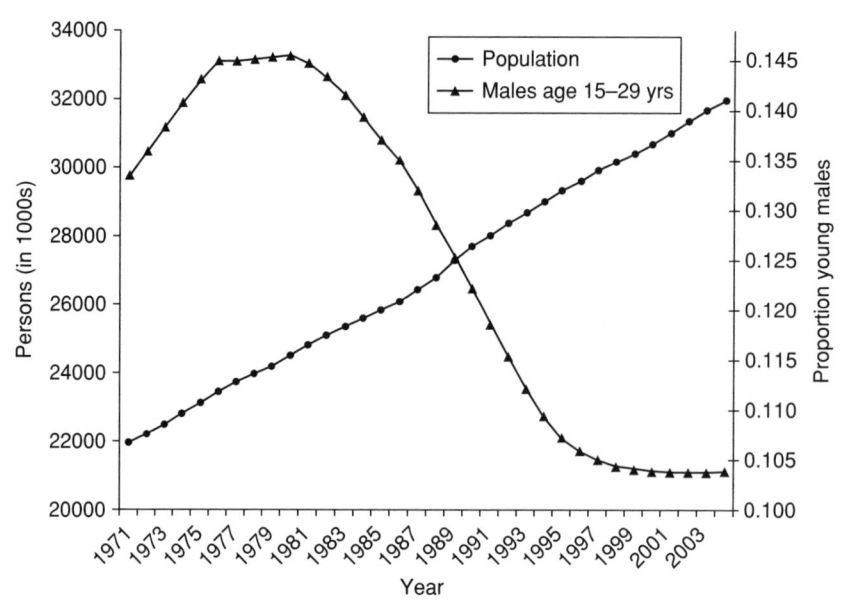

Fig. 24.2 Population of Canada from 1971–2003 and the proportion of males aged 15–29 years.

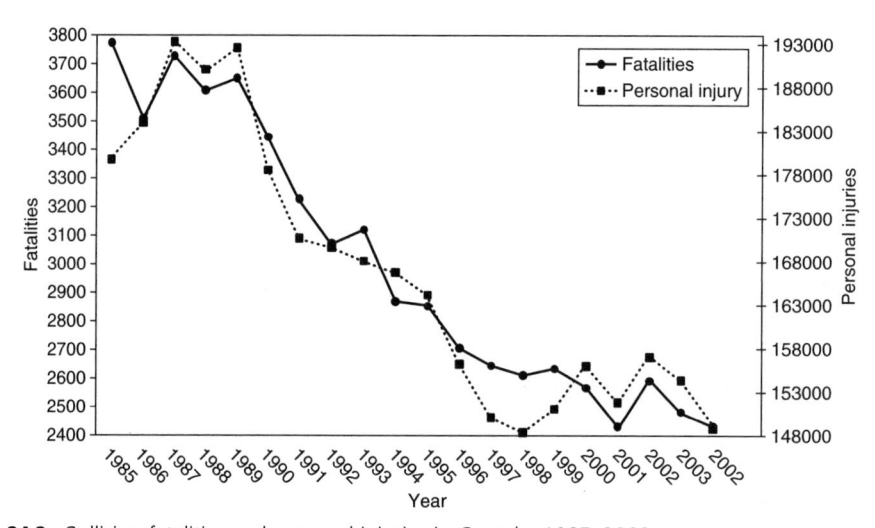

Fig. 24.3 Collision fatalities and personal injuries in Canada, 1985–2002.

The fatality decline also matches the decrease in the proportion of young males in the population. It should be noted that males aged 16–34 years represent 30.1% of male drivers and 16% of all Canadian drivers, yet they account for 39.5% of traffic fatalities and 43.6% of injuries.

Looking specifically at driver fatalities, Figure 24.5 shows that the number of driver fatalities for young men aged 15–29 has dropped considerably, regardless of whether they were drivers, passengers or pedestrians, and mirrors the decline in the proportion of young men in the

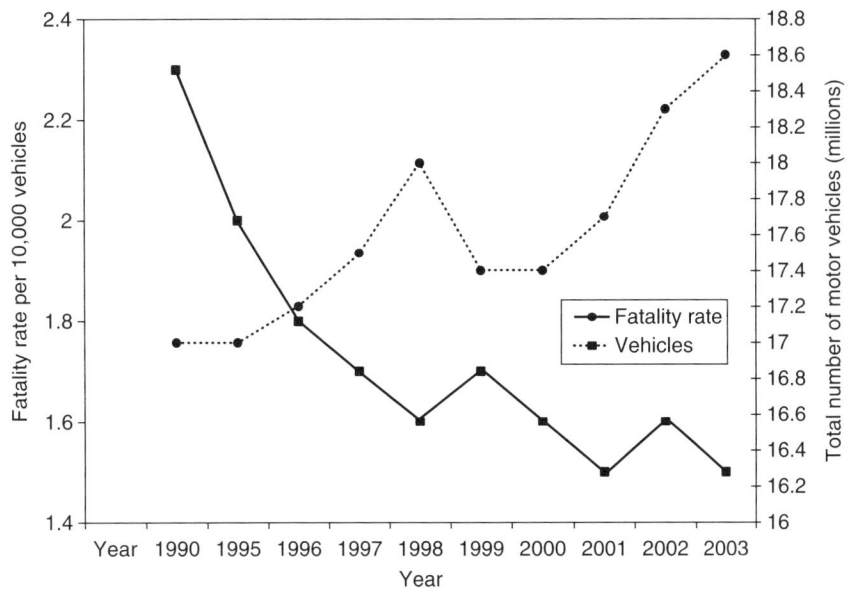

Fig. 24.4 Canadian fatality rates and number of registered motor vehicles.
Data source: North American Transportation Statistics (note that a gap exists on the x-axis).

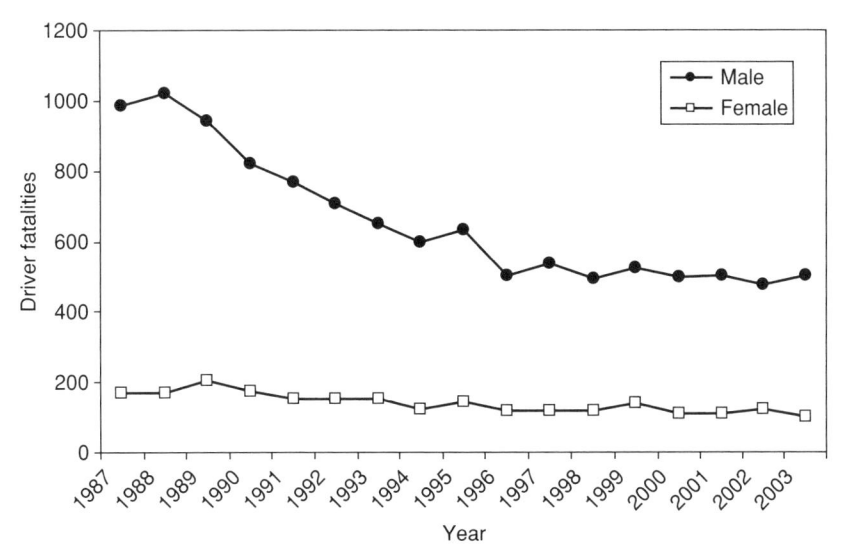

Fig. 24.5 Driver fatalities among 16–29-year-olds in Canada, 1987–2003.

population over the same time period. The number of driver fatalities for young females is lower than that for young males and has remained relatively constant over this same time period. Similarly, the Insurance Institute for Highway Safety reported that, in the United States in 2008, male passengers in passenger vehicles averaged a rate of 6.85 fatalities per 100,000 people compared to 4.55 deaths for female passengers (Insurance Institute for Highway Safety 2008).

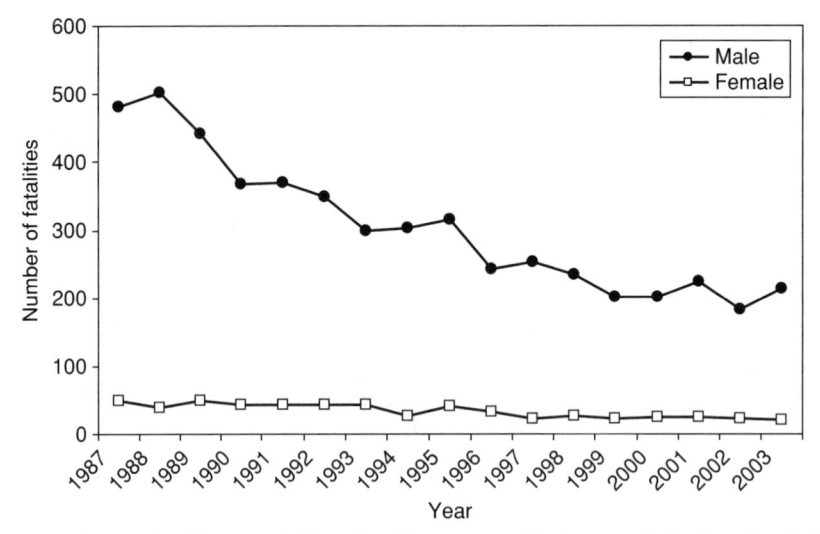

Fig. 24.6 Alcohol-involved (BAC > 0) driver fatalities among 16–29-year-olds in Canada, 1987–2003.

Transport Canada (2004) reported that for the 10-year period from 1992, males constituted 61% of pedestrian fatalities while females accounted for only 39%. Males are over-represented in fatalities, not only as drivers, but as passengers and pedestrians as well, lending support to the young male syndrome.

Alcohol-related fatalities have similarly declined over time for young male drivers in Canada (Figure 24.6). Macdonald (2003) noted the decline in these statistics for Ontario and suggested population changes in Ontario drivers, such as an increase in the ageing population of drivers and female drivers, as a contributing factor. Here, it is interesting to note that the national decline in fatalities of young men (aged 16–29) where alcohol has been consumed also coexists with the decrease in the proportion of young men over the same period. Young female driver fatalities, where alcohol has been consumed, remained at a lower, constant level from 1987 to 2003.

To summarize the Canadian population and road safety data, the following points were made:

- The population of Canada has increased almost 50% over three decades, while the number of young males has declined.
- Collision fatalities and injuries have both declined over the same three decades.
- Fatality rates have declined while the number of vehicles on the roads in Canada has increased.
- Men are disproportionately killed in roadway crashes.
- The number of male drivers, aged 16–29, that are killed on roadways has steadily decreased, compared to almost no change in the figures for women over the same period.
- Very few young women drivers are killed in drinking and driving incidents, compared to their male counterparts.
- The number of male drivers killed, who have consumed alcohol, has also been declining over this time period compared to women.
- The decrease in the fatality rate for young male drivers is mirrored in the decrease of their numbers in the Canadian population.

Canadian road safety statistics seem to be indicating that, as the proportion of young men in the population has decreased (hence, fewer young male drivers), so have injuries and deaths, whether attributable to alcohol consumption or not. The 'young male syndrome', described by Wilson and Daly (1985), receives support in these data. This must be recognized in explaining the improvement in road safety statistics, apart from governmental policy changes and automotive engineering improvements. While governments and auto manufacturers undoubtedly are pleased to accept credit for enhanced highway safety statistics (Gutoskie 2004), this state of affairs must be tempered by recognizing that changing demographic variables are also in operation.

Alternative explanations for greater risk-taking in young men

An alternative explanation for evolutionary pressures leading to risk-taking in the service of the formation of dominance hierarchies in young men has been the notion that the developing brains of young adults do not mature until the mid-20s or even in the third decade of life. Specifically, the frontal lobes have been chosen for study because this area of the brain is related to executive functioning (Bennett and Baird 2006; Romer and Hurt 2010; Sowell et al. 1999; Van Leijenhorst et al. 2009), which controls impulsivity and working memory. Romer and Hurt's (2010) research indicates that those scoring high on impulsivity engage in risky behaviours (e.g. fighting, gambling, alcohol consumption) at an early age. Surprisingly, not all those with weakness in working memory engaged in risky activities. These results contradict the belief that adolescent risk-taking is due to poor executive cognitive functioning.

Despite the controversy in the brain structure and functioning literature on the possible maturational differences between young men and women, and consequent behavioural differences, does the possible immaturity of the frontal lobes constitute a reasonable explanation for the greater number of fatalities and injuries experienced by male drivers? If one looks at distributions of fatalities and injuries of male drivers, the eventual decline from the adolescent and young adult years is a long time coming, but males still have a higher fatality and injury rate than females. Even if impulsivity might be responsible for fatalities/injuries of drivers and pedestrians, how can the higher fatalities and injuries experienced by *passengers* be explained? How does the presence of passengers enhance the probability of a crash? We might suspect that those young drivers with immature impulse control would get into trouble without passengers present. Passengers might elevate the risk for crashes by distracting the driver or goading the driver to engage in dangerous driving, but not to the extent actually seen in the Insurance Institute for Highway Safety (2008) data.

Another possible explanation for the heightened probability of young male driver crashes could be their aggressive tendencies being activated when they experience stress on the road. My (DLW) research exploring commuting stress resulting from the interaction of trait stress and situational factors (i.e. traffic congestion with its attendant frustrations) has found that aggressive behaviours are one of the most common reactions to stress. Interestingly, we have not obtained gender differences in aggressive driving behaviours, but only in violent behaviours, with males more prone to physical confrontations with other motorists (e.g. Hennessy and Wiesenthal 1997, 1999, 2001a,b; 2002a,b, 2004, 2005; Wiesenthal et al. 2000a,b, 2003;. Hennessy et al. 2004; Wickens and Wiesenthal 2005). While aggressive encounters may be responsible for fatality and injury of drivers and others on the roadway, stress and the frustration-aggression hypothesis cannot explain the higher fatality and injury statistics for young male passengers and pedestrians.

Finally, if enhanced automobile safety devices, enhanced enforcement of traffic regulations, graduated licensing initiatives, and better highway maintenance were responsible for the decline in fatalities, we still would be left with attempting to explain why males have higher fatalities and injuries, even when controlling for their greater exposure to driving risk. Both males and females would be expected to show a similar decline if these factors constituted the whole story, but the data fail to show a similar decrease.

Conclusions

The present research indicates that clear trends in Canadian data indicate the importance of considering age and sex as explanations for reductions in fatalities, injuries, and alcohol-related crashes. It is clear that the decline in the proportion of young males (ages 15–29) in Canada over the last three decades has resulted in appreciable decreases in traffic injuries and fatalities. Current work is aimed at producing a statistical model, based upon approaches such as time series and regression analysis, to identify the proportion of safety related increments solely attributable to population variation and the proportion that may be credited to engineering advances and policy implementations.

Unlike the usual evolutionary models that provide only post hoc explanations of behaviour, the Mesquida (2002) formulation offers predictions about the potentiality for conflict, while accounting for environmental resources. Its application in traffic safety offers broad predictive and explanatory powers relating risky driving behaviours to a variety of other inherently risky activities—criminality, coalitional aggression, gambling, and dare-devil sporting activities (so-called 'extreme sports')—all of which might serve partly to establish dominance over other males through competition. This biologically-based drive for male ascendance extends across species, but also enables societies to recognize, and thus regulate, the competitive drive, offering the possibility for diversion into socially acceptable venues (e.g. sports).

Although evolutionary psychology has not commonly been incorporated in the realm of traffic safety and driver statistics, it may well play a role in explaining why males are over-represented in collision fatalities and injuries, with or without the involvement of alcohol. Coupled with the changing demographics of young men, it offers a diverse and interactive approach to the interpretation of the changes in traffic statistics, suggesting that the commonly investigated reasons for improvement, such as graduated driver licensing or roadway upgrades, are not the sole factors contributing to traffic statistical trends.

Acknowledgements

The writing of the chapter was assisted by grant support from AUTO 21 and the Social Sciences and Humanities Research Council of Canada. The authors wish to thank Chris Mesquida, Tony Nield, Irwin Silverman, Jenn Steeles, Maggie Toplak, and Fred Weizmann for their advice and assistance.

References

Abdel-Aty, M.A. and Abdelwahab, H.T. (2000). Exploring the relationship between alcohol and the driver characteristics in motor vehicle accidents. *Accident Analysis and Prevention*, **32**, 473–82.

Anderson, K.L. (1999). Snowboarding: the construction of gender in an emerging sport. *Journal of Sport and Social Issues*, **23**, 55–79.

Arkes, J. (2006). Does the economy affect teenage substance abuse? *Health Economics*, **16**, 19–36.

Bateman, A.J. (1948). Intra-sexual selection in Drosophila. *Heredity*, **2**, 349–68.

Begg, D. and Langley, J. (2001). Changes in risky driving behavior from age 21 to 26 years. *Journal of Safety Research*, **32**, 491–9.

Bennett, C.M. and Baird, A.A. (2006). Anatomical changes in the emerging adult brain: a voxel-based morphometry study. *Human Brain Mapping*, **27**, 766–77.

Beullens, K., Roe, K., and Van den Bulck, J. (2008). Video games and adolescents' intentions to take risks in traffic. *Journal of Adolescent Health,* **43**, 87–90.

Beullens, K. and Van den Bulck, J. (2008). News, music videos and action movie exposure and adolescents' intentions to take risks in traffic. *Accident Analysis and Prevention,* **40**, 349–56.

Bruno, G. (2009). Finding a place for the 'Sons of Iraq'. Council on Foreign Relations, Washington, DC. Available at: http://www.cfr.org/iraq/finding-place-sons-iraq/p16088

Burns, R.G. (1999). Socially constructing an image in the automobile industry. *Crime, Law and Social Change,* **31**, 327–46.

Byrnes, J.P., Miller, D.C., and Schafer, W.D. (1999). Gender differences in risk taking: a meta-analysis. *Psychological Bulletin,* **125**, 367–83.

Cohen, L.E. and Land, K.C. (1987). Age structure and crime: symmetry versus asymmetry and the projection of crime rates through the 1990s. *American Sociological Review,* **52**, 170–83.

Courtenay, W.H. (2000). Constructions of masculinity and their influence on men's well-being: a theory of gender and health. *Social Science and Medicine,* **50**, 1385–401.

Daly, M. and Wilson, M. (1988). *Homicide.* A. de Gruyter, New York.

Dowling, J. (2003). MP demands ban on video hoons. The Sun-Herald, 16 November 2003. Available at: http://www.smh.com.au/articles/2003/11/15/1068674436769.html

Evans, L. (1991). *Traffic safety and the driver.* Van Nostrand Reinhold, New York.

Evans, L. (2002). Traffic crashes. *American Scientist,* **90**, 244–53.

Ferguson, S.A., Hardy, A.P., and Williams, A.F. (2003). Content analysis of television advertising for cars and minivans: 1983–1998. *Accident, Analysis and Prevention,* **35**, 825–31.

Fischer, P., Kubitzki, J., Guter, S., and Frey, D. (2007). Virtual driving and risk taking: Do racing games increase risk-taking cognitions, affect, and behaviors? *Journal of Experimental Psychology—Applied,* **13**, 22–31.

Fischer, P., Guter, S., and Frey, D. (2008). The effects of risk-promoting media on inclinations toward risk taking. *Basic and Applied Social Psychology,* **30**, 1–11.

Fischer, P., Greitemeyer, T., Morton, T., *et al.* (2009). The racing-game effect: why do video racing games increase risk-taking inclinations? *Personality and Social Psychology Bulletin,* **35**, 1395–409.

Fox, J.A. (2000). Epilogue, 2005: after the crime drop. In: A. Blumstein and J. Wallman (eds), *The crime drop in America,* pp. 288–317. Cambridge University Press, Cambridge.

Gartner, R. and Parker, R.N. (1990). Cross-national evidence on homicide and age structure of the population. *Social Forces,* **69**, 351–71.

Greenberg, D.F. (1985). Age, crime, and social explanation. *American Journal of Sociology,* **91**, 1–21.

Gutoskie, P. (2004). 'Global perspectives on road safety targets.' Canadian Multidisciplinary Road Safety Conference, June 28th, Ottawa, Ontario, Canada.

Harre, N., Brandt, T., and Dawe, M. (2000). The development of risky driving in adolescence. *Journal of Safety Research,* **31**, 185–94.

Hennessy, D.A. and Wiesenthal, D.L. (1997). The relationship between traffic congestion, driver stress, and direct versus indirect coping behaviours. *Ergonomics,* **40**, 348–61.

Hennessy, D.A. and Wiesenthal, D.L. (1999). Traffic congestion, driver stress, and driver aggression. *Aggressive Behavior,* **25**, 409–23.

Hennessy, D.A. and Wiesenthal, D.L. (2001a). Further validation of the Driving Vengeance Questionnaire. *Violence and Victims,* **16**, 565–73.

Hennessy, D.A. and Wiesenthal, D.L. (2001b). Gender, driver aggression, and driver violence: an applied evaluation. *Sex Roles,* **44**, 661–76.

Hennessy, D.A. and Wiesenthal, D.L. (2002a). Aggression, violence, and vengeance among male and female drivers. *Transportation Quarterly,* **56**, 65–75.

Hennessy, D.A. and Wiesenthal, D.L. (2002b). The relationship between driver aggression, violence and vengeance. *Violence and Victims,* **17**, 707–18.

Hennessy, D.A. and Wiesenthal, D.L. (2004). Age and vengeance as predictors of mild driver aggression. *Violence and Victims,* **19**, 469–77.

Hennessy, D.A. and Wiesenthal, D.L. (2005). Driving vengeance and wilful violations: clustering of problem driving attitudes. *Journal of Applied Social Psychology,* **35**, 61–79.

Hennessy, D.A., Wiesenthal, D.L., Wickens, C.M., and Lustman, M. (2004). The impact of gender and stress on traffic aggression: are we really that different? In: J.P. Morgan (ed.), *Focus on aggression research,* pp. 157–74. Nova Science Publishers, Hauppauge, NY.

Hirschi, T. and Gottfredson, M. (1983). Age and the explanation of crime. *American Journal of Sociology,* **89**, 552–84.

Hrdy, S.B. (1981). *The woman that never evolved.* Harvard University Press, Cambridge, MA.

Insurance Institute for Highway Safety (2008). Fatality facts 2008. Available at: http://www.iihs.org/research/fatality_facts_2008/teenagers.html

Jackson, T.T. and Gray, M. (1976). *Perceptual and Motor Skills,* **43**, 471–4.

Krahé, B. (2005). Predictors of women's aggressive driving behavior. *Aggressive Behavior,* **31**, 537–46.

Krahé, B. and Fenske, I. (2002). Predicting aggressive driving behavior: the role of macho personality, age, and power of car. *Aggressive Behavior,* **28**, 21–9.

Kruger, D.J. and Nesse, R.M. (2004). Sexual selection and the male:female mortality ratio. *Evolutionary Psychology,* **2**, 66–77.

Macdonald, S. (2003). The influence of the age and sex distribution of drivers on the reduction of impaired crashes: Ontario, 1974–1999. *Traffic Injury Prevention,* **4**, 33–7.

Mast, M.S., Sieverding, M., Esslen, M., Graber, K., and Jäncke, L. (2008). Masculinity causes speeding in young men. *Accident Analysis and Prevention,* **40**, 840–2.

McKenna, F.P., Waylen, A.E., and Burkes, M.E. (1998). *Male and female drivers: how different are they?* AA Foundation for Road Safety, Basingstoke.

Mesquida, C.G. (2002). 'Resources, mating, and male age composition: An evolutionary psychology perspective on coalitional aggression.' PhD dissertation, York University, Toronto, Canada.

Mesquida, C.G. and Wiener, N.I. (1996). Human collective aggression: a behavioral ecology perspective. *Ethology and Sociobiology,* **17**, 247–62.

Mesquida, C.G. and Wiener, N.I. (1999). Male age composition and severity of conflicts. *Politics and the Life Sciences,* **18**, 181–9.

Owens, I.P.F. (2002). Sex differences in mortality rate. *Science,* **297**, 2008–9.

Romer, D. and Hurt, H. (2010). 'No simple explanation for why adolescents take risks.' Paper presented at the American Academy of Pediatrics, Vancouver.

Rudin-Brown, C.M., Lavack, A.M., Watson, L.M., Mintz, J.H., and Colterman, B. (2008). Be driving excitement! Accelerate the future! Zoom-zoom! Unsafe themes in recent Canadian automotive advertisements. *Proceedings of the 18th Canadian Multidisciplinary Road Safety Conference* (pp. 1–14). Canadian Association of Road Safety Professionals, Whistler, BC, Canada.

Schonfeld, C., Steinhardt, D.A., and Sheehan, M. (2005). A content analysis of Australian motor vehicle advertising: effects of the 2002 voluntary code on restricting the use of unsafe driving themes. *Proceedings of the Australasian Road Safety Research, Policing and Education Conference.* Wellington, New Zealand.

Shin, P.C., Hallett, D., Chipman, M.L., Tator, C., and Granton, J.T. (2005). Unsafe driving in North American automobile commercials. *Journal of Public Health,* **27**, 318–25.

Simpson, H.M. (2003). The evolution and effectiveness of graduated licensing. *Journal of Safety Research,* **34**, 25–34.

Small, P. (2007). *Killer drivers won't do jail time.* The Toronto Star, 30 May 2007. Available at: http://www.thestar.com/News/GTA/article/219249

Social Issues Research Centre (2004). Sex differences in driving and insurance risk: an analysis of the social and psychological differences between men and women that are relevant to their driving behaviour. Social Issues Research Centre, Oxford.

Sowell, E.R., Thompson, P.M., Holmes, C.J., Jernigan, T.L., and Toga, A.W. (1999). *In vivo* evidence for post-adolescent brain maturation in frontal and striatal regions. *Nature Neuroscience,* **2**, 859–61.

Tilleczek, K.C. (2004). The illogic of youth driving culture. *Journal of Youth Studies,* **7**, 473–98.

Transport Canada (2004). *Pedestrian fatalities and injuries,* 1992–2001 (Fact Sheet TP 2436E RS-2004–01E). Transport Canada, Ottawa.

Trivers, R.L. (1972). Parental investment and sexual selection. In: B. Campbell (ed.), *Sexual selection and the descent of man 1871–971,* pp. 136–79. Heinemann, London.

Ulmer, R.G., Preusser, D.F., Williams, A.F., Ferguson, S.A., and Farmer, C.M. (2000). Effect of Florida's graduated licensing program on the crash rate of teenage drivers. *Accident Analysis and Prevention,* **32**, 527–37.

Van Leijenhorst, L., Zanolie, K., Van Meel, C.S., Westenberg, P.M., Rombouts, S.A.R.B., and Crone, E.A. (2009). What motivates the adolescent? Brain regions mediating reward sensitivity across adolescence. *Life Sciences and Medicine,* **20**, 61–9.

Waller, P.F., Elliot, M.R., Shope, J.T., Raghunathan, T.E., and Little, R.J.A. (2001). Changes in young adults offense and crash patterns over time. *Accident Analysis and Prevention,* **33**, 117–28.

Wasser, S.K. (1983). *Social behavior of female vertebrates.* Academic Press, New York.

Wickens, C.M. and Wiesenthal, D.L. (2005). State driver stress, traffic congestion, and trait stress susceptibility. *Journal of Applied Biobehavioral Research,* **10**, 83–97.

Wiesenthal, D.L., Hennessy, D.A., and Gibson, P.M. (2000a). The Driving Vengeance Questionnaire (DVQ): the development of a scale to measure deviant driver attitudes. *Violence and Victims,* **15**, 115–36.

Wiesenthal, D.L., Hennessy, D.A., and Totten, B. (2000b). The influence of music on driver stress. *Journal of Applied Social Psychology,* **30**, 1709–19.

Wiesenthal, D.L., Hennessy, D.A., and Totten, B. (2003). The influence of music on mild driver aggression. *Transportation Research Part F,* **6**, 125–34.

Williams, G.C. (1966). *Adaptation and natural selection.* Princeton University Press, Princeton, NJ.

Wilson, M. and Daly, M. (1985). Competiveness, risk taking, and violence: the young male syndrome. *Ethology and Sociobiology,* **6**, 59–73.

Wilson, M. and Daly, M. (1994). Evolutionary psychology of male violence. In: J. Archer (ed.), *Male violence,* pp. 253–88. Routledge, London.

World Health Organization (2002). Gender and Road Traffic Injuries. Department of Gender and Women's Health, Geneva. Available at: http://www.who.int/gender/documents/road_traffic/a85576/en/index.html

Chapter 25

Evolutionary robotics

Dylan Evans and Walter de Back

Introduction

Robotics has been heavily influenced by cognitive psychology and evolutionary biology, but the child prodigy of these two disciplines, evolutionary psychology, has not yet had such a large impact on the field. Here we describe how evolutionary psychology may come to have a greater impact on robotics in the near future. First, however, we will briefly outline the role that the two parent disciplines of evolutionary psychology have played in the development of robotics, and describe the few lessons that have been drawn so far by roboticists from evolutionary psychology itself.

Cognitive psychology

The cognitive revolution swept through psychology in the 1960s, displacing the behaviourist paradigm that had held sway since the 1920s. Its origins, however, lie in the 1950s. If one day had to be singled out as the birthday of cognitive science, it is surely 11 September 1956. It was on that day that three seminal papers were presented at a historic meeting at the Massachusetts Institute of Technology (MIT). Allen Newell and Herbert Simon spoke about a 'logic theory machine', inaugurating the modern discipline of Artificial Intelligence (Newell and Simon 1956). Noam Chomsky described 'three models for the description of language' in a paper that has been described as marking the birth of modern linguistics (Dennett 1995, p. 384; Chomsky 1956). Finally George Miller presented a paper about short-term memory that is now recognized as one of the foundational papers of cognitive psychology (Miller 1956).

It is somewhat harder to date the origins of robotics, but the pioneering work of Grey Walter and other figures in the cybernetics movement in the late 1940s has a good claim to be considered as laying the foundations for the discipline as we know it today. The mechanical tortoises built by Walter are noticeably different from the robots after the rise of cognitive science a decade later (Walter 1951). For one thing, Walter made no attempt to build a 'mind' out of various functional components such as memory, perceptual processing, and so on. Rather, his emphasis was on the behaviour of the robot as a whole, and on seeking the simplest internal structure that would enable complex behaviour to emerge when the robot was placed in the right kind of environment. This approach may have been largely due to the fact that the integrated circuit had not yet been invented, which meant that Walter had to work entirely with conventional analogue electronic components, large numbers of which would have been required to build components with a distinct psychological function.

The invention of integrated circuits, in which large numbers of components were packed onto a small silicon chip, gave roboticists the means, and the simultaneous rise of cognitive science, the motive, to pay more attention to the internal computational structure of their machines. An early example of the new generation of 'cognitive robots' was Shakey. Developed at the Artificial Intelligence Center at Stanford Research Institute between 1966 and 1972, Shakey had a complex cognitive architecture in which distinct functions such as perception, planning, and natural language processing were implemented by separate programs, which reflected the emphasis of cognitive psychology on the functional decomposition of mental processes.

Evolutionary biology

Biorobotics and evolutionary robotics

By the late 1980s, the cognitive approach to robotics was coming under increasing criticism from Rodney Brooks and his colleagues at MIT. Brooks argued that reasoning and planning via symbol systems, as in Shakey, was often unnecessary, since similar behaviours could be achieved more simply and economically by building robots out of distinct modules, each directly linking sensors to actuators. Unlike cognitive robots, in these 'behaviour-based' robots there was no central processing unit through which all information would be routed. Instead, complex behaviour emerged from the interactions between largely autonomous modules organized into a hierarchy where higher levels could inhibit lower ones.

Brooks and his colleagues were inspired directly by ideas from evolutionary biology. He drew a parallel between his approach to designing robots incrementally, and the stepwise manner by which natural selection builds complexity in nature:

> The advantage of this approach is that it gives an incremental path from very simple systems to complex autonomous intelligent systems. At each step of the way, it is only necessary to build one small piece, and interface it to an existing, working, complete intelligence.
>
> (Brooks 1991, p. 403)

Others have criticized the analogy between the construction of behaviour-based robots and the evolution of organisms. Paul Griffiths (1999, p. 51) is surely right when he states that it is 'implausible that our brains evolved by adding separate mechanisms subserving new functions'. The mind may be massively domain-specific as many evolutionary psychologists contend, but if it is, the various mechanisms surely evolved in parallel, just like the organs of the body.

Brooks also likes to compare the development of artificial intelligence with the history of life on earth, suggesting that it would not be possible to build robots with human-level intelligence without first endowing them with less complex capacities such as locomotion. From this perspective, the construction of Deep Blue, a chess-playing computer which won a six-game match against world champion Garry Kasparov in 1997, was an ambiguous achievement. Since many people find chess cognitively-challenging, it was perhaps natural for the pioneers of artificial intelligence to regard chess-playing as a legitimate goal for computers, but Brooks viewed such projects as dead ends. While IBM was building Deep Blue, other researchers were struggling to build robots that could walk over anything except a perfectly flat surface without falling over. Brooks argued that the flexibility of human intelligence would only be replicated in robots that had first acquired a bedrock of capacities, such as locomotion, that had evolved much earlier than abstract reasoning.

The robots built by Brooks were inspired by general principles of evolutionary biology, and sometimes by generic features of large classes of organisms such as hexapod locomotion, but were not generally intended to replicate particular species. During the 1990s, however, researchers began to use robots to test detailed hypotheses about the biological mechanisms of individual species. Barbara Webb, for example, developed a spiking neuron network to model the process by which crickets processed acoustic signals, and mounted the circuit on a mobile robot to test her hypotheses about how female crickets locate males by detecting their songs (Webb 1995). Other roboticists have used a similar approach to test hypotheses about whisking in rats and navigation in ants, giving rise to a dynamic subfield known as 'biorobotics' (Webb 2001).

A related area of research, known as 'evolutionary robotics' (Nolfi and Floreano 2000), uses genetic algorithms and other types of evolutionary computation to develop artificial neural networks (and sometimes robot morphology). Most experiments in evolutionary robotics follow the same basic format. First, a population of artificial chromosomes that code for the control system and/or the morphology of a robot are randomly generated. Next, these chromosomes are decoded and the resulting robot (which may be physical or simulated) is set free to act in a given environment while its performance on various tasks is automatically evaluated according to a predefined criterion known as the fitness function. The fittest robots are then allowed to reproduce by generating copies of their genotypes, with the addition of changes introduced by some genetic operators such as mutation and recombination. This process is then repeated for several generations, allowing selection to enhance the average fitness of the population in the same way as it has done so effectively in the history of life as we know it.

Most current research in biorobotics and evolutionary robotics tends to fall into one of the following categories:

Navigation

A capacity for navigation is one of the most basic requirements for an autonomous mobile robot, so it is unsurprising that this remains a focus of evolutionary robotics research. Typical tasks that robots are set include obstacle avoidance and navigating toward a target area. Since most animal navigation is visually guided, this field of research includes a great deal of work on visual sensing, including such areas as object recognition and discrimination, though many robots still use infrared sensors or sonar instead of, or in addition to, a camera.

Competitive co-evolution

Most experiments in the evolution of robot navigation involve testing each member of each generation, one-by-one, in a fixed physical environment. Experiments in co-evolution, by contrast, involve a continually changing dynamic environment consisting of other agents. Biologists have long speculated that such an environment may enhance the adaptive power of natural selection. For example, Dawkins and Krebs (1979) argued that competing populations may reciprocally drive one another to increasing levels of behavioural complexity by producing an evolutionary arms race. Researchers in artificial life began to develop virtual simulations of such arms races in the early 1990s (Miller and Cliff 1994), and more recently, researchers in evolutionary robotics have developed fully-embodied physical models of predator–prey dynamics.

Cooperative collective robots

Experiments in collective robotics need not be entirely competitive. A thriving strand of research studies the evolution of cooperation among teams of mobile robots. Such experiments often involve modelling the behaviour of social insects (e.g. ants, bees) which display sophisticated collective intelligence. This swarm intelligence (Bonabeau et al. 1999) emerges from the networks of interactions among agents that are individually quite simple. Research in this area includes work on social signalling; Luc Steels (1998), in a pioneering series of experiments, used robots to explore various hypotheses about the origins of language.

Evolvable hardware

Most research in evolutionary robotics takes the robot body as a fixed parameter, focusing exclusively on evolving robot control systems. However, a few projects have applied the evolutionary process to robot bodies. The most striking experiment so far in this field involved the evolution of simple locomotive systems composed of bars and actuators, and then used rapid-prototyping technology to produce the multi-linked structures, so that the only human input needed was the final attachment of snap-on motors (Lipson and Pollack 2000). Another approach in evolvable hardware is evolving robot controllers in reconfigurable electronic circuits such as field-programmable gate arrays (FPGAs). These are argued to be analogue dynamical continuous-time systems and thereby avoid the constraints of discrete digital design (Thompson 1997).

The explicit discussion of evolvable hardware draws attention to an aspect of evolutionary robotics that may not be apparent from the discussion so far: the fact that experiments can be done by means of computer simulations as well as with real, physical robots. The relative advantages and disadvantages of each approach are the subject of some debate in evolutionary robotics. Some argue that physical robots are better models since they incorporate real problems in coping with physical forces like sensory noise, energy consumption, damage, and inertia. However, experiments involving physical robots are very time-consuming. Simulated robots avoid this problem, but also circumvent the physical problems that may well be essential to the structure of natural minds. Another important problem with physical robots is the inability to evolve bodily structures (evolvable hardware as described above is a rare exception), which is essential for the modelling of co-evolution of body and brain. This is so far only possible in simulation. Hybrid approaches perform the most time-consuming periods of evolution in simulation, and subsequently evolve the controllers further in physical robots.

Evolutionary psychology

The previous two sections have provided a brief summary of the huge impact that the two parent disciplines of evolutionary psychology—cognitive psychology and evolutionary biology—have each had on the development of robotics. Given this legacy, it is somewhat surprising that evolutionary psychology has itself had little influence so far on robot research. In this section we discuss some ways in which evolutionary psychology may be applied to the development of robots in the near future.

Synthetic evolutionary psychology

The methods used by evolutionary psychologists so far may be described as analytic, in the sense that they start with a real system (the minds of modern humans and/or of various ancestral species) and attempt to collect data about this system that might permit us to infer the internal structure of this system. There is nothing inherently wrong with analytic methods—indeed, they are the backbone of most modern science—but researchers are increasingly aware of their drawbacks. One important drawback is that the difficulty of analysing a system grows exponentially with the complexity of the system. Since minds, especially the minds of advanced primates such as our own and those of our recent ancestors, are notoriously complex systems, it follows that all analytic methods are very hard to apply to the study of the human mind in a way that all researchers agree upon. One has only to look at the voluminous polemical literature that has arisen in response to the version of the Wason selection task originally described by Cosmides and Tooby, to realize how difficult it is to derive uncontroversial conclusions from analytic methods in evolutionary psychology (Cosmides and Tooby 1992).

Valentino Braitenberg (1984) has proposed, in respect of such difficulties, that when it comes to complex systems it is often easier to discover their internal structure by synthetic methods. In other words, if we wish to discover how some system works, it is often easier to do so by building successively more complex models, rather than by attempting to infer the mechanism from mere observation:

> It is pleasurable and easy to create little machines that do certain tricks. It is also quite easy to observe the full repertoire and behaviour of these machines even if it goes beyond what we had originally planned, as it often does. But it is much more difficult to start from the outside and to try to guess internal structure just from the observation of behaviour.
>
> (Braitenberg 1984, p. 20)

Braitenberg refers to this generalization as the law of uphill analysis and downhill synthesis. But this way of putting things is misleading to the extent that it suggests that the researcher is faced with a choice between analytic and synthetic methods. In reality, analytic and synthetic methods are not alternatives, but complementary aspects of a dialectic that involves moving back and forth between the analysis of empirical data and the construction of simple models of underlying mechanisms.

Synthetic methods involve building models of the system under investigation and then observing their behaviour. The more closely the behaviour of the model corresponds to the behaviour of the target system, the more confident we can be that the internal structure of the model corresponds to the internal structure of the target. Because we have built the model ourselves, its internal structure is transparent, and need not be inferred by analysis.

With a few notable exceptions (such as Geoffrey Miller, Gerd Gigerenzer, and Douglas Kenrick), most evolutionary psychologists have not attempted to translate their hypotheses about human mental structure into working machines. This might seem to exclude evolutionary psychology from cognitive science, a core feature of which is often taken to be the design-based approach. However, this conclusion is too quick, for emphasis on design does not imply that all cognitive scientists must take an active part in building artificial minds. It simply requires that cognitive scientists propose models of the mind that are computational enough to permit computer programs to be readily designed on the basis of the models. Many of the domain-specific mechanisms proposed by evolutionary psychologists take such a form; they are not specified in terms of any programming language, but they are often spelled out in a form that would be relatively easy to convert into a computer program. The models proposed by evolutionary psychologists thus count as fully cognitive, and evolutionary psychology is firmly within the fold of cognitive science.

Even so, it seems a shame that evolutionary psychologists have not taken more interest in translating their models into real machines. The tools of artificial intelligence and computational modelling might well offer them ways of testing their hypotheses and thus enable them to answer the common charge of telling 'just-so stories'. Critics of evolutionary psychology frequently dismiss it on the grounds that it promulgates untestable theories. Evolutionary psychologists acknowledge that there are methodological difficulties posed by investigating the history of the mind, but point out that most of these difficulties are not particular to their discipline. Most of them are common problems faced by all those who wish to investigate evolutionary hypotheses, so to be consistent the critics should also dismiss the whole of evolutionary biology. Such general arguments, however, would be strengthened if evolutionary psychologists could also point to experimental ways of testing their hypotheses. Robotics could supply evolutionary psychology with just such experimental techniques.

Computer simulations of evolutionary processes are already common. In one early example, Thomas Ray (1992) designed a virtual world called Tierra and populated it with a simple digital organism. This was a simple self-replicating program which occasionally made mistakes in the copying process. These 'mutations' led to an increasingly diverse population of digital organisms. Competition for limited memory space on the computer's hard disk ensured that there was differential survival. The conditions for natural selection were therefore all in place, and Ray was able to observe numerous cases of digital evolution complete with virtual viruses, parasite resistance, and other surprisingly 'natural' features.

Tierra is only a virtual world, and thus subject to the criticisms of roboticists like Rodney Brooks, who argue that it is all too easy in such simulations to make some crucial but unnoticed simplification that renders the simulation invalid. In order to avoid this potential danger, Brooks recommends that cognitive scientists work with actual physical robots. By using the techniques of evolutionary robotics, evolutionary psychologists could observe artificial evolution in the real world, giving rise to a genuine lineage of robots evolving by natural selection. Perhaps these methods can provide a way for evolutionary psychologists to test their hypotheses about mental evolution.

As an example of such an approach, den Dulk et al. (2003) studied the evolution of dual-route dynamics for affective processing. In particular, they used the standard methods of evolutionary robotics to examine the evolutionary justification given by LeDoux (1998) for his dual-route model of fear-processing. LeDoux has found evidence that, in many mammals, fear is processed simultaneously by two neural pathways, one subcortical and the other largely cortical. The subcortical route is faster but generates many false positives, while the cortical route is slower but more accurate. LeDoux argues that this dual-route mechanism evolved by natural selection because it allowed animals to get the best of both worlds by escaping quickly when necessary, but not wasting too much effort on false alarms. By allowing agents to evolve in a simple environment consisting of predators and food, den Dulk et al. (2003) found that agents did indeed evolve a dual-route mechanism similar to that proposed by LeDoux, but only when certain conditions were met: the food and the predator had to be relatively hard to distinguish, and information must take significantly longer to propagate via the cortical route than via the subcortical route.

A programme of research in synthetic evolutionary psychology might use much of the same tools as those used by mainstream research in evolutionary robotics, but the issues for investigation would be substantially different. As we have already seen, evolutionary robotics is generally concerned with navigation, co-evolution, cooperation, and evolvable hardware. Evolutionary psychologists would instead be more interested in using robots to explore psychological questions such as the history and structure of the various mental modules that comprise the human mind.

A representative sample of the kinds of problems that are thought by evolutionary psychologists to have led to the evolution of mental modules include: 1) finding food and discriminating between nutritious substances and toxins; 2) detecting and avoiding predators; 3) eliminating infectious agents from the body; 4) establishing and maintaining strategic alliances; 5) communicating with conspecifics; 6) finding and keeping mates; and 7) rearing offspring.

Experiments in evolutionary robotics have already begun to explore the first two problems in this list, and have touched on the problem of communication, but have left the other problems virtually unexplored. This is partly due to technological limitations, but such considerations apply only to conducting experiments with real physical robots. Computer simulations of evolving robot populations could explore the other areas without much difficulty. There is great scope, then, for a broad research programme to explore all of these problems in a systematic way, perhaps by a graded approach that tackles each problem in the order in which they were faced by our ancestors. Such an incremental approach might also throw light on the way in which prior adaptations may be co-opted by natural selection as the basis for solutions to later problems.

Open-ended evolution

A well-known and important difference between artificial and natural evolution is that the former is goal-directed, while the latter is open-ended. Artificial evolution is a technique originally devised to optimize certain parameters. Natural evolution, in contrast, does not converge to a single solution and then stop: it is open-ended and continuous. There are no optimal solutions in nature because the problems, arising from the biological context which includes many co-evolving creatures, are constantly changing. Evolutionary psychologists are interested in the mechanisms that result from natural, and therefore open-ended, evolution. A synthetic approach to evolutionary psychology should attempt to replicate this process. Over the years, various technical procedures have been proposed that partly overcome the major differences between artificial and natural evolution, such as massive co-evolution, variable genotype length, and complex genotype/phenotype mapping schemes.

The goal-directed nature of artificial evolution is mainly due to the use of fitness functions, which are mathematical formulae used to calculate the relative fitness of an individual. Selection mechanisms operate on the basis of the relative differences between fitness values: better individuals are selected for reproduction. After many iterations, evolution thus selects the individuals that optimize the components of the fitness function. In evolutionary robotics, the task of robot is thus implicitly coded in the fitness function (although the behavioural components are not described). Floreano and Urzelai (2000) propose the fitness space as a framework in which fitness functions can be positioned consisting of three dimensions: 1) *Functional/behavioural*: is fitness dependent on the internal functioning of the controller or the behavioural manifestation? 2) *Explicit/implicit*: are there many or very few variables included in the fitness function? 3) *External/internal*: are the variables in the fitness function only available to the fitness evaluator or to the evolving robot as well? According to this framework, the key to approximate open-ended evolution as found in nature is to make a fitness function as behavioural, implicit, and internal as possible.

Alternative evolutionary schemes

In evolutionary theory, fitness is defined in terms of relative reproductive success. This is incompatible with standard artificial evolution on pain of tautology: selection for reproduction cannot be based on relative fitness if this is itself based on reproductive success, because it implies that reproduction already took place. A more biologically plausible evolutionary scheme for experiments in synthetic evolutionary psychology would therefore be one in which selection is not based on explicit fitness values. Instead, selection should emerge from the interaction of the agents with their environment. Fitness values should be used solely for the purpose of cross-comparison and other analyses of the experimental results. This will have impact on the overall set-up of an experiment, which will render it more biologically plausible, without making it necessarily more detailed or complex.

An experiment of this kind should have populations of robots roaming the environment at the same time, in contrast to standard evolutionary robotics where individual controllers are often tested one at a time on the same physical robot. Having simultaneous individuals introduces problems of survival and reproduction if resources are limited. The adaptations to cope with such problems are precisely the ones in which we are interested in evolutionary psychology. Furthermore, having simultaneous populations of potential mates implies continuous reproduction. This results in a gradual generational change, in contrast to many evolutionary robotics experiments where generations are discrete in the sense that they are separated by a selection operation. Gradual generational change and simultaneous populations of individuals means having different (grand)parents and offspring in the same environment, which together with the problem of survival (and growth), introduces the problem of parental care. This would also allow adaptations for social transmission and cultural evolution to evolve. Even a few steps closer to a biologically plausible context should allow sexual selection to influence morphology, having the environment include predation, or even a full-fledged co-evolving food chain. The possibilities of this kind of experiment are of course limited by the constraints imposed by physical design (e.g. appropriate sensors and effectors), adaptability of robot controllers (e.g. sufficiently complex neural networks), mechanical issues (e.g. energetic autonomy), and many other technical limitations.

Experimental design in synthetic evolutionary psychology

Incorporating the alternative evolutionary scheme just described would have a substantial impact on experimental design. Because fitness functions are omitted in the experimental set-up, there are no means to formulate a specific task, which is the first step in designing a standard experiment in evolutionary robotics. Research in synthetic evolutionary psychology should proceed differently. Although specifying an entire experimental framework of synthetic evolutionary psychology is beyond the scope of this chapter, and involves long and thorough experimentation, a few important aspects are noted here.

A hypothesis about the emergence or structure of a particular mental module or behavioural trait is generally based on the following aspects: (changing) environmental conditions, evolutionary forces, and already existing mental structures. That is, an adaptation emerges from new or changed situations in the individual's environment; it benefits the organism's ability to cope with the evolutionary forces; and it is based on existing mental structures or neural substrates that facilitate the emergence of new abilities. Verifying such a hypothesis involves synthesizing these aspects in the experiment and analysing evolving individuals for emerging adaptations in coherence with changing environments.

Changing environmental conditions have played a central role in the evolution of all species. Either it gradually forces adaptation in a species by slow environmental change such as diminishing food resources, or it radically forces selection by historical natural accident like a meteor impact or epidemic. Researchers disagree about the relative importance of these two kinds of environmental pressures, but everyone agrees that both have played a part in the evolution of the human mind. It is relatively easy to synthesize both kinds of environmental changes in conditions in synthetic evolutionary psychological experiments. This can be used to test the robustness of certain adaptations, to cause the emergence of new adaptations, and to provide insight into the trade-off between the importance of robust processes and actual sequences in explaining the history of the mind. Experimenters could slowly increase or decrease the presence of predators, food resources, and mates. Historical accidents can be simulated through sudden changes in availability of resources or removal of individuals.

The existing mental structures that facilitate the emergence of a certain adaptation can be synthesized in two ways that resemble the different goals in evolutionary psychology. If one is particularly interested in exploring the historical trajectory of the mind, the mental structures that facilitate certain adaptations are themselves the results of earlier evolutionary experiments. That is, a graded or incremental approach is taken to evolve increasingly complex robot controllers. This is achieved by increasing the difficulty to survive and reproduce by adding more and more environmental complexity. On the other hand, if one is more interested in the structure of the mind, a somewhat more pragmatic approach can be adopted. In these experiments, the existence of particular structures can be assumed. These structures can, for example, be synthesized by means of neural controllers that are trained by using supervised feedback, without concern for the evolutionary plausibility of their existence. Although this undermines the historical account of adaptation, it can prove to be a fruitful and time-saving approach to investigate the emergence of higher-level adaptations.

Once these three conditions have been specified and synthesized, the individual creatures must be designed with the appropriate sensors, motors, and neural controllers. Experimenters subsequently look for evidence for adaptations in changing behaviour, examine internal functioning of the controller and robustness to changing conditions, analyse fitness changes, make comparisons with naturally evolved creatures, or perform lesions and look for dissociations. Plausible stories about the historical evolution of situated, embodied minds can then be supported or eliminated by hard evidence.

Synthetic evolutionary psychological research has the potential to yield new, or confirm existing, hypotheses about the evolution of specific adaptations. Furthermore, it can provide insight about the importance of the different variables in the evolution of natural minds. It can provide us with increasingly complex machines with needs comparable with those of living creatures. And perhaps these machines will exhibit cognitive abilities, evolved out of the most primal forces in nature, comparable with ours.

The application of evolutionary psychology to human–robot interaction

Some might argue that the research programme we have outlined in the previous section is out of place in a book about 'applied evolutionary psychology', on the grounds that this programme is more about the application of robotics to evolutionary psychology than vice versa. We would disagree with such a claim, since our suggestions about open-ended evolution and alternative evolutionary schemes, which are directly inspired by evolutionary theory, would clearly modify the existing experimental protocols in evolutionary robotics. Furthermore, some roboticists are already carrying out experiments along the lines we have described, and have been directly inspired by ideas from evolutionary psychology in doing so. For example, Alan Winfield and Frances Griffiths are building an artificial society of robots in order to model the mechanisms by which a group of organisms might make the transition from the merely social to the cultural (Winfield and Griffiths 2010). The project is inspired by a memetic theory of culture (Blackmore 1999), itself inspired by Richard Dawkins' speculations about non-genetic replicators at the end of *The Selfish Gene* (Dawkins 1976).

Nevertheless, if we want a less ambiguous example of how robotics might benefit from ideas drawn from evolutionary psychology, we should turn to the study of human–robot interaction. When robots are designed specifically to have rich psychological interactions with humans, as in pedagogical robots and artificial companions (Wilks 2010), rather than merely physical interactions such as surgical robots and military robots, the designers ignore evolutionary psychology at their peril, or must laboriously rediscover for themselves what they could have learned more quickly from evolutionary psychologists. A case in point is the so-called 'uncanny valley hypothesis' first advanced by the Japanese roboticist Masahiro Mori (1970). He proposed that the relationship between a robot's degree of resemblance to a human and a human observer's emotional response was highly non-linear. Robots lacking humanoid features would, he speculated, produce a neutral response in observers, but as robots become progressively more humanlike, we would experience a more positive emotional reaction to them. However, at the point just before it became impossible to distinguish between robots and humans, the robot would evoke in us a sense of revulsion and horror. In reference to the shape of the graph he produced to illustrate his hypothesis (see Figure 25.1), Mori dubbed this zone 'the valley of the uncanny'.

Whether Mori's use of the word 'uncanny' was intended to evoke Freud's earlier essay on this particular feeling is unclear, but in that essay Freud (1955 [1919]) specifically refers to 'ingeniously constructed dolls and automata' as prime examples of things that seem spooky. Citing a

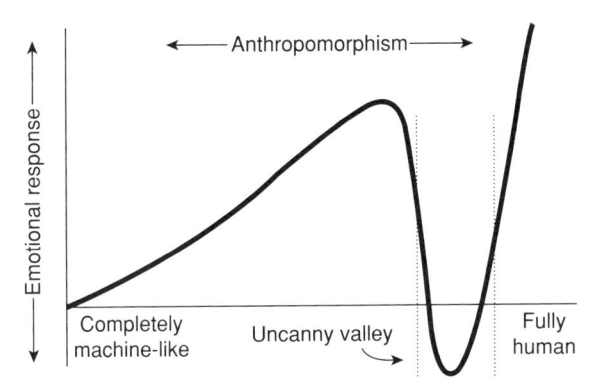

Fig. 25.1 The valley of the uncanny.

1906 essay by Ernst Jentsch, Freud notes that 'in telling a story one of the most successful devices for easily creating uncanny effects is to leave the reader in uncertainty whether a particular figure in the story is a human being or an automaton'. (However, Freud's explanation for this phenomenon, which traces it to the fear of castration, is bizarre and unconvincing.)

Various evolutionary explanations for the uncanny valley have been proposed, all of which are more convincing than Freud's, and which are not mutually inconsistent. One explanation focuses on the psychological mechanisms for mate selection. According to this view, automatic, stimulus-driven appraisals of uncanny stimuli elicit aversion by activating an evolved cognitive mechanism for rejecting potential mates with low fertility, poor hormonal health, or ineffective immune systems, based on visible features of the face and body that are predictive of those traits (Green et al. 2008). Another explanation focuses on pathogen avoidance: uncanny stimuli may activate a cognitive mechanism that originally evolved to motivate avoidance of potential sources of pathogens by eliciting a disgust response. Thus, the visual anomalies of android robots have the same effect as those of corpses and visibly diseased individuals: the elicitation of alarm and revulsion (Green et al. 2008).

Those who build robots intended for rich psychological interactions with humans could also benefit from evolutionary psychology when writing the software for such machines. For example, evolutionary psychology has focused attention, in a way that other branches of psychology have not, on the cognitive mechanisms that underlie reciprocity and fairness. Those who wish to build robot companions might glean much from this literature, perhaps yielding some counter-intuitive guidelines for robot design. For example, it may be that a robot companion who does not punish its owner (in some appropriate way) when the owner treats it unfairly is less engaging than one who does (Evans 2010).

Conclusion

During its relatively brief history, evolutionary psychology has used many different methods, from experimental manipulation of human behaviour in the laboratory to observation of indigenous peoples and analysis of archaeological data. All these methods may be called analytic, in the sense that they collect data about already-existing systems and then analyse them. We have argued that evolutionary psychologists could benefit from extending their methodological repertoire to include synthetic methods, which involve constructing artificial systems. Such artificial systems can provide useful models of evolved minds and evolutionary histories that might provide evolutionary psychologists with additional data to test hypotheses about mental structure and evolutionary trajectories. One kind of synthetic method in which evolutionary psychologists have so far shown little interest, is evolutionary robotics. We have argued that, by ignoring this field, evolutionary psychologists are missing out on a valuable research tool, and we have sketched out a research programme involving the use of robots to test evolutionary psychological hypotheses.

We have referred to some initial work that uses robots to test hypotheses about human evolution. At the moment, however, this work is too rare and patchy to warrant description as a full-fledged research programme. Our manifesto for a new field of synthetic evolutionary psychology remains, for the moment at least, more of a vision than a reality. Time will tell whether this vision is embraced by enough researchers to make it a permanent and fruitful addition to the evolutionary psychologist's repertoire of methodologies.

Acknowledgements

We are grateful to Geoffrey Miller, Verena Hafner, Lars Zwanepol, Tijn van der Zant and Rens Kortmann for their comments. Dylan Evans' work was supported by a Platform Grant from the Engineering and Physical Sciences Research Council, EPSRC GR/M97503.

References

Blackmore, S. (1999). *The meme machine*. Oxford University Press, Oxford.

Bonabeau, E., Dorigo, M., and Theraulaz, G. (1999). *Swarm intelligence: from natural to artificial systems*. Oxford University Press, New York.

Braitenberg, V. (1984). *Vehicles: experiments in synthetic psychology*. MIT Press, Cambridge, MA.

Brooks, R.A. (1991). Intelligence without representation. In: J. Haugeland (ed.), *Mind design II: philosophy, psychology, artificial intelligence*, pp. 395–420. MIT Press, Cambridge, MA.

Chomsky, N. (1956). Three models for the description of language. *IRE Transactions on Information Theory*, **2**, 113–24.

Cosmides, L. and Tooby, J. (1992). Cognitive adaptations for social exchange. In: J. Barkow, L. Cosmides, and J. Tooby (eds), *The adapted mind: evolutionary psychology and the generation of culture*, pp. 163–228. Oxford University Press, Oxford.

Dawkins, R. (1976). *The selfish gene*. Oxford University Press, Oxford.

Dawkins, R. and Krebs, J.R. (1979). Arms races between and within species. *Proceedings of the Royal Society B*, **205**, 489–511.

den Dulk, P., Heerebout, B.T., and Phaf, R.H. (2003). A computational study into the evolution of dual-route dynamics for affective processing. *Journal of Cognitive Neuroscience*, **15**, 194–208.

Dennett, D. (1995). *Darwin's dangerous idea: evolution and the meanings of life*. Allen Lane, London.

Evans, D. (2010). Wanting the impossible: the dilemma at the heart of intimate human-robot relationships. In: Y. Wilks (ed.), *Close engagements with artificial companions. Key social, psychological, ethical and design issues*, pp. 75–87. John Benjamins, Amsterdam.

Floreano, D. and Urzelai, J. (2000). Evolutionary robots with on-line self-organization and behavioral fitness. *Neural Networks*, **13**, 431–43.

Freud, S. (1955 [1919]). The 'Uncanny'. In: J. Strachey (ed.), *The standard edition of the complete psychological works of Sigmund Freud*, Vol. XVII, pp. 217–56. Hogarth Press, London.

Green, R.D., MacDorman, K.F., Chin-Chang, H., and Vasudevan, S. (2008). Sensitivity to the proportions of faces that vary in human likeness. *Computers in Human Behavior*, **24**, 2456–74.

Griffiths, P. (1999). Author's response. *Metascience*, **8**, 49–62.

LeDoux, J. (1998). *The emotional brain: the mysterious underpinnings of emotional life*. Weidenfeld and Nicholson, London.

Lipson, H. and Pollack, J. (2000). Automatic design and manufacture of robotic lifeforms. *Nature*, **406**, 974–8.

Miller, G.A. (1956). The magical number seven, plus or minus two: some limits on our capacity for processing information. *Psychological Review*, **63**, 81–97.

Miller, G.F. and Cliff, D. (1994). Protean behaviour in dynamic games: arguments for the co-evolution of pursuit-evasion tactics in simulated robots. In: D. Cliff, P. Husbands, J. A. Meyer, and S. Wilson (eds), *From animals to animats 3: Proceedings of the Third International Conference on Simulation of Adaptive Behaviour*, pp. 411–20. MIT Press, Cambridge, MA.

Mori, M. (1970). Bukimi no tani. [The uncanny valley] (translated by K.F. MacDorman and T. Minato). *Energy*, **7**, 33–5.

Newell, A. and Simon, H.A. (1956). The logic theory machine- a complex information processing system. *IRE Transactions on Information Theory*, **2**, 61–79.

Nolfi, S. and Floreano, D. (2000). *Evolutionary robotics: the biology, intelligence and technology of self-organizing machines*. MIT Press, Cambridge, MA.

Ray, T.S. (1992). An approach to the synthesis of life. In: M.A. Boden (ed.), *The philosophy of artificial life*, pp. 111–45. Oxford University Press, Oxford.

Steels, L. (1998). The origins of syntax in visually grounded robotic agents. *Artificial Intelligence*, **103**, 133–56.

Thompson, A. (1997). Artificial evolution in the physical world. In: T. Gomi. (ed.), *Evolutionary robotics: from intelligent robots to artificial life*, pp. 101–25. AAI Books, Ottawa, ON.

Walter, W.G. (1951). A machine that learns. *Scientific American*, **185**, 60–3.

Webb, B. (1995). Using robots to model animals: a cricket test. *Robotics and Autonomous Systems*, **16**, 117–34.

Webb, B. (2001). Can robots make good models of biological behaviour? *Behavioral and Brain Sciences*, **24**, 1033–50.

Wilks, Y. (ed.) (2010). Close engagements with artificial companions. *Key social, psychological, ethical and design issues*. John Benjamins, Amsterdam.

Winfield, A.F.T. and Griffiths, F. (2010). Towards the emergence of artificial culture in collective robot systems. In: P. Levi and S. Kernbach (eds), *Symbiotic multi-robot organisms*, pp. 431–9. Springer, Berlin.

Index

abandonment rage 103
academic learning
  learning in school 79, 86–7
  motivation 79–80, 85–6
  see also educational psychology
addiction 103
adultery 94–7, 98
  see also monogamy
advertising see marketing
affiliation response 316
Afghanistan 228
age-crime curve 207
  see also young male syndrome
age differences 10
aggression 115–16, 123–4
  coalitional aggression 122–4, 399, 405
  frustration-aggression 103
  life history (LH) theory 210–15
  young male syndrome 399
    societal and evolutionary process model 404–5
al-Betawi, Hamed 225
alcohol 261
  drink driving 162, 403–4
Allport, G.W. 194
Al Qaeda 228, 232
altruism 9, 19, 42, 45–6
  competitive altruism 45, 159–60, 175–7
  evolutionary theory 173–4
  foster care 244–5
  'green' behaviour 160, 161, 165, 179–81
  prestige and reputation-based helping 174–5
    policy implications 181–2
  self-sacrifice and war 223–5
  sexual advantage 177–9
  see also charitable behaviour; cooperation; help-giving
American Civil War 234–5
ancestral environment 7, 8, 11–12
  coalitional conflict 122–4
  counter-exploitation 120–1
  help-giving 118–19
  hierarchies 121–2
  pair-bonding 97
  resource conflicts 119–20
  social organization 20–2, 115, 116
androgens 96, 290
anomie 203
antidepressants 104, 283
Apple computers 321, 322
appointed leaders 27, 28
arousal 389–90
artificial intelligence 414
  see also robotics
association value 121
attachment behaviours 93–4
attachment theory 205, 285–6
autonomy 39

Bandura, A. 352
bargaining 11, 119
Bartholet, Elizabeth 243
Bartlett, F. 388
Beccaria, Cesare 201
beggars 178
behaviour
  mental health and 279–80
  ultimate and proximate explanations 1–2, 36, 312–13
Belsky, J. 69, 70
Bentham, Jeremy 202
biases 8–9, 18
  folk heuristics 83
  intergroup prejudice 192–3
  same-race 59–60, 122–3, 189, 190, 191
    life history (LH) theory 211
  sex biases 131–2
    sexual attribution bias (SAB) 44, 55
  status quo bias 8, 316
Bin Laden, Osama 228
biological sensitivity to context 68, 69
biorobotics 415–16
bipedalism 97, 331
body odour 331, 333
  biological signals
    cues of complementary genes 337–8, 343
    cues of current state 334–5
    indicators of genetic quality 337
    individual recognition 334
    mediation of reproductive physiology 335–7
Boehm, Christopher 24
Bowlby, John 205, 286
Boyce, W.T. 69, 70
brain chemistry 94, 96, 99, 100–2, 103, 104, 266
brain size 37, 55–6, 78
brain structure
  communication apparatus 385, 387, 388
  left and right hemispheres 296
  maturation and gender differences 409
  old brain 362, 363, 366
  visual information 362–4, 365
Braitenberg, Valentino 417–18
brand positioning 321–3
breastfeeding 261
bride price societies 133, 135

Brooks, Rodney 415
Brosnan, Sarah 250
bubble psychology 17, 18, 19
bullying 43, 44
burnout 42–3, 44
business 16
   Credit Crunch (2008) 17–20
   see also management

Canadian driver safety data 405–9
capital goods 12
car manufacturers 322–3, 403
cave paintings 363, 364
celebrities 350–1, 353
Central Intelligence Agency (CIA) 232
charitable behaviour 44, 149
   visible donations 176–7
   see also altruism; help-giving
cheating 9, 11, 118, 208–9
Chechnya 229, 230
chess 415
Chevrolet 322, 323
Chicago school 203
child development
   accelerated pubertal development 58
   attachment theory 205, 282–6
   cognitive and behavioural characteristics 57
   conditional adaptations 57–8
   core knowledge 59
   deferred adaptations 56–7
   differential susceptibility to experiences 66–70
   education *see* educational psychology
   environmental influence 66–7
   extended juvenile period 55–6
   facultative adaptations 57–8
   false-belief tasks 59
   family investment 61–2
      criminal behaviour and 207
      facial resemblance and 244, 252
      foster care *see* foster care
      grandparents 64–5, 242
      healthy psychological development 65–6
      maternal 62–3, 151, 242
      paternal 63–4, 242
      step-parents 65
   gender differences 56, 59
   healthy baby hypothesis 63
   immature appearance 57
   in-group/out-group distinctions 59–61, 190–1
   intentional agency 59
   language acquisition 87, 190–1
   natural selection 55
   ontogenetic adaptations 57
   peer relationships 81, 85, 86
   predictive adaptive responses 58
   shared attention 59
   skeletal competencies 59
   social orientation 58–61
   undernutrition 260–1
child maltreatment 246–7
Chomsky, Noam 414
CIA (Central Intelligence Agency) 232
circumcision 136

clandestine adultery 94–7, 98
   see also monogamy
climate change 162
clitoridectomies 136
coalitional aggression 122–4, 399, 405
Coca-Cola 322
co-evolution 16–17
   self-regulation 22–3
      leader effectiveness 28–9
cognitive development 83–4
cognitive robots 414–15
   see also robotics
cognitive theories of crime 204–5, 209–10
colour 297–301
communication
   media naturalness 386, 390–2
      ambiguity 388–9
      cognitive effort 386–7
      human evolution and 384–6
      implications for managers 393–4
      implications for researchers 392–3
      physiological arousal 389–90
   media richness theory 381–2, 391
      evidence against 382–4
      evidence for 382
competitive altruism 45, 159–60, 175–7
competitive behaviour
   conspicuous consumption 36, 43–4, 46
   gossip 43–5
   intragroup and intergroup 41–2
   intrasexual competition 39–40, 43–4, 46
   rivalry 43–4
   sport *see* sport
competitive disadvantage theory 208
computer games 402–3
   Deep Blue chess computer 415
conditional adaptations 57–8
conspicuous altruism 165, 176–7, 178
conspicuous consumption 36, 43–4, 46
consumer behaviour 7, 311–12
   evolutionary approach 312, 318, 320, 324
      forming friendships 316
      gaining status 316–17
      proximate and ultimate levels of explanation 312–13
      romantic partner attraction 314–15
      self-protection 315–16
      solving adaptive challenges 313–14
   'green' behaviour 160, 161, 165, 179–81, 317
   kin care 317
   marketing applications 312, 318
      brand positioning 321–3
      consumer targeting 319–20
      customer segmentation 318–19
      product development and strategic messaging 320–1
      situational influences on social motives 323–4
contact hypothesis 194
control theory 203–4
cooperative behaviour 9, 11, 149
   group competition 231
   party politics 123, 124
   see also altruism; charitable behaviour; help-giving

cooperative breeding 62
costly signalling 43
  evidence for 158
    'green' behaviour 160, 161, 165, 179–81
    informational value 157
    reputational benefits 158–9
counter-exploitation 120–1
courtship 99
Credit Crunch (2008) 17–20
criminal behaviour 201
  age-crime curve 207
  evolutionary social science theories
    behavioural genetics and evolutionary psychology 205
    cognitive theories 209–10
    competitive disadvantage theory 208
    disease stress 209–10
    frequency-dependent theories of psychopathy 208–9
    life history (LH) theory 210–15, 217
    reactive heritability theory and epigenetic influences 206–7
    sexual selection theory 207–8
  left-handedness and 295
  psychopathy 208–9, 212, 214, 216
  rape 136, 247–8
    see also sexual coercion
  reputation and culture of honour 207–8, 209, 214, 215
  self-control theory 212, 216
  standard social science theories 201–2
    cognitive theories 204–5
    control theory 203–4
    cultural, sub-cultural and social learning theories 203
    functionalist theories 202–3
    personality theories 205
    positivist theories 202
  young male syndrome 400–1
criminal justice 120–1, 126
crowd behaviour 19
  sport 292
cuckoldry 61, 64
cultural collectivism 209
cultural evolution 17, 22–3

Darwin, Charles 1, 8n, 39, 202, 222, 223, 224, 229, 299, 312
Darwinian psychiatry 276–7, 278, 282, 285, 286
  see also mental health
Dawkins, Richard 416, 422
Deep Blue (chess computer) 415
deferred adaptations 56–7
den Dulk, P. 419
deodorants 333, 336
  see also body odour
disabled children 247
disease-avoidance 188, 192
diseases of civilization 259
diseases of poverty 260, 271
disease stress 209–10
disgust 268
depression 103, 104, 278–9, 280, 281, 283

developmental psychology *see* child development
diathesis stress model 67
differential susceptibility hypothesis 66–70
disposition effect 9
division of labour 11–12
divorce 97
dominance 23–6, 60
dopamine 94, 96, 99, 100, 101, 102, 266
drink driving 162, 403–4
driver safety
  Canadian data 405–9
  demography and 401–2, 403–4
  drink driving 162, 403–4
  gender difference 404, 409–10
  governmental initiatives 401
  media influences and 402–3
Dual Inheritance Theory 17
Dunbar's number 19, 37
Durkheim, Emile 202–3

eating 1
eco-friendly products 160, 161, 165, 179–81, 317
economics
  age differences 10
  ancestral environment 7, 8, 11–12
  capital goods 12
  consumer behaviour *see* consumer behaviour
  Credit Crunch (2008) 17–20
  disposition effect 9
  division of labour 11–12
  endowment effect 8, 9, 18, 250–1
  folk theories 11n, 12–13, 19
  frequency-dependent selection 9, 19–11, 20
  gender differences 10
  income inequality 12
  individuality versus heterogeneity 10
  markets 11
  psychology and 212–2, 30–1
  rationality and biases 8–9, 18
  selfish and pro-social behaviour 9–10, 19
  social policy 13
  social problems and 21–2, 30–1
  taxation 12
  trade 11
  utility function 7
  willingness to pay/accept (WTP/WTA) 8
  zero-sum 12
educational psychology
  cognitive development 83–4
  evolution of the human mind 78–9
  fluid intelligence 79, 80
  folk competencies 79, 80–3, 84–5
    limitations 87–8
    rules of thumb 83
  learning in school 79, 86–7
  modular domains of the mind 79, 80–1
  motivation 79–80, 85–6
  reading skills 87
  social learning 79, 85–6
  writing skills 86
elected leaders 27, 28

electronic communication
 media naturalness 386, 390–2
  human evolution and 385–6
  implications for managers 393–4
  implications for researchers 392–3
  physiological arousal 389–90
 media richness theory 381–2, 391
  evidence against 382–4
  evidence for 382
Ellis, B. J. 69, 70
emergent leaders 27, 28
endowment effect 8, 9, 18, 250–1
entertainment *see* media research; television programming
environmental adversity 67–9
 life history (LH) theory 210, 217
environmental behaviours 160, 161, 165, 179–81, 317
envy 36, 37–9
equilibrium
 Evolutionarily Stable Strategy (ESS) 11
 punctuated equilibrium 17
error management theory 188–9, 191
Evolutionarily Stable Strategy (ESS) 11
evolutionary developmental psychology *see* child development
evolutionary educational psychology *see* educational psychology
evolutionary medicine 2–3
 Darwinian psychiatry 276–7, 278, 282, 285, 286
 see also mental health
 public health *see* public health
evolutionary psychology
 application 2–4
 history of the discipline 1–2
evolutionary robotics *see* robotics
executive functioning 212, 216
exploitation 120–1
extra-pair copulation 95
Eysenck, Hans 205

face processing
 choice of political leaders 122
 emotional and social signals 83, 382, 384, 385, 389
 facial resemblance and parental investment 244, 252
 infants 59
facultative adaptations 57–8
families
 investment in children 61–2
  foster care *see* foster care
  facial resemblance and 244, 252
  grandparents 64–5, 242
  healthy psychological development 65–6
  maternal 62–3, 151, 242
  paternal 63–4, 242
  step-parents 65
family businesses 317
family honour 207–8, 209, 214, 215, 228
family size 137, 144
fear reactions 351
 over-perception of threat 189, 193, 195, 369–70
 self-protection 315–16
Ferri, Enrico 202
financial crisis (2008) 17–20
Fiorina, Carly 25

Floreano, D. 420
fluid intelligence 79, 80
folk biology 81, 82, 83, 84–5
folk competencies 79, 80–3, 84–5
 limitations 87–8
 rules of thumb 83
folk economics 11n, 12–13, 19
folk physics 82, 83, 84–5
folk psychology 81, 84
football 292–3, 297, 299
foster care 239–40
 adoptive versus foster placements 245
 child maltreatment 246–7
 facial resemblance and 244, 252
 kin as foster parents
  application of research findings 242–3
  background to current US practice 240–1
  insights from evolutionary theory 241–2
 race considerations 243–6
free-riders 118, 176, 317
frequency-dependent selection 9, 10–11, 20, 208–9
Freud, Sigmund 422–3
friendship 316, 354–6
frustration-aggression 103
functional flexibility 189, 191
functionalist theories 202–3
future discounting 18

game theory 9, 10–1
gang members 42, 60–1
gender differences
 child development 56, 59
 driver safety 404, 409–10
 hormones and competition 290–1
 intrasexual competition 39–40, 43–4, 400
 leadership 41–2
 maturation and brain structure 409
 media research 370–2
 organizational models of management 24–6
 parental investment 61–4, 241–2
 physical strength 119–20, 124
 risk preference 10
 see also risk-taking behaviours
 romantic rejection 103
 suicide bombers 229–31
gender equity 131
 attitudes and behaviours 143–4
 convergence of gender roles 136–7
 control of fertility 145–6
 correlates of women's power in traditional societies 133–4
 cultural influences 139
 economic costs of parenthood 144
 evolved sex biases 131–2
 gendered work patterns 134–5
 hidden costs of equal opportunities 144, 145
 Human Development Index (HDI) 138
  economic and political opportunities 141–2
  gender empowerment measure 141–2
  gender ratio 138, 139–41, 142, 143
  world values 142
 importance of resources 132–3
 infanticide 143, 145
 modern developed nations 138

politics 135
religious influences 139, 142, 143
sex and marriage 136
sexual transgression 136, 145
smaller families 136–7
voting rights 145
warfare 135
work schedules 144–5
World Economic Forum (WEF) 139
gene-environment interactions 66–7, 206
General Motors 322–3
genocide 224
Genocide Watch 190
global threats 30
glory 222–3
gossip 19, 43–5, 355, 356–7
grandparents 64–5, 242
'green' behaviour 160, 161, 165, 179–81, 317
Griffiths, Frances 422
Griffiths, Paul 415
group competition 231
groups
    in-group/out-group distinction 59–61, 81, 190–1
        see also intergroup prejudice
    party politics 122–4
group selection 9
    see also multi-level selection
group solidarity 60–1
Guantánamo Bay 232

Hamas 225, 226, 235
Hamilton, William 61
handicap signalling 43
    see also costly signalling
hawk-dove game 10–11
health promotion see public health
healthy baby syndrome 63
help-giving 118–19, 149–50
    costly signalling 43
        evidence for 158
        informational value 157
        reputational benefits 158–9
    direct reciprocity 154
        conditional cooperation 155–6
        structural incentives 154–5
    evolutionary theory 149–51
        adaptive on average 151–2
    inclusive fitness theory 152–3
    indirect reciprocity 156–7
    reputational benefits 45, 159–60, 175–7
        consequences of non-reinforcement 163
        costs of bad reputation 157
        creating reputational pressures 161–2
        'do-gooder derogation' 165
        effectiveness of identifying and promoting benefits 165
        explicit incentives versus intrinsic motives 164–5
        future benefits versus immediate costs 164
        harnessing the power of reputation 161
        implicit cues of reputation 160–1
        lack of concern for 163–4
        outweighed by costs of helping 162–3
        preconditions and evidence 159
        situational potency 163

vested interests 153–4
see also altruism; charitable behaviour; cooperation
Henry, Patrick 233
herd behaviour 19
hereditary leaders 27, 28
heritability 205, 206–7
heterogeneity 10
hierarchy 23–6
    ancestral environment 121–2
    male and female status hierarchy 41
higher-order reasoning 126–7
Hirschi, Travis 203–4
Hitler, Adolf 234
hominid evolution 97, 364
honour revenge ideology 207–8, 209, 214, 215
    sacred values 225, 228
    suicide bombers 229–30
hormones 96, 100, 101, 102, 290–1, 292–3, 299–300, 301, 337
    see also testosterone
human attachment 93–4, 205, 285–6
Human Development Index (HDI) 138
    gender empowerment measure 141–2
    gender ratio 138, 139–41, 142, 143
    women's economic and political opportunities 141–2
    world values 142
human needs 36
human traits 115
hunter-gatherer societies 20, 21, 22
hygiene 264, 267–70

Ibn Khaldûn 224
imaginative abilities 126
Implicit Association Test 60
inclusive fitness theory 61, 152–3
income inequality 12
individualism 202, 203
individuality 10
Indonesia 227
infanticide 143, 145, 246, 247
inference mechanisms 126–7
infidelity 94–7, 98
    see also monogamy
in-group favouritism 59–61, 81, 190–1
    see also intergroup prejudice
intergroup prejudice 186–8
    contact hypothesis 194
    disease-avoidance 188, 192
    evolutionary approaches 188
        error management theory 188–9, 191
        functional flexibility 189, 191
        specific threats posed by out-groups 188
    life history (LH) theory 211
    over-perception of threat 189, 193, 195
    promoting tolerance 194–6
    psychological processes 189–90
        biased attitudes and cognitions 192–3
        in-group/out-group categorization 190–1
    race-bias 59–60, 122–3, 189, 190, 191
    stereotyping 126, 193
Internet 381
    see also electronic communication
Interpersonal Relations Rating Scale (IRRS) 212–14

intoxicants 261, 264
intrasexual competition 39–40, 43–4, 46
Iran 227
Iraq 229, 230, 399
Israel 225, 226–7

James, William 23
Jentsch, Ernst 423
Jihad 225, 230, 231
Jones, Owen 250
journalism 367–8, 370, 371, 372, 373

Kanazawa, S. 354–5
Kashmir 227
kinship 152–3
  consumer behaviour 317
  foster care placements 240–1
    race considerations 244
Krebs, J. R. 416

language acquisition 87, 365
leadership
  dominance 23–6
  effectiveness 28–9
  gender differences 41–2
  paths to leadership 23–6
  politicians 121–2
  self-regulation 28–9
Le Doux, J. 419
left-handedness 293–7
Lewin, Kurt 286
libertarian paternalism 13
life history (LH) theory 210–15, 217
Lombroso, Cesare 202
loneliness 254–6
loss aversion 18

Machiavelli, Niccolò 28
major histocompatibility complex (MHC) 96, 337, 339
male-warrior hypothesis 42
Malthus, Thomas 7, 8
management 16
  electronic communication 383–4, 393–4
  organizational model
    dominance 23–6
    gender differences 24–6
    human group sizes 37
    leader effectiveness 28–9
    paths to leadership 26–8
  traditional and modern societies 20–2
  workplace envy 36, 38–9
see also business
market bubbles 17, 18, 19
marketing
  brand positioning 321–3
  cars 322–3, 403
  consumer targeting 319–20
  customer segmentation 318–19
  evolutionary needs of customers 312, 318, 320, 324
  product development and strategic messaging 320–1
  situational influences on social motives 323–4
market theory 11

marriage 93, 133, 135, 136, 144
  see also monogamy
martyrdom 225
  see also self-sacrifice
mass society 115, 116–17
  anonymity of mass politics 124–5
    emotional reactions 127
    inference mechanisms 126–7
    processing without cues 125–6
    stereotyping 126
mate choice 99–102
  altruism and 177–9
mate poaching 95
maternal investment 62–3, 151, 242
mating aggression 210, 211, 212, 213, 215, 248
McCain, John 223
media naturalness 386, 390–2
  ambiguity 388–9
  cognitive effort 386–7
  human evolution and 384–6
  implications for managers 393–4
  implications for researchers 392–3
  physiological arousal 389–90
media research 361, 372–3
  driver safety and media influences 402–3
  gender-specificity 370–2
  journalism 367–8, 370, 371, 372, 373
  media equation 361–2
  news media 369–72
  old brain responses 362, 363, 366
  sensationalism 349, 353, 367, 369, 370
  survival-relevancy 369–70
  television see television programming
  visual information 351, 362–4, 365
  will to mediate 364–7
media richness theory 381–2, 391
  evidence for 382
  evidence against 382–4
menstrual synchrony 334–5
mental health
  attachment theory 205, 285–6
  Darwinian psychiatry 276–7, 278, 282, 285, 286
  depression 103, 104, 278–9, 280, 281, 283
  diagnosis 277
    behavioural analysis 279–80
    functional assessment 278–9
    how to collect 279–80
    what to collect 277–9
    where to collect 281
  post-traumatic stress disorder (PTSD) 283–4
  prevention 284–6
  schizophrenia 279
  therapy
    delineating the approach 281–2
    suppression of adaptive symptoms 282–4
Monahan, John 251
monogamy
  clandestine adultery 98
  mate choice 99–102
  neurobiology of human attachment 93–4, 96, 99, 100–2, 104
  phylogenetic origin 97
  romantic attraction 98–9

romantic rejection 102–3
serial monogamy 97–8
social monogamy 95, 132
Mori, Masahiro 422
Moss, Kate 350
Mullah Omar 228
multi-level selection 16–17
Mussolini, Benito 233

narcissism 42
naturalistic fallacy 3
needs 36
Nesse, Randolph 276, 281, 286
networks 19
neurobiology of pair-bonding 93–4, 96, 99, 100–2, 104
Newell, Alan 414
news media 369–72
non-verbal behaviour 280
nutrition
   pregnancy 58
   risk factors 260–1, 264, 265, 270–1, 285

obesity 264, 270–1, 285
occupational hazards 264
oestrogens 96, 102
olfaction 330–1
   see also body odour; perfume design
ontogenetic adaptations 57
Organizational Citizen Behaviours (OCB) 45–6
organizational models 23–6
   gender differences 24–6, 41–2
   human group size 37
   leader effectiveness 28–9
   paths to leadership 23–6
   traditional and modern societies 20–2
ostracism 44
out-group hostility 59–61, 81, 190–1
   see also intergroup prejudice
oxytocin 94, 96, 102

pair-bonding 93–4, 96, 99, 100–2, 104
   see also monogamy
Palestine 226–7, 229
parental investment 61–2
   criminal behaviour and 207
   facial investment and 244, 252
   foster care see foster care
   grandparents 64–5, 242
   healthy psychological development 65–6
   maternal 62–3, 151, 242
   paternal 63–4, 242
   step-parents 65
parenting behaviour 67–8
Partner Bonding Scale 95–6
party politics 122–4
paternal investment 63–4, 242
perfume design
   biologically-informed design
      amplifying quality perception 338–9
      constant fertility enhancement 338
      cueing complementary genes 339–41

deodorants 333, 336
   see also body odour
effects on the wearer 333, 336
human olfaction 330–1
origins of perfume use 331–2
perfume diversity 332
psychological effects 332–3
undesirable side effects
   disruption of function 341–2
   ethical concerns 342
   product works too well 341
personality theories 205
   life history (LH) theory 210–12
physical strength 119–20, 123–4
physiological arousal 389–90
Pliny the Elder 331–2
policy-making see social policy
political behaviour 116
   anonymity of mass politics 124–5
      emotional reactions 127
      inference mechanisms 126–7
      processing without cues 125–6
      stereotyping 126
   criminal justice 120–1, 126
   gender equity 135
   leaders 121–2
   misapprehensions 117, 118–19, 127
   party politics 122–4
   propaganda 195
   redistribution 119–20
   social welfare 118–19
polyandry 132
polygyny 93, 131, 132, 133, 134
pornography 117
positive discrimination 46
positivist theories 202
post-traumatic stress disorder (PTSD) 283–4
poverty 260, 271
predictive adaptive responses 58
pregnancy
   body odour 337
   nausea and vomiting 2–3
   nutrition 58
   timing 1–2, 144, 145–6, 210
   unplanned 136
prejudice see intergroup prejudice
preventive medicine
   mental health 284–6
   public health 266
primates 23, 24, 25, 26, 37, 56, 64, 116
   biological reactivity in rhesus monkeys 68
   colouration 299
   social learning 79
prisoner's dilemma 9, 154
propaganda 195
pro-social behaviour 9–10, 19, 45
   see also altruism; cooperation; help-giving
prospect theory 18
proximate explanations of behaviour 1–2, 36, 312–13
prudence 18
psychiatry see mental health
psychoactive drugs 264, 265, 266

psychopathy 208–9, 212, 214, 216
puberty 58
public goods 9
public health
　diseases of civilization 259
　diseases of poverty 260, 271
　handwashing 267–70
　health promotion 267, 272
　high blood pressure 261
　indoor smoke 264
　intoxicants 261, 264, 266
　lack of exercise 264
　lead exposure 264
　low vegetable and fruit intake 264
　micronutrient deficiencies 264
　mismatch between ancestral and modern environments 259–60, 265–7, 269–71, 285
　obesity 264, 270–1, 285
　occupational hazards 264
　prevention versus treatment 266
　risk factors for global burden of disease 262–3
　sanitation and hygiene 264, 267–70
　smoking 162, 261, 266, 270
　undernutrition 260–1
　unsafe sex 261
public policy see social policy
punctuated equilibrium 17
punishment 120–1, 157

race-bias 59–60, 122–3, 189, 190, 191
　foster care placements 243–6
　life history (LH) theory 211
rage 103
rape 136, 247–8
　see also sexual coercion
Rappaport, Roy 224
rationality 8–9, 18
Ray, Thomas 418–19
reading skills 86–7
reciprocal altruism 9, 45–6
reciprocity
　direct reciprocity 154
　　conditional cooperation 155–6
　　structural incentives 154–5
　indirect reciprocity 156–7
　　punishment 157
redistribution 119–20
rejection 102–3
religion
　gender equity and 139, 142, 143
　sacred values and self-sacrifice 224–8
reproduction 1–2
　body odour and 335–7
　neurobiology of human attachment 93–4, 96, 99, 100–2, 104
　timing of pregnancy 1–2, 144, 145–6, 210
reputation
　competitive altruism 45, 159–60, 175–7
　costs of bad reputation 157
　creating reputational pressures 161–2
　criminal behaviour and honour 207–8, 209, 214, 215
　'do-gooder derogation' 165

　effectiveness of identifying and promoting reputational benefits 165
　evolutionary theories of altruism 174–5
　explicit incentives versus intrinsic motivation for helping 164–5
　future benefits versus immediate costs 164
　harnessing the power of reputation 161
　implicit 160–1
　lack of concern for 163–4
　non-reinforcement 163
　outweighed by costs of helping 162–3
　preconditions and evidence 159
　situational potency of reputational cues 163
resource conflicts 119–20
respiratory illness 67
revenge 207–8, 209, 214, 215
　sacred values 225, 228
　suicide bombers 229–30
reward systems 38–9
　brain chemistry 94, 96, 99, 103, 261, 265–6
risk preference 10
risk-taking behaviours 399
　driving see driver safety
　young male syndrome 400–1
　　maturation and brain structure 409
　　societal and evolutionary process model 404–5
rivalry 43–4
road safety see driver safety
Robertson, James 285
robotics
　cognitive psychology 414–15
　evolutionary biology
　　biorobotics and evolutionary robotics 415–16
　　competitive co-evolution 416
　　cooperative collective robots 416
　　evolvable hardware 416–17
　　navigation 416
　evolutionary psychology 423–4
　　alternative evolutionary schemes 420–1
　　application to human-robot interaction 422–3
　　experimental design 421–2
　　open-ended evolution 419–20
　　synthetic evolutionary psychology 417–19
romantic attraction 96, 98–9, 103, 314–15
　altruism and 177–9
　consumer behaviour and 413–15
romantic rejection 102–3
rules of thumb 83

sacred values 224–8
sacrifice see self-sacrifice
sanitation 264, 270
Scarr, Sandra 65–6
schizophrenia 279
schooling 78, 79–80, 86–7
　see also educational psychology
scientific management 21
Seaman, Julie 249
self-control theory 212, 216
self-deception 31
self-efficacy 39
self-esteem 36

selfish behaviour 9–10
self-protection 315–16
 fear reactions 351
 over-perception of threat 189, 195, 315
self-regulation 22–3
 leader effectiveness 28–9
 long-term and short-term interests 31
self-sacrifice 222–3
 sacred values 224–8
 status and 317
 suicide bombers 225, 229–31
 war 223–5, 228–31, 233–6
serial monogamy 97–8
serotonin 100, 101, 104
sex biases 131–2
 sexual attribution bias (SAB) 44, 55
 see also gender differences; gender equity
sex drive 96
sexism 211, 212
sex-offending behaviour 207
sexual attraction 96, 98–9, 103
 altruism and 177–9
 consumer behaviour and 314–15
 odour 330, 338, 339
sexual attribution bias (SAB) 44, 59
sexual coercion 210, 211, 212, 213, 215
sexual dysfunction 104
sexual equality *see* gender equity
sexual harassment 249–50
sexually-transmitted diseases 261
sexual selection theory 207–8
Simon, Herbert 414
smell 330–1
 see also body odour; perfume design
Smith, Adam 11, 232
smoking 162, 261, 266, 270
soap 267–70
social brain hypothesis 36–7, 55–6
social comparison 37–9
social competencies 58–61, 278–9
social contracts 201–2
social decision-making *see* political behaviour
social exclusion 44
social groupings 37
social identity theory 187
social influence 383–4
social learning 79, 85–6, 203, 352
socially responsible products 160, 161, 165, 179–81, 317
social monogamy 95, 132
social organization 20–2, 115, 116
social policy 13, 46–7, 239, 251–2
 child maltreatment 246–7
 foster care legislation *see* foster care
 property and contract law 250–1
 reducing incidence of rape 247–8
 sexual harassment legislation 249–50
social pressures 162
social problems 21–2, 30–1
social welfare 118–19
socio-economic status 119–20
speech 384
sport 290, 301

 colour influences 297–301
 football 292–3, 297, 299
 home advantage 291–3
 left-handedness 293–7
 testosterone and human competition 290–1, 292–3, 301
Stanton, Gregory 190
status 23–6, 36
 burnout and 43
 conspicuous consumption 36, 43–4
 consumer behaviour 316–17
 envy 36, 37–9
 intrasexual competition 39–40, 46
 rivalry 43–4
status quo bias 8, 316
Steels, Luc 416
step-parents 65
 child maltreatment 246–7
 see also foster care
stereotypes 126, 193
strain theory 203
stressful environments 67–9
 life history (LH) theory 210, 217
subcultures 19
suicide bombers 225, 229–31

Taliban 228
taxation 12
technological change 12
television programming
 artificial friendships 354–6
 celebrities 350–1, 353
 credibility of information 351
 gossip and 356–7
 sex and crime 349, 353, 367
 social learning 352–4
 visual information 351
 see also media research
territoriality 292
terrorists 236
testosterone
 dominance and 339
 fertility and 337, 338
 mate choice and 96, 100, 101, 102
 sport and competition 290–1, 292–3, 299–300, 301
Tinbergen, Niko 1
tobacco 261, 266, 270
tolerance 186, 194–6
torture 232
trade 11
tragedy of the commons 30, 179–81
Trivers, Robert 61

ultimate explanation of behaviour 1–2, 36, 312–13
ultimatum game 9
undernutrition 260–1
United Nations Development Program (UNDP) 139
United Nations Human Development Report 138
Urzelai, J. 420
utility function 7

vasopressin 94, 95, 96
vested interests 153–4

victimization 43
video games 402–3
Vietnam war 233
virtual worlds 418–19
   see also media research
Visa 322
visual information 35, 362–4, 265
Volvo 321, 322
Vonnegut, Kurt 355

Wal-Mart 321, 322
Walter, Grey 414
warfare 123–4, 135
   evolutionary drives 231
   genocide 224
   moral norms 232–3
   over-perception of threat 189, 195
   self-sacrifice 223–5, 228–31, 233–6
Warhol, Andy 350

Wason selection task 417
Webb, Barbara 415
Weber, Max 23
Vietnam war 233
Winfield, Alan 422
workaholism 42–3
workplace envy 36, 38–9
World Economic Forum (WEF) 139
world values 142
World War II 234
writing skills 86, 364, 365

York, Alvin 232
young male syndrome 400–1
   driving behaviour see driver safety
   maturation and brain structure 409
   societal and evolutionary process
      model 404–5

zero-sum mentality 12